Blue versus Orange

U.S. GOVERNMENT
OFFICIAL EDITION NOTICE

AUTHENTICATED
U.S. GOVERNMENT
INFORMATION
GPO

Use of ISBN Prefix

This is the Official U.S. Government edition of this publication and is herein identified to certify its authenticity. ISBN 978-1-935352-02-0 (e-book ISBN 978-1-935352-09-9) is for this U.S. Government Printing Office Official Edition only. The Superintendent of Documents of the U.S. Government Printing Office requests that any reprinted edition clearly be labeled as a copy of the authentic work with a new ISBN.

THE UNITED STATES NAVAL WAR COLLEGE
VIRIBUS MARI VICTORIA

Legal Status and Use of Seals and Logos

For Sale by the Superintendent of Documents, U.S. Government Printing Office
Internet: bookstore.gpo.gov Phone: toll free (866) 512-1800; DC area (202) 512-1800
Fax: (202) 512-2104 Mail: Stop IDCC, Washington, DC 20402-00001

ISBN: 978-1-935352-02-0 e-book ISBN: 978-1-935352-09-9

NAVAL WAR COLLEGE HISTORICAL MONOGRAPH SERIES NO. 21

The historical monographs in this series are book-length studies of the history of naval warfare, edited historical documents, conference proceedings, and bibliographies that are based wholly or in part on source materials in the Historical Collection of the Naval War College.

The editors of the Naval War College Press express their special gratitude to all the members of the Naval War College Foundation, whose generous financial support for research projects, conferences, and printing has made possible the publication of this historical monograph.

To my siblings,
for all of the help along the way

Blue versus Orange: The U.S. Naval War College, Japan, and the Old Enemy in the Pacific

Hal M. Friedman

NAVAL WAR COLLEGE PRESS
NEWPORT, RHODE ISLAND
2013

The contents of this volume represent the views of the author. His opinions are not necessarily endorsed by the Naval War College or by any other agency, organization, or command of the U.S. government.

Printed in the United States of America

Historical Monograph Series

NAVAL WAR COLLEGE PRESS
Code 32
Naval War College
686 Cushing Road
Newport, R.I. 02841-1207

The Library of Congress has cataloged the printed edition as follows:

Friedman, Hal M., 1965-
Blue versus Orange : the U.S. Naval War College, Japan, and the old enemy in the Pacific, 1945-1946 / Hal M. Friedman. — U.S. Government official edition
pages cm
Includes bibliographical references and index.
ISBN 978-1-935352-02-0 (pbk.)
1. United States. Navy—Maneuvers—History—20th century. 2. Naval maneuvers—United States—History—20th century. 3. War games, Naval—United States—History—20th century. 4. Pacific Area—Defenses. 5. Japan—Strategic aspects. 6. Pacific Area—Strategic aspects. I. Title. II. Title: U.S. Naval War College, Japan, and the old enemy in the Pacific, 1945-1946.
V245.F75 2013
359.4'8097309044—dc23

2013013623

TABLE OF CONTENTS

LIST OF FIGURES AND MAPS

Figures

Maps

FOREWORD

The study of the history of naval warfare is an integral part of the Naval War College's educational programs. The importance of the discipline was firmly established with the foundation of the College in 1884 by the initial contributions of both the College's founder, Stephen B. Luce, and his successor as its president, Capt. Alfred Thayer Mahan. Historical research and analysis has continued as a recognized element of the academic life of the institution for over a century and a quarter. Nowhere has the history of warfare at sea been more thoroughly investigated and analyzed for the professional purposes of the U.S. Navy than at the Naval War College. Nowhere is there a more logical requirement for a corpus of relevant source materials or for an academic research department devoted to new research in naval history.

On 1 January 2003, the College's Maritime History Department was established as part of the Center for Naval Warfare Studies to carry out this function. Predating this, a program for the publication of books and source materials on the history of naval warfare was formally established by the College in 1975, with the series of books known as the Naval War College Historical Monographs. In order to encourage and make more widely known the College's extensive collections for historical research, including its archives, historical manuscripts, and associated materials in the Naval War College Museum's collection, the series has been restricted to publications of book-length works that deal with the history of naval warfare and that are based wholly or in part on the source materials in the Naval War College library's Naval Historical Collection and in the Naval War College Museum. As the series has developed over more than thirty-five years, these works have taken a variety of forms, including bibliographies and conference proceedings; many of them have been edited historical documents from the College's rich historical collections. This series is now managed by the Maritime History Department in collaboration with the Naval War College Press and the head of the Naval War College's Naval Historical Collection.

Consistent with the earlier books that have appeared in this series since 1975, this study of naval war gaming at the Naval War College between 1945 and 1946 is based primarily on materials in the Naval War College's archives and historical collection. With this volume, Dr. Hal Friedman, of Henry Ford Community College in Dearborn, Michigan, directly complements his earlier study in this same series, *Digesting History: The U.S. Naval War College, the Lessons of World War Two, and Future Naval Warfare, 1945–1947,* published in 2010. While the earlier volume looked at the broad range of the educational activities at the Naval War College under its twenty-sixth president, Adm. Raymond Spruance, this work focuses on the specific activity of war gaming, which the College had been developing since William McCarty Little delivered the first lecture in Newport on the subject in 1886. By 1946 the Naval War College had, over a period of sixty years, developed the art of war gaming into one of its signature teaching and research tools for thinking about naval warfare.

Today, after a span of another sixty years, the Naval War College has developed that art even further. In many ways, the war gaming conducted there today, with the assistance of computers and in a very modern building, may seem completely different from and foreign to what is described in this book. Yet Friedman's close analysis of the procedures and practice reveals that the fundamental purpose was much the same—an attempt to understand human reactions and decision making. Up to now, few naval historians have been able to penetrate the process and explain clearly the practices of war gaming in this period. The Naval War College Museum and the Naval War College library have collections of both paper records and three-dimensional objects associated with war gaming from earlier periods. While experts in war gaming clearly understand these processes, explanations that are accessible to a broader audience interested in naval history have been lacking.

With this volume, Dr. Friedman has made a significant contribution both to the history of the Naval War College and to the history of naval war gaming. His book evokes for the reader the laborious and detailed process of war gaming in this era and helps us to understand why an automated approach was attractive to officers engaged in this activity in the 1950s. It is particularly interesting to see the character of naval thinking in 1945–46, when the U.S. Navy was in the midst of transition from total war to peace. At the same time, one can begin to find traces of preparation for the beginning of the tensions that would emerge in the Cold War era. A clearer transition to the new era would become evident in the following year, the academic year 1946–47, which will be the subject of Friedman's next volume. In the earlier span of time that this volume covers, 1945–46, a number of officers on the Naval War College's staff and faculty had very recently seen action in the Pacific War and were using that experience to instruct a younger generation of officers in what they had learned directly about the nature and character of war at sea. Clearly, the American officers who played these war games retained respect for—and even wariness of—the former enemy that they had met in combat not long before.

JOHN B. HATTENDORF, D.PHIL.
Ernest J. King Professor of Maritime History
Chairman, Maritime History Department

PREFACE

This book is about the U.S. Naval War College's war-gaming activities in the academic year 1945–46. It is the second book in a planned trilogy on the immediate postwar Naval War College. My earlier work focused very generally on how the College viewed future naval warfare in 1945–47. This study looks more specifically at the war gaming that went on at the College in the same period, as will a planned third book. More particularly, while my first look at the Naval War College was a general exploration of the College's reaction to the end of the Second World War and the beginning of the Cold War, here I direct my attention to the Pacific, as I have done in previous monographs, because of the importance of that region to the Navy for the half-century prior to the period of the present book. In effect, I want to look at how the end of World War II and the beginning of the Cold War impacted the Naval War College in terms of its change of focus from Japan to the Soviet Union as the primary enemy in the Pacific Basin. This exploration may also prove especially fruitful in that the 1940s were precedent setting with respect to the origins of the Navy's 1980s Maritime Strategy.[1]

Accordingly, this book's prequel set the context by looking at Adm. Raymond Spruance's public addresses, lectures by instructors and guests, and student theses to determine what kind of future naval world these officers perceived and how the United States would have to conduct its peacetime and wartime naval operations. This second book looks at the war games held in the fall of 1945 and the winter of 1946 in which Japan, or "Orange," was still the primary enemy. The last book in the trilogy will explore how and why the Soviet Union, or "Purple," became in the 1946–47 academic year the hypothetical enemy.[2] These studies are particularly significant in illuminating the years between the end of the war and the formulation of Containment. The importance of the events between the fall of 1945 and the beginning of Containment as a coherent American foreign policy in the fall of 1947 is why I have chosen those dates as the chronological parameters for this book. This transition period is especially valuable as a window through which to explore institutions like the Naval War College in periods of transition from a hot war to a cold one.[3]

The immediate postwar period entailed the reconstitution of the College on a full-time basis after its reduced wartime status, first under the presidency of Vice Adm. William Pye and then under Admiral Spruance from March 1946 onward. Spruance was charged by Fleet Adm. Chester Nimitz, Chief of Naval Operations (CNO), with the strategic reformulation of American naval doctrine for the atomic and Cold War contexts of the postwar period. Some of these reforms had begun with Vice Admiral Pye even before the war ended. Pye, for instance, called for an expanded institution capable of teaching a tenfold increase in officers by way of a three-tiered educational program consisting of a Command and Staff Course, the War College Course, and an Advanced Course.[4]

Adm. Raymond Spruance, President of the Naval War College, 1946–48

Pye continued until the end of his tenure to preside over the six-month courses that had become the order of the day since before the war. He had begun, however, to make preparations for returning to a full course. By the time the war ended, the Naval War College had also started to consider joint service education for officers from the other services, as well as personnel from the State Department. Most important for purposes of this study, however, because of the sudden end of the war, the 1945–46 curriculum could not be changed to reflect the complete defeat of Japan or the rise of the Soviet Union as the probable next enemy of the United States. That would have to await the 1946–47 academic year.[5] This inability to completely update the curriculum goes far toward explaining some of the war games' major aspects, as do wartime Pacific Fleet practices. In general, the war games reflected strategic scenarios from the late Pacific War, and the operational and tactical aspects of the Exercises heavily reflected a combination of both interwar and wartime naval doctrine, with an equal emphasis on surface warfare and naval aviation. If any major naval combat arm was relegated to a lesser status in the war games, it was the submarine, especially in its maritime interdiction role. One major purpose of this book will be to explain the pedagogical, institutional, and operational reasons for this phenomenon in the war games.[6]

I have been selective in analyzing the war games, or "Operations Problems," as they were called. For the fall term—when two classes, the Command and Staff and Preparatory Staff classes of December 1945, were present—I have covered only

Vice Adm. William Pye, President of the Naval War College, 1942–46

Operations Problems 4, 5, and 6. I chose not to study Operations Problem 3 from the fall, since it dealt with operations in the Indian Ocean, not the Pacific. I also looked at Operations Problems 1 and 2 in the winter term, when the two groups were combined into the Command and Staff class of June 1946, but I did not do so for the fall, since they repeated the same Problems. There was no Operations Problem 6 in the winter, Operations Problem 5 that term had to do with the Soviet Union in the North Atlantic, and Operations Problem 4 in the winter was a repetition from the fall. Also repeated from the fall were various Search, Communication, and Board Maneuver Exercises; the reader will find these analyzed only once.

Not all of the Exercises were composed of identical sections. Some of the Operations Problems contained what the Naval War College called "Statements of the Problem," "Staff Solutions" for both "Blue" (the United States) and Orange, a record of the actual board game (called the "History of Maneuver"), and in a few cases Critiques. Most of the Operations Problems, however, lacked at least some of these elements. The chapters that follow record the Operations Problems in detail, but they are not identical in organization, because of the nature of the surviving records.[7] In addition, I have devoted several chapters to describing in detail what the Naval War College called "Maneuver Rules." These were the procedures by which Exercises and Operations Problems were conducted, including such details as the computation of damage to ships and planes and the impact of weather. While previous studies have outlined the maneuvers and the rules in basic form, I have devoted more space to detail in this monograph.[8]

Fleet Adm. Chester Nimitz, Chief of Naval Operations, 1945–47

The reader will also find this book a straightforward, chronological narrative. I have not attempted to analyze deeply, or "deconstruct," what the historical actors have said but instead have taken great pains to record what was done at the time, while creating a readable narrative. Even the occasional apparent contradictions in the sources have been preserved, unresolved—there being no external "reality" to which to appeal. I think the value of this type of account is to allow readers to see what these war games then represented so as to make their own judgments about the ideas enunciated at the time. Moreover, I have found myself a bit reluctant to judge these officers harshly if their ideas about future naval warfare turned out to be wrong or even wrongheaded. This was a very fluid time, and there was no quick formula to produce the "answers." These officers were preparing strategic, operational, and tactical doctrine for the security of the Republic against a threat they knew almost nothing about, with potentially revolutionary new weapons that they knew almost nothing about either—an awesome challenge indeed. Having to provide those answers at the conclusion of a major war as a new type of limited war was facing the country, in a context of limited resources, must have made the task even more daunting.

A narrative account is also in order since this work marks out fairly new territory, historiographically speaking. While there are seminal works about the Naval War College in the interwar period and its impact on the Pacific War, there have not been many on the post-1945 period.[9] Some of the works available are organizational histories and studies of the College;[10] others, strategic in nature, focus on the College's role in a limited chronological sense, very early in the Cold War;[11] yet others look at its role much later in that era.[12] In addition, there were articles written at the time under Naval War College auspices, lectures by guests and instructors, and theses by students on American naval strategy during the Cold War.[13] So far, however, there seem to be no monographs covering the Naval War College's contribution to strategic policy throughout the entire Cold War. This work does not fill that niche, nor will my future monograph. By focusing on the immediate postwar period, however, I hope to complement the prewar studies as well as those that have been done on the Naval War College during the Cold War. I also hope this study is a guide for other scholars who are seeking a more complete explanation of this time period and who plan on future research into the Naval War College's role during the Cold War.

NOTES 1 Historians such as Michael Palmer, Richard Hegmann, and John Hattendorf have illustrated that the Maritime Strategy was not created by Secretary of the Navy John Lehman in the early 1980s but originated in the late 1940s and was developed throughout the remainder of the Cold War. The strategy included relegation of the Pacific theater to a subordinate status, after the Atlantic, the Mediterranean, and the Persian Gulf. The Pacific was nevertheless seen as important, since American naval forces were to keep Soviet forces in East Asia occupied so they could not reinforce Russian forces in Europe and the Middle East; see Palmer, *Origins of the Maritime Strategy: The Development of American Naval Strategy, 1945–1955* (Annapolis, Md.: Naval Institute Press, 1990). The strategy continued to develop in the 1950s and 1960s; see Richard Hegmann, "Reconsidering the Evolution of the US Maritime Strategy, 1955–1965," *Journal of Strategic Studies* 14 (September 1991), pp. 299–336. By the late 1970s, the Strategic Concepts Branch in the Office of the Chief of Naval Operations (OPNAV), as well as OPNAV's Strategic Studies Group and the Naval War College's Center for Naval Warfare Studies—the latter two groups established in Newport in 1981—resurrected the northern and western Pacific as regions of American naval-force power projection that would be critical in preventing the movement of Soviet reinforcements from the Far Eastern Maritime Provinces of the Union of Soviet Socialist Republics (USSR) to Europe in case of a war with the North Atlantic Treaty Organization (NATO); see John Hattendorf, ed., *U.S. Naval Strategy in the 1970s: Selected Documents,* Newport Paper 30 (Newport, R.I.: Naval War College Press [hereafter NWCP], 2007); and Hattendorf, *The Evolution of the U.S. Navy's Maritime Strategy, 1977–1986,* Newport Paper 19 (Newport, R.I.: NWCP, 2004).

2 See Hal M. Friedman, *Digesting History: The U.S. Naval War College, the Lessons of World War Two, and Future Naval Warfare, 1945–1947* (Newport, R.I.: NWCP, 2010). For NWC's color coding of countries, see Michael Vlahos, *The Blue Sword: The Naval War College and the American Mission, 1919–1941* (Newport, R.I.: NWCP, 1980), pp. 97–130.

3 For the concept of hot war to cold, see Jeffrey Barlow, *From Hot War to Cold: The U.S. Navy and National Security Affairs, 1945–1955* (Stanford, Calif.: Stanford Univ. Press, 2009).

4 John Hattendorf, B. Mitchell Simpson, and John Wadleigh, *Sailors and Scholars: The Centennial History of the U.S. Naval War College* (Newport, R.I.: NWCP, 1984), pp. 175–77. See also Nepier Smith, "Historical Analysis of the Organizational Success of the Naval War College during the Twenty-Five Years following the Second World War" (Naval War College Advanced Research Project, Naval Historical Collection, Naval War College, Newport, R.I., 1974), pp. 11–67.

5 Hattendorf et al., *Sailors and Scholars,* pp. 175–77; Smith, "Historical Analysis," pp. 11–67.

6 For detailed accounts of interwar American naval doctrine, see Joel Davidson, *The Unsinkable Fleet: The Politics of U.S. Navy Expansion in World War II* (Annapolis, Md.: Naval Institute Press, 1996), pp. 11–12, 14–16, 19–21, 23–24, 32, 34, 60, 96–97; Thomas Wildenberg, *Destined for Glory: Dive Bombing, Midway, and the Evolution of Carrier Airpower* (Annapolis, Md.: Naval Institute Press, 1998), pp. 48–64, 83–98, 126–28, 141, 155, 157–60, 163–64, and 170–71; William McBride, *Technological Change and the United States Navy, 1865–1945* (Baltimore, Md.: Johns Hopkins Univ. Press, 2000), pp. 111–212; Thomas Hone and Trent Hone, *Battle Line: The United States Navy, 1919–1939* (Annapolis, Md.: Naval Institute Press, 2006), pp. 110–25; Craig Felker, *Testing American Sea Power: U.S. Navy Strategic Exercises, 1923–1940* (College Station: Texas A&M Univ. Press, 2007), pp. 61–75; John Kuehn, *Agents of Innovation: The General Board and the Design of the Fleet That Defeated the Japanese Navy* (Annapolis, Md.: Naval Institute Press, 2008), pp. 87–92, 173–75, and 205; Joel Holwitt, *"Execute against Japan": The U.S. Decision to Conduct Unrestricted Submarine Warfare* (College Station: Texas A&M Univ. Press, 2009), pp. 22, 62–83; and Albert Nofi, *To Train the Fleet for War: The U.S. Navy Fleet Problems, 1923–1940* (Newport, R.I.: NWCP, 2010), pp. 25, 51–55, 60, 62, 64–66, 74, 76–77, 80, 85–86, 93–94, 100–102, 104–105, 110–17, 121–25, 129–36, 139–46, 151, 155–59, 169, 197–203, 207–16, 219–27, 229–37, 253–63, and 287–88. For an excellent account demonstrating how wartime practices were really a combination of interwar doctrine and wartime lessons, see Trent Hone, "U.S. Navy Surface Battle Doctrine and Victory in the Pacific," *Naval War College Review* [hereafter *NWCR*] 62, no. 1 (Winter 2009), pp. 67–105.

7 Dr. Evelyn Cherpak, curator of NWC's Naval Historical Collection, e-mails to author, 9 June, 18 June, 22 June, and 19 July 2010.

8 Ronald Spector, *Professors of War: The Naval War College and the Development of the Naval Profession* (Newport, R.I.: NWCP, 1977), pp. 71–87; Vlahos, *Blue Sword,* pp. 131–56.

9 Edward Miller, *War Plan Orange: The U.S. Strategy to Defeat Japan, 1897–1945* (Annapolis, Md.: Naval Institute Press, 1991), and Vlahos, *Blue Sword*. For a recent work that looks briefly at the interwar period as a prelude to the Pacific War, see Douglas Smith, *Carrier Battles: Command Decision in Harm's Way* (Annapolis, Md.: Naval Institute Press, 2006).

10 J. S. Hurlburt, "War Gaming at the Naval War College, 1969–1989," *NWCR* 42, no. 3 (Summer 1989), pp. 46–51; Hattendorf et al., *Sailors and Scholars;* Smith, "Historical Analysis," cited in Hattendorf et al., *Sailors and Scholars,* p. 248; Philip Crowl, "Education versus Training at the Naval War College: 1884–1972," *NWCR* 26, no. 3 (November–December 1973), pp. 2–10; James Barber, "The School of Naval Warfare," *NWCR* 21, no. 8 (April 1969), pp. 89–96.

11 Robert Fisher, "The U.S. Navy's Search for a Strategy, 1945–1947," *NWCR* 48, no. 3 (Summer 1995), pp. 73–86.

12 Hattendorf, ed., *U.S. Naval Strategy in the 1970s;* Hattendorf, *Evolution of the U.S. Navy's Maritime Strategy;* Bud Hay and Robert Gile, *Global War Game: The First Five Years,* Newport Paper 4 (Newport, R.I.: NWCP, 1993); Robert H. Gile, *Global War Game: Second Series, 1984–1988,* Newport Paper 20 (Newport, R.I.: NWCP, 2004); David Rosenberg, "Being 'Red': The Challenge of Taking the Soviet Side in War Games at the Naval War College," *NWCR* 41, no. 1 (Winter 1988), pp. 81–93; Robert Wood, "The Conceptual Framework for Strategic Development at the Naval War College," *NWCR* 40, no. 2 (Spring 1987), pp. 4–16; James Barber, "Mahan and Naval Strategy in the Nuclear Age: A Lecture Delivered at the Naval War College," *NWCR* 24, no. 7 (March 1972), pp. 78–88; Donald White, "Admiral Richard L. Conolly: A Perspective on His Notions of Strategy," *NWCR* 24, no. 3 (November 1971), pp. 73–79. For an example of this type of literature from the post–Cold War period, see Kenneth Watman, "Global 2000," *NWCR* 54, no. 2 (Spring 2001), pp. 75–88.

13 These articles numbered in the dozens during the Cold War, but a sample from primarily the 1950s includes Robert Carney, "Role of the Navy in a Future War," *NWCR* 6, no. 10 (June 1954), pp. 1–12; George Phelan, "Sea Power and Strategies for the Control of the Seas," *NWCR* 6, no. 10 (June 1954), pp. 15–36; James Field, "Seapower and Military Strategy Today," *NWCR* 8, no. 8 (April 1956), pp. 21–39, and "Origins of Maritime Strategy and the Development of Sea Power," *NWCR* 7, no. 7 (March 1955), pp. 1–23, and "The Influence of Sea Power on Modern Strategy," *NWCR* 10, no. 1 (September 1957), pp. 31–52; J. F. McInteer, "The Significance of Seapower to the United States," *NWCR* 12, no. 1 (September 1959), pp. 1–32; Joseph Bredestege, "Limited Nuclear War at Sea," *NWCR* 19, no. 6 (February 1967), pp. 4–31; and Harry James, "An Analysis of Limited Maritime War," *NWCR* 19, no. 6 (February 1967), pp. 33–74.

ACKNOWLEDGMENTS

I have incurred many debts in the process of producing this monograph. None of the individuals or organizations mentioned here is in any way responsible for the opinions asserted in this work. Any accountability for interpretations or errors is mine alone. However, each of the people cited assisted me in a very significant way and deserves thanks for that assistance.

Like my earlier works, this book would not have been possible without my wife Lisa. I cannot imagine what things would be like without her. She has been not only my soul mate in terms of our life together but also the most active supporter of career aspirations that any spouse could ask for. This support has come while she has pursued her own career and has taken the central role in our son's activities. I simply do not know what I would have done without her during these nearly twenty-five years of personal and professional life.

My son, Jeffrey, is now an adult. My hope is that this work will add further to his appreciation of studying the past. More exactly, I hope this work will further demonstrate to him what his grandparents' world was like. I also hope that this book gives him some idea of what I did during all of those hours in front of the computer and what contributions I am trying to make to my profession. Finally, I hope that he personally will never have to experience war or its aftermath firsthand, and I hope that his country's future imperial adventures are kept to a minimum in his lifetime.

My parents, Irving and Elaine Friedman, continue to represent my personal connection with the past. It has been nearly a decade since my father's death and over seven since my mother's. It is strange for my siblings and me to be without parents. For me, studying 1940s America brings them a bit back to life, if only momentarily and in certain dimensions. Their lessons to me about the Great Depression and the war taught me more about history—both writing it and teaching it—than they probably knew. In all kinds of ways, my parents' absence in my life will be particularly felt as I write additional monographs in the years to come.

I continue also to gain great support from my in-laws, Ronald and Carolyn Sampsell, who are fortunately both still with us and hold special places in the creation of my books, my thesis, and my dissertation. They provided Lisa and me with various kinds of support during graduate school. Since then, they have been the best kind of inspiration to our son. I am so glad that he has been able to grow up in close physical proximity to at least one set of grandparents, as I did. They have also provided me with many historical lessons as well, since they were part of the last, sizable generation in the United States to grow up on farms. My in-laws also continue to be advisers to me concerning my career, as they were throughout the 1990s.

I continue to benefit as well so much from my interaction with my siblings, and this book is dedicated to them. My sister Karen and my brother-in-law Doug have been an example to me through the hard work they have demonstrated in their careers and their attention to professionalism and detail. My sister Nancy was particularly important in instilling me the idea of attending college and obtaining a professional position in whatever field I chose. I especially thank my sister Margaret and my brother-in-law Willard for providing encouragment early in my professional life and for continuing to give me career advice even as they transitioned to new careers. I've also been helped immensely by my sister Phyllis and brother-in-law Mark through their support, which included a day on the nuclear-powered aircraft carrier USS *Ronald Reagan,* thanks to their friend, then the executive officer, Capt. (now Rear Adm.) Herman Shelanski. Days spent like that, and walks around the fast battleship USS *Wisconsin* in downtown Norfolk, truly inspired me to start writing about American naval officers in the postwar period and their ideas about future wars. I also want to thank my brother Alan for introducing me to Professor John Bowditch (now deceased), formerly chair of the History Department at the University of Michigan, and for always listening when I talked about how difficult it is to publish professionally while teaching full-time at a community college.

Two individuals who should have been thanked for helping me with my first book but were not, through my own oversight, were my godparents—my great aunt and great uncle, Rose and Mac Carp. They have, like my parents, been particular examples to me about pursuing a career as a historian. Both are now, unfortunately, deceased, but they provided me with financial support through my undergraduate years as well as moral support afterward. My Uncle Mac came to Canada from imperial Russia in 1913; with his thick accent, his ability to converse in several languages, and his constant interest in historical and political affairs, he always reminded me of the typical—or perhaps stereotypical—European intellectual now living in the United States. He was, in fact, the intellectual of our family. Similarly, my Aunt Rose, raised in a Jewish farming colony in Bad Axe, Michigan, and then educated as a school librarian, was the teacher of the family. She was also a feminist in her own time—or, as one cousin said, she was the "Katharine Hepburn" of the family. Standing six feet tall, deciding to attend college and then graduate school in the 1920s and 1930s, establishing her career, wearing slacks, and marrying late, Aunt Rose was a unique example of living history in and of herself. I greatly miss our conversations about the past, but I'm glad she was alive during my early years.

I must also thank many colleagues at Henry Ford Community College (HFCC). Several stand out: President Andrew Mazarra, now retired; his successors, Interim President Sally Barnett and President Gail Mee; Vice President/Dean of Academic Education Edward Chielens, now retired; his successors Interim Vice President/Dean of Academic Education Lynne Hensel and Vice President of Academic Affairs for Arts and Sciences Reginald Gerlica; Social Science Associate Deans Henry Bowers and Kim Schopmeyer; and John McDonald, president of the Henry Ford Community College Federation of Teachers (American Federation of Teachers Local 1650). John McDonald is deserving of a special thanks, since he has been a champion of faculty contractual access to professional development opportunities, particularly at community colleges, where time for research and scholarship is at a premium. He negotiated the provision in our contract for faculty sabbaticals, and Presidents Mazarra's and Mee's support in approving the sabbaticals for the fall semester of 2003 and the winter semester of 2011, respectively, are greatly appreciated.

The Executive Board and rank and file of Local 1650 have also gone far in creating numerous sources of funding for instructors to pursue research for professional development, especially conference presentations. Vice President/Controller Marjorie Swan and (now retired) Vice President/Dean of Student Services Michael Meade additionally approved several of my requests for such funding, during which ideas for this book were presented and refined. I have also received helpful hints and ideas from our college library staff, especially Kathy Cunningham, Pat Doline, Dan Harrison, Barb Lukasiewicz, Terrence Potvin, and Nancy Widman. A special thanks goes to Terry Potvin for major assistance with interlibrary loan requests for the research on this book.

Maggie Anderson, Tom Anderson, Rodney Barnhart, Hani Bawardi, Shatha Baydoun, Kathryn Beard, John Burks, Mario Di Ponio, Kimberly Dyer, Michael Johns, Saeed Khan, Cris Lewandowski, Lynda Litogot, Gary Mitchell, Devissi Muhammad, Wendy Osthaus, Sam Plaza, Pamela Sayre, Bill Secrest, Ken Shepherd, Reginald Witherspoon, and Sue Zimmerman, as well as retirees Virginia Caruso, Bill Hackett, Robert Spiro, Morris Taber, the late Raye Howe, the late Armen Ovhanesian, the late Berny Rogers, the late Vince Solomon, the late Art Thomas, and the late Bill Watson—all these colleagues have gone even farther in fostering this type of professional environment. Whether they realize it or not, they have created and sustained a History Department at Henry Ford Community College that is the most conducive environment to unemcumbered exchange about pedagogy, research, writing, and historiography that I have ever encountered in the field. I value our comradeship more than I can express in words, and I hope it continues for a very long time. I have to particularly thank Pamela Sayre for taking on much of the day-to-day departmental administration that is so typical in small colleges. Her willingness to take on these tasks freed much of the time I needed to write my books. I have also especially valued the mentoring provided to me by Mario Di Ponio and Robert Spiro and their advice to keep teaching, service, scholarship, and life "in perspective."

To the list of HFCC historians must be added the names of four now-retired HFCC political scientists, Brian Coyer, the late Tom Payette, John Smith, and Barbara Suhay. In particular, my conversations with Brian about the literature on bureaucratic rivalry were quite useful, and John lectured me—and still does—about the importance of publishing as a community-college scholar. In addition, I have to acknowledge several colleagues with whom I have had general conversations about political and international affairs that have added to this book, especially Nabeel Abraham, Henry Bowers, Pete Cravens, Kalvin Harvell, Tarek Joseph, Dan Kearney, Greg Osowski, Tony Perry, Eric Rader, Kim Schopmeyer, Brian Smith, and Cynthia Stiller. Special thanks are due to Tony Perry and Kim Schopmeyer for helping me with computer problems that developed during the preparation of the manuscript, as well as to Dr. Vivian Beaty, professor of Instructional Technology, and Glenn Gaddis, instruction technician, for assistance with the photographs in the book.

Numerous people at the Naval History and Heritage Command (formerly the U.S. Naval Historical Center) in Washington, D.C., helped me get a start on this project in the 1980s and 1990s and continued on to help me finish it in the early twenty-first century. These individuals include Dr. Dean Allard, Bernard Cavalcante, and Kathy Roar. More particularly with this project, as well as my previous ones, I must thank Kathy Lloyd, now-retired head of the Operational Archives Branch, for her flexibility in providing me with research facilities. Included also are Dr. Regina Akers, John Hodges, Ken Johnson, Allen Knechtmann, Tim Pettit, and Mike Walker, all of whom have made my research excursions to the Operational Archives so successful. In addition, this last bit of research would not have been very successful if it were not for now-retired Senior Historian Ed Marolda, former Contemporary History Branch chief Gary Weir, Contemporary Branch historians Jeff Barlow and John Sherwood, and former Contemporary Branch historian Randy Papadopoulos (now the secretariat historian in the Office of the Under Secretary of the Navy).

The staff at the U.S. Naval War College were indispensable to the creation of this book. Thanks go first to philanthropist Edward Miller for generously establishing the Edward S. Miller Research Fellowship in Naval History. That award allowed me to travel to Newport, Rhode Island, in October 2003 and obtain the materials noted above. Thanks must also go to Rear Adm. Joseph Strasser, USN (Ret.), then executive director of the Naval War College Foundation, and Dr. John Hattendorf, the Ernest J. King Professor of Maritime History, chair of the Maritime History

Department, and director of the Naval War College Museum, for coordination of the award and especially for encouragement in my research efforts. Professor Hattendorf was particularly helpful in the logistics of administrative support as was his assistant in 2003, Patricia Cormier, who did more photocopying than I had a right to ask of her. Putting in an equally wonderful effort during the week when I was in Newport were Evelyn Cherpak, curator of the War College's Naval Historical Collection, and Teresa Clements, archival technician, both of whom pulled numerous records for me as I raced against the research clock. At subsequent meetings and conferences, as well as through numerous emails, Drs. Hattendorf and Cherpak were extraordinarily supportive of my work. The staff at the Naval War College have similarly made this endeavor a very fruitful and easy professional project. Thanks go especially to Dr. Carnes Lord, director of the Naval War College Press; Pelham Boyer, managing editor; Phyllis Winkler, book review editor; Robert Cembrola of the Naval War College Museum; John Kennedy, Director of Museum Education and Public Outreach; Art Lamoureux, visual information specialist; Robin Lima, Interlibrary Loans at the Naval War College Library; and Lori Almeida, the Naval War College Press administrative assistant. Critical to the book's production have been Jerry Lenihan, John Lanzieri, Shannon Cole, and Albert Fassbender of the Desktop Publishing Division at the Naval War College.

At Michigan State University and the University of Michigan, a number of people again need to be mentioned. To Don Lammers, thanks are again due for being such a professional example, not only as a dissertation director but as a teacher, scholar, and colleague as I embarked on writing these additional monographs. To John Shy at the University of Michigan, I still have to thank you for helping me get my career off the ground, even though you don't want to be thanked. In the end, suffice it to say that you continue to inspire me with your example as a scholar and that I am paying you back by "paying forward."

Other individuals were also key to this study being completed. Chris Hamel taught me what being a historian is all about. Not only did he instruct me in how and why historians analyze and write the way they do, but his help in classifying my early work into strategic and bureaucratic dimensions was the starting point for my postdissertation monographs. Our long conversations about these matters were a basic building block on which this work flourished, and his assistance during graduate school was vital to the beginning of my career. I continue to admire him as well for being able to put up with the often ridiculous issues of day-to-day academic administration while still staying focused on the Big Picture.

Along these same lines, several colleagues have been instrumental in providing me with career advice and assistance at numerous points. Michael Palmer of East Carolina University, Gary Weir, now chief historian of the National Geospatial-Intelligence Agency, and Carol Reardon of Penn State University have been informal mentors to me in the last twenty years, either because of the books they wrote that helped matters gel for me or for pointers they provided me in numerous conversations over the phone, by e-mail, or in person at professional conferences having to do with teaching, publishing, or job searches. In addition, I have to thank Jonathan Winkler of Wright State University for numerous conversations about the strategic and national-security history of the United States, and especially the logic of the Naval War College exercises and maneuvers in the 1940s. I also have to thank James Levy of Hofstra University for his feedback and amplification about research on U.S. Navy surface warfare in the early Pacific War, and Trent Hone, director of Softwar Engineering at Trimble Mobile Resource Management. In particular, Trent's articles on U.S. Navy tactical surface doctrine in the interwar and Pacific War periods, as well as our e-mail conversations, literally put everything into perspective for this book. Most especially, I need to belatedly thank Marc Gallicchio, current chair of the History Department at Villanova University. Marc not only wrote a letter of recommendation for my application for the Miller Fellowship in 2003 but was courteous enough to look the application package over before I submitted it. In addition, he has always been ready with advice about my place in the profession whenever I've asked in the last twenty-plus years.

My professors at Eastern Michigan University (EMU), my undergraduate alma mater, are perhaps even more central to this study, since they first introduced me to serious historical study and taught me that college teachers can also be first-rate scholars. I must thank the late Richard Abbott for first discussing the idea of the United States in the 1940s Pacific with me during the 1980s and for continuing to discuss this project with me as I pursued it in graduate school and as he scuba-dived in Micronesia. I still find it ironic and more than a bit intimidating that I taught his course in U.S. military history on an occassional basis during and after his terminal illness, and I am trying to follow in his footsteps as much as possible. He continues to be greatly missed.

I must also thank Don Briggs, George Cassar, Robert Grady, Mike Homel, Jim Johnson, Roger King, Karen Linderman, Jim Magee, Walter Moss, Joe Ohren, James Pfister, Leonis Sabaliunas, Les Shearer, Janice Terry, Jiu Upshur, Reinhard Wittke, and the late Richard Goff for instilling in me the professionalism and high standards which I was able to hone in graduate school and am now able to impart to students at Henry Ford Community College. The History and Political Science Departments at Eastern Michigan in the 1980s were the most encouraging environments I can think of in which a first-generation college student from a working-class background could get started in the profession.

Similarly, I have to thank several individuals from the current EMU History and Philosophy Department who have been instrumental in my graduate education, my career as an instructor, and my development as a historian. These include—but are not limited to—past and present members Marsha Ackerman, Kathy

Chamberlein, Mark Higbee, Jim Holoka, Richard Nation, Gersham Nelson, Philip Schmitz, and Matt Schumann. Especially helpful in these matters has been Michael Homel, whose EMU History Reader's Group has been a wonderful outlet of professional activity for me and a magnificent release from the rigors of teaching.

Most especially at EMU, I have to thank Rob Citino, now at the University of North Texas. Rob and I have become colleagues in the last few years, and he has become a professional model for me to emulate. In just over twenty-five years of teaching, always with heavy classroom and administrative loads, Rob has managed to publish nine monographs, is now working on additional ones, and has become a recognized specialist in his area, all while maintaining an extraordinary reputation for teaching excellence in the classroom. I'll never catch up, Rob, but I am sure going to try!

I would also like to acknowledge the assistance I received from my membership in the Michigan War Studies Group at the University of Michigan's Department of History. Since I joined that informal weekly group in January 1991, MWSG has been a forum for communicating my ideas, my hopes, and my fears about professional life in general. Special thanks go to past and present members Saiful Islam Abdul-Ahad, Keith Arbor, Wil Blythe, Bill Boardman, Denver Brunsman, George Cassar, Rob Citino, Tom Collier, David Fitzpatrick, Paul Forage, Jim Hill, Jim and Chris Holoka, Bob Jefferson, Doron Lamm, Sheldon Levy, Gerald Lindermann, Jonathan Marwil, Dennis Ringle, Mike Riordon, Pamela Sayre, Matt Schumann, Stanley Shapiro, Jack and Alan Sherzer, Ken Slepyan, and Bruce Zellers for constantly providing me with new perspectives on military history and strategic thought.

In addition, the Metro Detroit Historians Collegium has been another outlet for my professional activities. A group of historians from small, two-year and four-year college history departments in the Detroit area, the collegium has become a local forum for history instructors at institutions that have minimal resources for professional development. The collegium was kind enough to hear a presentation on my first book and continues to be significant to my professional development. Within the collegium, I especially have to thank Jayne Morris-Crowther, formerly of Madonna and Oakland Universities; Tom Klug from Marygrove College; Tony Baracco, Shawn Dry, Valerie Emanoil, Ed Gallagher, Tim Koerner, Marilynn Kokoszka, Nancy Shockley, and Mike Vollbach from Oakland Community College; Steve Berg, Evan Garrett, Michael Johns, and Michael Swope from Schoolcraft College; Roy Finkenbine from the University of Detroit-Mercy; and Duane Ashley from Wayne County Community College. In addition, I have to thank my colleagues from HFCC who helped start the Collegium, namely, John Burks, Virginia Caruso, Mario Di Ponio, Michael Johns, Richard Marquis, Devissi Muhammad, Pamela Sayre, Bill Secrest, Ken Shepherd, Bob Spiro, and Wendy Osthaus.

ACRONYMS

AA	antiaircraft
AD	destroyer tender
AE	ammunition ship
AF	provision ship
AG	miscellaneous auxiliary ship
AGC	amphibious force flagship
AGS	survey ship
AH	hospital ship
AK	cargo ship
AKA	attack cargo ship
AKS	general stores issue ship
AM	minesweeper
AN	net layer
AO	fleet oiler
AOG	gasoline tanker
AP	transport
APA	attack transport
APD	high-speed transport
APL	barracks ship
AR	repair ship
ARB	battle-damage repair ship
ARG	internal combustion engine repair ship
ARL	landing craft repair ship
ARS	salvage vessel
ASW	antisubmarine warfare
ATCOM	Atoll Commander
ATF	fleet tug
ATO	ocean tug
ATR	rescue ocean tug
ATS	Army Transport Service
BATDIV	battleship division
BB	battleship

BUPERS	Bureau of Naval Personnel
CA	heavy cruiser
CAP	combat air patrol
CARDIV	Carrier Division
CARGRU	Carrier Group
CB	battle cruiser
CCS	Combined Chiefs of Staff
CG	Commanding General
CGC	Coast Guard cutter
CIC	Combat Information Center
CINC	Commander-in-Chief [game]
CINCPACFLT	Commander-in-Chief, Pacific Fleet
CINCPOA	Commander-in-Chief, Pacific Ocean Areas
CINCSWPA	Commander-in-Chief, Southwest Pacific Area
CL	light cruiser
CLAA	antiaircraft cruiser
CNO	Chief of Naval Operations
COM 14	Commandant, Fourteenth Naval District [or any numbered naval district]
COM5FLEET	Commander, (Fifth) Fleet [or any numbered fleet]
COM7PHIBFOR	Commander, Amphibious Force, Seventh Fleet
COMAIR	Commander, Air Forces
COMAIRPAC	Commander, Air Forces, Pacific
COMBATDIV	Commander, Battleship Division
COMCARDIV	Commander, Carrier Division
COMCARGRU	Commander, Carrier Group
COMCRUDIV	Commander, Cruiser Division
COMDESDIV	Commander, Destroyer Division
COMDESPAC	Commander, Destroyers, Pacific
COMDESRON	Commander, Destroyer Squadron
COMFAIRWING	Commander, Fleet Air Wing
COMFWDAREA	Commander, Forward Area, Central Pacific
COMINCH	Commander in Chief, U.S. Fleet
COMMINDIV	Commander, Mine Division
COMNORPACFOR	Commander, Northern Pacific Force
Como.	Commodore
COMSERPAC	Commander, Service Force, Pacific
COMSERRON	Commander, Service Squadron
COMSOPAC	Commander, South Pacific Force
COMSOWESTPAC	Commander, Southwest Pacific
COMSUBPAC	Commander, Submarine Force, Pacific Fleet
CORTRON	escort squadron
CRUDIV	cruiser division
CTF	Commander, Task Force
CTG	Commander, Task Group
CV	fleet aircraft carrier

CVB	battle carrier
CVE	escort carrier
CVL	light carrier
DD	destroyer
DE	destroyer escort
DESDIV	destroyer division
DESRON	destroyer squadron
DL	destroyer leader
DMS	destroyer minesweeper
FAIRWING	Fleet Air Wing
GCT	Greenwich Civil Time
HF	high frequency
IFF	Identification Friend or Foe
IJN	Imperial Japanese Navy
ISCOM	Island Commander
JCS	Joint Chiefs of Staff
JICPOA	Joint Intelligence Center, Pacific Ocean Areas
JM	utility aircraft
KCS	kilocycles per second
LCI	landing craft, infantry
LCS	landing craft, support
LCT	landing craft, tank
LSM	landing ship, medium
LST	landing ship, tank
MCS	megacycles per second
MESA	Maximum Engine Speed Allowed
MINRON	Mine Squadron
MN	very-high-frequency voice radio
MTB	motor torpedo boat
NATO	North Atlantic Treaty Organization
NHC	Naval Historical Collection [at NWC]
NSD	Naval Supply Depot
NWC	Naval War College
NWCP	Naval War College Press
NWCR	*Naval War College Review*
OCL	prewar modernized light cruiser
ODD	prewar modernized destroyer
OPNAV	Office of the Chief of Naval Operations
OTC	Officer in Tactical Command
PACOM	Pacific Command
PATDIV	patrol division
PC or SC	submarine chaser
PE	patrol vessel
RG	record group
SC or PC	submarine chaser
SERVRON	service squadron

SOPA	Senior Officer Present Afloat
SS	submarine
SUBDIV	submarine division
SWPA	Southwest Pacific Area
TBS	Talk Between Ships
TF	task force
TG	Task Group
TU	Task Unit
UN	United Nations
USSR	Union of Soviet Socialist Republics (Soviet Union)
VAC	Fifth Amphibious Corps [or any numbered amphibious corps]
VF	fighter aircraft
VFB	fighter-bomber
VG	light transport plane
VHF	very high frequency
VOS	observation scout aircraft
VP	naval patrol aircraft
VPB	naval patrol bomber
VS	scout aircraft
VSB	scout bomber
VSO	scout observation aircraft
VTB	torpedo bomber
YF	provision store lighter
YMS	district motor minesweeper
YP	yard patrol boat

I *Context at the College*

The Setting at the U.S. Naval War College, 1945–1946

The Naval War College's curriculum, like the Navy as a whole and in fact the entire American national security establishment, was in great flux in 1945–46. Some of the change at the College had begun before the end of the war, during the tenure of Rear Adm. (later full admiral) Edward Kalbfus, President of the Naval War College from 1934 to 1936 and again from 1939 to 1942. It continued in the tenure of Vice Adm. William S. Pye, who was President from November 1942 until March 1946. Kalbfus's second term as President entailed ensuring that the Naval War College (NWC) stayed open during the war and remained vital to the Navy's mission even as most officers were sent to the various combat theaters. Toward the end of staying operational while the Navy geared up for and then began fighting the war, Kalbfus took on several additional, administrative positions as the commander of Naval Operating Base Newport. Because he feared that the wartime drawdown at NWC would reduce the President's billet below flag level, Kalbfus also took on command of the base's Naval Training Center, the Naval Net Depot, the Naval Torpedo Station, Naval Air Station Quonset Point, Naval Fuel Depot Melville, and the Naval Hospital.[1]

The Naval War College Curriculum, 1945–1946

At NWC, Kalbfus presided over an institution whose traditional three resident courses—Advanced, Senior, and Junior—conducted over an eleven-month period had been curtailed in May 1941 in favor of five-month courses. One of these, the Command Course, was for more senior regular Navy officers with more than six years of commissioned service; it would focus largely on the strategic aspects of naval warfare. The second course, the Preparatory Staff Course, was for more junior officers—including Naval Reserve, Army, and Marine Corps officers, State Department Foreign Service Officers, and foreign military officers—and would address the fundamentals of naval operations, such as staff procedures, and the relationship between the tactical and strategic aspects of naval warfare. A third course, the Correspondence Course, was to be retained in its previous form throughout the war and the early postwar period.[2]

By 1946, under Admiral Pye's direction, a new organization was in place at the Naval War College. Between 1941 and 1943, for instance, the College was organized into departments of Operations, Intelligence, and Administration. The Department of Operations was further divided into sections for the Correspondence Course, Command Course Strategy and Tactics, and Preparatory Staff Course Tactics and Minor Strategy. By 1946, there were separate departments of Operations, Intelligence, and Research & Analysis, as well as the Correspondence Course. In addition, Operations was now divided into Strategy, Tactics, Logistics, and Aviation sections.[3]

Adm. Edward Kalbfus, President of the Naval War College, 1934–36 and 1939–42

Kalbfus's other major accomplishment was to expand the Naval War College's pamphlet *Estimate of the Situation* into a more holistic publication, *Sound Military Decision. Estimate of the Situation* dated from 1910, was derived from lectures on how American naval officers were to sum up operational situations, and outlined the procedures—including a standardized format for orders that became known as the "Order Form"—by which they were to make strategic and tactical decisions, first in Naval War College war games, later in fleet exercises known as Fleet Problems, and then in actual operational situations. *Estimate of the Situation* had been revised several times by the late 1930s when Admiral Kalbfus decided to expand it. Specifically, Kalbfus sought to communicate to American naval officers a philosophy about naval warfare and a training manual for the operational conduct of naval warfare, one that focused on various fundamentals without locking officers into a too-rigid schematic. Officers were to be taught, for instance, that naval warfare was both an art and a science, in the sense that while Kalbfus thought naval warfare had various fundamentals, these might be changed by technology or other factors, such that officers had to be ready for imprecise situations.[4]

In the new manual's terminology, Kalbfus saw officers, in operational planning at the Naval War College or in the fleet, as needing first to understand the "General Situation" in which they would be operating. Next, officers had to understand what their "Objective" was in the context of the "Strategic General Situation." "The Estimate of the Situation" followed, in which officers were to ascertain their own and their enemies' various capabilities and the tasks that were feasible given those capabilities. The Estimate of the Situation would also assist in creating an operational plan that took into account "Task Organization," as well as

Adm. Ernest King, Commander-in-Chief of the U.S. Fleet, 1941–45 and Chief of Naval Operations, 1942–45

"Annexes" that were to cover particular aspects of the "Operation Plan" and its contingencies. Kalbfus went farther than previous versions of the *Estimate of the Situation* by creating processes and procedures whereby officers learned to exercise their Operation Plans and practiced actually doing so and supervising subordinates in carrying out the various stages.[5]

When Vice Admiral Pye became President, he added to *Sound Military Decision* a publication entitled *The Operational Function of Command, Including Sound Military Decision,* which Adm. Ernest J. King, Commander-in-Chief, U.S. Fleet (COMINCH) and CNO, had issued as *Naval Directives and the Order Form.* These pamphlets were still in use as the war ended and as Admiral Pye and his staff were working on a new curriculum that took into account the realities of World War II. Yet the war had ended so abruptly that the academic year 1945–46 was still divided into the two six-month Command and Preparatory Staff classes that ran from July to December 1945. From January to June 1946 an even more abbreviated, five-month course would be combined, as the Command and Staff Course.[6]

As a result, classes in this academic year lacked a fully updated curriculum, and the three classes pursued a course partially based on the prewar and wartime pamphlets. As during the war, there would be course work in the major aspects of naval warfare fundamentals, as well as guest and instructor lectures. In fact, at times students themselves would be allowed to lecture on topics that were their specialties. There was a great deal of emphasis on guest lecturers, since so many senior officers were available to teach about their wartime experiences and views on lessons for the present and future. There were, however, already some changes. For instance, there was now a much greater emphasis on carrier and submarine warfare, logistics, communications, amphibious warfare, international law, international relations, military government, atomic weapons, intelligence, naval organization, and how the Navy would fit into a large future national security structure that would also encompass the Army and the State Department.

There was also a continuation of the curriculum requirement that students write theses, either on fairly general topics, such as the potential of Russian sea power or the postwar U.S. naval establishment, or more detailed topics such as the need for postwar overseas bases or greater attention to logistics. In general, there was consensus that the next war in which the U.S. Navy fought would be a repetition of World War II to the extent that a balanced fleet would be needed to project American power to the shores of Europe and Asia; deliver that power ashore if necessary to carry out U.S. national security objectives; and operate globally and self-sufficiently for long periods of time, much as the Navy had had to do in the Pacific between 1943 and 1945, especially off Okinawa in the spring of 1945.[7]

There was not much concern about the Soviet Navy, but there was a great deal of attention to how a balanced fleet would operate in a probable atomic environment. The fleet was obviously to still be built around aircraft carriers that projected tactical airpower. In addition, battleships (BBs) and cruisers were still seen as needed to provide antiaircraft (AA) and surface protection to the carriers, as well as fire support for amphibious operations. Destroyers (DDs) would still be necessary for the myriad of tasks they had been used for during the war, while submarines were still to employed in scouting, attack, and the interdiction of sea lines of communication. Moreover, there was now a clearly recognized need for an afloat "train" to keep the fleet at sea over long deployments, a shore-base system also meant to ensure that the fleet could stay at sea, and amphibious assault forces for power projection.[8]

An example of the curriculum can be seen in the Course Prospectus prepared in June 1942 and retained throughout the war. A June 1943 foreword signed by Admiral Pye described the primary mission of an officer assigned as a Naval War College student as strengthening his own professional judgment and building his own style of leadership, based on a foundation of character, experience, and knowledge. The Prospectus itself spoke to these matters in a more detailed manner. Specifically, it declared, the College had a mission related to "tuning" the minds of subordinate commanders so as to make possible a harmonious exercise of command through "unity of thought and effort." This "tuning" would require a fundamental understanding of naval strategy and tactics—as well as of joint operations with the Army—"fortified" by a background of history, world politics, economics, and international law. In addition, each officer was to possess "sufficient" knowledge about arriving at sound decisions, transmitting his intent to subordinates, and interpreting the strategic and tactical dispositions of his superiors. The Naval War College represented an "uninterrupted opportunity" to direct officers into "definite paths of thought" that were seen as correct, in that they would lead the individual to a "high state" of mental development. All of this, of course, was designed to prepare officers for the exercise of command.[9]

After an administrative description of the College, the Prospectus stated that free and full discussion between faculty and students would be fostered, with ample time for questions during lectures, staff presentations, and Maneuver Critiques. This emphasis on constructive criticism, within the ethics of the military profession, was coupled with the assertion that the Naval War College maintained a close liaison with the fleet and fleet methods. In addition, officers fresh from the fleet were to make their knowledge of current procedures and operations available to both staff and students. In Chart and Board Maneuvers, for example, students were to employ current fleet methods, but they could also employ methods of their own that they thought might be superior to the fleet's. "The Naval War College, itself, advocates no particular strategical or tactical methods or doctrines, but endeavors to encourage sound progress in the development of the science of naval warfare." After giving a brief description of the Correspondence, Command, and Preparatory Staff Courses, the Prospectus again stressed the role of the latter in furthering an officer's understanding of the "fundamental considerations" governing the successful conduct of war. Much of the emphasis here was on logical reasoning to arrive at sound decisions, to plan operations, and to issue effective orders. Lectures by staff instructors and guest speakers would give students broad introductions to topics, which students in turn would further explore in individual reading and class discussions. These topics, however, would also be supplemented by Exercises and Maneuvers, with Chart Maneuvers addressing strategic problems and Board Maneuvers addressing tactical issues.[10]

The Chart and Board Maneuvers, the Prospectus emphasized, accounted for actual conditions, such as weather, communications, and damage, as well as sophisticated equipment, on the basis of the latest available data and the "best known practices" in the fleet. However, it also warned the student officer to understand fully the "motives" behind these strategic and tactical maneuvers and not to read "anything more" into the motives than actually existed. In particular, the Chart Maneuvers were exercises in the solution of "assumed" military problems, the formulation of directives, and familiarization with the composition of existing fleets and important strategic areas. These factors were what the Naval War College considered the fundamentals for sound judgment in actual warfare. This practice in "increased mental facility" would help in dealing with situations encountered in the Operations Problems, which in turn were to provide lessons for real situations that might later present themselves. Student officers were also warned not to assume that the final results of strategic or tactical maneuvers, if carried out to a final result, were to be considered conclusive with respect to actual naval warfare. The Naval War College saw the final result as strictly an opportunity for analysis of "correct" and "incorrect" applications of naval warfare principles: "The maneuver is the means to an end and not the end itself."[11]

Combined with these lectures and Maneuvers, as mentioned above, would be courses and discussions in aspects of international relations, international law, economics, and American foreign policy. Economics especially was to be studied, because of the central role of finance in the conduct of national security affairs, the Navy's mission of keeping sea routes open, and the weaknesses of the United States itself in strategic raw materials. Along these lines, students were to demonstrate familiarity with subjects of their selection by way of a thesis, supported by additional reading. In addition, while the Naval War College was not going to "check up" on officers, it was assumed that they were also doing a significant amount of reading from prescribed book lists. These lists included a section on Strategy and Tactics, including naval history, especially works of classic authors such as Rear Adm. Alfred Thayer Mahan. There were also, however, books on more recent changes to naval warfare, such as submarine and air warfare. There was also a section on Foreign Policy and International Relations, with memoirs by politicians and diplomatic histories by such historians as Charles Beard and Samuel Bemis. Additionally, there were a reading list on World War I, with a strong focus on submarine and surface warfare, especially the battle of Jutland; historical studies of sea power, including Japanese practice; books on the British Empire and Europe, including the Nazi period; studies on East Asia, especially as it impacted American foreign policy; and a contemporary section on leadership in war, propaganda, and American military policy. These final sections saw the students reading books by Winston Churchill, German admiral Reinhard Scheer, Sir Julian Corbett, Basil Liddell Hart, John Foster Dulles, Adolf Hitler, and U.S. Army colonel Emory Upton.[12]

A detailed synopsis of the Command Course informed students that they would need a thorough knowledge of *Sound Military Decision,* as they would have to solve an Operations Problem, section by section, following the procedure of the manual's Estimate Form. There would also be Demonstrative Search Exercises, in which the students would become familiar with the various search methods employed by aircraft and surface ships. There would additionally be exercises in Developing Task Force Dispositions, as well as Demonstrative Board Exercises and Chart Maneuvers to familiarize students with the mechanics of these Maneuvers and the applicable Naval War College publications. Students were also scheduled for Seminars on Current Events, primarily with presentations by staff, but possibly by students, as noted above, on topics on which they were experts. These Seminars on Current Events usually involved strategic and tactical Operations Problems on paper and then Chart and Board Maneuvers in which students took roles as commanders carrying out their written solutions. Critiques of these Maneuvers would follow, in the form of a detailed History of the Maneuver by the staff, with input from student officers. What the College called "Quick Decision Problems" would also take place, various strategic and tactical scenarios in which student commanders would have to make decisions in time periods more compressed than for Operations Problems and then have their decisions critiqued. Further, there were to be more specialized Amphibious Exercises and studies of Cruiser Warfare on Trade, focusing on recent and wartime developments.[13]

The synopsis of the Preparatory Staff Course was very similar, except that given the more junior and reserve status of its members, there was a greater emphasis on such basics as Employment of Ships and Weapons, Maneuvers in Connection with Study of General Signal Book and General Tactical Instructions, and Organization of the Navy Afloat and Ashore. The relatively elementary nature of this course could also be seen in such subjects as Employment of Local Defense Forces, Historical Maneuvers of the Maneuver Board, and Exercises in the Formulation of Directives. When Chart and Board Maneuvers took place, members of the Preparatory Staff class took roles as Subordinate Commanders to the Student Commanders from the Command Class or assisted as members of the Maneuver Staff.[14]

Cdre. George Bowdey, Chief of Staff to the President of the Naval War College, 1940–41 and 1945

Another aspect of the curriculum could be seen beginning in January 1946, when mine seminars were scheduled. On 16 January, Como. George Bowdey, Chief of Staff to the President of the Naval War College, directed numerous officers from the Command and Staff Course to these seminars on the naval mine. A section on Mine Construction broke down the various types of mines and especially the means by which they were armed and fired. The lecture was then to proceed to Mine Countermeasures, in particular how important it was to degauss ships (i.e., neutralize their magnetic fields), employ paravanes (towed minesweeping gear), and hunt for mines using a combination of sonar, lookouts, and aircraft. Finally, the lecture covered the means for Minelaying, whether by surface ships, aircraft, or submarines, addressing the weight of mines that could be carried by aircraft; the purposes for which the mines were deployed; and factors that had to be taken into account, such as current, tide, weather, depth, and enemy defenses. The point of the lecture was that World War II had demonstrated how potent a weapon the mine was and the vital importance of careful planning and exact knowledge of the weapon.[15]

A major part of the curriculum comprised fairly detailed "hints" for students writing their theses. In early March 1946, the Command and Staff class was told by Rear Adm. Allan Smith, Bowdey's successor, that the purpose of the thesis was practice in research, collation, evaluation, and condensation of material in support of "definite" conclusions—tasks similar, he pointed out, to staff duties students would one day perform. The students were to employ books, news magazines, and newspapers available in the Naval War College Library. There were tips on "skim reading" books (what would be called "graduate-school reading" in this author's day). The instruction went on: creation of an outline; note taking on index cards; formulation of the thesis, evidence, and conclusion; the importance of remaining within the assignment's parameters; and the use of research material. To Smith, all this constituted a method of "approach and attack" for the project.[16]

Rear Adm. Allan Smith, Chief of Staff to the President of the Naval War College, 1946–48

In early March, Smith repeated this guidance, with the proviso that if a student desired to write on a different subject the change would have to be justified, with a research strategy, to the Chief of Staff and the Academic Section. Other aspects of the thesis were addressed, such as length, exclusion of classified material, and points for emphasis for various subjects. For example, students writing on the foreign policies of and relations between the United States and Great Britain, Russia, or China were to include geographic, economic, climatic, "racial," social, and military factors as well as background on any conflicts these nations had had with each other. Students looking at historical influences on naval strategy and tactics were directed to pay attention to ship propulsion, armor, armament, naval bases, aircraft, and the "employment of nuclear energy in missiles." Substantial lists of books by noted scholars like Owen Lattimore, Nicholas Spykman, A. Whitney Griswold, Edward Earle, George Blakeslee, and Bernard Brodie were provided; also listed were works by such noted journalists as Walter Lippmann, recent Army and Navy publications, and works by former policy makers such as Sumner Welles. The students were also directed to congressional sources, foreign-policy journals, news magazines, speeches, State Department bulletins, and writings of military theorists like Mahan, Corbett, and Army Air Forces major Alexander De Seversky.[17]

By early May, Smith was detailing procedures whereby students would turn in two copies of their theses by 21 May, after which selected students would read and comment on theses in conjunction with instructors. Staff and student readers were not to assign marks or even comments on the copies of the thesis but were to provide evaluations for the Chief of Staff on separate sheets. These comments were not to be made known to the writer. Smith again listed a number of thesis subjects from which the students could choose: Blockade; the Postwar U.S. Naval Establishment; the Foreign Policies and International Relations between the United States and China, the U.S. and Russia, or the U.S. and the U.K.; the Influence of Industrial Power on Naval Warfare; and Outstanding Factors influencing Naval Strategy and Naval Tactics throughout History.[18]

War Gaming and the Naval War College

In addition to lectures, discussions, readings, and a thesis, war gaming was an integral part of the Naval War College curriculum, and it had been since the 1890s. The officer most responsible for the introduction and integration of naval war gaming at the College was Lt. William McCarty Little, a medically retired officer on the Naval War College staff. McCarty Little was not the first military officer to lecture, write about, or advocate war gaming; the British, German, and then American armies had been doing so since the 1870s. McCarty Little began lecturing on the value of war gaming in 1886 at the College, where gaming was inaugurated in 1895. Games were played in Luce Hall until 1934, when they were moved to the Maneuver Room in the newly constructed Pringle Hall. The primary purpose was not to replay previous battles or campaigns or even to create conditions for predictions. Instead, the games were meant as analytical tools to teach naval officers about giving clear and concise orders, understanding the value of time, prioritizing, and—even at this early stage in the history of the College—teaching the American public about the naval profession and its importance to the nation.[19]

Lt. William McCarty Little, first Naval War College war gaming specialist, 1887–1915

By 1906, the Naval War College was war gaming using Japan as the primary hypothetical enemy, though that did not preclude the possibility of Great Britain or Germany fulfilling the role. In the coming years, the College often replayed Jutland as its Annual Problem, even with Japan as the primary hypothetical enemy. As Michael Vlahos argues, eventually "the game" at times became an almost mystical and religious spectacle; often the playing of the game and learning of the rules were much more important than the strategic and tactical scenarios employed, the historical reality, or even the eventual outcome.[20] This approach to gaming, to some extent, went along with mainstream thinking at the Naval War College. By the interwar years, war gaming—though obviously not a religious exercise—was perceived as more than the analytical tool it had been in McCarty Little's day. War gaming in Pringle Hall after 1934 was seen as the primary way to teach naval officers about decision making and to give them decision-making experience. The primary sponsor of this line of thinking about war gaming was Francis McHugh, a civilian Board Plotter and later Operations Research Analyst who served at the Naval War College from 1934 to 1974. McHugh wrote a number of books and articles on naval war gaming, in all of which he emphasized gaming as a way to learn and exercise decision making. Particularly, McHugh held that war gaming was a simulation, governed by predetermined rules, data, and procedures for selected aspects of a conflict situation.[21]

Fig. 1
The Pringle Hall Game Board

McHugh went on to argue that numerous naval situations could be war-gamed, including landing operations or forces consisting of air, surface, or subsurface units. Providing military commanders with decision-making experience or information was, to him and to the Naval War College, the "ideal" aim of every war game. Individual games were played to give decision-making experience at some specified level or with some particular type of force. Games might also evaluate certain weapons systems, types of operations, or tactical doctrine. An additional purpose was to expose officers from one specialist area of the Navy to another. McHugh denoted experiential games as "educational" and informational games as "analytical." He also emphasized, however, that the focus of both on understanding

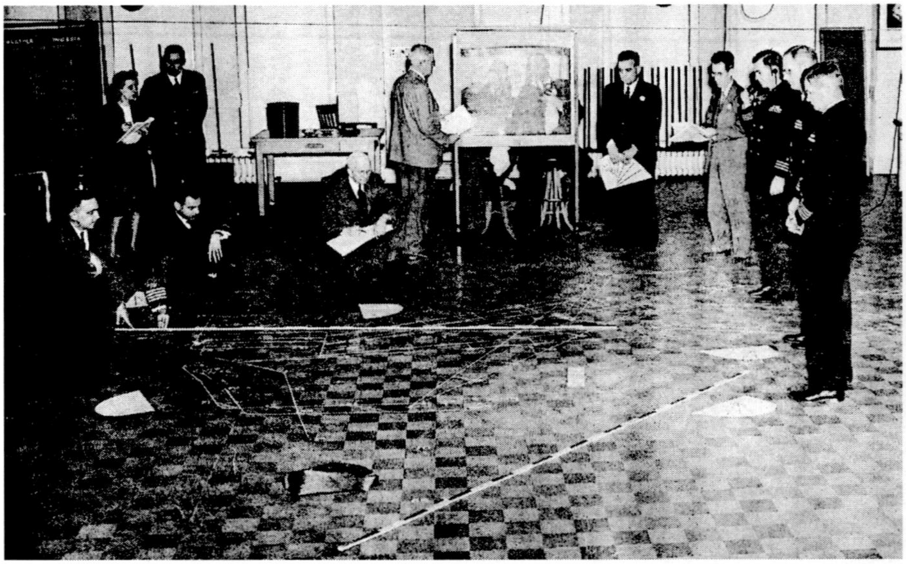

Fig. 2
A Post–World War II Game Board

Mr. Francis McHugh, Operations Research Analyst, 1934–74

plans, command structures, forces, weapons systems, and areas of operation would be of value in planning and conducting "real-world operations." McHugh further asserted that war games possessed the "inherent" advantage of simulating almost any type of military operation, including those too remote, too expensive, or too complicated to exercise in actuality. Moreover, he thought that these simulations could be used to explore and shape organizations and systems of the future as well as to develop operational concepts and plans.[22]

As noted above, the emphasis on the war game as an educational or analytical teaching tool meant that the scenario and its outcome did not have to conform to real-world situations or even result in a winner. Most games, McHugh argued, were not played out to a logical end, because that was not the purpose. McHugh thought that even if games did not match what happened in the real world, the simulation was not faulty if the officers involved learned lessons about decision-making capability and limitations. Indeed, McHugh contended that it might be not only difficult to pick a winner in a war game but even undesirable. He envisioned scenarios, for instance, in which it was not known which side might have the capacity to win until the game was actually played. He noted that while early Naval War College war games had been conducted to develop strategic and tactical naval doctrine, more recent games had been chiefly educational.[23]

From Pye to Spruance

Another part of the context of the 1945–46 academic year—in addition to the institutional limitations on the Naval War College at the end of World War II, the curriculum changes that had already been brought about by the summer of 1945, and the pedagogical purposes of the war games—was the relief of Admiral Pye by Adm. Raymond Spruance as President of the Naval War College. The change of command did not take place until March 1946, and by then Admiral Pye and his staff had created the curriculum and organization for the following year. Much of the new curriculum was based on the fact that naval warfare had changed so much during the war. There was increased emphasis now on joint operations, because of the clear direction the war had taken in these terms. Moreover, the Naval War College was affected by new developments in the postgraduate education of American military officers—notably the wartime establishment of the Army-Navy Staff College, in Washington, D.C. (the predecessor of the postwar National War College and today's National Defense University). Also, present at military postgraduate institutions were not only officers from each of the U.S. military branches but also State Department Foreign Service Officers and even foreign military officers.[24] These developments, in turn, had already exerted significant influence on Admiral Pye, who became a member of a board (since known as the Pye Board) to recommend a postwar reorganization of naval officer education. Its January 1944 report detailed what the Board thought education and training for American naval officers should comprise, from precommissioning to preparation for the highest levels of command.[25] In terms of postgraduate education and training for American naval officers, the Pye Board saw the need for officers at all stages—but especially more senior ones—to understand the general principles, capabilities, and limitations of combat aviation, surface, submarine, ground, and amphibious forces, as well as the logistical support of those forces. Larger numbers of officers, the Board thought, should be undergoing education at various points in their careers, and this additional education would have to be increasingly specialized, though not to the detriment of the more general aspects of the profession that the Board thought so necessary.[26]

The 1944 Pye Board considered the basic tenets of the 1920 Knox-King-Pye Board, which had broken down requirements for postgraduate education for naval officers at specific stages of their careers, to be still sound. For instance, all officers at the junior-division-officer level were to have tours of postgraduate education that related directly to their combatant branches or technical specialties. Next, officers would undertake an additional educational tour at the department-head or minor-staff level. Line officers would undertake a third postgraduate tour as they became commanding officers of ships or took on greater responsibilities as staff members; line and staff officers would have a fourth as they became commanders of, respectively, groups of ships or aviation forces or major shore establishments. A final tour of postgraduate education would take place when officers reached senior flag rank, as fleet or force commanders. The Board specified the approximate ranks at which officers would undergo these various levels of postgraduate education. There were also provisions made for specialized courses for officers from branches other than the line; the specialty schools—including those of other services—that existed or should be established for all technical areas; and the years in officers' career when they would be expected to attend these schools.[27]

The reflection in the Naval War College's curriculum of Admiral Pye's thinking can be seen in his welcome address to the Command and Preparatory Staff classes as they began their tours in early July 1945 (when, of course, Japan had not yet surrendered and an invasion of the home islands was still in prospect). Pye talked about the need for better trained naval officers in the "far-flung" operations of the U.S. armed forces and about the joint nature of operations—especially amphibious and air, involving the Army, Navy, and Marines. He additionally talked about the classified nature of the past and present naval affairs the officers were about to study, the abbreviation of courses to five instead of eleven months, and the need for the students to read the *Prospectus of the Naval War College Courses: Command and Preparatory Staff.* More generally, Admiral Pye described the purpose of the courses as increasing student officers' technical knowledge of naval weapons, types of ships and aircraft, and "standard practice" for the employment of such weapons and platforms in attaining objectives in naval warfare. The courses were also to increase students' understanding of the attributes and characteristics of command and the methods by which it was most "effectively" exercised, as well as to teach student officers the "mental process" that many years' experience "has proved to be the most effective in reaching sound military decisions." Another aim was to instruct students in the formulation of directives by which military decisions could be translated into effective operations; finally, it was intended to teach student officers an effective method for supervising planned operations. Without an understanding of these elements of naval command, Pye did not think a successful outcome in an operation could be attained. He concluded with a general breakdown of the Command and Preparatory Staff Courses. He also mentioned the need for Naval Reserve officers and officers from other branches to acquaint themselves with naval affairs as quickly as possible; for all officers to begin reading Navy publications such as *The Operational Function of Command* and *Sound Military Decision,* as well as library materials on history and warfare; and to begin preparing for lectures and reading materials on international law.[28]

Vice Adm. Louis Denfield, Chief of Naval Personnel, 1945–47

Pye demonstrated concern about the Naval War College being able to stay current with fleet doctrine and operational practices and procedures just a few weeks after the Japanese surrender. In late September 1945, he sent a memo to Fleet Admiral King via Vice Adm. Louis Denfield, Chief of the Bureau of Naval Personnel (BUPERS), about the terms of reference of the College. Pye recounted the myriad of Navy agencies that had controlled the Naval War College since 1884, including the Bureau of Navigation, the Office of the Assistant Secretary of the Navy, and the General Board of the Navy. He pointed out that at one time the President of the Naval War College had been a member of the General Board, and that, Pye felt, had kept the Naval War College current on developments in the fleet. Since 1934, he reminded King, the College had been brought back under the jurisdiction of the Bureau of Navigation (later renamed BUPERS); Pye was not

Capt. Bern Anderson, Instructor, Department of Strategy, Naval War College

certain why, since numerous studies had indicated that it should be under the Office of the Chief of Naval Operations (OPNAV). Pye saw that as especially advisable since the College had recently been asked to provide new courses in communications, intelligence, weapon development, and modern electronics equipment—though it did not have officers well versed in such a wide variety of subjects. Moreover, he asserted, changes in weapons, strategic and tactical doctrine, and fleet plans had so drastically changed during the war that the curriculum—which basically dated from the prewar era—required "sweeping changes," not least since its wartime mission, focused on training Naval Reserve officers, was coming to an end. Accordingly, Pye recommended that the College be transferred from BUPERS to OPNAV and that the President become again a member of the General Board. Pye's request was turned down by King, but the latter did agree that the Chief of Naval Operations would exercise close supervision over the College's mission, policy, and course of study and that the President would be able to refer directly to the CNO on matters pertaining to curriculum, future plans, and war studies.[29]

When Admiral Spruance took command in March 1946, he was just as supportive of the Naval War College as he had always been, and he agreed with many of Pye's ideas, but he wanted to see the College go even farther with the study of logistics, lessons from the war's naval battles, and a new naval operational planning manual. He disliked *Sound Military Decision* as too complicated for operational situations such as he had encountered during the war; he thought that Kalbfus's work had been too heavily revised and supplemented. Along with Capt. Bern Anderson, an instructor in the College's Department of Strategy at this time, Spruance created a manual that was essentially a standardized check-off sheet for estimating an operational situation—the *Naval Manual of Operational Planning,* later adopted by OPNAV.[30]

Conclusion

The Naval War College by the summer of 1945 was emerging from serious institutional constraints in terms of personnel, mission, and the time the President and staff had to change the curriculum to match the naval developments of World War II. These institutional factors should be kept in mind as later chapters detail the rules and procedures by which the College's war games were conducted. In any case, the pedagogical concept was that the war games did not necessarily have to match fleet doctrine or operational reality in all aspects to be worthy educational exercises, though the Exercises and Operations Problems were quite similar to fleet doctrine and operational experience. As chapters on the Maneuvers themselves will suggest, there still were good reasons for practicing aspects of surface warfare along with naval aviation, submarine warfare, amphibious assault, and logistics. They will also suggest that because they were so complex in design and implementation, changing war games came last in any curriculum revision, after reading lists, lecture topics, and thesis projects.[31]

NOTES

1 John B. Hattendorf, B. Mitchell Simpson III, and John Wadleigh, *Sailors and Scholars: The Centennial History of the U.S. Naval War College* (Newport, R.I.: NWCP, 1984), pp. 164–69.

2 Ibid.

3 "Naval War College Administrative Organization: Departments & Divisions," attached to Capt. Gail Morgan, Secretary, Naval War College, Memorandum for Files, 13 August 1947, Papers of Raymond A. Spruance, Naval Historical Collection [hereafter NHC], folder 14, box 2, series 1, collection 37, Naval War College, Newport, R.I. [hereafter Spruance Papers].

4 *Sailors and Scholars,* pp. 155–61; Frank Snyder, ed., *Sound Military Decision: U.S. Naval War College* (Annapolis, Md.: Naval Institute Press, 1992), pp. xii–xv, xl–xliii, 1–4.

5 Snyder, *Sound Military Decision,* pp. xl–xliii, 219; and Douglas Smith, *Carrier Battles: Command Decision in Harm's Way* (Annapolis, Md.: Naval Institute Press, 2006), pp. 33–36.

6 Hattendorf et al., *Sailors and Scholars,* pp. 160, 177, 180–83. For the organization of the courses in 1945–46, see NWCP reader's report, 4 January 2011, in possession of the author.

7 See Hal M. Friedman, *Digesting History: The U.S. Naval War College, the Lessons of World War Two, and Future Naval Warfare, 1945–1947* (Newport, R.I.: NWCP, 2010), pp. 1–138, 349–50; and Nepier Smith, "Historical Analysis of the Organizational Success of the Naval War College during the Twenty-Five Years following the Second World War" (Naval War College Advanced Research Project, NHC, Naval War College, Newport, R.I. [hereafter NWC], 1974), pp. 11–19, 76–77.

8 See Friedman, *Digesting History,* pp. 143–49, 156–61, 195–96, 199–207.

9 "Prospectus of the Naval War College Courses: Command and Staff," June 1942, NHC, Record Group [hereafter RG] 19: College Publications, 1 and 2, NWC.

10 Ibid., pp. 3–8.

11 Ibid., pp. 8–10.

12 Ibid., pp. 10, 15–19.

13 Ibid., pp. 11–12.

14 Ibid., pp. 13–14.

15 George Bowdey, "Memorandum for Command and Staff Class" and "Digest of Presentation on the Naval Mine," 14 and 16 January 1946, NHC, folder 2565, box 131, RG 4: Publications, NWC.

16 Allan Smith, "Hints on Thesis Writing," 28 February 1946, NHC, folder 2559-A, box 130, RG 4, NWC, pp. 1–5.

17 Allan Smith to Command and Staff Class, "Directive for Thesis," 1 March 1946, pp. 1–3 and annexes A–F, NHC, folder 2559, box 130, RG 4, NWC.

18 Allan Smith, "Memorandum for Officers of the Naval War College Staff and Theses Reading Detail," 6 May 1946, pp. 1–4, NHC, folder 2559-B, box 130, RG 4, NWC.

19 Ronald Spector, *Professors of War: The Naval War College and the Development of the Naval Profession* (Newport, R.I.: NWCP, 1977), pp. 71–87.

20 Michael Vlahos, *The Blue Sword: The Naval War College and the American Mission, 1919–1941* (Newport, R.I.: NWCP, 1980), pp. 131–56.

21 Francis McHugh, *A Manual of War Games* (Newport, R.I.: Naval War College, 1960), pp. 1–3; McHugh, *Fundamentals of War Gaming* (Newport, R.I.: Naval War College, 1966); McHugh, "War Gaming and the Navy Electronic Warfare Simulator," *NWCR* 19, no. 7 (March 1967), pp. 22–32; and McHugh, "Eighty Years of War Gaming," *NWCR* 21, no. 7 (March 1969), pp. 88–90.

22 McHugh, *Manual of War Games,* pp. 1-5, 1-8, 1-10, 1-12, 1-20, 1-21.

23 Ibid., pp. 1-25, 3-1, 3-2, 3-7.

24 Hattendorf et al., *Sailors and Scholars,* pp. 173–77.

25 "Report of Board to Study the Methods of Educating Naval Officers," encl. A, 18 January 1944, Naval War College Library, Newport, R.I.

26 Ibid., pp. 6–7.

27 Ibid., pp. 16–21, 21–37, 84–100.

28 William S. Pye, "Incoming Address to Students," 2 July 1945, pp. 1–2, 4, 7–13, NHC, folder 8, box 1, RG 11: Addresses, NWC.

29 Pye to Ernest J. King, 24 September 1945; Louis Denfield to Pye, 29 September 1945; and King to Pye, 15 October 1945; all in Papers of Richard W. Bates [hereafter Bates Papers], folder 2, box 13, Naval War College Library, Newport, R.I., available at www.batespapers.org.

30 Hattendorf et al., *Sailors and Scholars,* pp. 183–93.

31 John B. Hattendorf, e-mail to author, 8 February 2011.

II *Maneuver Rules*

The Conduct of the Maneuver, June 1945

The Maneuver Rules, or the procedures by which the Maneuvers were conducted, began with a foreword asserting that they depended for success on simulating "dynamical" aspects of the realities of war and on the attitudes of the student officers. Realities, it stated, were represented in the Rules, but when the Rules did not fully cover a situation, recourse was made to the "imagination, judgment, and loyalty to the aims of the Naval War College." Accordingly, students were to be "scrupulous" in avoiding transmitting information during a Chart or Board Maneuver except by methods that would be available to them at sea under similar circumstances. A Student Commander, for instance, was to avoid giving verbal instructions if enemy force commanders, who would at sea not be in earshot, would nevertheless be in hearing distance (perhaps in the next room) during the Maneuver. "Unauthorized conversations between student commanders tend to destroy the value of a maneuver, and to relieve a subordinate of making decisions on his own responsibility and privilege." Only by strict self-discipline, it was noted, could students gain the full value from the Naval War College Maneuvers.*

The Chart Maneuver

There were two types of Maneuvers, the Chart Maneuver and the Board Maneuver. The purpose of the Chart Maneuver was to simulate the many and varied strategic situations that arose in war so that they could be studied and operations could be "visualized." The Chart Maneuver, it was thought, had an advantage in that it was simulated and could thus create a greater number of situations for study than was possible in the fleet. The Chart Maneuver, however, had the disadvantage in that the results obtained and the conclusions drawn were "correct" only as far as the Rules and Special Conditions on which the Maneuver was based were correct. Intangible factors, such as the efficiency of personnel, indoctrination, and morale, were considered equal for both sides, which might not be the case in war. Accordingly, every effort was made to draw up the Maneuver Rules in the context of existing

* "The Conduct of the Maneuver," June 1945, p. a, NHC, folder 2525-D, box 124, RG 4, NWC. Subsequent page-number references in this chapter are to this source.

conditions so that inaccuracy and artificiality did not violate the soundness of the conclusions. (Page 2.)

Preliminary to a Maneuver, a specific Problem was drawn up and issued to all participants. To derive the most benefit, each student was required to study the Problem and submit a Solution. Each Solution was to state the Motives (i.e., pedagogical purpose) of the Problem and then the General Situation. The latter was a statement of the general causes of the war, the attitudes of neutrals and allies, and the progress of operations to date. The General Situation would also include orders from higher authority, the composition of the officer's own forces, and the state of his own defenses and outposts. There would additionally be the latest information on enemy forces, especially size, location, bases, composition, and logistics. Finally, the General Situation would provide pertinent meteorological, hydrographic, and topographic data. A Special Situation, with Special Conditions, might be required, supplementing or modifying the data and the Maneuver Rules. The Problem then informed students what was required of them in producing the Solutions. Solutions would then be selected for the Chart Maneuver. (Pages 2–3.)

The Discussion of Rules

The Problem's Rules for the Conduct of the Chart Maneuver served two specific purposes. One was to prescribe methods and conventions for conducting the Maneuver. The second purpose was to impress on student officers the constraints imposed on naval operations by materiel limitations. The Rules stated what could be accomplished with the instruments of naval warfare; students were warned that distinctions had been made between assumptions on which a Rule was based and the terms of the Rule itself. In making assumptions regarding possibilities and restrictions actually existing, all available data were considered and a decision reached that represented a "conservative" average of data and opinion. Once the facts about actual practice had been determined, it was necessary to render them in a form simple enough to prevent "unnecessary" delay in the Maneuver. The student was cautioned that the Rules might seem arbitrary or approximate, but as long as they were based on "substantially" correct premises and produced a "good resemblance" of wartime conditions, they served their purpose. Also, unless otherwise stated in the Problem, data on displacement, speed, drive, type of fuel, fuel capacity, fuel expenditure, and fleet organization were found in Naval War College publications of the latest issue. When data in those publications differed, the fact was indicated in the Statement of the Problem. (Page 4.)

Chart Maneuver Materials

Various implements were employed in a Chart Maneuver. Strategic plotting charts, paper forms called "flimsies," drawing instruments, red and blue pencils, and various forms were drawn by students from racks outside the Maneuver Room in Pringle Hall. The Maneuver Staff, consisting of the Naval War College staff assisted by students, kept the Master Plot of the Maneuver. Each Student Commander handling a force also kept a Plotting Chart to track the movements of all elements of his force as well as those of other friendly and enemy forces with which he was concerned. There was also a Communication Detail that used a "ditto machine" (spirit duplicator) to make copies of messages that were delivered. The forms themselves included Message Blanks, Record of Previous Move Blanks, Aircraft Flight Forms, Aircraft Flight Diagrams, and Fuel Account Blanks, as well as Fuel Work Sheets. Documents for use by the Maneuver Staff included Aircraft Casualty Forms, Information Blanks, Air Damage and Umpire Communication Records, Umpire Damage and Move Umpire Check-Off Sheets, the Ship Damage Table, and Radar Contact Reports. (Page 5.)

The Conduct of the Maneuver

After the Problem had been solved on paper by the student officers, the Director of the Maneuver would designate the various Student Commanders for the forces concerned. Time was then allotted in which the Student Commanders studied their tasks and prepared directives. Should the circumstances of the Problem be such that a Student Commander could realistically confer with Subordinate Commanders, such conferences might be authorized by the Director before the start of the Maneuver. (Page 6.)

The Maneuver Staff itself was to conduct the Maneuver as "expeditiously" and with as little "artificiality" as possible. It was also to expect full cooperation from all of the students. The number of officers assigned to the Maneuver Staff was governed by the extent of operations, the forces involved, and the number of officers available. If the Staff itself was small, one officer might be detailed to perform several functions. The Director of the Maneuver was the Naval War College staff member in charge of the Maneuver's conduct and the coordination of the Maneuver Staff. The Director was the arbiter in all cases; his decisions were final. There was, additionally, a Plotting Officer, who was in charge of the Master Plot, announced the lengths of the moves, and announced when flimsies on moves were handed in. The Plotting Officer also announced the Maneuver Time when changes were made; his decisions were subject to the Director. This officer additionally determined contacts and wrote them up; transmitted information to the Liaison Officers and students; and informed the Damage Umpire of all damage assigned, except from air attacks. Finally, the Plotting Officer supervised the preparation of the Maneuver's record, called the "History of the Maneuver," and the diagrams of moves for the post-Maneuver Critique. (Pages 6–7.)

Liaison Officers were also assigned as intermediaries between the Director and the Student Commanders. They cleared up any uncertainties about contacts by telephone or personal conversations with the students and then made decisions about the results, also subject to approval by the Director. In addition, Liaison

Officers were to inform the Plotting Officer, the technical assistants, and the Maneuver Recorder—who compiled the History of the Maneuver—about all decisions. It was generally the Liaison Officers' duty to make the Maneuver as realistic as possible for the students while expediting its progress. Relatedly, there was an Air and Weather Umpire, possibly with one or more assistants. Specifically, the Air and Weather Umpire had general oversight over the operation of aircraft and kept records of aircraft available to each side. The Air and Weather Umpire also received and checked Aircraft Flight Forms; assisted the Plotting Officer, especially with respect to aircraft operations and contacts; and determined the results of aircraft attacks, the latter subject to change by the Director. Further, the Air and Weather Umpire informed the Director, the Plotting Officer, the Damage Umpire, and the Recorder of the damage done by air attacks. Additionally, the Air and Weather Umpire maintained a separate plot of aircraft operations. Finally, he prepared Weather Predictions for the next move, submitted these to the Director for approval, and transmitted copies to each player and the Plotting Officer when so directed. Subject to the approval of the Director, the Air and Weather Umpire also set general and local weather conditions. (Page 7.)

The Move Umpire, one for each side in the Maneuver, was to receive flimsies of ship and aircraft movements as well as the Record of Previous Moves. He transmitted these flimsies to the Master Plot and Records of Previous Moves to the Recorder, and he assisted the Plotting Officer. The Communication Umpire was to supervise the application of the Communication Rules. He was to teach the two students detailed as Assistant Communication Umpires their duties before the Maneuver began and advise the two Student Commanders about their Communication Plans. His assistants were responsible respectively for the communications of the two opposing sides. They were to receive and examine Message Forms from the Student Commanders, determine from the Student Commanders' Communication Plans which addressees of a given message could be reached on the circuits used and which required relays, and record each message on the Communication Record Form. They were also to determine whether the Student Commanders had taken time delays—such as for preparation of the messages—into account when transmitting and receiving messages. The Assistant Communication Umpires were additionally to compute the time of delivery of messages, have orderlies prepare copies of the messages and deliver them to the originator and addressees via pneumatic tubes when the Maneuver Time (that is, scenario time, usually much faster than actual clock time) reached the calculated time of delivery, and keep the Master Plot informed about the position of forces so that transmission distances could be checked. Assistant Communication Umpires also had the responsibility of delaying delivery of messages to submerged submarines, deciding the impact of jamming, and turning over all messages files and Communication Records to the Recorder at the end of the Maneuver.

Other technical assistants included a Land Operations Umpire, a Damage Umpire, and a Logistics Umpire. In addition, the Recorder prepared the History of the Maneuver under the supervision of the Plotting Officer. (Pages 7–9.)

When the Maneuver started, stress was placed on expediting it as much as possible by promptness in handling flimsies, messages, and decisions. Resolution of disagreements was reserved to the Director, during final discussions and the Critique. When the Maneuver started, the Air and Weather Umpire distributed copies of the Weather Prediction. In addition, the Plotting Officer posted and announced information concerning the first move and set the game-floor clock dial to indicate the Maneuver Hour for the beginning of the move. The Director or Plotting Officer then announced the length of the move and that weather conditions as distributed prevailed unless students were otherwise notified. Variations were introduced as desired by the Air and Weather Umpire, subject to the Director; in unsettled weather conditions, the probability of contacts was determined by the throw of dice. Having noted the announced length of a move and the weather conditions, each Student Commander plotted on his work chart the intended movements of his vessels and aircraft for the entire period of the move. (Pages 9–10.)

Conventions followed in plotting included indicating in all plots the position of the Main Body of ships by a dot within a small square and of detachments or single ships by dots within small circles. In addition, a wavy line was drawn over a circle to indicate a submerged submarine's position; tracks of vessels proceeding on the surface were drawn with full lines. Tracks of submerged submarines and aircraft in flight were drawn with dotted lines. All tracks of ships and aircraft were also labeled to identify the ship, aircraft, or force; the unit's course and speed; and the direction of its movement, using an arrowhead for ships and a *T* for aircraft (with the crossbar in the direction of flight). Each point at which a change of speed or course was made was indicated and labeled with the time. In addition, each point where a submarine left or returned to the surface or an aircraft landed or took off also had to be shown. Tracings of data on the flimsies were made in black lead pencil. All flimsies indicated proposed tracks completed at the end of the current move, and each flimsy had two reference points, usually latitudes and longitudes. A legend was also displayed in the upper right-hand corner of the flimsy in colored pencil, with blue for U.S. forces and red for opposing forces. This legend also gave the number and Maneuver Time of the move; the composition of the force, specifically its task designator and units assigned; the identity of the Student Commander, giving both his Maneuver title and his actual name; and the student's room number. Each flimsy contained all the information with respect to the composition and movements of forces that was required by the Master Plot to determine contacts and the result of contacts. If radar or antisubmarine listening devices were used during the Maneuver, that was stated, giving names and identification numbers of the ships involved as well as the sensors' arcs of coverage. (Pages 10–11.)

So that the Plotter could quickly determine the position of a ship or unit at any given time, ground speed and times of course and speed changes were entered on the flimsy. Zigzagging was indicated with a specific symbol, and a notation was made about actual speed, speed of advance on the base course, and the zigzag plan in use. In addition, when vessels were accompanied by air patrols, a note was made on the flimsy about the composition, altitude, station, and purpose of the patrol, and its presence was indicated on the track. The presence of an antisubmarine screen—especially for lone vessels, such as a fleet aircraft carrier (CV)—was indicated by dots or a broken circle around a circle showing the initial position of the screen; the number of dots represented the number of screening vessels. Student Commanders were at liberty to change their intended movements at any time, for "good reason." In such a case they submitted a "Modified Move," notifying the Director as soon as possible, before the Director advanced the Maneuver Time. The flimsy for a Modified Move was consecutive to its number, such as "Move 1, Mod. 2." In addition, the flimsy was to indicate the tracks of the forces whose movements had been changed—from the time of modification to the end of the current move—with the original track from the time of Modification to the end of the move crossed out. (Pages 11–12.) It was understood that Student Commanders might quite frequently change plans of operations or send out messages that would cause other forces to make their own changes. These changes in plans might be planned by the Student Commander any time after the announced Maneuver Time or after a contact or message delivery brought new information or orders.

Flimsies were submitted to the Master Plot when called for, accompanied by a Record of Previous Moves, as well as Aircraft Flight Forms if aircraft flights were planned. Records of Previous Moves were submitted only at the end of a full move, not for a partial one. In addition, at the beginning of each move, Aircraft Flight Forms and flimsies were submitted for all flights that were planned for that move. If flights were not planned but became necessary at some point during the move, the Air Umpire was notified and the Aircraft Flight Form and flimsy were prepared and submitted. The same procedure was used if a flight was modified during a move. In order to expedite computation of the times required for aircraft to complete various legs of their flights, the lengths of legs were always rounded to the nearest ten miles and times of the flight to the nearest five minutes. When flights were repeated—provided the weather did not change—a single drawing of the airplane tracks for the first flight recorded subsequent flights as well. When this procedure was used, the airplane tracks were placed on a flimsy separate from the ship flimsy, but the latter showed the points and times of the beginnings of each repeated flight. The airplane flimsy was also marked "Repeated Flight" and there was inserted on that flimsy a table showing the individual planes of each flight, start times, course changes, and the time of completion of the flight. The ship flimsy was also to bear the notation "Plane flights on separate sheet." (Pages 13–14.)

Continuing on about the recording of flights, the Rules noted that if the flight legs were "very short," as in the case of antisubmarine patrols, a figure was to be drawn around the ship or formation at the position of the first launch. The figure was to show the outline of the patrol, the number of planes in the patrol, and the hours during which it was to be flown. Aircraft tracks were dotted and indicated on the flimsy with a horizontal *T.* The course did not need to be indicated, but the ground speed did. Similarly, flimsies for submarines were to show whether the boats were surfaced or submerged. If the latter, the flimsy was to show the depth of submergence; the exact times of employing the periscope, if one was being employed; and the periods spent listening for radio transmissions, with or without a vertical antenna exposed. Students filled out records for all moves at the end of the full move for surface ships and submarines and forwarded them to the Master Plot with the flimsy for the next move. (Records of Previous Moves were not required for aircraft.) On Message Forms, the originator of a message was to include himself as an information addressee, in order to obtain a copy. A student could file messages with the Assistant Communication Umpire to be delivered at times subsequent to that which the Maneuver had reached. Moreover, the student could ask the Master Plot to hold the Maneuver Time to permit him to get the message off. (Pages 14–15.)

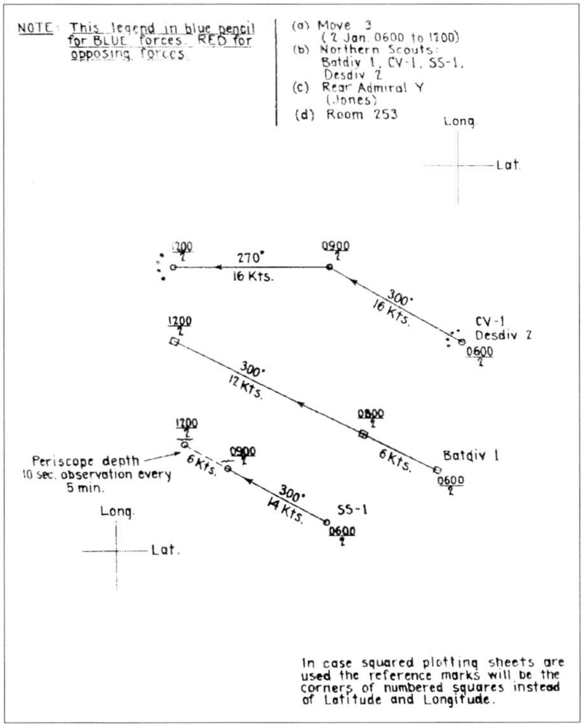

Fig. 3
Example of tracing a move

To simulate the uncertainty of wartime conditions, the end of the Maneuver Room housing the Master Plot, or the headquarters of the Maneuver Staff, was screened off. Students not detailed to the Maneuver Staff were not to enter this area. Flimsies, Records of Moves, messages, and other papers were sent to the Master Plot by pneumatic tube. The Plotter, a civilian draftsman or technical assistant, transferred the movements shown on flimsies to the Master Plot and entered course, speed, and times of positions. After all ship and aircraft movements had been plotted, the Plotting Officer examined the Master Plot for any contacts that might have occurred. If none had occurred and all communications had been cleared, he advanced the Maneuver Clock to the end of the move and called for the next one. If contact had been made or an important message was still to be delivered, the Plotting Officer advanced the clock to or beyond the time of the contact or delivery of the message and announced the new Maneuver Time over a loudspeaker. Information concerning contacts and action subsequently taken was conveyed to and from the Student Commanders on "appropriate" forms in duplicate. Student Commanders then decided their new actions, wrote synopses of their decisions on the forms, and returned one copy to the Master Plot. If desired, the Student Commanders then changed the planned movements of their forces and submitted flimsies for a Modified Move. If contacts were important or complex, Liaison Officers, as noted above, would consult the students on both sides by message, in person or by phone. Once results were decided, the Maneuver continued. When engagements became general or, in the opinion of the Director, all lessons derived from the Maneuver had been clearly developed, the evolution terminated—it was not designed to demonstrate "tactical situations." (Pages 18–19.)

Time (Min.)	Ground Speed in Knots Distance to Nearest Ten Miles										Time (Min.)
	60	70	80	90	100	110	120	130	140	150	
5	5	10	10	10	10	10	10	10	10	10	5
10	10	10	10	20	20	20	20	20	20	30	10
15	15	20	20	20	30	30	30	30	40	40	15
20	20	20	30	30	30	40	40	40	50	50	20
25	25	30	30	40	40	50	50	50	60	60	25
30	30	40	40	50	50	60	60	70	70	80	30
35	35	40	50	50	60	60	70	80	80	90	35
40	40	50	50	60	70	70	80	90	90	100	40
45	45	50	60	70	80	80	90	100	110	110	45
50	50	60	70	80	80	90	100	110	120	130	50
55	55	60	70	80	90	100	110	120	130	140	55
60	60	70	80	90	100	110	120	130	140	150	60
65	65	80	90	100	110	120	130	140	150	160	65
70	70	80	90	110	120	130	140	150	160	180	70
75	75	90	100	110	130	140	150	160	180	190	75
80	80	90	110	120	130	150	160	170	190	200	80
85	85	100	110	130	140	160	170	180	200	210	85
90	90	100	120	140	150	170	180	200	210	230	90
95	95	110	130	140	160	170	190	210	220	240	95
100	100	120	130	150	170	180	200	220	230	250	100
200	200	230	270	300	330	370	400	430	470	500	200
300	300	350	400	450	500	550	600	650	700	750	300
400	400	470	530	600	670	730	800	870	930	1000	400
500	500	580	670	750	830	920	1000	1080	1160	1250	500

Fig. 4
Airplane speed-time-distance tables

Various records were kept by the Student Commanders and the Maneuver Staff: Orders, Plans, Doctrine, and Special Instructions; track charts; a file of messages sent; a file of messages received; a file of Records of Moves; Fuel Accounts; a file of Aircraft Flight Forms; a location for all carrier planes; and copies of diagrams showing cruising and other dispositions.

Maneuver Staff members had other specific duties as well. For instance, the Recorder determined the length of and conditions during each move. He also figured Blue dispositions and movements as well as those of the enemy. Additionally, the Recorder reported engagements, the results, and damage received on both sides. Finally, he recorded comments by Maneuver Staff that were used during the Critique. The History of the Maneuver itself was divided by the Recorder into sections that were to correspond to the periods covered by each move; all moves were identified in the History. (Pages 19–20.)

As soon as possible after the completion of the Maneuver, all student officers and Maneuver Staff assembled for the Critique. This began with comments on student Solutions of the Problem, followed by a brief History of the Maneuver, in which Master Plots for each move were shown on slides. During the display

Time (Min.)	Ground Speed in Knots Distance to Nearest Ten Miles										Time (Min.)
	160	170	180	190	200	210	220	230	240	250	
5	10	10	20	20	20	20	20	20	20	20	5
10	30	30	30	30	30	40	40	40	40	40	10
15	40	40	50	50	50	50	60	60	60	60	15
20	50	60	60	60	70	70	70	80	80	80	20
25	70	70	80	80	80	90	90	100	100	100	25
30	80	90	90	100	100	110	110	120	120	130	30
35	90	100	110	110	120	120	130	130	140	150	35
40	110	120	120	120	130	140	150	150	160	170	40
45	120	130	140	140	150	160	170	170	180	190	45
50	130	140	150	160	170	180	180	190	200	210	50
55	150	160	170	170	130	190	200	210	220	230	56
60	160	170	180	190	200	210	220	230	240	250	60
65	170	180	200	210	220	230	240	250	260	270	65
70	190	200	210	220	230	250	260	270	280	290	70
75	200	210	230	240	250	260	280	290	300	310	75
80	210	230	240	250	270	280	290	310	320	330	80
85	230	240	260	270	280	300	310	330	340	350	85
90	240	260	270	290	300	320	330	350	360	380	90
95	250	270	290	300	320	330	350	360	380	400	95
100	270	280	300	320	330	350	370	380	400	420	100
200	530	570	600	630	670	700	730	770	800	930	200
300	800	850	900	960	1000	1050	1100	1150	1200	1260	300
400	1070	1130	1200	1270	1330	1400	1470	1530	1600	1570	400
500	1330	1420	1500	1580	1670	1800	1830	1900	2000	2080	500

Fig. 5
Repeated airplane flights flimsy

of the slides, the Director or one of the Assistant Directors discussed "interesting" features and situations, merits and "defects" of particular decisions and movements by each side, and any strategic and tactical points demonstrated. During the discussion, student officers and the Maneuver Staff were given opportunities to make such comments as they desired. It was assumed that many questions would indicate a "healthy difference" of professional opinion. However, the discussion was not to stray from the point at issue. (Pages 20–21.)

The Board Maneuver and Its Purpose

Board Maneuvers were tactical problems prepared with definite aims in mind. For each Problem, a Commander-in-Chief (CINC) was selected for each side from the class. Each Commander-in-Chief then assembled his Subordinate Commanders, selected by the staff, for a conference, when the Director of the Maneuver deemed that such a meeting would have been "practicable" in the conditions of the scenario. There was an important distinction between the Problem and the Maneuver. The Problem set forth conditions requiring action—a Situation. Study of the Situation and of reasons to change or preserve it was to result in a decision and then directives to be put in effect—that is, played—in the Maneuver. As the Maneuver proceeded, students were to note how the static initial Problem was left behind, how new Situations developed, and how those new Situations presented their own problems that required new Estimates and Solutions. The Maneuver continued until a point had been reached where no further value could obtained from it. At this point, a Commander-in-Chief might be required to prepare a brief Estimate of the Situation as it then existed. The Director would present a History and Critique of the Maneuver, as with Chart Maneuvers, and call on the Commanders-in-Chief and Subordinate Commanders for comment and explanation. The one exception was the "Quick Decision" Maneuver, limited in time and scope, with an expedited Critique. (Page 23.)

Tactical Problems were framed to develop tactical phases of naval warfare, and they used a format similar to Chart Maneuvers, addressing strategic issues. First, the Motives of the Problem were indicated, followed by the General Situation, the Special Situation, and Student Requirements. A Tactical Problem was to be solved in an Estimate of the Situation, leading to a Decision in which the detailed directives were formulated and issued as Plans and Orders. The Motive of a Tactical

Maneuver was twofold. First, the Maneuver was to deduce, develop, and illustrate tactical principles. Second, it was to afford opportunity for the practical application of such principles. Students were instructed to study *Sound Military Decision* accordingly. (Pages 23–24.)

Board Maneuver Materials

The list of equipment and administrative materials required for a Board Maneuver was extensive: a Maneuver Board, chalk or crayon for plotting, model ships, strips for forming groups of ships, Turning Cards, Torpedo Fire Cards, Range Wands and Tapes, a plotting table with drawing instruments and protractors, screens, a ditto machine, smoke templates, and Tactical Plotting Sheets. Forms used by students in Board Maneuvers (kept in racks outside the Maneuver Room) included Message Blanks, Aircraft Flight Forms and Diagrams, Move and Gunfire Sheets, Torpedo Fire Blanks, Mine Laying Blanks, and Tactical Plotting Sheets. Each student had a copy of a naval manual referred to as *Fleets,* describing Blue and opposing-force ship characteristics. Forms used by the Maneuver Staff included Damage and Radar Contact Reports; Aircraft Casualty and Target Indicator Forms; Information, Gunfire Scorer's Memo, and Submarine Information Blanks; Umpire Torpedo Fire, Air Damage, and Communication Umpire Records; Ship Damage and Gunfire Penalty Tables; Force Damage Summaries; and Umpire Damage, Fire Distribution, Ammunition Expenditure Tally, and Move Umpire Check-Off Sheets. (Pages 24–25.)

The Maneuver Board was to represent a portion of the earth's surface, either sea or land, depending on the natural features of the area under consideration. The board was divided into large squares, sixty inches on a side, and each large square was subdivided into small, six-inch squares (black and white floor tiles). The Junior Maneuver Board was divided by white lines into large squares forty inches on a side, subdivided into four-inch squares. Each small square was to represent a thousand yards on a side. The boards were lettered in the same manner as Tactical Plotting Sheets. When it was necessary to indicate latitude and longitude, it was customary to equate two thousand yards in the north–south direction of the board to a minute of latitude. Longitude parallels were then marked off in chalk to represent a Mercator projection. Letters were also placed on the balcony that overlooked the gaming floor in Pringle Hall or on the margins of the board to indicate true directions of north, south, east, and west; these could be changed to suit the needs of particular Problems. (Page 26.)

The tracks of forces on the board were shown by chalk lines; surface forces were indicated in white chalk and aircraft in yellow. Tracks of single ships were drawn in dotted lines, while tracks of two or more ships in column were solid lines. Hulls of model ships were painted blue or red to indicate friendly and opposing forces, respectively, and numbered (e.g., BB-1, CA-1) as in the manual *Fleets.* Submarines and the locations of aircraft were indicated as prescribed by the Director of the Maneuver.

Class Designation	Model Size	Distinctive Features
BB	Large	1 Mast, 1 Stack
CC	do	1 Mast, 1 Stack White Stripe
CV	do	Flat Flight Deck
CA	Intermediate	1 Mast, 2 Stacks
CL	do	1 Mast, 4 Stacks
DL	Small	1 Mast, 2 Stacks
DD	do	1 Mast, 2 Stacks
DM	do	1 Mast, 2 Stacks
SS)) SM)		As indicated by the Director
Train	Intermediate	1 Stack

Other types as designated by the Director.

Each vessel on the maneuver board will be given a serial number corresponding to the number in the "Fleets". Destroyer divisions will be numbered according to fleet organization or as indicated in the statement of the problem.

Fig. 6
Model ship type designations and features for the Maneuver Board

Model ships were made in several sizes and shapes to indicate various ship types, as shown in figure 6. Other types of ships were indicated by the Director. Destroyer divisions were numbered according to fleet organization or as indicated in the Statement of the Problem. These rules also applied for ships on the Junior Maneuver Board, which had a longer list of ship types than did the main Maneuver Board. (Pages 26–29.)

Model ships were also attached to metal strips so fitted as to permit the formation of a division, squadron, or fleet of any size at three, six, or seven hundred or a thousand yards' distance on the Maneuver Board and 250 or 500 yards, respectively, on the Junior Maneuver Board. The freedom of movement of the strips when joined permitted close approximation to ship's tracks even when making significant changes of course in column formation. Turning Cards were used to plot the track of a ship during a move; the cards were constructed to the scale of the Maneuver Boards. The straight edge of the card represented the distance that a ship would move on a straight course in three minutes at thirty knots; movement at speeds under thirty knots was indicated by figures along the edge. A circle on the Turning Card represented the turning circle of the ship, assuming a tactical diameter of a thousand yards. The retardation of speed due to the turn was incorporated both in the card and the straight lines tangent to the turning circle that represented the new path of a ship after course changes, in multiples of fifteen degrees. The numbers along the tangent sides represented the position the ship would reach in three minutes at the indicated speed during the Maneuver Move. To use this card, the straight edge was laid in the direction of the actual course at the beginning of the move, with the zero point at the position of the ship at that time. If a turn had begun in a previous move but was still in progress, the zero point was laid from the original point of the beginning of the turn and moved to the point in the turn the ship would have reached. (Pages 30–31.)

Torpedo Fire Cards were small, blue prints bound in book form, each type of torpedo on a separate card, except for variable-speed torpedoes, which had three cards to cover their variations in speed and range. Torpedo Fire Cards also included target speeds, from six to twenty-three knots, and gave effective ranges and sight (target) angles. Additionally, the students used protractors of transparent celluloid to determine target angle (that is, the relative bearing of the torpedo as seen from the target), roll, visible surface spray, pitch, and yaw, in accordance with the Maneuver Rules. In addition, Range Wands—strips of wood painted black and white in alternating sections—were employed. In the middle of each white section was a figure indicating thousands of yards from the origin to the middle of the section. The wand was used by placing the zero end at the center of one ship and noting in which section lay the ship whose range was being measured. (Pages 31–32.)

As in Chart Maneuvers, Student Commanders handling forces were not permitted to see such portions of the Maneuver Board as were beyond effective visible range from their ships. Screens were employed for this purpose, or the Student Commander could be placed out of the sight of the board. When a Student Commander so separated was moving his force on a Plotting Sheet, his moves were handled in the same way as those of a Chart Maneuver, except that the Director might cause them to be transferred by the staff either to the Maneuver Board, the Master Plot, or both. Student Commanders thus separated from the board were given information in the same way as in a Chart Maneuver. A Submarine Commander could see the board only by permission of the Director, as could an Aircraft Group Commander. The Director would limit the length of observations by Aircraft Commanders. (Page 32.)

A ditto machine was used to reproduce clear dispatches rapidly. Fire Effect Tables gave average expected results of fire at various ranges for one gun of each type used, as well as rate of fire. Fire Effect Diagrams gave average expected fire effects for one gun of each type of ship against specific target-ship types, taking account of armor, penetration, and firing rate. Students were to obtain these tables from the Naval War College Archives. Gun Scorers used Gunfire Penalty Tables. Message Blanks, Aircraft Flight Forms, and Umpire Communication Records were used as for Chart Maneuvers; the students indicated their fire distributions (i.e., which weapons on which ships were to engage which targets) and movements on Move and Gunfire Sheets. These were the forms from which the Move Umpire would obtain data for the Fire Distribution Sheets and from which the Gun Recorder would record gunfire data as determined on the board. These forms were also used by the Chief Scorer to compute gunfire effect. Each Move Umpire had a Fire Distribution Sheet to record fire distribution of the ships he was umpiring and to record changes in course and speed, as well as the final speeds of all these ships. When all gunfire for the move had been indicated on the sheet, it was delivered to the Chief Scorer. (Pages 33–34.)

Class Designation	Color of Sail
BB	Green
BM	As designated by Director
CC	Green and White
CA	White
CL	Yellow
CV	Brown
CM	As designated by Director
DL	No sail
DD	No sail
DM	No sail
BS)	
SM)	
PE)	
PC)	
PG)	
PS)	
PY)	
AD)	
AS)	
ASR)	
AV)	
AZ)	
AR) - - - - - - - - - - -	As designated by Director
AF)	
AC)	
AO)	
AE)	
AK)	
AP)	
AH)	
AT)	
AM)	
AG)	
Y)	

Each vessel used on the maneuver board will be given a serial number corresponding to the number in the Fleets. Destroyer divisions will be numbered according to fleet organization or as indicated in the statement of the problem.

Fig. 7
Model ship type designations and features for the Junior Maneuver Board

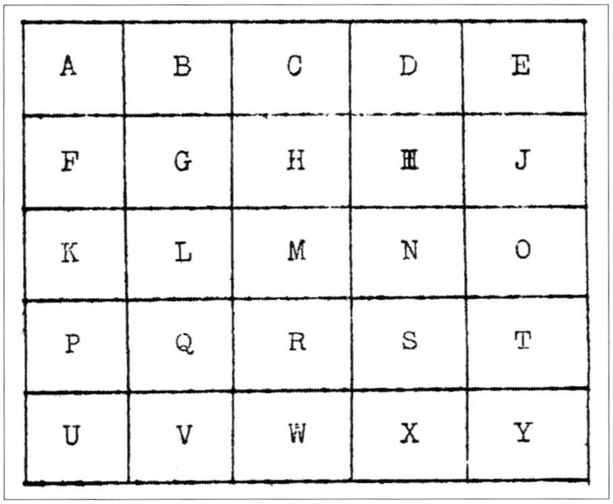

A	B	C	D	E
F	G	H	I	J
K	L	M	N	O
P	Q	R	S	T
U	V	W	X	Y

Fig. 8
Tactical plotting sheet

Students familiarized themselves with Umpire Damage Sheets, Torpedo Fire Blanks, and Mine Laying Blanks, as well with the sections of the Maneuver Rules that pertained to these forms. Information Blanks were given to Student Commanders of submerged submarines to record information they obtained by their periscopes. Additionally, Information Blanks were given to Student Commanders of aircraft and surface ships that were not in plain sight of each other because of distance, poor visibility, or darkness. Student Commanders of these units were to record on the blanks smoke or ships they saw or heard. Damage Reports, printed on blue or red paper, were prepared by the Chief Scorer, with one copy prepared by his assistants. The original was delivered to the Director or the Assistant Director for relay to the unit commander concerned only; the duplicate was filed for record. Aircraft Casualty Forms were used by the Assistant to the Air Umpire to inform students of aircraft casualties. Tactical Plotting Sheets—square prints divided into a hundred blocks, each with a number at the center, "00" in the northwest corner of the grid, and "99" in the southeast corner—represented the hundred stations on the board, each station representing a thousand square yards. When more than one sheet was required to cover the plot, individual sheets were identified by letters, *M* designating the primary Maneuver Board area and other letters indicating adjacent areas. (Pages 34–35.)

The Maneuver Staff and the Conduct of the Maneuver

Just as in a Chart Maneuver, members of the staff for a Board Maneuver were appointed from the College staff and student officers. In the case of a Board Maneuver, however, the Maneuver Staff was formed under the direction of the head of the Department of Operations. The number of officers detailed was governed by the forces involved in the Problem as well as the number of officers available. Here again, the Maneuver was to be conducted as expeditiously but also as realistically as possible. (Page 36.)

The Director was charged with the conduct of the Maneuver, the History of the Maneuver, its Record, and the Critique. He was to decide the length of moves, weather and visibility conditions, and the applicability of Maneuver Rules, and decide on situations not covered by the Rules. There were also two Assistant Directors. The senior of the Assistant Directors was appointed the Executive Officer of the Maneuver and coordinated its progress. Each Assistant Director was in charge of the operations on the Maneuver Board for his respective side. They were to announce the length of the moves as well as the weather and other general conditions and keep those matters updated on the move and weather blackboards. They also called for the Move Blanks for their sides and directed that the Maneuver Moves be made on the board once all the blanks had been checked by the Move Umpires. The Assistant Directors also required students to correct any discrepancies noted in their Moves and Gunfire Sheets in regard to the capabilities of the ships concerned. These officers additionally checked the correctness of the laying and moving of smoke screens, and they notified the Plotting Officer if they gave students permission to modify a move. Finally, the Assistant Directors were to supply the Recorder with notes and comments concerning the operations of their sides for the History of the Maneuver. (Page 37.)

The Communication Umpire was responsible for all communication between students, ensuring that they were in accordance with the Maneuver Rules and generally supervising communication systems. The Communication Umpire familiarized himself with the Communication Plans so as to ensure that messages were transmitted in accordance with the Rules. Additionally, he was to stay informed about damage to ships and other circumstances impacting communications and charge time delays for special conditions not covered by the Rules. In a similar fashion, the Gunnery Umpire was responsible for the measurement and recording of gunfire. Before a move had been made on the board, he was to measure ranges and target angles, decide whether end-on or broadside fire was involved, and determine penalties due to smoke, gas, sun glare, roll, pitch, yaw, masked fire, or battery interference. The Gunfire Recorder for each side then recorded the data on the Gunfire Sheet for the Gunfire Scorers and delivered this information to the Chief Scorer. (Pages 37–38.)

The Torpedo and Mine Umpire was responsible for the conduct of torpedo fire according to the Rules. He saw that all torpedoes were properly plotted on the Master Plot and acted on all torpedo situations that showed up on the Master Plot. He also determined all torpedo hits by surface ships, submarines, and aircraft; informed endangered ships about their situations; and decided damage, including that from depth charges and ramming. He was to require students handling torpedo craft to record torpedoes and depth charges on hand and their expenditure, and he was to inform the Chief Scorer of all damage inflicted by torpedoes, mines, depth charges, or ramming. This umpire was additionally responsible to see that all mine operations took place according to the Rules. He was to see that all minefields were properly plotted on the Master Plot, clear up all mine-related situations that arose, and assess mine hits. He also required students to keep records of their mine allowances on hand and the mines expended, informed the Chief Scorer of all mine hits, supplied the Recorder with his notes and comments concerning torpedoes and mines for the History of the Maneuver, and kept the Director informed of the entire torpedo and mine situation. (Pages 38–39.)

The Submarine Umpire was responsible for ensuring that submarines operated according to the Maneuver Rules. He kept a check-off list of submarine actions and saw that submarine Maneuver Moves were plotted on the Master Plot. He resolved submarine contacts, kept in constant contact with the Master Plot, and gave Submarine Commanders such information as they were entitled to under the Maneuver Rules. He was to decide, for instance, when and by whom periscopes were sighted and what results were obtained by listening equipment. He was to keep the Director informed of all submarine matters, require the Submarine Commanders to keep records on battery life, and supply the Recorder with notes and comments on submarine operations for the History of the Maneuver. (Page 39.)

The Air Umpire was responsible to see that aircraft were "gamed" in accordance with the Rules and that the movements of aircraft, aircraft carriers, or other vessels operating planes appeared on the Master Plot. He received Aircraft Flight Forms from Aircraft Commanders and recorded combat and operational losses. In addition, the Air Umpire was to decide results of aerial engagements and resolve uncertainties in aerial contacts. He decided what damage was inflicted by aerial bombs and torpedoes and informed the Chief Scorer. He also kept the Gunnery Umpire up to date about the locations and capabilities of spotting planes and the Director informed as to the air situation, and supplied the Recorder with notes and comments regarding air operations. Finally, he was responsible for that part of the History of the Maneuver that concerned aircraft operations. (Page 39.)

Similarly, the Plotting Officer was responsible for plotting the Maneuver Moves of all forces on the Master Plot, informing the Director and Submarine and Air Umpires of all contacts between forces on the Maneuver Board, notifying the Director when any uncertainties about each move had been cleared up, and supervising the preparation of "blueprints" and slides for use in the Critique. The Recorder kept the Record of the Maneuver and accordingly kept himself current of situations on the Maneuver Board and the Master Plot. He obtained from the Assistant Directors and the various umpires such notes and comments as they desired included in the History of the Maneuver. In order to facilitate the preparation of the Critique, the Recorder divided the data on the Maneuver into periods that corresponded to the moves. (Page 40.)

Just as in a Chart Maneuver, there was a Move Umpire for each side, acting as aides to the Assistant Directors. Each umpire was to receive all forms submitted by Student Commanders from his side, scrutinize the forms "carefully" for compliance with Maneuver Rules, keep check-off lists for all moves made by his side, and report any discrepancies to his respective Assistant Director. Once fire had been commenced by either side, he was to ensure that Fire Distribution Sheets showed fire distribution, changes of course, changes of speed, and final average speed of all units on his side. These forms were then sent to the Chief Scorer and the Move and Gunfire Forms to the Gunfire Recorder. Each Move Umpire reported to his Assistant Director when all Maneuver Moves for his side had been turned in and checked. At the end of the Maneuver, each umpire delivered all of his forms to the Recorder. (Page 40.)

The Communication Umpire had two Assistant Communication Umpires, one for each side. They had numerous duties. They were to receive all Message Forms from students and use the Maneuver Rules to check the forms for errors. Using the Communication Plan, they were to check which addresses could be reached directly on various circuits employed and which required that delays be imposed to account for relays. For visual communications, they were to determine whether the positions of the addressees on the board required delays. They were additionally to note whether time of preparation had been taken into account by the Student Commander when a message responded to a message or information slip. Further, the Assistant Communication Umpires filled in on the Message Forms time delays charged for such activities as coding and decoding and computed the times of delivery. They also recorded each message on the Communication Record Form, by call sign and heading as well as text times (the time it would take the compose and transmit the message—in Maneuver Time) to determine the transmission period the circuit in use would need to handle it. In addition, these officers would record whether or not a circuit was already in use and how many circuits could be used simultaneously. They could also delay transmissions and deliveries if appropriate. (Pages 40–41.)

The Assistant Communication Umpires had orderlies prepare the requisite numbers of copies of messages—including one for the Master Plot—and deliver them to the addresses via pneumatic tube at the predetermined times of delivery. The assistants' other duties were to stay up to date from the Master Plot or the board on the position of forces, so that transmission distances might be checked and to assist the Director and Communication Umpire in the application of communication aspects of the Maneuver Rules. The Assistant Communication Umpires delayed the delivery of messages to submerged submarines to conform to radio listening periods and conferred with each other to decide and apply the effect of jamming if used. At the end of the Maneuver, they turned in all of their forms and files to the Recorder. (Pages 41–42.)

The Chief Scorer supervised gunfire scoring, the expenditure of ammunition, and the recording and reporting of all damage and ammunition expenditure. He sent damage and ammunition reports to the Director or Assistant Directors and informed the Communication Umpire of damage to ships, when that damage reached 30 percent or higher. Five or more Gunfire and Damage Scorers assigned as aides to the Chief Scorer computed damage inflicted by gunfire, recorded the damage sustained, and kept a record of ammunition expenditure. All members of the Maneuver Staff were expected to work in close cooperation, especially the Move, Communication, and Gunnery Umpires. Umpires, however, were to refrain from rendering decisions outside of their own particular jurisdictions, though the free interchange of information between these three umpires was considered vital to the success of the Maneuver. (Page 42.)

The Board Maneuver itself usually began with the opposing forces sufficiently far apart to permit maneuvering for advantageous positions, thus affording the students practical experience with different methods of approach. The length of each move was at the discretion of the Director but was three minutes of actual clock time unless otherwise announced. Aircraft moves were designated by letter instead of number and might be of different length than those for surface ships. After announcing the length of a move, the Director or Assistant Director called out, "Hand in Move No. ——." Each officer commanding a force was at once to submit various forms to the appropriate umpires. First came a Move and Gunfire Sheet, showing exactly what movement his vessels were to make during the impending move and what main-battery gunfire, if any, they were to deliver. The Student Commander was to indicate on an additional form if he used secondary-battery fire, filling out columns marked at the top with asterisks. Also submitted were any Message Blanks showing any communications coming from his ships during the move as well as any Torpedo Fire, Mine Laying, or other forms to cover any other action taken during the move. Forms thus submitted were complete and "definite," not dependent on what any other forces might do. In addition, copies of all forms submitted were retained, except Message Forms. (As noted, Student Commanders received file copies of messages on which they had included themselves as information addressees.) (Page 43.)

When Student Commanders had turned in their Move and Gunfire Forms, and the forms had been "properly" verified, Move Umpires informed the Assistant Directors, the Director, and the Communication Umpire that the Student Commander in question was ready to "make" the move—that is, physically change the position of pieces on the board. However, if a Move Blank or other form that had been turned in called for performance that was impossible under the Rules, the Move Umpire returned the form to the student for correction. Ships on the board would not be moved until the Director ordered the move made. Also, though, as noted above, a Move and Gunfire Form could be altered with the approval of an Assistant Director, forces had to move at least one minute's worth in accord with the form already handed in. After receiving "Ready" reports from the Communication and Move Umpires, the Director broadcast, "Make Move No. ——," whereupon students shifted the models of their units, strictly in accordance with the Move and Gunfire Sheet, and the Maneuver Time was advanced to that representing the end of the move. (Page 44.)

Should a new move be ordered before damage from that just completed had been determined, the Director would decide when to apply its effects as to sinking or loss of speed. As in Chart Maneuvers, tracks of vessels proceeding on the surface were marked in white chalk, those of single ships as broken lines, and those of groups on the same track as solid lines. Aircraft and Submarine Maneuver Moves, however, were called for separately and were usually half an hour or more in length. For each aircraft move, all flights were specified on the Aircraft Flight Form. Aircraft moved on the board were represented in a manner determined by the Director, using yellow lines.

The Maneuver was closed when the Director deemed that the tactical lessons to be derived had been developed. The Director then called on the respective Commanders-in-Chief for written Estimates of the Situation, along with any comments they might want to make. Other unit commanders could also be called on for similar estimates and comments, and later the students would assemble for the delivery of the History and Critique of the Maneuver by the Director and the Maneuver Staff. At "appropriate" times during the Critique, students were given opportunities to ask questions and make observations pertinent to the Maneuver, in an environment of free discussion. Once the Critique was finished, the Record of the Maneuver was completed and turned in. The record included the Statement of the Problem, the History of the Maneuver, and the Critique of the Maneuver, with prints of the entire record. Messages, Move and Gunfire Sheets, Damage Records and Reports, Torpedo Fire Blanks, Mine Laying Blanks, Aircraft Flight Forms, and other forms, all of which had been preserved and turned in to the Recorder at the close of the Maneuver, were used in preparing the record. (Pages 44–45.)

Scoring Gunfire Effects

The Rules noted that the method of scoring gunfire effects in the Board Maneuver had to take into consideration all the factors and conditions affecting gunfire. The satisfactoriness of scoring depended on whether rates of gunfire were "reasonably" correct under the conditions existing at that moment and the Staff's ability to evaluate with "reasonable" correctness the effect of damage to the target as a result of the hits made. Moreover, the methods needed to be simple and expeditious. The Rules for the Conduct of the Maneuver discussed three general methods that could be used to evaluate either the number of hits made in a given time under existing conditions or the damaging effect on the target of the hits. (Page 46.)

The first method was the arbitrary decision. To produce a "correct" interpretation of a tactical situation, the method was based on a sound knowledge of the qualities and capabilities of the forces engaged and a sound estimate of the conditions existing at the time. "That is, it must be a mental application of the chances and rules that would govern a more formal and accurate investigation." This type of decision was frequently necessary to avoid excessive delay in settling a minor tactical situation but was not considered suitable for an accurate decision on the results of gunfire that was the principal factor in a battle. (Page 46.)

The second method, chance, played a considerable part in actual battle, the authors of the June 1945 "Conduct of the Maneuver" document noted. Of two salvos fired, exactly similar except with regard to the accidental distribution of splashes, one might make several hits and the other none at all. It was further pointed out that of two salvos making the same number of hits on a target ship, one might only do superficial damage while the other might put the ship out of action, depending on the location of the hits. It was believed appropriate to leave to chance, by using dice or other means, the number of hits being made in a given time as well as the effect of these hits. If the chance was "properly" applied, in accordance with an accurate study of the existing conditions, the results might be those that would be obtained in battle. It was further held that if each Tactical Problem were repeated a sufficient number of times using the chance method of scoring, the probability could be deduced for the given situation that a particular side would win. The primary objection to the chance method, however, was that time was not available to reenact each Tactical Problem, whereas any one iteration might lead to false conclusions about the relative strength of forces or merits of different formations and maneuvers. (Page 47.)

The third method was known as the "Invariable Average Method." Since the principal purpose of the Maneuver Board was to develop tactical knowledge and skill by showing the advantages of a force, position, formation, or maneuver over another, it was "desirable" that the results shown on the board be those that would, on average, be obtained in actual battle. This rule of applying the Invariable Average, it was argued, had been used for tactical maneuvers at the Naval War College since 1916. It was also noted that gunfire was reckoned according to the Maneuver Rules and that the effect of gunfire was derived from the actual Fire Effect Tables and Fire Effect Diagrams used by the fleet. For these tables, students consulted a confidential pamphlet entitled *Construction of Fire Effect Tables,* which could be obtained from the Naval War College Archives. (Pages 47–48.)

Damage to ships was categorized as Above Water Damage (due to gunfire or bombs) and Under Water Damage (from torpedoes, depth charges, bombs, or ramming). These two types of damage obviously differed, but the survival times of ships and the absolute amounts of damage suffered were expressed numerically, using a unit of value equivalent to the effect of a penetrative hit by a fourteen-inch shell. Damage was expressed in percentages of original life of the ship, to the nearest 10 percent. As soon as damage was determined, each Student Commander was notified of that received by vessels under his immediate command. A Student Commander could ascertain the damage to ships of his own side but not under his personal command only from the Student Commander of the damaged ship, by signal or dispatch. The state of damage to ships of one side could be made known to the Student Commanders of the other side only if the Director so desired. In addition, all damage during a three-minute Maneuver Move was to become effective at the end of the move in which it was received. (Page 48.)

Torpedoes, Mines, Submarines, Aircraft, Bombs, and Communications

The Maneuver Rules directed students to the *Manual of Torpedo Fire,* a confidential fleet publication on the theory of torpedo fire and methods of firing. Torpedo Fire Cards had instructions for use, with a sketch to assist in interpretation. In salvo firing, "dispersion," or spread, was horizontal and was expressed as the angle between the paths of the two exterior torpedoes. It was, in fact, the difference in "offset" between these two torpedoes, though the students were cautioned that in actual practice dispersion might exist even when all of the shots were fired with the same offset. For efficient salvo firing, it was necessary for the students to prescribe the course of the torpedoes as if a single weapon. Dispersion was deliberately introduced when multiple weapons were fired at a single ship in order to compensate for errors in estimates of its course and speed. In firing against formations, dispersion was used to cover more than one ship of the targeted formation. The plotting of torpedoes fired and the determination of hits were governed by the Maneuver Rules. (Page 49.)

Mine operations were in accordance with standard publications, and the procedures for them on the Maneuver Board were in the Rules. Similarly, provisions for submarines were in Maneuver Rules; the officer detailed to command was assigned a separate room in Pringle Hall and, as noted, could not see the Maneuver Board or Master Plot unless authorized by the Director. He submitted flimsies for his moves instead of Move and Gunfire Sheets. When the Submarine Commander wished to use his periscope or come to the surface, this was indicated on the flimsy; the Assistant Director for Submarines furnished information that would have been obtained under actual conditions. (Page 50.)

The Aircraft Commander was also, as previously mentioned, assigned a separate room during the Board Maneuver and not permitted to see the Maneuver Board or Master Plot unless allowed by the Director under "special" circumstances.

Three-minute Maneuver Moves required all communications started to be submitted when the move was turned in. After the move had been made, no additional communications for it could be received except those permitted by the Communication Umpire as being based on unforeseen circumstances that developed during the move. With other than three-minute moves, students submitted messages at such Maneuver Times as the Communication Umpire decided. (Pages 50–51.)

Conclusion

The distributed "Conduct of the Maneuver" instruction was an introduction for the student to the basic procedures that would govern Chart and Board Maneuvers. Obviously quite intricate and involved, these procedures delved in even greater detail than mentioned above into matters such as weather, visibility, and how various weapons platforms and systems could be employed in the war games.

EXIT

III *More "Rules of the Road," June 1945*

As noted in the previous chapter, the provisions of the Chart and Board Maneuver Rules continued beyond their Conduct of the Maneuver section.[1] Rules applying to both Chart and Board Maneuvers began with a number *1* following the letter that designated the Rule (for example, A-1). Next, those that only applied to Chart Maneuvers began with the number *100* following the letter designating the Rule. Rules that only applied to Board Maneuvers began with *200* following the letter designating the Rule. The Conduct of Maneuver cited in the previous chapter was issued separately to student officers and studied in preparing for Chart and Board Maneuvers.*

General Rules

Actual types of ships and aircraft were used in the Operations Problems and designated by standard nomenclature. Data for ships, aircraft, and weapons were given in current Naval War College publications or in the Statement of the Problem. The sizes of combatant vessels were expressed in tons standard displacement (actual weight, in tons of 2,240 pounds), those of merchant vessels in gross tons (cubic capacity in tons of one hundred cubic feet). There were five categories of size. "Large ships" had a displacement or tonnage of eighteen thousand tons or more, while "intermediate ships" were between 8,000 and 17,999 tons. "Small ships" had displacements or tonnage of between 3,000 and 7,999, while "destroyers" were placed in a 200–2,999 category. A fifth category, "submarines," was not given a displacement or tonnage; all surface craft of 199 tons or less were classed with submarines, due to the similarity of their silhouettes. All aircraft carriers were classed as "large ships" for gunfire and visibility purposes, but only battleships and battle cruisers (CBs) were classified as capital ships. (Page a-2.)

In the Maneuver Rules, "bunker fuel capacity" (that is, for a ship's own use, not cargo for transfer or delivery) referred either to coal- or oil-burning ships. For coal, "Full Load" or "Maximum Bunker Capacity" meant the ship's bunkers were full;

---

* "Maneuver Rules," June 1945, pp. I–XXI, a-1, NHC, folder 2525-E, box 124, RG 4, NWC. Subsequent page-number references in this chapter are to this source.

"Normal Fuel Capacity" was two-thirds of Full Load. In oil-burning surface ships, "Full Load" or "Emergency Full Load" was the amount estimated to be required under normal conditions for the steaming radius prescribed in the ship's Navy Department design characteristics; "Normal Fuel Capacity" was two-thirds that load. For submarines, "Normal Load" was the "economical" peacetime load, using only tanks specifically designed as fuel bunkers; an "Emergency Load" was that used in addition to ballast tanks designed for carrying either water or fuel. (Pages a-2 to a-3.)

The unit of distance used in the Chart Maneuver was the nautical mile, taken as two thousand yards; the unit of range in the Board Maneuver was one thousand yards, and the unit of speed was the knot, one nautical mile per hour. Wind force was expressed as 0 to 12 on the Beaufort scale; the velocity of wind in knots was taken to be five times the given figure. The state of the open sea corresponded to wind force: the sea in a wind force of 0 to 3 was considered "smooth," from 4 to 6 "moderate," from 7 to 9 "rough," and from 10 to 12 "heavy." The direction of the sea was the same as that of the wind. Unless otherwise stated, sunrise was always 0600 and sunset at 1800. (Pages a-3 to a-4.)

Position reports were subject to error, depending on whether they came from surface ships or aircraft and on conditions affecting the report. Positions given to students by the Director of the Maneuver were rarely exact positions but were subject to errors at the discretion of the Director. Tables of Errors were issued as an approximate guide to the limits that could be introduced by the Director. For naval patrol planes (VP) in weather that was suitable for obtaining fixes the errors were assumed to be only 50 percent of the values in the error tables (Table 2[a]), but for aircraft flying at night without fixes the values would be doubled, due to the difficulty of making drift observations at night. Vessels in sight of each other for whom a reference position had been sent out within a four-hour period were usually not subject to an error of position relative to each other greater than two miles. A position relative to something in view of both the reporter and the receiver of the report, such as a searchlight or a flare, would ordinarily not be subject to an error greater than two miles. It was explained to students that the Maneuver Rules were empirical, intended to describe actual practice, on the average. Decisions by the Director were made in that spirit. As an example, if a certain penalty was attached to making a course change of thirty degrees or more, the Director might "properly" apply it to a change of twenty-nine degrees. (Pages a-4 to a-5.)

(1) Vessels

Time since weather was suitable for obtaining a fix	Error in Miles
0 hours	2
4	4
12	8
24	14
36	22
48	30

(2) (a) Aircraft during daylight for extended flight over water or strange terrain.

Time since leaving ship or landing field (no fix during the flight)	Error in Miles
0 hours	0
1	4
2	8
3	12
4	16
8	32
12	48
16 plus	60

(b) VP planes - Weather suitable for sights

For VP planes, with weather conditions suitable for obtaining fixes, the errors will be assumed to be 50% only of the values of Table 2 (a) above.

Fig. 9
Tables of Errors

Classification Rule A-2 (a)		Angle between sea and course: 0 - 30	31-80	81-120	121-180
Moderate Sea					
Large	MESA	(M)	(M)	(M)	(M)
	Loss	(2)	(1)	(0)	(0)
Intermediate	MESA	(M)	(M)	(M)	(M)
	Loss	(2)	(1)	(0)	(0)
Small	MESA	(31)	(M)	(M)	(M)
	Loss	(3)	(2)	(0)	(0)
DD & DL	MESA	(29)	(32)	(34)	(M)
	Loss	(4)	(3)	(1)	(0)
Old Destroyer	MESA	(26)	(30)	(30)	(M)
	Loss	(6)	(5)	(2)	(1)
Submarine	MESA	(M-2)	(M-1)	(M)	(M)
	Loss	(3)	(2)	(1)	(0)
Rough Sea					
Large	MESA	(20)	(26)	(M)	(M)
	Loss	(4)	(3)	(1)	(0)
Intermediate	MESA	(18)	(23)	(28)	(M)
	Loss	(4)	(3)	(2)	(1)
Small	MESA	(18)	(20)	(25)	(30)
	Loss	(5)	(4)	(3)	(1)
Destroyer type	MESA	(17)	(19)	(23)	(27)
	Loss	(7)	(5)	(3)	(2)
Submarine	MESA	(M-3)	(M-2)	(M-2)	(M-2)
	Loss	(4)	(3)	(2)	(2)
Heavy Sea					
Large	MESA	(10)	(12)	(15)	(18)
	Loss	(10)	(5)	(3)	(2)
Intermediate	MESA	(10)	(10)	(12)	(16)
	Loss	(10)	(5)	(3)	(2)
Small	MESA	(10)	(10)	(12)	(16)
	Loss	(10)	(8)	(5)	(3)
Destroyer type	MESA	(10)	(10)	(12)	(15)
	Loss	(10)	(10)	(6)	(3)
Submarine	MESA	(10)	(10)	(10)	(12)
	Loss	(9)	(9)	(8)	(7)

NOTE: "M" denotes designed maximum engine speed (from "FLEETS") (See Rule C-7).

Fig. 10
MESA Table

Speed and Fuel

The Maneuver Rules addressed speed and fuel. Specified "ideal" underway conditions included a ship being undamaged, not going against a current, and having a clean bottom; with no paravanes streamed, nothing under tow, and not towing another ship; and not keeping station in formation, and so not having to maintain what was called "reserve speed." Engine speed corresponded to the engine revolutions being made. Under ideal underway conditions, engine speed was the same as speed through water or over ground. Ground speed was engine speed corrected for deviations from ideal underway conditions and was the speed plotted on the Maneuver Board or charts. "Reserve speed" was the difference between current speed and the highest immediately available. Engine speed being made at the Maximum Engine Speed Allowed (MESA) and penalties due to sea conditions were shown in the MESA Table. Conditions at the end of a move governed for that move. Speed losses due to damage were in percentages of designated maximum engine speeds, to the nearest half-knot, except for aircraft, whose speeds were taken to the nearest five knots. (Pages b-1, c-1 to c-2.)

Speed losses from fouled bottoms—in percentages of designated maximum engine speed per month out of dock—were 1 percent for battleships; 3 percent for carriers, cruisers, and destroyers; and 2 percent for other ships. Four months was the maximum loss applied.

Towing speed in smooth sea was the towing vessel's speed multiplied by the ratio of the towed vessel's displacement to the sum of displacements of the towing vessel and the towing gear. The strength of the towing gear, however, limited speed, so that large vessels could be towed at no more than ten knots, intermediate and small vessels at twelve knots, destroyers at fourteen knots, and submarines at eleven knots. In other than smooth seas, towing speed was determined by reductions indicated in the MESA Table, using the larger of the losses shown for the towing and towed vessels. (Page c-4.)

Vessels in formation, except the Guide (the ship on which others took station), had to have a minimum reserve speed for station keeping: at a Formation Engine Speed of ten knots or less, one knot of reserve speed; eleven to twenty-two knots, two knots; twenty-three to thirty knots, three knots; and thirty-one knots or more, four knots. Moreover, the Formation Engine Speed plus the required reserve speed had to be within the MESA range; if Formation Engine Speed was so increased that a vessel could not make the reserve speed required, it was to break formation. With a paravane streamed, a vessel could not make more than twenty-eight knots without losing the paravane itself, and a vessel operating a paravane was not to run at over twelve knots engine speed for more than six minutes, to ensure it could stream or recover the paravane. Laying smoke screens did not reduce the speeds of ships. (Pages c-4 to c-5.)

For changes of speed in a Chart Maneuver, two hours was required to raise steam in additional boilers of coal-burning ships, forty-five minutes in oil-burning ships. In addition, a steam vessel of destroyer size or less under Port Condition was to have enough steam up for an engine speed of one-half maximum (not to exceed fifteen knots) and be able to get under way thirty minutes after commencing preparations. Steam vessels of intermediate and small sizes under Port Condition were to have enough steam for ten knots and be able to get under way forty-five or thirty minutes, respectively, after commencing preparations to do so. Steam vessels of large size in port could not use their main engines for one hour after commencing preparation to do so and before the necessary steam had been raised. A vessel with an internal combustion engine, on the other hand, was able to get under way within ten minutes. When under way, if no additional steam was being maintained, coal-burning steam vessels could increase speed immediately by two knots if engine speed was three or more knots less than maximum; if engine speed was within three knots of maximum, a one-knot increase was allowed. Oil-burning ships could increase immediately to 75 percent of maximum speed if the engine speed was ten knots or more, or to 85 percent of maximum if within two knots of 75 percent of maximum. If the engine speed was within two knots of 85 percent of the maximum, speed could immediately be increased to maximum. When maintaining additional steam, engine speed could be immediately increased to that for which steam was being maintained, except that coal-burners required half an hour to work up to speed when increasing to maximum from a speed below 90 percent of maximum. When working up to speed, the Mean Engine Speed during this half hour was the Average Engine Speed (the average of the speeds at the beginning and the end of the half hour). Fuel expenditure was that corresponding to the speed to which the ship was increasing. (Pages c-5 to c-6.)

Unless otherwise stated, all ships were assumed to have steam for the maximum allowed under the MESA Table. Surface ships could not increase speed, however, during a three-minute Maneuver Move more than the number of knots given in the Change of Speed Table. Submarines on the surface could increase speed eight knots in a three-minute move, and those submerged or awash could increase speed five

Size Class (Rule A-2)	*Maximum Speed 19 or less	20-29	30 or more
Large	3	4	5
Intermediate	4	5	6
Small	5	6	8
Destroyer or Submarine	6	8	10

*(as given in "FLEETS" or as corrected for damage).

Fig. 11
Change of speed: Board Maneuvers

knots in the same three-minute move. By backing engines, any ship could be stopped in three minutes from any speed. Unless engines were reversed, however, vessels could not lose speed at a greater rate than those shown in the Change of Speed Table. Speed changes were considered to take place at a uniform rate during the move. The mean speed during the move was used in plotting. (Pages c-6 to c-7.)

All losses of speed were effective at the end of the move in which the damage was received, and all damage was permanent. Loss of speed was expressed in a percentage of original designed speed, to the nearest half-knot. Under Water Damage to any ship, by bomb, torpedo, or mine, reduced that ship's designed speed in proportion to the damage received. The reduction was also in addition to any speed losses due to Above Water Damage or other reasons. Above Water Damage above 29 percent—49 percent in battleships, heavy cruisers (CAs), or battle cruisers—inflicted a loss of speed. Thirty to 49 percent (except for the heavy ships listed) resulted in a 10 percent loss of speed; 50 to 69 percent Above Water Damage in battleships, heavy cruisers, or battle cruisers resulted in 20 percent loss, or 30 percent in other types; and 70–79 percent Above Water Damage caused a 50 percent loss of speed in all types. A ship that received Total Damage of 80 percent or more could not make more than five knots. (Page c-7.)

Fuel: Max Speed	Engine Speed Made: Maximum	1 kt. below maximum	2 kts. below maximum	3 kts. below maximum
Coal: 12-17	4	24	X	X
:18-23	4	8	X	X
:24-29	4	8	16	X
:30 plus	4	6	8	24

Fuel: Max Speed	Engine Speed Made: Maximum	1 kt. below maximum	2 kts. below maximum	3 kts. below maximum
Oil :12-17	24	X	X	X
:18-23	6	12	X	X
:24-29	6	12	24	X
:30 plus	6	12	16	24

Fig. 12
Engine speed being made

The number of hours that a steam vessel could maintain various engine speeds at or near its designed maximums is shown in figure 12. When these limits were reached, a steam vessel was to slow for six hours to a speed it could maintain indefinitely. After this period, it could return to the original high speed. In addition, a vessel running at or near maximum speed would be subject to Chances of Breakdown for each hour of operation at such speed. When moving a vessel at a high speed involving a Chance of Breakdown, the student was to report the fact at once to the Director, giving maximum speed, the speed being made, and the number of hours at that speed. The Director would then decide whether a breakdown was to occur and assign a penalty accordingly. (Pages c-8 to c-9.)

Drive as given in FLEETS	Max. Speed	Engine Speed Made: Maximum	1 kt. below maximum	2 kts. below maximum	3 kts. below maximum
R	12-17	1/48	0	0	0
	18-23	1/24	1/96	0	0
	24-29	1/18	1/48	1/96	0
	30 plus	1/18	1/24	1/48	1/96
T	12-17	1/48	0	0	0
G	18-23	1/36	1/144	0	0
E	24-29	1/24	1/72	1/144	0
T.G.	30 plus	1/18	1/36	1/72	1/144
C	1-11	1/24	1/96	0	0
	12-17	1/18	1/48	1/96	0
C.E.	18 plus	1/12	1/24	1/48	1/96

R - Reciprocating.
T - Turbine
G - Geared turbine
E - Electric
C - Internal Combustion
T.G. - Turbine. Geared cruising turbine
C.E. - Internal-combustion electric drive

Fig. 13
Chances of Breakdown

For computing fuel expenditure, "Port Condition" was to mean that main engines were not in use and steam was being maintained for in-port use only. "Stand-By Condition" meant that the main engines were ready for use, and Underway Condition meant that they were in actual use. Moreover, fuel expenditures for Port, Stand-By, and Underway Conditions were computed from data given in the Navy publication *Fuel Expenditure Tables of Fleets.* A steam vessel burning either coal or oil was classified, for determining fuel expenditure, according to the type being used. (Such a vessel could use either type of fuel as long as it was available, but not both simultaneously.) A hundred tons of oil was equal to 150 tons of coal. While raising or keeping steam for more than the engine speed being made, fuel expenditure was increased as given in the *Fleet Expenditure Tables.* If the speed of the ship was less than that shown in the *Tables,* the student was to pick out expenditure for the lowest speed shown. There was no increase for coal-burners if raising or keeping steam for no more than two knots above engine speed, but 15 percent for more than two knots above engine speed. (Pages c-10 to c-11.)

For oil-burners, there was no increase for raising or keeping steam up to five knots above engine speed but a 10 percent increase if over five knots. For vessels keeping station in formation at a distance of a thousand yards or less—except for the Guide—fuel expenditure was increased by 3 percent, 7 percent for ships with paravanes out. For ships with fouled bottoms, fuel expenditure increased by twice the percentage calculated for loss of speed due to fouled bottoms. Unless it was otherwise stated, it was assumed that at the beginning of an operation bunkers of all ships were at Full Load and that all fueling ships had full cargoes. Coal-burners

In Port

Vessel Supplying Coal: Type	Vessel Coaled: Type	Max. number at one time	Max. rate per hour one vessel can receive. (Fraction of bunker capacity.
AC	Any	2	1/12

At Sea

Vessel Supplying Coal: Type	Vessel Coaled: Type	Max.number at one time	Max. rate per hour one vessel Frac. of bunker capacity	Sea	Speed while coaling	Time of maneuver: Hrs. required to take in tow, rig & cast off
AC	Any	1	1/20	S	5	2
				(M (R (H	Coaling impossible	

NOTE: Coal burning vessels may take on board an excess of 10% of coal over their allowed coal bunker capacity, assuming this excess to be stored in fire rooms and other below-deck spaces and on deck.

Fig. 14
Coaling: in port and at sea

Vessel Supplying Oil: Type	Vessel Supplying Oil: Max.Delivery (Tons per hr.)	Vessel oiled: Type	Vessel oiled: Max.number at one time	Vessel oiled: Max.rate one vessel can receive(Tons per hour).	Time of Maneuver: Hrs. required to come alongside connect up, and disconnect.
AO	500-1800	DD	6	200	½ per DD N.B. Use both sides of oiler simultaneously
AD	15-180	DD	4	200	do
Other than AO & AD	80-150 BB 60-230 CL 15 Misc.	DD	2	200	1/2
AO	500-1800	SS	6	15	1/2 per SS each side
AS	6-180	SS	4	15	do
Other than AO & AS	80-150 BB 60-230 CL 15 Misc.	SS	2	15	1/2
AO	500-1800	(Large (Int. (Small	2	250	1
Other than AO	80-150 BB 60-230 CL 15 Misc.	do	1	do	1

Fig. 15
Oiling in port

Vessel Supplying oil		Vessel oiled (B-Broadside A-Astern)					
Type	Max. Delivery (Tons per hr.)	Type	Max. Number at one time	Max. Rate one vessel can receive (Tons per hour).	Sea	Speed while oiling	Time in hrs. exclusive of pumping
AO	500-1800	DD	2(B)	120	S	10	1/2
"	"	"	"	"	M	7½	1/2
"	"	"	1(A)	50	R	6	3/4
"	-	"	-	-	H	Oiling impossible	
AD	15-180	"	2(B)	100	S M	10 7½	1/2 1/2
"	"	"	1(A)	50	R	6	3/4
"	-	"	-	-	H	Oiling impossible	
AO	500-1800	SS	1(A) or 2(B)	10	S	7½	1/2
"	"	"	1(A)	10	M	6½	1/2
"	-	"	-	-	R) H)	Oiling impossible	
AS	6-180	"	Same as with AO	15	S.M.R	Same as with AO	
"	-	"	-	-	H	Oiling impossible	
Other than AO, AD, AS	80-150BB 60-230CL 15-Misc.	DD or SS	1(A) 1(B)	100 or 15	S.M.	5	1/2
"	-	"	-	-	R) H)	Oiling impossible	
AO and other than AO	15 to 1800	(Int (Small	2(B)	100	S.M.	5	3/4
"	"	"	1(B)	"	R	3	1
"	"	"	1(A)	50	S.M.	10	1
"	"	"	1(A)	50	R	5	1¼
"	-	"	-	-	H	Oiling impossible	

Fig. 16
Oiling at sea

could take on board an excess of 10 percent of coal over their allowed bunker capacity (in the presumption that this excess could be stored in fire rooms and other below- and on-deck spaces). (Pages c-11 to c-12.)

Visibility, Audibility, Detection, and Smoke Screens

"Natural Visibility" was that obtained without the use of artificial illumination, under conditions classed as day, night, or twilight. Day Condition was defined as sunrise to sunset; Twilight Condition was the half hour (unless otherwise stated) following sunset and an equal period before sunrise. Night Condition was the end of evening twilight to the beginning of morning twilight. Day, night, or twilight visibility was classified as "high," "normal," "low," or "fog." Visibility other than normal might be in the Statement of the Problem or announced at any other time by the Director of the Maneuver. The range of visibility varied according to a number of factors. Normal visibility was dependent on the height of the observer, the number and efficiency of lookouts, the character of the object being observed, the clarity of the atmosphere, and the amount of light. The height of the observer was assumed to be the height of the ship's bridge at night and the ship's mast during the day. Ranges of visibility of ships on the surface and aircraft in the air—under conditions other than fog or twilight—can be seen in figures 17–22. In all of these cases, visibility was in miles, and where only one set of figures was given, the object seen was not only visible at that range but recognizable as to general type. Where a second set of figures was given, the lesser one was the range at which recognition as to general type was possible. (Pages d-1 to d-7.)

For purposes of visibility, coal-burning ships were considered to make smoke at all speeds. Oil-burning ships were considered to emit gas but not funnel smoke at all speeds two or more knots below maximum. Even at those lower speeds, however, the stack gases of groups of eight or more oil-burning ships proceeding together were considered just as visible as the smoke of a group of three or fewer coal-burners. At speeds within two knots of maximum speed, oil-burners made the same smoke as coal-burners, but vessels with internal combustion engines were considered not to make gas or smoke at all. This last point is an oddity of the Maneuver Rules, since in actual operations vessels with internal combustion engines did make smoke. (Page d-8.)

Seen from

Object seen	Vessels on surface: Large Int.	Small	Dest.	Sub.	Sub.# periscope	Aircraft at Alt. indicated.	Ft.
SMOKE - 8 plus (See Rule D-5)	40	40	40	35	25	35	any
4 - 7	35	35	35	30	20	30	any
1 - 3	30	30	30	25	15	30	any
VESSELS ON SURFACE							
Large, Int.	25	22	20	16	13	25	(up
	19	16	14	10	8	15	(
Small	22	20	18	14	12	20	(to
	18	15	13	9	7	15	(
Dest.	20	17	15	11	9	20##	(10,000*
	17	14	12	8	6	15	(
Sub.	8	7	7	4	3	10##	5,000*
	6	5	5	3	3	8	
AIRCRAFT IN THE AIR	See Rule J-10.						

#For visibility from periscopes when unrestricted. Otherwise, use Rule D-10.

##NOTE: The figures given above for destroyers and submarines observed from aircraft apply with a condition of "smooth sea"; otherwise an aircraft can see and recognize destroyers and submarines on the surface at 15 and 7 miles respectively.

*NOTE: Distances that vessels may be seen by aircraft at greater altitudes than 10,000 feet are reduced one mile for each increase of 1000 feet in altitude, but distances shall not be reduced below two miles.

- -

For Classification of Vessels, see Rule A-2.
For Visibility of Periscopes and Submerged Submarines, see Rules D-8 and D-9. Also Rule J-13(a).

Fig. 17
Day: high visibility

In Miles

Seen from

Object seen	Vessels on surface: Large Int.	Small	Dest.	Sub.	Sub.# periscope	Aircraft at Alt. indicated.	Ft.
SMOKE - 8 plus (See Rule D-5)	25	25	25	25	20	25	any
4 - 7	20	20	20	20	16	20	any
1 - 3	18	18	18	18	12	18	any
VESSELS ON SURFACE							
Large, Int.	14	13	12	10	8	15	(
	12	11	10	9	7	13	(
Small	13	12	11	9	7	14	(10,000*
	11	10	9	8	6	12	(
Dest.	12	11	10	8	6	13##	(
	10	9	8	7	5	11	(
Sub.	7	6	6	4	3	8##	5,000*
AIRCRAFT IN AIR	See Rule J-10.						

For visibility from periscopes, when unrestricted, otherwise use Rule D-10.

NOTE: The figures given above for destroyers and submarines observed from aircraft apply with a condition of "smooth sea"; otherwise an aircraft can see and recognize destroyers and submarines on the surface at 8 and 5 miles respectively.

* NOTE: Distances that vessels may be seen by aircraft at greater altitudes than those listed in last column are reduced one mile for each increase of 500 feet.

- -

For classification of vessels, see Rule A-2.
For visibility of periscopes and submerged submarines, see Rules D-8 and D-9. Also Rule J-13(a).

Fig. 18
Day: normal visibility

In Miles

Seen from

Object seen	Vessels on surface: Large Int.	Small	Dest.	Sub.	Sub.# periscope	Aircraft at Alt. indicated.	Ft.
SMOKE - 8 plus (See Rule D-5)	7	7	7	7	6	7##	(
4 - 7	7	7	7	7	6	7##	(6,000*
1 - 3	6	6	6	6	5	6##	(
VESSELS ON SURFACE							
Large, Int.	6	6	6	6	5	6##	(
	5	5	5	5	4	5##	(
Small	6	6	6	6	5	6##	(5,000*
	5	5	5	5	4	5##	(
Dest.	5	5	5	5	4	5##	(
Sub.	4	4	4	4	3	4##	2,000*
AIRCRAFT IN AIR	See Rule J-10.						

#NOTE. For visibility from periscopes when unrestricted. Otherwise use Rule D-10.

##NOTE. Distances noted in table for visibility from aircraft and altitudes indicated in last column are variable and dependent on clouds and atmospheric conditions.

* Note. Distances that vessels may be seen by aircraft at greater altitudes than those listed in last column are reduced one mile for each increase of 500 feet.

- -

For classification of vessels, see Rule A-2.
For visibility of periscopes, and submerged submarines, see Rules D-8 and D-9. Also Rule J-13 (a).

Fig. 19
Day: low visibility

With moonlight, clear sky, clear atmosphere, and horizon

In Miles

Seen from

Object seen	Vessels on surface: Large Int.	Small	Dest.	Sub.	Sub. periscope	Aircraft # Alt. 1000 ft.
SMOKE - 8 plus (See Rule D-5)	5	5	5	5	0	3
4 - 7	4	4	4	4	0	3
1 - 3	3	3	3	3	0	3
VESSELS ON SURFACE:						
Large, Int.	3	4	5	5	0	3
Small	$2\frac{1}{2}$	3	4	4	0	2
Dest.	2	$2\frac{1}{2}$	3	3	0	2
Sub.	1	$1\frac{1}{2}$	2	2	0	1
Periscopes	0	0	0	0	0	0
Aircraft in air (at 3000 ft. or less)	$\frac{1}{4}$	$\frac{1}{4}$	$\frac{1}{4}$	$\frac{1}{4}$	0	See Rule J-10

With moon less than 45° high, objects bearing in the moon's reflection can be seen twice the above distance. Aircraft in path of moon $\frac{1}{2}$ mile only.

For classification of vessels, see Rule A-2.

NOTE: Distances that vessels may be seen by aircraft at greater altitudes than those listed in last column are reduced one mile for each increase of 500 feet.

Fig. 20
Night: high visibility

With clear sky, clear atmosphere and horizon, but no moon

In Miles

Object seen	Seen from: Vessels on surface: Large Int.	Small	Dest.	Sub.	Sub. periscope	Aircraft Alt. 1000 ft.
SMOKE - 8 plus) See	3	3	3	3	0	1/2
4 - 7) Rule	2	2	2	2	0	1/2
1 - 3) D-5	1½	1½	1½	1½	0	1/2
VESSELS ON SURFACE						
Large, Int.	1½	2	3	3	0	1/2
Small	1¼	2	2½	2½	0	1/4
Dest.	1	1½	2	2	0	1/4
Sub.	½	½	¾	¾	0	0
Periscopes	0	0	0	0	0	0
Aircraft in air (at 1000 ft. or less)	¼	¼	¼	¼	0	See Rule J-10

For classification of Vessels see Rule A-2.

#NOTE: Distances that vessels may be seen by aircraft at greater altitudes than those listed in last column are reduced 1/4 mile for each increase of 500 feet.

Fig. 21
Night: normal visibility

With sky overcast, moist atmosphere, and no horizon

In Miles

Object seen	Seen from: Vessels on surface: Large Int.	Small	Dest.	Sub.	Sub. periscope	Aircraft 1000 Ft. Alt.
SMOKE - 8 plus) See	½	½	½	½	0	0
4 - 7) Rule	½	½	½	½	0	0
1 - 3) D-5	½	½	½	½	0	0
VESSELS ON SURFACE						
Large, Int.	½	½	½	½	0	0
Small	1/3	1/3	1/3	1/3	0	0
Dest.	¼	¼	¼	¼	0	0
Sub.	1/10	1/10	1/10	1/10	0	0
Periscope	0	0	0	0	0	0
Aircraft in air (at 3000 ft. or less)	0	0	0	0	0	See Rule J-10

For classification of Vessels, see Rule A-2.

Fig. 22
Night: low visibility

During twilight periods, visibility ranges gradually increased or decreased according to the following rules. Twilight was divided into five equal parts, with each six minutes in duration; visibility during each period was based on certain Day and Night Visibility Conditions. For instance, twenty-five to thirty minutes before sunrise or after sunset, visibility was 1.5 times night visibility, whereas at from nineteen to twenty-four minutes it was two times night visibility. For thirteen to eighteen minutes before sunrise or after sunset, visibility was a quarter of day visibility, half of day visibility at seven to twelve minutes, and three-quarters of day visibility at one to six minutes. The Director, however, was free to decide differently on these matters. In fog, visibility from surface craft was limited to distances determined by the Director; from aircraft, it was the same distance as from surface craft if the aircraft was on or very close to the water. Otherwise, aircraft could see nothing of surface or subsurface craft in fog. If they were above the fog, however, they could see other aircraft above the fog according to visibility conditions pertaining there. (Pages d-8 to d-9.)

At the same time, a submarine could be seen by any observer within six thousand yards unless it remained submerged at periscope depth or greater and did not expose its periscope making more than three knots, or within a thousand yards of surface ships or aircraft, or longer than specified in figure 23. The exposed periscope of a submerged submarine, however, could not be observed at all at night or if it was more than six thousand yards away. The snorkel of a submarine could be sighted as if it were a periscope. Also, when a submarine fired its torpedoes at periscope depth, the swirl from the torpedoes could be seen by any surface or aircraft within 2,500 yards in smooth sea or within 1,500 yards in moderate sea. If less than one minute intervened between exposures, the duration of the two exposures plus the intervening time was regarded as one continuous exposure. One minute, therefore, had to intervene between periscope exposures, except for the interval before a firing exposure. Additionally, the Director could introduce false sightings for up to 50 percent of all periscope or snorkel contacts. (Pages d-9 to d-10.)

Sea Smooth, no white caps Wind 0-3		Sea Choppy with white caps Wind 4 and above	
Distance (yds)	Time (sec.)	Distance (yds)	Time (sec.)
1000-2000	10 sec.	1000-2000	30 sec.
2000-4000	20 sec.	2000-4000	60 sec.
4000-6000	30 sec.	4000-6000	90 sec.

Fig. 23
Visibility distances of submarine periscopes

A Submarine Commander who had his boat submerged was to enter on his flimsy the Maneuver Time at which his periscope was raised, whether he observed on a specified bearing or took an "all around look," and the duration of periscope exposure. A look on a specified bearing was to take ten seconds; ten seconds was also allotted to shift from observation of one ship to another. As an example, a thirty-second exposure permitted the Submarine Commander to look at each of two bearings. A ten-second "all around look," on the other hand, would give information only on the bearings of vessels within ten thousand yards; discern whether the vessels were nearer or farther away than six thousand yards; and determine whether a sighted vessel was large, intermediate, small, or less in size. After a look of ten seconds on a designated bearing, information was given to the Submarine Commander by the Director: this information was to include—within seven thousand yards actual range—the type of vessel; whether it was friendly or enemy; the exact compass bearing; and the target angle on the center of the ship, with inserted ambiguity. For instance, if the actual target angle was 345–015 degrees (i.e., the submarine was in an arc between fifteen degrees on the target's port bow and its starboard bow), the student would be furnished with information that the Correct Target Angle was 165–195 (meaning that the submarine was on the target's port or starboard quarter, in an arc the same size). In general, target angles were more accurate the greater the range, up to seven thousand yards. Other ranges to targets for submarines included seven to five thousand, given within 1,800 yards of actual range; five thousand to 2,500, within one thousand yards of actual range; 2,500 to 1,500 yards, within five hundred yards; and 1,500 yards or less, range given as accurately as could be measured on the Master Plot. (Pages d-10 to d-11.)

The speed of a target was normally determined by the Student Commander from his own plot. If given to the Student Commander, the accuracy of the information was governed by the range and the number of observations taken. For instance, for a target seven to ten thousand yards distant in actual range, the Student Commander would be able to ascertain the approximate ship type and, if large, whether friendly or enemy. At that range, it would also be possible to obtain an exact compass bearing; target angles as noted above, for ranges within seven thousand yards; and ranges within two to four thousand yards of the actual figure, inaccuracy varying with the range. Speed, however, was not obtainable at this range. Above ten thousand yards actual range, it would be possible to obtain an exact compass bearing but only "very meager" information about the composition of the force, and some indication of range, as above. Target angles, though, could not be obtained. (Page d-11.)

Visibility from a ship using searchlights under conditions of Night High and Night Normal Visibility were the same as Night High Visibility from a large or intermediate-sized ship. In Night Low Visibility, the ranges were those of Night Normal Visibility as seen from a large or intermediate-sized ship. Under Fog Conditions, searchlights did not improve visibility. For other, nearby ships not using searchlights and one side of the searchlight beam, the ranges of visibility were increased by 50 percent. For game purposes, in conditions of Night High or Night Normal Visibility a searchlight could silhouette a ship six thousand yards away, or three thousand yards away in Night Low Visibility, but not at all under Fog Conditions. If the vessel was silhouetted by (projected against) the searchlight beam within the ranges prescribed above and was not more than eight thousand yards from the projected beam, its visibility for gunfire and recognition purposes was as shown in figure 24. (Pages d-11 to d-12.)

Each searchlight used for searching for airplanes in the air had one chance in fifty of picking up and holding any single plane or formation of planes within three thousand yards and of holding it until the plane or formation had passed to six thousand yards. However, an airplane held by a searchlight beam could be seen by all observers within six thousand yards. In addition, under Night High and Night Normal Visibility, searchlights were visible from thirty miles and gun flashes from ten miles. Under Night Low Visibility, searchlights were visible from fifteen miles and gun flashes from five miles. Under Fog Conditions, either day or night, searchlights were visible to an observer in the direct beam for a distance of a mile greater than that at which the ship using the searchlight could be seen. The range of visibility of a ship in fog, however, was determined by the Director of the Maneuver. (Pages d-12 to d-13.)

Star shells fired by guns from three-inch to six-inch caliber could also be used in Maneuvers. They were considered to be burst by time-fuse action; the burst was at an altitude of 1,500 feet. Parachute flares dropped from airplanes would descend to 1,500 feet before attaining full illumination. During illumination, star shells and flares dropped at a rate of four hundred feet per minute and drifted

Searchlight Silhouette

Illumination for Gunfire Purposes. Type can be made out.

Night High or Night Normal Visibility	
Smoke and Large or Int. Vessels	9000 yards
Small	7500 yards
Dest.	6000 yards
Sub.	3000 yards
Night Low Visibility	
Smoke and Large or Int. Vessels	4500 yards
Small	3700 yards
Dest.	3000 yards
Sub.	1500 yards

Fig. 24
Searchlight silhouette

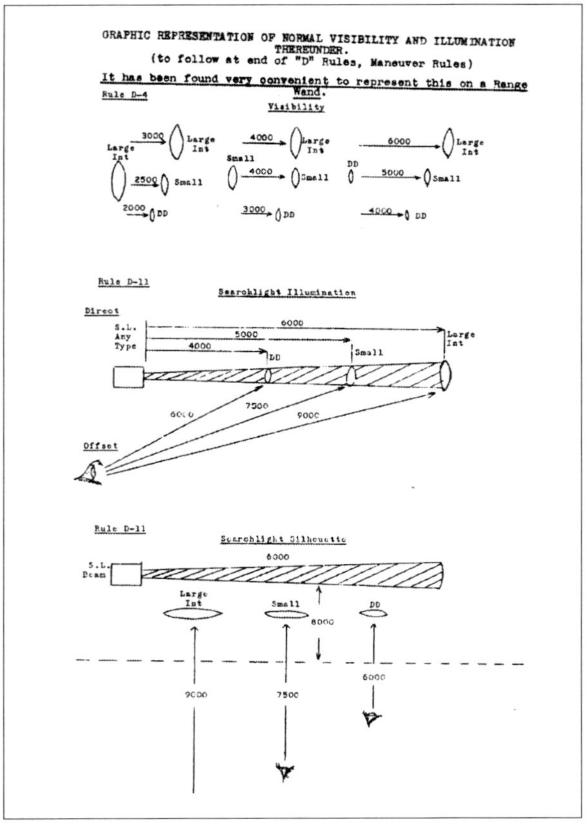

Fig. 25
Normal visibility and illumination

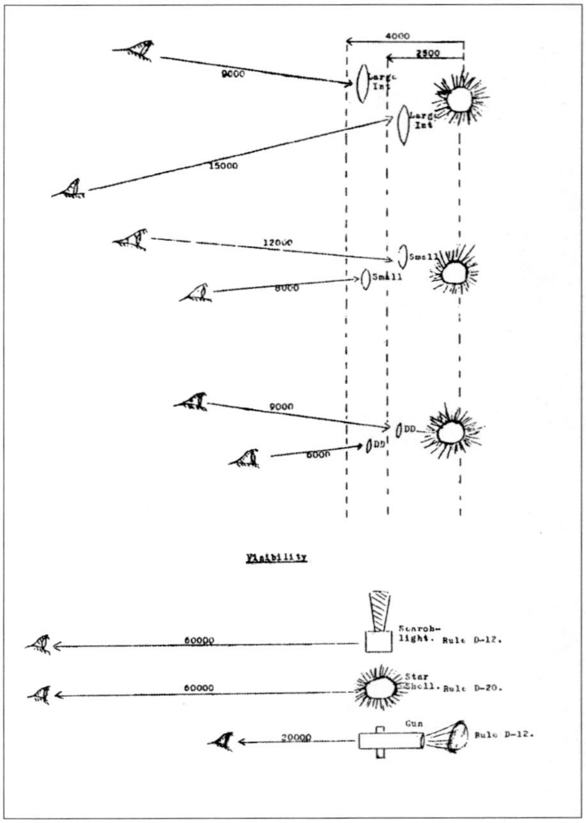

Fig. 26
Star shell illumination

with the wind. The parachute flares weighed 19.5 pounds each, and planes could carry a specific number as determined in the fleet manual *Aircraft Characteristics.* A flare could be carried in place of a hundred-pound bomb, and twenty flares could be carried in a naval patrol plane. Flares were supplied by ship or aircraft tenders. While a star shell or flare was burning, its light was assumed to illuminate a circular area of water beneath it. Characteristics for the flares and shells can be seen in figures 27 and 28; a higher rate of fire was needed if the target was moving at high speed. The rate of fire and horizontal range used were stated by the student, as were the bearing limits of any search spread. (Pages d-13 to d-14.)

If star shells or flares were being used to illuminate a target for gunfire or for searching an area, they had to burst no more than four thousand yards beyond the target and, while burning, had to cross the target's line of sight. If the burst was from zero to 2,500 yards beyond the target, the target was considered to be projected against the circle of illumination, in which case conditions for gunfire or illumination were considered most favorable. If the burst was between 2,500 and

Gun	Range (yds.)	Diameter (yds.)	Time (Min.)	Burst Per min
6"1 6" 5"9	17,000	1,000	1	4
5"5	15,500	800	1	5
5"1 5"	14,500	800	.8	6
4"7 4"1 4" 3"9	8,000	600	.7	7
3"5 3" 2"9	6,500	500	.4	8
Flares	---	2,000	3.0	2

Fig. 27
Star Shell and parachute flare characteristics

Star Shell or Parachute Flares

	(a) Burst 00 to 2500 yards over Target	(b) Burst 2501 to 4000 yards over Target
	Illumination for gunfire purposes. Type can be made out.	Illumination for gunfire purposes. Type can be made out.
Night High or Night Normal Visibility		
Object Seen		
Smoke and Large or Int.	15,000 yds.	9,000 yds.
Small	12,000 "	8,000 "
Dest.	9,000 "	6,000 "
Sub.	5,000 "	5,000 "
Night Low Visibility		
Smoke and Large, or Int.	6,000 yds.	4,000 yds.
Small	5,000 "	4,000 "
Dest.	4,000 "	3,000 "
Sub.	2,000 "	1,000 "

Fig. 28
Star Shell and parachute flare night characteristics

four thousand yards beyond, the conditions were considered less favorable. Beyond four thousand yards the burst was considered to give no illumination, while bursts short of the target obscured it if the burst was in the line of sight. If properly illuminated, however, a target was visible for gunfire and recognition purposes under Night Conditions—other than fog—as shown in the Star Shell Illumination characteristics, figures 27 and 28. (Pages d-14 to d-25.)

Ships other than the target could be illuminated or obscured by accident or design, subject to the above conditions. Under Fog Conditions, however, the use of star shells and flares did not improve visibility, and Star Shell Illumination characteristics set the outside range limits of gunfire when using star shells or flares; the effectiveness of gunfire was determined by range, the size of the firing ship, and the caliber of the gun. When using star shells or flares for searching, an arc was considered thoroughly illuminated only if the centers of bursts were not farther apart than the width of illumination of an individual burst. When a vessel was under satisfactory illumination by a parachute flare under Night Conditions (other than fog), a plane could bomb it with an accuracy 50 percent that of Day Conditions. Satisfactory illumination by flares required that two or three flares be constantly in the air within two hundred yards of the target and between altitudes of five hundred and 1,500 feet during a bombing period. The time of the burning of the flares was considered three minutes. Illumination had to be provided by a special group of airplanes, since flares could not be dropped by bombing aircraft. (Pages d-15 to d-16.)

Under Night Conditions other than fog, star shells and parachute flares were visible for thirty miles in Night High and Night Normal Visibility and fifteen miles in Night Low Visibility. Under Fog Conditions, star shells and flares were visible at a distance a mile greater than the natural range of visibility of a ship, as determined by the Director. Moreover, searchlight beams, if trained directly on an observer, would cause "serious" interference with vision, adversely affecting gunfire spotting, torpedo control, and maneuvering. The Director decided the extent of this interference in all cases and could even decide that a star shell burst or flare completely impeded vision under certain conditions, if a student's plan for such action fully satisfied him. (Page d-16.)

Normal ranges of audibility of gunfire and of destroyers proceeding at high speed were calculated in terms of miles and thousands of yards. For instance, gunfire twelve-inch or higher in caliber could be heard from a distance of twenty miles, or forty thousand yards. Eight-to-eleven-inch gunfire could be heard fifteen miles, or thirty thousand yards, while six-to-7.5-inch gunfire could be heard at ten miles, twenty thousand yards. Four-to-5.5-inch gunfire could be heard from seven miles, or fourteen thousand yards, and calibers three-inch or smaller could be heard from five miles, ten thousand yards. Destroyers proceeding at high speed could be heard a mile away, two thousand yards. The distance from which an airplane could be heard was, when it was heading directly toward or away from the listener, much less than when it was passing at right angles; the ratio between these distances was assumed to be approximately one to three. Heavier-than-air craft could be heard by the unaided ear within two miles and at six thousand feet of altitude if the plane was a fighter (VF), observation scout (VOS), or scout observation (VSO) plane. A scout bomber (VSB) or torpedo bomber (VTB) could be heard within 2.5 miles and at eight thousand feet. A naval patrol plane could be heard within three miles at up to ten thousand feet. For between two and seven planes flying together, two thousand feet would be added to these altitude figures and a mile to the horizontal distance. In the case of more than seven planes, three miles and two thousand feet were added, respectively. These figures could also vary according to wind and any noises near the listener. If there was any wind, the ellipse marking the limit of audibility would be shifted downwind a distance equal to its major axis multiplied by the wind force and divided by thirty. A vessel firing or under fire would be able to hear through the air (as opposed to underwater sound on hydrophones) only one-fourth as far as normal, and a destroyer proceeding at high speed would be able to hear half as far as normal. The direction of a persistent noise could be told to the nearest forty-five degrees. (Pages d-17 to d-18.)

Underwater sound equipment was assumed to be carried by all ships, and students were to inform the Plotting Officer when the ships they were commanding were using it for the detection of other ships. A sketch or memorandum of the search plan was submitted, and the Director would use various rules to determine the outcome. Ranges and bearings could be obtained by echo-ranging (i.e., active sonar), within a basic limiting range of three thousand yards. In antisubmarine screens, the effective limiting echo range would determine the maximum distance between screen units for fully effective search. The operation area, the season of the year, special local conditions, and other factors affecting physical conditions would establish the effective limiting range. The greatest factor here was the variation of water temperature with depth—that is, the temperature gradient. Measured variation per hundred feet of depth was used to predict the effective range. The higher the water temperature, the smaller the effects of temperature would be. In high latitudes and cold water, the sun's heating effect was small. In deep seas, the large masses of water, wind, and sea were considered to hold down temperature effects. Students, however, were to expect temperature effects from time to time. Figure 29 gives approximate effective limiting detection ranges for an area in which the average water temperature is about sixty-five degrees Fahrenheit at the surface against a submarine operating at one-hundred-foot submergence. Students would need tables that gave five-degree surface temperature variations so they could make predictions in all possible operating areas. (Pages d-18 to d-19.)

Temperature Change with Depth	Result of Dice Roll: Inshore Operations	Result of Dice Roll: Offshore Operations	Effective Limiting Range
0-1°F/100 ft.	1, 7	All odd	3000 yds.
2°F/100 ft.	3, 5, 8	All even	2000 yds.
3°F/100 ft.	4, 6, 9	- - - -	1500 yds.
4°F/100 ft.	10, 12	- - - -	1200 yds.
5°F/100 ft.	11	- - - -	1000 yds.
6°F/100 ft. (or more)	2	- - - -	Ineffective

Fig. 29
Water temperature variations

For purposes of underwater sound detection, "inshore operations" were defined as within one hundred miles of a continental body of land, while "offshore operations" were more than a hundred miles out. An effective limiting echo-range factor was set in a "fairly realistic" fashion with a pair of dice; the roll would be used to pick the effective limiting echo range, which had to be taken into account in screening formations. The limiting range of any particular sets of conditions was determined by multiplying the effective range established above by successive coefficients from figure 30. Echo-ranging was not considered to be prevented by the sound of a ship's own propellers or wake. However, the general noise level was "somewhat" higher when the projector was pointed abaft the beam. When the projector was pointed directly aft, the noise level would be high, and propeller sounds would interfere with ranging. (Pages d-19 to d-20.)

Factors for speeds marked with asterisks in figure 30 applied only when such speeds were used prior to detection. Once contact had been established, it was presumed, a competent operator could compensate for the effect of increased speed. However, even with new modulated-wave equipment, smooth seas were considered less favorable for echo-ranging than moderate seas, since smooth surfaces produced false echoes. Since echo-ranging was not greatly affected by the moderate pitching and rolling of the ship, there was no need for students to distinguish between detections by a submerged submarine or surface ship in these cases. There were also factors considered for shallow water, as seen in figure 30, but speed of the target did not affect echo-ranging. The target's wake close astern, however, would reflect sound waves, and there was no way to distinguish between the echoes returned by the wake and those from the hull. (The effective length of that wake might be as much as fifty yards per knot of target speed, but its reflecting character diminished with time, depending on the state of the sea and the speed at which the wake was formed.) Initial contact was developed by determining the limiting target bearings and the mean of those bearings. The mean bearing would be displaced aft of the target due to the effect of the wake from the target, but not much for surface ships, and the low speeds of submerged submarines tended to minimize this effect. The mean bearing, however, would change regularly and definitely relative to the target's movement to the ranging ship. In practice, students were to use the limiting bearings, smooth relative movement of the target, propeller sounds, and discrimination between good echoes (from the hull) and increasingly poor ones aft of the target to determine that observations were being made on the target's hull rather than its wake. (Pages d-20 to d-22.)

(1) Coefficient I - (Relative Bearing)

Angle of Train	Value of Coefficient I
0 - 90 (360 - 270)	1.0
90 - 160 (270 - 200)	0.9
160 - 200	0.3

(2) Coefficient II - (Speed of Ranging Ship)

Own Speed	Value of Coefficient II
0 - 12	1.0
13 - 15	0.9
16 - 20**	0.5
21 - 25**	0.4
26 plus **	0.3

(3) Coefficient III - (State of Sea)

State of Sea	Value of Coefficient III
Smooth	0.8
Moderate	1.0
Rough	0.8
Heavy	0.5

(4) Coefficient IV - (Depth of Water)

Depth	Value of Coefficient IV
0 - 100 fathoms	0.7
101 fathoms plus	1.0

Fig. 30
Echo-ranging coefficients

Turbulent masses of water in the line of the sound beam, such as those caused by depth-charge explosions, would interrupt echo-ranging through them for about twenty minutes. Other limited-area turbulence—such as water masses churned up by destroyers making sharp turns at high speed—would cause similar interference. In general, determinations of target range and bearing (or of the fact that a target did not exist in the ten-degree segment centered on the selected projector bearing) could be made every five seconds. The projector could be trained twenty degrees per second, but no observations could be made when training so quickly. Detection was possible while training at a rate of five degrees per second, but five seconds had to elapse after detection before the range and bearing could be determined. Bearings were usually correct to two degrees and ranges to twenty-five yards. Only after the target itself had been located could range and bearing be assumed to be those of the target. (Page d-22.)

Bearings and some information about ship type could be obtained by listening. The basic limiting range assumed for underwater sound listening was fifteen thousand yards. The limiting range for any particular set of conditions was determined by multiplying successively this value by the listening range coefficients from figure 31. The bearing of the target—or the fact of its absence—in the ten-degree segment centered on the selected bearing of the listening device could be established every twenty seconds, within the range determined. The listening device was trained twenty degrees per second, and at that rate or less detection was possible. Twenty seconds, however, had to elapse after detection before the bearing was determined. Bearings obtained were correct to plus or minus one degree for a supersonic device, plus or minus five degrees for a sonic device. (Pages d-23 to d-24.)

Examples for determining limiting ranges for both echo-ranging and listening were supplied, and they make clear how this detailed feedback was supplied to players. In one case, a destroyer in the screen has one supersonic projector fixed at 060 degrees relative and the other at 300 degrees relative. The speed of the destroyer is twelve knots, the depth of the water is one thousand fathoms, and the seas are moderate. A submerged submarine at a speed of five knots is attempting to penetrate the screen. Given that Coefficients I–IV (from figure 30) were all 1.0, the basic range for echo-ranging was three thousand yards, and the limiting range was the same. In a second example, a submarine is detected by echo-ranging at three thousand yards. Coefficients I–V from figure 31 are 1.0, 0.3, 0.9, 1.0, and 0.7, respectively, and the basic range for listening is fifteen thousand yards; the limiting range is therefore 2,835 yards. The submarine's propellers cannot be heard by this listening device until the range closes to 2,835 yards. (Pages d-24 to d-25.)

After the limiting range had been determined by these methods, the percentage chance of picking up a target vessel that came within range could be obtained from the tables in figure 32. Once a target had been picked up, it could be tracked effectively by echo-ranging and listening out to the limiting ranges determined above. False submarine contacts from echo-ranging were frequently made and reported. At the discretion of the Director, false contacts as high as 50 percent of the contacts given could be introduced. After time for development, from thirty seconds to two minutes, a "considerable" number of

(1) Coefficient I - (Relative Bearing)

Angle of Train	Value of Coefficient I	
	Supersonic	Sonic
0-20	1.0	0.7
21-80	1.0	1.0
81-120	0.9	0.7
121-160	0.8	0.5
161-180	0.6	0.1

(2) Coefficient II - (Speed of Listening Ship)

Own Speed	Value of Coefficient II	
	Supersonic	Sonic
0-2	0.8	0.8
3-4	0.6	0.2
5-10	0.5	0.0
11-15	0.3	0.0
16-20	0.2	0.0
21-25	0.1	0.0

(3) Coefficient III - (State of Sea)

State of Sea	Value of Coefficient III	
	Listening ship on surface	Listening ship submerged
Smooth	1.0	1.0
Moderate	0.9	0.9
Rough	0.6	0.7
Heavy	0.5	0.7

(4) Coefficient IV - (Depth of Water)

Depth	Value of Coefficient IV	
	Supersonic	Sonic
0-50 fathoms	0.7	1.0
51-100 "	0.8	0.8
101-200 "	1.0	0.2
201 fathoms plus	1.0	0.0

(5) Coefficient V - (Speed and Type of Target)

Target Speed	Large	Type of Target Int. & Small	DD	SS Surface	SS Submerged
1-3	0.1	0.05	0.05	0.1	0.2
4-6	0.3	0.1	0.1	0.[illegible]	0.7
7-10	0.5	0.3	0.3	.5	1.0
11-15	0.7	0.6	0.5	0.7	-
16-20	0.9	0.7	0.6	0.9	-
21-25	1.0	0.9	0.7	1.0	-
26 plus		1.0	1.0	-	-

Fig. 31
Listening range coefficients

false contacts could be eliminated. About 15 percent of the false contacts, however, could not be eliminated and were taken as probable submarines. Vessels equipped with these underwater sound devices could use them to maintain contact during both fog and darkness. (Pages d-26 to d-27.)

The results of depth-charge attacks against submarines as a result of sound contact were decided by the Director, who was to base his decision on the Maneuver Rules as well as the capabilities of the attacking vessels for tracking by sound. Partial identification of type of ship and approximate speed (from counting revolutions of the propeller) could be obtained by listening, and this information could be given to students at the discretion of the Director. Propeller revolutions could be counted as far away as three-quarters of the chance-detection distances given by the Maneuver Rules. No information as to type of ship or number of revolutions could be obtained by echo-ranging, but the number of vessels could be counted by either listening or echo-ranging if the contacts were not less than ten degrees apart. Submarines, it was ruled, could make torpedo attacks entirely on the basis of their underwater sound equipment. (Pages d-27 to d-28.)

Radio direction finders were carried as specified in the Maneuver Rules, and shore radio-direction-finder stations listed in Navy Department publications could be specified in an Operations Problem. Reception was on frequency bands D–N (as defined in the Rules), except for designated stations that could operate on Bands D–Z. Subject to certain qualifications, a direction finder could obtain the bearing of any radio transmission provided it was listening on the frequency used. A radio direction finder had to be within the limiting distance of the transmission; if on Frequency Band N or below, the transmission had to last at least thirty seconds, or if on Frequency Band O or above, forty-five seconds; and the listening station could not be transmitting while the bearing was being taken. Operators not listening on the exact frequency but searching in a fixed band not more than two hundred kilocycles in width could detect the transmission, tune to it, and obtain a bearing if the transmission lasted at least three minutes. Shore-station nets could listen only on a designated frequency until shifted to another frequency; it required three minutes to shift the direction finders of a net to a new frequency. Except in aircraft, this multilateral method of radio direction finding could be used to locate a transmitter; when such a method was not used, a bearing reciprocal to the actual bearing was furnished to the student. In either case an error was factored in, determined by casting a die twice and picking out the error in a table (figure 33). The bearing given the student would include the error. (Pages d-28 to d-29.)

	Percent of Determined Limiting Range			
	1 - 75%	76 - 95%	96 - 100%	101% plus
Percent chance of detection	100%	75%	50%	10%

	Percent of Determined Limiting Range			
	1 - 75%	76 - 95%	96 - 100%	101% plus
Percent chance of detection	100%	75%	50%	10%

Fig. 32
Limiting range determination on chance targets

		Second die					
		1	2	3	4	5	6
First Die	1	0	0	0	0	0	0
	2	0	0	+1	+1	+1	+1
	3	+1	-1	-1	-1	-1	-1
	4	+2	+2	+2	-2	-2	-2
	5	+3	+3	-3	-3	+4	+4
	6	-4	+5	-5	-6	-7	+10

Fig. 33
Table of direction finder errors

Radar equipment for detection, fire control, and recognition was installed on all ships, shore stations, and aircraft in the Maneuver in accordance with certain policies. Radars in surface ships included medium-range ship and aircraft detectors; main, secondary, antiaircraft, and dual-purpose battery director radars; and identification equipment (i.e., Identification Friend or Foe, or IFF) used with radar. Submarines were equipped with aircraft-warning devices as well as radar designed to detect surface craft. Aircraft were equipped with detection radars of various types and identification equipment. Vessels, shore stations, and aircraft were equipped as noted below. To guard against interference between radars, not more than six air-detection radars operating on approximately the same frequency could be used simultaneously by any one disposition of ships. The six might include one radar for the Officer in Tactical Command (OTC), one for the Fighter Director, and one for each of four designated "guard" ships. (Page d-29[a]-1.)

	CB CC BB	CV	CL CA	ODD DD DL	SM OSS SS	Large Unt. Auxiliaries	Aircraft	Shore Stations and other Vessels.
Detection:	3	1	-	1	1	1	(As stated in problem)	(As stated in problem)
Fire Control:	C	4	4	1 or 2	1	1	None	"
	(One for each Director, Main, Secondary and AA)							
Automatic Recognition:	2	2	2	1	1	1	1 per plane	"

Fig. 34
Radars by ship type, shore station, and aircraft

Within certain technical limitations, detection radars used by surface units could be expected under the Maneuver Rules to give accurate ranges and bearings of surface units at distances up to thirty thousand yards—depending on the size of the target—and of aircraft above the optical horizon out to ninety-five miles. Under some circumstances, the altitude of planes could be determined with usable accuracy by detection radars; at short ranges, altitude could be determined with greater accuracy by antiaircraft fire-control radar. Detection of submerged submarines with exposed periscopes was doubtful, but submarines on the surface could be detected, though not beyond seven thousand yards, because of their smaller radar cross sections. Radars used by planes in flight could be expected to give accurate ranges and good bearings of units. Fire-control or gunnery radar was normally used for fire control purposes during periods of low visibility. This radar might also be used to supply tactical information, such as the position of other ships in one's own force, to facilitate station keeping in low visibility or at night, and to direct star-shell and searchlight illumination. (Pages d-29[a]-1 to d-29[b].)

Results obtained from the use of radar equipment would be decided by the Director using the limitations above as a guide as modified to suit the purposes of the Maneuver. Students informed the Plotting Officer when ships, aircraft, or shore stations under their command were using radar equipment, for either detection or fire control. Students would also furnish a sketch or memorandum of the plan for radar use in Mooring Diagrams, Aircraft Flight Diagrams, or Aircraft Flight Forms. The use of radar would also be indicated on the Move and Gunfire Sheet. Limiting range and bearing were not affected by local conditions, the annual season, the temperature, or the time of day, but a radar was affected by rain on the antenna, humidity, cold fronts, the proximity of adjacent land to the instrument's beam, and the radar's calibration and adjustment. The use of lower power at frequencies above sixty megacycles per second (MCS), especially above 250 MCS, was thought to reduce the chances of enemy interception. (Page d-29[b].)

For detection radars, a pair of dice would be used as shown below to obtain limiting ranges. At all ranges, however, accuracy was assumed to be within five hundred yards. In addition to the limiting ranges for detection radar shown in figure 37, planes in flight had to be above the minimum altitudes shown in figure 35 to be detected from ships on the surface or to detect ships on the surface. For fire-control radars, limiting ranges were as shown in figure 36. For all of these sensors, frequency ranges were assumed to be above 60 megacycles, unless otherwise indicated. (Pages d-29[c] to d-29[d].)

Radar could itself be detected and its direction determined by "suitably" equipped stations up to 250 miles away. In order to penalize excessive use of radar in some manner approximating the harmful effects that might result in war, the approximate bearings from a receiving station of ships operating radar within direction-finding distance could be, at the discretion of the Director, given to the enemy commander. It was also possible to create radar interference; the Director would also decide when this was possible and what sectors it was being done in. In general, the Director supplied radar data as he saw necessary or desirable. It was not considered practical to be given continuous information, as would be the case in actual service; instead, "pertinent and timely" information, such as marked changes of course and changes in disposition—conditions that would be disclosed by radar—was issued. (Pages d-29[d] to d-29[e].)

Distance Nautical Miles	Minimum Altitude Thousands of Feet.
10	½
40	2
60	4
75	6
85	8
95	10

Fig. 35
Minimum altitudes for aircraft radar detection

Finally, tactical smoke screens could be laid by any oil-burning ship, by any ship or aircraft carrying special smoke apparatus, or by aircraft carrying smoke bombs. To be so dense as to prevent vision, a smoke screen had to be laid by at least one ship of light cruiser (CL) size or larger or by three smaller ships that were not more than five hundred yards apart. In order for smoke to lie on the water and not rise at once, atmospheric conditions had to be favorable, and the relative wind as felt by the smoke-laying ships had to be fifteen knots or more. Atmospheric conditions at night were considered always favorable for smoke screens,

Transmitter	Target	Effective limiting range.
Surface Ship or Station, not including SS on surface.	Surface Ship not SS.	(Range accuracy - 25 yds.) (Bearing accuracy - ¼ deg. 1/2 to 10 miles.
"	SS on surface	1/2 to 5 miles.
"	Plane, if altitude places it above transmitter's horizon.	1/2 to 15 miles.

Fig. 36
Fire control radar limiting ranges

8. For detection radar, a pair of dice will be used as shown in the following tables to obtain the limiting ranges:-

Transmitter	Target	Sum of dice rolls	Effective limiting Range, yards.
(a) Surface ship or station, including SS on surface.	Surface ship, not SS.	2	20,000
		3	21,000
		4	22,000
		5	23,000
		6	24,000
		7	25,000
		8	26,000
		9	27,000
		10	28,000
		11	29,000
		12	30,000
(b) "	SS on surface	2	2,000
		3	3,000
		4	4,000
		5	5,000
		6	6,000
		7	7,000
		8	8,000
		9	9,000
		10, 11, 12	10,000
(c) Submerged SS, not below periscope depth	Surface ship, not SS.	2, 9	10,000
		3, 10	11,000
		4, 11	12,000
		5, 12	13,000
		6	14,000
		7	15,000
		8	16,000
(d) "	SS on surface.	2, 8	2,000
		3, 9	3,000
		4, 10	4,000
		5, 11	5,000
		6, 12	6,000
		7	7,000

Note: A submerged SS, even with periscope exposed, may not be detected at any range.

Transmitter	Target	Sum of dice rolls	Effective limiting Range
(e) Surface ship, station, SS on surface, SS submerged not below periscope depth.	Plane, if altitude places it above the transmitter's visual horizon.	2	35 miles
		3	40 miles
		4	45 miles
		5	50 miles
		6	55 miles
		7	60 miles
		8	65 miles
		9	70 miles
		10	75 miles
		11	85 miles
		12	95 miles
(f) Plane in flight.	Plane in flight.	As designated by Director.	

Transmitter	Target	Sum of dice rolls	Effective limiting range
(g) Plane in flight.			
Small(Fighter)	(No detection radar carried in these planes)		
Patrol-Bomber (Altitude over 300 feet.)	SS on surface	2 to 12	3 to 15 miles (Min.range 1/4 miles.)
" "	Other single surface units.	2 to 12	4 to 25 miles (Min.range 1/4 miles.)
" "	Dispositions of two or more vessels.	2 to 12	7 to 45 miles (Min.range 1/4 miles.)

Fig. 37
Detection radar limiting ranges

but whether or not they were during the day was determined by the Director or asserted in the Statement of the Problem. When conditions permitted the laying of smoke, the screen always lay properly at first, but the Director would decide whether atmospheric conditions changed so as to obviate its effectiveness. The effect of this decision would be applied by one of the Assistant Directors, and the decision would not be announced to the students. Students told the Assistant Director in each instance of the emission of smoke to form a screen and when the smoke stopped. Moreover, a student was required to mark the intention to smoke in the Move Blanks of the Move and Gunfire Sheet. A smoke screen could be started at will or by signal. It could also be stopped at will, if the smoke-laying ships were clear of the screen; if they were operating in the screen, the emission of smoke could not be stopped except by signal. A smoke screen did not protect a ship laying it during the three-minute Maneuver Move in which the smoke screen was begun; it could, however, interfere with the gunfire of the smoke-laying ships and of other ships between which it lay. The smoke-laying ship in this case would be subject to a funnel-smoke penalty. After the emission of smoke had ceased, destroyers and other ships would be protected, but only by the screen already laid. Smoke emitted during a move would be carried directly to leeward at the speed of the wind during the three subsequent moves. During these three moves the smoke was impenetrable, but at the end of the fourth—including that in which the smoke was first emitted—it dissipated entirely. (Pages d-30 to d-31.)

The smoke area at the end of a move during which smoke was begun would be determined by drawing a line in the direction in which the wind was blowing from the position of the lead ship at the beginning of the move, a line equal in length to the distance a particle of smoke would travel during the move. The student drew a second line from the downwind end of that line to the position of the leader at the end of the move. This line was the forward edge of the smoke area and was designated as line *b*. In the same manner, the student was to determine a line for the rear ship and label it *c*, the after edge of the smoke area. The outer ends of lines *b* and *c* were connected by a line parallel to the shape of the formation at the beginning of the move, *d*. Another line was then drawn connecting the inner ends of lines *b* and *c* through the formation as it existed at the end of the move; this was line *e*. The smoke area was that bounded by lines *b, c, d,* and *e;* it was usually a parallelogram, but might appear as two triangles with apexes touching if the smoke-laying ship or ships made a significant change of course during the move. In plotting smoke screens, it was usually "most convenient" to lay out the smoke for one move as described above, then to translate the block of smoke bodily in each succeeding move the distance it would travel by the existing wind. It was found that the blocks of smoke laid out on the second and subsequent moves would connect with the first and form a continuous smoke screen. (Pages d-31 to d-32.)

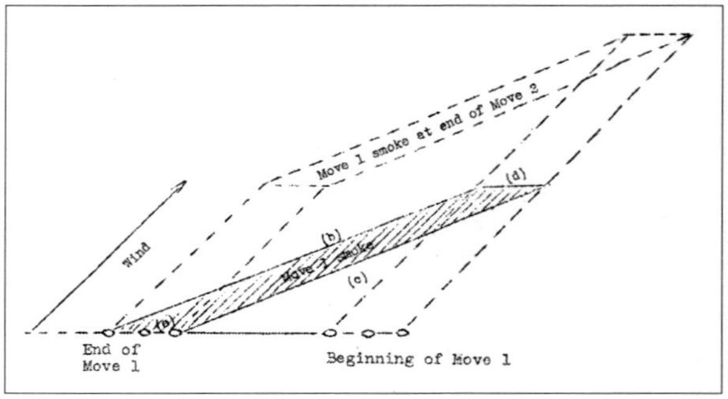

Fig. 38
Beginning of move 1

Students were asked to label each block of smoke with the move in which it was laid. They had to bear in mind that each block of smoke represented the situation at the *end* of the move. The use of a "smoke template" assisted in plotting smoke screens: students cut a piece of paper into a triangle to the scale of the Maneuver Board, such that the hypotenuse gave the direction of the wind and the distance that smoke particles would move along it in a single three-minute move. For example, if the wind was northwest at Force 4, the smoke template would appear as in figure 40. If *AB* was kept in a north–south direction and *A* was placed on the leading or rear ship, the points in lines *a* and *c* could be "readily" determined. The part of the template to the left of the line *A–B* was to assist in alignment with north on the Maneuver Board. A distance of 1,500 yards at fifteen knots for a three-minute move was indicated along a "convenient" edge of the template in order to facilitate the employment of the smoke. The template was then marked with the direction and the force of the wind for identification. Having determined the smoke area, the smoke screen was considered effective if line *b* was equal to or greater in length than fifteen knots on the scale of the Maneuver Board, and ineffective if less. (Pages d-32 to d-33.)

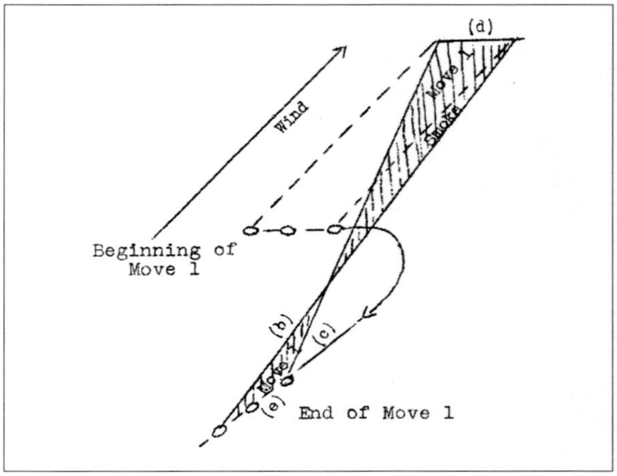

Fig. 39
End of move 1

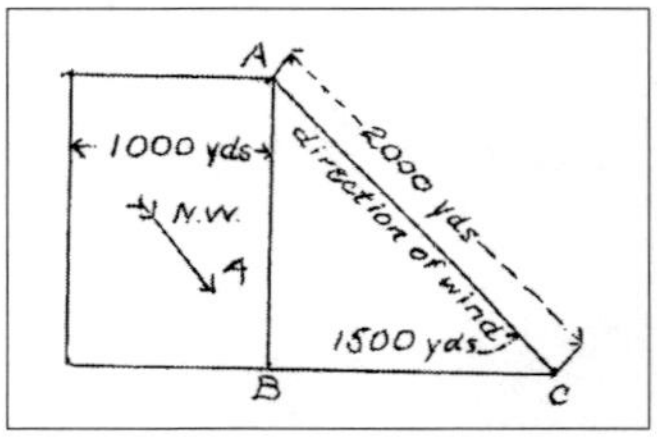

Fig. 40
Smoke template

NOTES

1 For previous looks, in less detail, at how the Maneuvers were conducted, see James Barber, "The School of Naval Warfare," *NWCR* 21, no. 8 (April 1969), pp. 89–96; Barber, "Mahan and Naval Strategy in the Nuclear Age: A Lecture Delivered at the Naval War College," *NWCR* 24, no. 7 (March 1972), pp. 78–88; Philip Crowl, "Education versus Training at the Naval War College: 1884–1972," *NWCR* 26, no. 3 (November–December 1973), pp. 2–10; Hal M. Friedman, *Digesting History: The U.S. Naval War College, the Lessons of World War Two, and Future Naval Warfare, 1945–1947* (Newport, R.I.: NWCP, 2010); Robert Gile, *Global War Game: Second Series, 1984–1988,* Newport Paper 20 (Newport, R.I.: NWCP, 2004); John B. Hattendorf, *The Evolution of the U.S. Navy's Maritime Strategy, 1977–1986,* Newport Paper 19 (Newport, R.I.: NWCP, 2004); John B. Hattendorf et al., *Sailors and Scholars: The Centennial History of the U.S. Naval War College* (Newport, R.I.: NWCP, 1984), pp. 175–77; Hattendorf, ed., *U.S. Naval Strategy in the 1970s: Selected Documents,* Newport Paper 30 (Newport, R.I.: NWCP, 2007); Bud Hay and Bob Gile, *Global War Game: The First Five Years,* Newport Paper 4 (Newport, R.I.: NWCP, 1993); J. S. Hurlburt, "War Gaming at the Naval War College, 1969–1989," *NWCR* 42, no. 3 (Summer 1989), pp. 46–51; Edward Miller, *War Plan Orange: The U.S. Strategy to Defeat Japan, 1897–1945* (Annapolis, Md.: Naval Institute Press, 1991); David Rosenberg, "Being 'Red': The Challenge of Taking the Soviet Side in War Games at the Naval War College," *NWCR* 41, no. 1 (Winter 1988), pp. 81–93; Douglas Smith, *Carrier Battles: Command Decision in Harm's Way* (Annapolis, Md.: Naval Institute Press, 2006); Nepier Smith, "Historical Analysis of the Organizational Success of the Naval War College during the Twenty-Five Years following the Second World War" (Naval War College Advanced Research Project, NHC, NWC, 1974), pp. 11–67; Ronald Spector, *Professors of War: The Naval War College and the Development of the Naval Profession* (Newport, R.I.: NWCP, 1977), pp. 71–87; Michael Vlahos, *The Blue Sword: The Naval War College and the American Mission, 1919–1941* (Newport, R.I.: NWCP, 1980), pp. 131–56; Kenneth Watman, "Global 2000," *NWCR* 54, no. 2 (Spring 2001), pp. 75–88; Donald White, "Admiral Richard L. Conolly: A Perspective on His Notions of Strategy," *NWCR* 24, no. 3 (November 1971), pp. 73–79; and Robert S. Wood, "The Conceptual Framework for Strategic Development at the Naval War College," *NWCR* 40, no. 2 (Spring 1987), pp. 4–16.

IV *Maneuver Rules Continued, June 1945*

Communications during the Chart and Board Maneuvers were effected by visual means, radio, sound, wire, and various carriers, such as postal, messenger, and plane. The main weapons of surface ships were guns and torpedoes.

Communications

Unless otherwise indicated, students in Maneuvers transmitted all communications to each other by means of Message Blanks, indicating on the form what means of communication was being simulated. Students filled out transmission instructions on the form and wrote the message in the space provided, using a hard pencil or typewriter and special carbon provided. The Message Blank was then sent to the Assistant Communication Umpire, who would—using values in Table E-4 on the guide to the Message Blank (figure 42)—compute the times of delivery for the messages and fill out that part of the form. Student Commanders were to be considered present in all parts of their commands for which other students were not assigned as Subordinate Commanders. It was therefore not necessary to use the Message Blank for communication within a command, but "proper regard" was to be given to actual existing capabilities and conditions when considering internal communication to have been accomplished.*

Whenever time—other than references to dispatches, identified by Date-Time Groups using Greenwich Civil Time (GCT)—was expressed in a message text, a letter suffix indicating the zone description was to be used. Thus, −1 to −12 were represented as letters *A* to *M*, except *J*, respectively, and positive 1 to 12 were letters *N* to *Y*, respectively. For example, the suffix for plus 5 was *R* while *Z* was used for Zone 0. To express time between zones, students used both letters, such as *WX* for plus ten and a half. The system was used exclusively in Navy and Blue–Red (i.e., communications to the United Kingdom) but not in Army–Navy communications or in messages to civilian agencies. (Page e-7[a].)

---

* "Maneuver Rules," June 1945, pp. e-1 to e-7(a), NHC, folder 2525-E, box 124, RG 4, NWC. Subsequent page-number references in this chapter are to this source.

Also, when time—again, other than in references to dispatches—was expressed in a message text, it was to be followed by time zone suffix letters, as provided for in article 935 of the 1939 *Communication Instruction, United States Navy.* This instruction defined the Time of Origin as the time at which the message was released by the originator; if that time did not appear on the message itself, the Time of Origin was that at which the message was filed for transmission. The Time of Origin Group was to be a four-digit group, with the first two digits indicating the hour, from 00 to 24, inclusive. The last two digits indicated the minutes after the hour, from 00 to 59, inclusive. Normally, GCT was to be employed to indicate the Time of Origin of all communications throughout the Navy. There was also to be a Date Group, consisting of two digits indicating the day of the month. When employed, the Date Group was to be combined with the Time of Origin Group. If the date was less than ten, it was to be preceded by a zero. Thus, the first day of any month was 01, and the tenth day was 10. The date had to agree with GCT or with whatever time was to be employed in the Time of Origin Group. The Date Group, when combined with the former, formed what the Navy called the "Date-Time Group." (Page e-7[a]-1.)

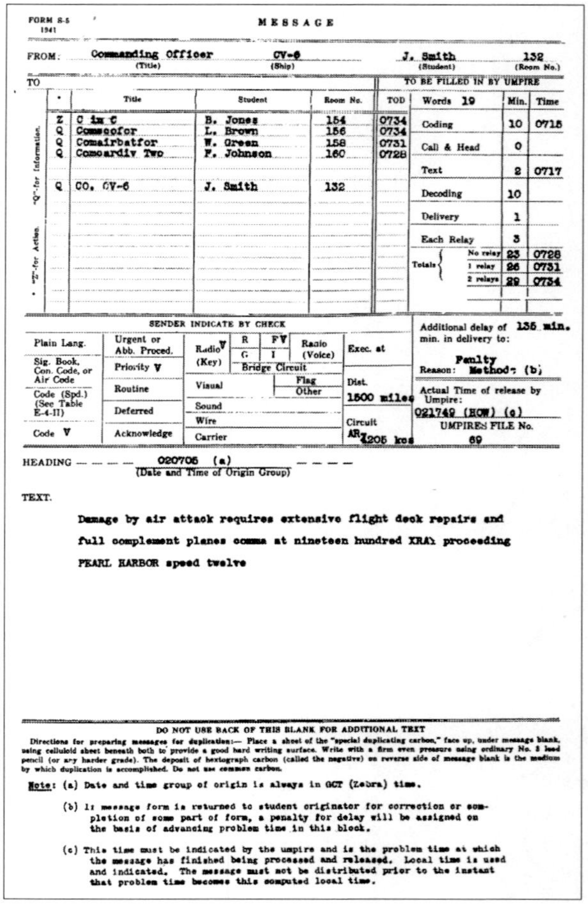

FORM S-5
1941

MESSAGE

FROM: Commanding Officer (Title) CV-6 (Ship) J. Smith (Student) 132 (Room No.)

TO

"Z"-for Action. "Q"-for Information.

•	Title	Student	Room No.	TOD
Z	C in C	B. Jones	154	0734
Q	Comscofor	L. Brown	156	0734
Q	Comairbatfor	W. Green	158	0731
Q	Comoardiv Two	F. Johnson	160	0728
Q	CO. CV-6	J. Smith	132	

TO BE FILLED IN BY UMPIRE

Words 19	Min.	Time
Coding	10	0715
Call & Head	0	
Text	2	0717
Decoding	10	
Delivery	1	
Each Relay	3	
Totals: No relay	23	0728
Totals: 1 relay	26	0731
Totals: 2 relays	29	0734

SENDER INDICATE BY CHECK

Plain Lang.	Urgent or Abb. Proced.	Radio V (Key): R, F V, G, I, Bridge Circuit	Radio (Voice)	Exec. at
Sig. Book, Con. Code, or Air Code	Priority V			
Code (Spd.) (See Table E-4-II)	Routine	Visual	Flag / Other	Dist. 1500 miles
	Deferred	Sound		
		Wire		Circuit AR 4205 kcs
Code V	Acknowledge	Carrier		

Additional delay of 135 min. min. in delivery to:

Reason: Penalty Method (b)

Actual Time of release by Umpire: 021749 (HOW) (c)

UMPIRES FILE No. 69

HEADING ———— 020705 (a) ———— (Date and Time of Origin Group)

TEXT.

Damage by air attack requires extensive flight deck repairs and full complement planes comma at nineteen hundred XRA1 proceeding PEARL HARBOR speed twelve

DO NOT USE BACK OF THIS BLANK FOR ADDITIONAL TEXT

Directions for preparing messages for duplication:— Place a sheet of the "special duplicating carbon," face up, under message blank, using celluloid sheet beneath both to provide a good hard writing surface. Write with a firm even pressure using ordinary No. 2 lead pencil (or any harder grade). The deposit of hextograph carbon (called the negative) on reverse side of message blank is the medium by which duplication is accomplished. Do not use common carbon.

Note: (a) Date and time group of origin is always in GCT (Zebra) time.

(b) If message form is returned to student originator for correction or completion of some part of form, a penalty for delay will be assigned on the basis of advancing problem time in this block.

(c) This time must be indicated by the umpire and is the problem time at which the message has finished being processed and released. Local time is used and indicated. The message must not be distributed prior to the instant that problem time becomes this computed local time.

Fig. 41
Form S-5 Message Blank

The Month Group was the name of the month, which could be abbreviated. This was a Text Group to be used to identify messages for reference purposes when the month of origin was other than the current month. The month of the referenced message, however, did not have to be indicated if it was the month preceding the current month and the date of the preceding month was greater than the present date of the current month. When required, the Month Group also followed reference numbers. The time recorded in message texts, except for references to other messages, was to be expressed in four- or six-digit numbers, with a letter suffix chosen from the table in figure 43, the letter suffix indicating the zone description. Students were warned that discretion was to be used when employing these suffixes to nonnaval addressees who might not understand them. Students were also referred to Time Zone Chart H.O. 5192 for exact zone boundaries, as these sometimes deviated to accommodate national boundaries. Both letters in the suffixes would be used when indicating time midway between zones. Given all of these stipulations,

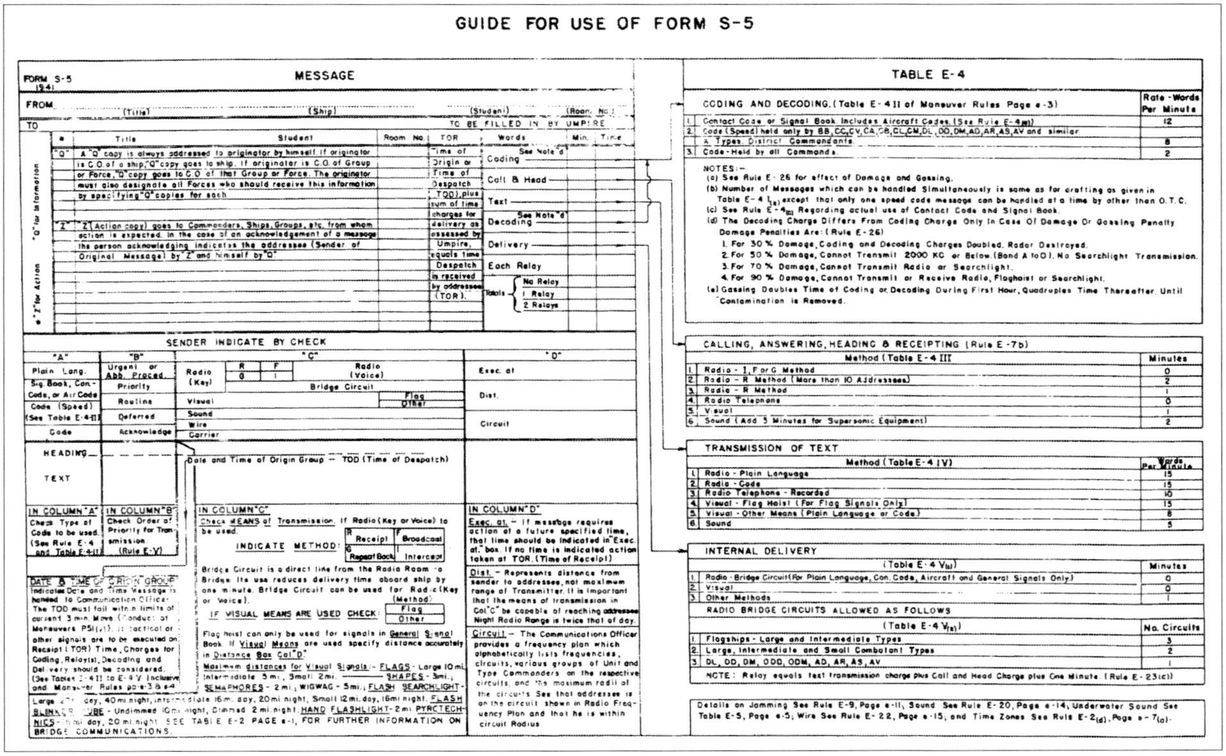

GUIDE FOR USE OF FORM S-5

FORM S-5 (1941)

MESSAGE

FROM (Title) (Ship) (Student) (Room No.)

TO — TO BE FILLED IN BY UMPIRE

*	Title	Student	Room No.	TOR	Words	Min.	Time
"Q"	A "Q" copy is always addressed to originator by himself. If originator is C.O. of a ship, "Q" copy goes to ship. If originator is C.O. of Group or Force, "Q" copy goes to C.O. of that Group or Force. The originator must also designate all Forces who should receive this information by specifying "Q" copies for each			Time of Origin or Time of Despatch (TOD), plus sum of time charges for delivery as assessed by Umpire, equals time Despatch is received by addressee (TOR).	Coding (See Note "d")		
					Call & Head		
					Text		
"Z"	"Z" (Action copy) goes to Commanders, Ships, Groups, etc. from whom action is expected. In the case of an acknowledgement of a message the person acknowledging indicates the addressee (Sender of Original Message) by "Z" and himself by "Q"				Decoding (See Note "d")		
					Delivery		
					Each Relay		
					Totals: No Relay / 1 Relay / 2 Relays		

"Q" for Information — * "Z" for Action

SENDER INDICATE BY CHECK

"A"	"B"	"C"	"D"
Plain Lang.	Urgent or Abb. Proced.	Radio (Key) R F G I; Radio (Voice)	Exec. at
Sig. Book, Con-Code, or Air Code	Priority	Bridge Circuit	
Code (Speed) (See Table E-4 II)	Routine	Visual — Flag / Other	Dist.
	Deferred	Sound	Circuit
Code	Acknowledge	Wire	
		Carrier	

HEADING — Date and Time of Origin Group — TOD (Time of Despatch)

TEXT

IN COLUMN "A" — Check Type of Code to be used. (See Rule E-4 and Table E-4 II)

IN COLUMN "B" — Check Order of Priority for Transmission. (Rule E-V)

IN COLUMN "C" — Check MEANS of Transmission. If Radio (Key or Voice) to be used. INDICATE METHOD: R Receipt; F Broadcast; G Repeat Back; I Intercept. Bridge Circuit is a direct line from the Radio Room to Bridge. Its use reduces delivery time aboard ship by one minute. Bridge Circuit can be used for Radio (Key or Voice). (Method) IF VISUAL MEANS ARE USED CHECK: Flag / Other. Flag hoist can only be used for signals in General Signal Book. If Visual Means are used specify distance accurately in Distance Box Col. "D". Maximum distances for Visual Signals:- FLAGS - Large 10 mi., Intermediate 5 mi., Small 2 mi. —— SHAPES - 3 mi.; SEMAPHORES - 2 mi.; WIGWAG - 5 mi.; FLASH SEARCHLIGHT - Large [illegible] day, 40 mi. night, intermediate 16 mi. day, 20 mi. night, Small 12 mi. day, 16 mi. night. FLASH BLINKER TUBE - Undimmed 10 mi. night, Dimmed 2 mi. night. HAND FLASHLIGHT - 2 mi. PYROTECHNICS - [illegible] mi. day, 20 mi. night. SEE TABLE E-2 PAGE e-1, FOR FURTHER INFORMATION ON BRIDGE COMMUNICATIONS.

DATE & TIME OF ORIGIN GROUP — Indicates Date and Time Message is handed to Communication Officer. The TOD must fall within limits of current 3 min. Move. (Conduct of Maneuvers P51(a)). If tactical or other signals are to be executed on Receipt (TOR) Time, Charges for Coding, Relay(s), Decoding and Delivery should be considered. (See Tables E-4 II to E-4 V Inclusive and Maneuver Rules pp. e-3 & e-4)

IN COLUMN "D" — Exec. at. - If message requires action at a future specified time, that time should be indicated in "Exec. at." box. If no time is indicated action taken at TOR. (Time of Receipt). Dist. - Represents distance from sender to addresses, not maximum range of Transmitter. It is important that the means of transmission in Col. "C" be capable of reaching addressee. Night Radio Range is twice that of day. Circuit - The Communications Officer provides a frequency plan which alphabetically lists frequencies, circuits, various groups of Unit and Type Commanders on the respective circuits, and the maximum radii of the circuits. See that addressee is on the circuit shown in Radio Frequency Plan and that he is within circuit Radius.

TABLE E-4

	CODING AND DECODING. (Table E-4 II of Maneuver Rules Page e-3)	Rate - Words Per Minute
1.	Contact Code or Signal Book. Includes Aircraft Codes. (See Rule E-4(g))	12
2.	Code (Speed) held only by BB, CC, CV, CA, CB, CL, CM, DL, DD, DM, AD, AR, AS, AV and similar A Types. District Commandants.	8
3.	Code - Held by all Commands.	2

NOTES:-
(a) See Rule E-26 for effect of Damage and Gassing.
(b) Number of Messages which can be handled Simultaneously is same as for drafting as given in Table E-4 I(a) except that only one speed code message can be handled at a time by other than O.T.C.
(c) See Rule E-4(g) Regarding actual use of Contact Code and Signal Book.
(d) The Decoding Charge Differs From Coding Charge Only In Case Of Damage Or Gassing Penalty. Damage Penalties Are: (Rule E-26)
1. For 30 % Damage, Coding and Decoding Charges Doubled. Radar Destroyed.
2. For 50 % Damage, Cannot Transmit 2000 KC or Below. (Band A to D). No Searchlight Transmission.
3. For 70 % Damage, Cannot Transmit Radio or Searchlight.
4. For 90 % Damage, Cannot Transmit or Receive Radio, Flaghoist or Searchlight.
(e) Gassing Doubles Time of Coding or Decoding During First Hour, Quadruples Time Thereafter. Until Contamination is Removed.

CALLING, ANSWERING, HEADING & RECEIPTING (Rule E-7b)

	Method (Table E-4 III)	Minutes
1.	Radio - I, F or G Method	0
2.	Radio - R Method (More than 10 Addresses)	2
3.	Radio - R Method	1
4.	Radio Telephone	0
5.	Visual	1
6.	Sound (Add 5 Minutes for Supersonic Equipment)	2

TRANSMISSION OF TEXT

	Method (Table E-4 IV)	Words Per Minute
1.	Radio - Plain Language	15
2.	Radio - Code	15
3.	Radio Telephone - Recorded	10
4.	Visual - Flag Hoist (For Flag Signals Only)	15
5.	Visual - Other Means (Plain Language or Code)	8
6.	Sound	5

INTERNAL DELIVERY

	(Table E-4 V(b))	Minutes
1.	Radio - Bridge Circuit (For Plain Language, Con. Code, Aircraft and General Signals Only)	0
2.	Visual	0
3.	Other Methods	1

RADIO BRIDGE CIRCUITS ALLOWED AS FOLLOWS

	(Table E-4 V(a))	No. Circuits
1.	Flagships - Large and Intermediate Types	3
2.	Large, Intermediate and Small Combatant Types	2
3.	DL, DD, DM, DDD, DDM, AD, AR, AS, AV	1

NOTE: Relay equals text transmission charge plus Call and Head Charge plus One Minute. (Rule E-23(c))

Details on Jamming See Rule E-9, Page e-11; Sound See Rule E-20, Page e-14; Underwater Sound See Table E-5, Page e-5; Wire See Rule E-22, Page e-15; and Time Zones See Rule E-2(d), Page e-7(a).

Fig. 42
Guide for use of Form S-5

GCT was to be used in the Date-Time Group in dispatches, and appropriate zone letter suffixes to be used whenever time was mentioned in a message. Time for drafting messages and the number of messages that could be drafted simultaneously were stipulated as well, as shown in Table E-4. (Pages e-7[a]-1 to e-8.)

In initiating a message as a result of receiving a message or an information slip, students were to take drafting time into account in assigning the Date and Reference Group. With respect to encryption, messages could be sent in various ways, such as Plain Language, Contact Code, Aircraft Code, Signals, codes held only by various types of commands, or codes held by all commands. The rates used in the computation of coding and decoding times by the Assistant Communication Umpires were in Table E-4; the number of messages that could be coded and decoded simultaneously on various types of ships was the same that could be drafted simultaneously. Students were assumed to possess the necessary code and signal books, and the Director could instruct that appropriate messages be encoded from the Contact Code or Signal Book; in that case, the computed time for coding and decoding

Table of Time Zones, Zone Descriptions and Suffixes.

Zone	Description	Suffix	Zone	Description	Suffix
7½W to 7½E	0	Z	----------	----------	----------
7½E to 22½E	-1	A	7½W to 22½W	1	N*
22½E to 37½E	-2	B	22½W to 37½W	2	O
37½E to 52½E	-3	C	37½W to 52½W	3	P
52½E to 67½E	-4	D	52½W to 67½W	4	Q
67½E to 82½E	-5	E	67½W to 82½W	5	R
82½E to 97½E	-6	F	82½W to 97½W	6	S
97½E to 112½E	-7	G	97½W to 112½W	7	T
112½E to 127½E	-8	H	112½W to 127½W	8	U
127½E to 142½E	-9	I	127½W to 142½W	9	V
142½E to 157½E	-10	K	142½W to 157½W	10	W
157½E to 172½E	-11	L	157½W to 172½W	11	X
172½E to 180	-12	M	172½W to 180	12	Y

NOTE: GCT is indicated by suffix Z. Reference to Time Zone chart H.O. 5192 should be made to learn the exact zone boundaries as they sometimes deviate slightly to accommodate national boundaries. etc. For time midway between zones use both letters.

*Letter N is also used to designate -13; this is to provide for a ship in zone -12 keeping Daylight Saving Time."

Fig. 43
Table of time zones, zone descriptions, and suffixes

was zero. When a Contact Code, Signal Book, or Aircraft Code was designated in the transmission instructions, the form and subject matter were to be such that the code designated could actually be used for it. Aircraft were assumed to have only the Aircraft Code and Contact Code, and the time taken to code and decode messages in aircraft was to be double that in ships. If special signals for radio direction finding were to be made, the text on the Message Blank was to be "Test for ___ minutes for direction finding." Students were to indicate in all cases on the Message Blank which code was being used. (Pages e-8 to e-9.)

Messages were handled over all channels in their Order of Precedence—Urgent, Priority, Routine, and Deferred. "Specially important" signals or dispatches concerning battle or impending battle were to be marked Urgent. This designation was to be used at first contact with any "important" enemy force and on all subsequent reports that represented vital new information rather than mere "amplification." The term "abbreviated procedure" was to be used with the Contact, Signals, and Aircraft Codes and was to be used in all messages of "great, immediate" importance. The Priority designation was to have precedence over Routine traffic and was not to be used without good reason. The Deferred designation was to be used in messages that did not require delivery before the beginning of office hours on the following day. These messages, however, were to be transmitted as soon as the circuit was cleared of traffic of higher precedence and, at any rate, within the time for delivery as indicated above. Students were therefore also to indicate precedence desired on all Message Blanks and indicate whether an acknowledgment was desired. (Page e-9.)

Communications in the various methods could be conducted simultaneously during maneuvers in airplanes where the same personnel handled different methods of communication, such as visual and radio. The methods employed for the transmission of messages

			Words per minute
I.	DRAFTING (Rule E-3)		
	(1) All messages including recorded telephone messages		10
	(2) Number of messages which may be drafted simultaneously:		
	Fleet and Force Flagships	3	
	Large and intermediate combatant type	2	
	All others	1	
II.	CODING AND DECODING (Rule E-4)		
	(Rate given is for each process)		
(1)	ECM.(and similar mechanical coding and decoding units)		14
(2)	Contact Code or Signal Book. (Includes Aircraft Codes. See Rule E-4(e))		12
	Code (speed) Hold only by DB,CC,CV,CA, CB,CL,CM,DL,DD,DM,AD,AR,AS,AV, and similar X-Types. District Commandants.		8*
	Code Held by all commands		2
(3)	See Rule E-26 for effect of damage and gassing.		
(4)	Number of messages which can be handled simultaneously same as for drafting.		
	*(For Code (spd)		
(5)	See Rule E-4(c) concerning actual use of Contact Code and Signal Books.		
III.	CALLING, ANSWERING, HEADING, RECEIPTING.		
	Method		Minutes
	Radio - I, F or G method		0
	Radio - R method (more than 10 addressees)		2
	Radio - R method		1
	Radio telephone		0
	Visual		1
	Sound		2**
	* Only one message can be handled at a time except by O.T.C.		
	** Add 5 minutes for supersonic equipment.		
IV.	TRANSMISSION OF TEXT.		
	Radio telephone - plain language		14
	Radio - plain language		15
	Radio - code		15
	Radio telephone - recorded		10
	*Visual - Flag hoist		15
	Visual - other means - plain language or code		8
	Sound		5
	*For Flag Signals only.		
V.	INTERNAL DELIVERY.		Minutes
	(1) *Radio - bridge circuit		0
	Visual		0
	Other methods		1
	*For plain language, contact code, aircraft code and general signals only.		
	(2) Bridge radio circuits allowed as follows:		
	Flagships, large and intermediate types	-3	
	Large, intermediate, and small combatant type	-2	
	DL, DD, DM, ODD, ODM, AD, AR, AS, AV	-1	

NOTE: For each relay, in addition to repeating the times for call, heading and transmission of text, add one minute (Rule E-23(c)).

Fig. 44
Table E-4

by radio were *I* for Intercept, *F* for Broadcast, *R* for Receipt, and *C* for Repeat Back. The last designation was normally used between shore stations sending messages for ships to copy in using the Intercept method. Radio telephones were also to be available for voice communication; Table E-4 gave time delays that were charged for calling, answering, heading, and receiving messages. Rates of transmission were also given in both plain language and code, as were times for the internal delivery of messages in this manner and the number of bridge radio circuits that were permitted. Accordingly, students indicated on message forms whether they were using radio or bridge radio as their method of communication. Table E-3 illustrates the number of radio transmitters and receivers that could be carried by various types of craft, as well as their frequencies. Subject to special rulings by the Communication Umpire, the number of radio circuits that could be fully guarded was equal to the number of transmitters on board. Other frequencies could be guarded for listening only, until all receivers on board were in use. In Chart Maneuvers, because of limited personnel, the maximum number of circuits that could be guarded continuously by vessels other than flagships for more than twenty-four hours was considered to be five for aircraft carriers; three for large-, intermediate-, and small-sized ships; and one for destroyers and submarines. These limitations were in addition to those imposed on Table E-3, though fleet flagships were allotted three additional circuits and other flagships could add one. Again, students in all cases indicated on the radio message the circuit designation to be used by the originating ship or shore station. (Pages e-9 to e-11.)

In order for jamming to be effective, the Maneuver Rules stipulated, each ship was to search a frequency band not broader than two hundred kilocycles per second (KCS) for frequencies in Band O and below, 1,000 KCS for frequencies in Band P and above. After two transmissions by an enemy within a frequency band that was being searched, any further transmissions two or more minutes in length on the same frequency could be jammed. The effectiveness of this jamming would depend on the strength of the jamming signals and would be decided by the Communication Umpire. The effect of interference was to vary in accordance with the relative strengths of the interference signals as heard by the receiving ship. If a transmission and the interference signals were of equal strength, reception was impossible. A lesser degree of interference was to have the effect of doubling the time for transmitting messages. If it was included in the Statement of the Problem for the purpose of illustrating a possible situation, one side might be given a specially designated transmitter that could be used to intercept 20 percent of the enemy's spotting frequencies for a period of twelve minutes from the time the interference started. The interference could not be started earlier than six minutes after the commencement of gunfire; the Director would decide whether the spotting frequencies were interfered with during this period. This interference was assumed

41-15 Mod. 6/44

TABLE E-3 RADIO EQUIPMENT

Line No.	Column	1	2	3	4	5	6	6	7	8	9	10	11	12	13	14	15	16
		TRANSMITTERS K-Key only KV-Key and Voice V-Voice only																
	Figures indicate distance in miles (Daylight) (a) Higher Frequency Bands	(b)(c) R-Y 4000 Key 600 Voice	(c) P-W 3000 Key 500 Voice	(c) P-N 3000 Key 500 Voice	(c) P-W 2000 Key 200 Voice	(c) P-U 500 Key 100 Voice	(c)		(c) P 500 Key 100 Voice	(c)	(c)	(c)	(c) Q-R or T-U 200 Key 50 Voice	(c) Q-U 500 Key 100 Voice	(c) Q-V 3000 Key 500 Voice	(c) Q-R or T-U 200 Key 50 Voice	(c) P-U 300 Key 50 Voice	(c) R 2000 Key 200 Voice
	Lower Frequency Bands (Key only)			I-O 800 Key	E-L 300 Key	J-O 200 Key	L-O 500 Key	L-N 300 Key		E-L 800 Key	E-L 500 Key	C-L 100 Key		J-N 200 Key	I-L 300 Key	N 100 Key		I-L 300 Key
1	Fleet Flag (BB)	2K	2K,1KV		2KV		1K	1K	1K	2K		1K						
2	Fleet Flag (not BB)	2K	1K,1KV				1K	1K		1K	2K							
3	BB,CB,CC	1K	1K,1KV		1KV			1K	1K	2K		1K						
4	CV	1K	3K,2KV	2KV						2K	1K							
5	CVS, XCV	1K	1K,2KV	2KV						2K	1K							
6	CA, CL (large)		3K	1KV						1K	2K							
7	BM, CL (small), CM		2K	1KV						1K	1K							
8	DL, DD (e)		1K		1KV						1K							
9	DDc, ODD, DE, DM, DMS, APD (e)				1KV	1KV			1K(f)									
10	AVD				1KV	2KV												
11	SS, SM (e)				1KV													
12	SSc, OSS, SMc, OSM, ASR (e)																	1K
13	PC, PE (e)					1KV												
14	PT, PTC (e)																1KV	
15	AO, AR, AS, XAO, XAR, XAS	1K (g)	1K							1K								
16	AV, XAV	1KV	1K,2KV	1K (p)						1K								
17	AC, AE, AF, AG, AH, AK, AO, AP, AZ and similar converted types (e)		1K							1K					1K(q)			
18	AM, AN, AT, PG, PR, PY and similar converted types (e)				1K													
19	AVP, XAVP				1K													1K
20	District Craft (e)								1V (l)									
21	ZN, ZR												1KV	1KV(m)				
22	VB(A), VP												1KV		1KV			
23	VB, VJ, VM, VR, VS, VT													1KV(m)				
24	VF												1KV					
25	VO															1KV		
26	Advanced Bases	Sets other than portable will be indicated in statement of Problem.																

Line No.	17	18	19	20	21	22	23	24	25	26	27	28	29	30	31	32	33	34	Line No.
	RECEIVERS														PORTABLE SETS				
	(c)	(c)	(c) P-Z	(c) P-X	(c) P-X	(c) P-Z	(c) P-Q	(c) P-Y	(c)	(c) Q-U	(c) P-V	(c) Q-R or T-U	(c) P-U	(c) Direction finder	(c) TRANSMITTER P-R 30 Key, voice	(c) RECEIVER P-R	(c) RECEIVER P-T	(c) TRANSMITTER P-T 30 Key, voice	
	A-M	B-L	N-O	I-O	B-O	F-O	L-O	K-O	O-J	J-N	H-O			O-N		I-L			
1	6	1	8	1	1	1								1(d)	1K	1			1
2	6	2	9	2		1								1(d)	1K	1			2
3	5	1	6	1	1									1	1K	1			3
4	6	3	9	3										3	2K	2			4
5	4	3	7	3										1	1K	1			5
6	5	2	6	2										1	2K	2			6
7	4	2	4	2										1	1K	1			7
8	1	2	1	2										1			1	1KV	8
9	2		1(j)			1								1					9
10	3		2			1								1					10
11					1									1					11
12					1									1(f)					12
13						1													13
14						1							1						14
15		3		1			2(h)	2	1					1	1K		1		15
16		3		1			4	5	5					2	1K		1		16
17		1		1		1(i)								1	1KV		1		17
18					1									1					18
19					2									1					19
20																			20
21											2(h)								21
22											3(h)								22
23										1(n)									23
24												1(n)							24
25										1(r)									25
26																			26

USE OF FREQUENCY BAND LETTERS

The frequency bands are indicated in Table E-3 by letters which show the complete range of the set. For example; The set in Column 1 has a range of "R-Y". This is inclusive, i.e., includes Bands R, S, T, U, V, W, X and Y. Suppose it is desired to establish a 2000 mile (maximum) circuit on frequency Band R; any of the transmitters in Columns 1, 2, 3, 4, 13 and 16 could be used on this circuit. Transmitters in Columns 5, 11, 12, 14 and 15 can operate on frequency "R", but do not have the required power (distance).

BAND LETTER	FREQUENCY				
R/T (r)	3000	-	30000	mgcls.	Super-high
	300	-	3000		Ultra-high
	30	-	300		Very-high
Z	26000	-	30000	kcs.	High
Y	23000	-	26000		
X	18100	-	23000		
W	13575	-	18100		
V	9050	-	13575		
U	8000	-	9050		
T	6000	-	8000		
S	4525	-	6000		
R	4000	-	4525		
Q	3000	-	4000		
P	2000	-	3000		
O	1500	-	2000		
N	1000	-	1500		Intermediate
M	600	-	1000		
L	500	-	600		
K	400	-	500		
J	350	-	400		
I	300	-	350		
H	224	-	300		
G	200	-	224		
F	190	-	200		
E	175	-	190		
D	100	-	175		
C	75	-	100		Low
B	15	-	75		
A	10	-	15		

For a brief general description of the Naval Employment of Frequencies, see Article 138, Communication Instructions, 1939.

(a) Distance is average reliable transmission in miles under favorable conditions by day. At night for Frequency Bands A-O maximum distance is double that given above. At night use of Frequency Bands P-Z and at all times use of Frequency Bands R-Z is attended by danger of abnormally long distance transmission. Consider distances given as approximations only.

(b) Reliable use of Frequency Bands Y-Z does not extend beyond 10 to 15 miles, but transmissions may be intercepted at great, though uncertain and variable distances beyond the skip distance. Ranges for R/T transmission will be the optical range corresponding to the height of the transmitting antenna above sea level (may be several hundred miles for planes.)

(c) Only one frequency at a time on each set.
(d) Also has Direction Finder set for Frequency Bands G-W.
(e) Only Squadron Flags can man more than one circuit continuously during Chart Maneuvers.
(f) Squadron Flagship two.
(g) AR none.
(h) AR one.
(i) AP, AH, AP two.
(j) ASR only.
(l) Combined transmitter and receiver.

(m) Plug in tuning circuits may allow only part of this frequency range on flight.
(n) May be used for homing.
(p) Bands I-O only (For aircraft homing).
(q) AF, AH, AP only (Bands I-L only).
(r) These three bands are assumed to fall under the general heading of R/T (radio-telephone), and shall be so designated. Key transmission may not be used with these circuits. The numbers of transmitters, receivers and the range shall be in accordance with Table E-2, page e-1.

Fig. 45
Table E-3: Radio equipment

to eliminate all of one's own radio communications. Moreover, if the interference was not being made by the Officer in Tactical Command or on a time schedule, a visual signal would be needed to stop it. (Pages e-11 to e-12.)

Shifting frequencies would also involve delays. For ships with spare equipment tuned in advance, there would be no delays, but if spare equipment was not tuned, a three-minute delay would ensue. For aircraft, any shift would entail a five-minute delay. Table E-3 gave the average reliable daylight transmission distance for each transmitter, as well as transmission distances during darkness. The Director could increase or decrease these distances at his discretion and without notice. Also, for frequencies in Bands A–O, distances could be decreased by proportionate reduction in transmitter power. A student could indicate such reduction by the distance he prescribed on his Message Blank. It would then be assumed that the message could be read only at a distance that was 10–20 percent greater and heard at a distance 20–30 percent greater than the distance prescribed by the student. Students were again warned, as they had been in Section D of the Maneuver Rules, that radio-direction-finding bearings of all ships sending or acknowledging messages, receiving messages, or testing equipment could be given to the opposing Student Commander by the Director in order to simulate real wartime situations. Only one set of bearings, however, would be given on any one transmission. Additionally, in order to emphasize the danger of indiscreet radio transmissions that could furnish information to the enemy, the Director could supply the opposing Student

Commander with plain-language messages, after a five-to-sixteen-minute delay; translations of approximately 5 percent of messages in Contact, Aircraft, or Signals Code after a delay of an hour or an hour and a half; and 2 percent of messages in codes held by various or all commands after a delay of two to eight hours. (Page e-12.)

Submarines submerged to a depth greater than periscope depth could not transmit or receive radio messages. At periscope depth, when its vertical antenna was exposed, a submarine could transmit and receive messages on frequencies in Bands B to P and higher. When a submarine used the vertical antenna while submerged at periscope depth, the effective distance of transmission was to be "about ¼" of the distance specified in Table E-3. If a student wanted to arrange for submarines to hear radio calls when submerged, listening periods had to be planned and indicated on flimsies and in the Communication Plan. The Communication Umpire would decide whether submarines would be allowed to hear their radio calls on the basis of existing conditions and the length of the listening periods arranged. (Page e-13.)

For Chart Maneuvers, when respective forces were in company students could conduct direct oral communication instead of simulating visual communications. The means and distances for visual communication and the number of simultaneous transmissions permitted were specified in Table E-2. In addition, flag hoists could be used for signals only; students had to indicate on their message forms whether the flag hoist or some other method was being used. The time charged for visual calling, answering, heading (addressing), and receiving was specified in Table E-4, as were the rates of transmission of text by visual means. The time of internal delivery for visual messages was to be zero. Concerning the Semaphore, Blinker, and Searchlight methods of communication, the Assistant Communication Umpire would judge from the positions of the addressees on the Maneuver Board whether the number of transmissions required exceeded the number that could be made simultaneously by the sending unit. If so, he would delay transmissions accordingly. For flag signals, if there was not enough relative wind to make flags fly visibly, and for flashing-light signals, if the sending vessel was rolling and pitching considerably, the Assistant Communication Umpire would delay delivery of the message. Enemy units in proper relative position to the sender would be permitted to intercept plain-language visual transmissions. (Pages e-13 to e-14.)

As for sound transmissions, underwater sound equipment on various types of ships and aircraft and the distances at which it could be used for communication are shown in Table E-5 below and the effective distance of sound in the air in Table E-2. In underwater sound communication, five minutes were charged for calling and other actions in addition to the values shown in Table E-4. Vessels using supersonic equipment for communication were also to be charged five minutes for these actions in addition to those in Table E-4; transmission by audio-frequency oscillators could not be received through interference by other audio-frequency oscillators.

Cable and land-wire messages, however, could be sent over existing lines, provided they did not traverse enemy territory. A ship could send such a message one hour after arrival at a port or send it by any method available to a Naval Communication Shore Station. This message had to be addressed to a ship or station at a port or place named in the address and was to be received at that address two hours after being sent. The message was delivered to the senior U.S. naval officer present one hour after the message was sent. If the ship addressed was present, the senior officer present would forward the message to that ship immediately. If the ship addressed was not present, the senior officer present could relay it as if it were an original message. (Pages e-14 to e-15.)

2955-1340
21/Jun/44

e-1

No.	Means of Transmission	(1)* Maximum Distance: Day	(1)* Maximum Distance: Night	Max. number simultaneous Hoists or Circuits by each Means (2) — Combatant types sizes **: Large	Combatant: Intermediate	Combatant: Small	Combatant: Destroyer	Combatant: Submarine	Auxiliary** Types, size: Large - inter	Auxiliary: Small, DD, SS	Aircraft in air: ZR, ZN	Aircraft in air: VP, VT, VS	Aircraft in air: VO, VF
1	FLAGS, large	10(d)	-	5	4	4			4				
	intermediate	5(d)	-				2			2			
	small	2(d)	-										
2	SHAPES	3	-	(a)	(a)	(a)	(a)	(a)	(a)	(a)			
3	SEMAPHORES	2(d)	-	4	3	3	2	1	2	1			
4	WIGWAG	5(d)	-	3	3	2	1	1					
5	FLASH, searchlight, large	20(d)	40	2	2				2				
	inter.	16(d)	20			2	3			1			
	small	12(d)	16	2			1	1					
6													
7	blinker light	-	10	1	1	1	1	1	1	1	1	1	
8	blinker tube, undimmed	-	10(d)	2	2	2	2	1	2	1			
	dimmed	-	2(d)	2	2	2	2	1	2	1			
9	hand flashlight	-	2(d)	-								1	1
10	PYROTECHNICS	16	20	1	1	1	1	1	1	1	1	1	1
11	HAND SIGNALS	1	-	-								1	1
12	MOVEMENTS OF PLANES	(x)	-									1	1
13	DECK & GROUND SIGNALS	(x)(d)	-	1S(b)	-	-	-	-			1R	1R	1R
14													
15	GUN	(x)	(x)	1	1	1	1		1	1			
16	WHISTLE	2	2	1	1	1	1	1	1	1			
17	VOICE	(x)	(x)	2	2	2	2	1	1	1			
(3)	Max. number simultaneous Bridge Messages by any or all means			6	3	3	2	1	2	1	1	1	1

Notes:- (x) = As decided by umpire. The ranges for R/T transmission will be the optical range corresponding to height of transmitting antenna above sea level (may be several hundred miles for planes).

S = sending only.

(b) = Can be made only by carriers.

(d) = more or less directional.

* = Max. distances in thousands of yards and are estimated for favorable conditions (see Rule E-2).

R = receiving only.

(a) = speed indicators allowed in addition to flag hoists. Other shape hoists may be made in lieu of flag hoists.

** = for class sizes (see Rule A-2).

TABLE "E-2" BRIDGE COMMUNICATIONS

LIMITING DISTANCES & NUMBERS

Fig. 46
Table E-2: bridge communications

The total delivery period of a message—as determined by the Assistant Communication Umpire by adding the various time charges for activities, such as coding—represented the number of minutes from the time that the message reached the communication office of the transmitting ship (i.e., Time of Origin) until it reached the action officer of the receiving ship (Time of Delivery). Communication rates and times specified in Table E-4 were designed to represent actual ones under average conditions. When unfavorable conditions were considered by the Assistant Communication Umpire to exist, he would, with the approval of the Communication Umpire, charge appropriate delays in the delivery of the messages. Each relay was to entail a delay equal to the time of transmission of the text plus time for activities such as the call and the heading. In addition, one minute was to be added if the heading and recording were changed. Messages designated to be sent on a busy circuit would receive additional delays until the circuit was clear; the next highest

TABLE E-5

UNDERWATER SOUND

	EQUIPMENT		
	Supersonic echo-ranging & listening (b)	Supersonic listening	Sonic transmitter
Distance (miles) for Communication	2		1
DL and DD	2		
ODD, DM, DMS	1		
SS (1100 tons and over)	1	1	
SS (under 1100 tons)	1		
OSS		1	1 (a)
ASR, PC, PE, PTC, PY YMS, YP	1		

(a) Any vessel can hear this transmitter through the hull.

(b) Set cannot be used simultaneously for echo-ranging and listening.

Fig. 47
Table E-5: underwater sound

whole minute would be used for each computation of time charged. If tactical or other signals were not to be executed on receipt, the student was to place the desired time of execution in the space provided on the Message Blank. (Pages e-15 to e-16.)

The Student Commander for each side was responsible for preparing a Communication Plan for his force. This plan would include the frequency plan for all circuits and which frequencies would be guarded on the bridge. The number of such circuits possible on various types of ships was to be obtained from Table E-4. Only messages in Plain Language, Contact Code, Aircraft Code, or Signals Code could be delivered on bridge radio circuits. The Communication Plan was also to include arrangements for communication with the shore. If intercept schedules were provided, the stations participating and the times and circuit designations were specified. Also, for the purposes of the Maneuver, probable addressees could be grouped into distribution lists, labeled *A, B,* etc. These would be issued as a separate and unclassified annex to the Communication Plan. If each list had more than one frequency, the circuit designation was to be two letters, the first of which would indicate the list, while the second would indicate the frequency band. Short-range tactical control circuits on very-high, ultrahigh, and superhigh frequencies would be indicated by the general designation "R/T." These lists would also include the Maneuver title, the actual student names, and the room numbers of the addressees. When appropriate, these lists would be designated in the address of a message instead of individual titles, student names, and room numbers. (Pages e-16 to e-17.)

When it came to shipboard damage impacting communications, 30 percent Above Water Damage would destroy a ship's radar and double the time needed to code outgoing messages or decode incoming messages. Fifty percent damage meant that the ship could not transmit radio messages in Frequency Bands A to O, inclusive, and could not transmit messages by searchlight. Seventy percent damage meant that ships could no longer transmit by radio, and 80 percent damage—Total Damage—meant that ships could not transmit or receive messages by radio or flag hoist. The gassing of a vessel in chemical attack would also double the time for coding outgoing messages or decoding incoming ones during the first hour after the attack and quadruple it thereafter until the chemical effects had been neutralized. (Page e-17.)

FREQUENCY PLANS

The making of frequency plans is the joint function of Command and radio engineering experts. It is the duty of Command to indicate:

(1) Which flags, ships, or shore stations are to be grouped together on each circuit. (The following is a guide and not a rule. The number of radio sets which can transmit on a single circuit, without undue traffic delays, varies with the amount of traffic and has been found to be from 6 or 8 for very busy circuits, to not over 15 ships or one carrier's planes. The number which can receive is limited only by the availability of apparatus and personnel.

(2) What daylight range it is desired to insure.

(3) Whether a key or voice circuit is desired.

It is the duty of radio engineering experts to recommend the frequency(ies) to be used on a given circuit, provide the necessary alternate frequency(ies), the necessary separation between frequencies, etc. It will be assumed that this has been provided in all student plans made at the War College.

Frequency plans prepared for use during Operations Problems will provide one or more key and voice cirucits for each of the following:

(1) OTC to his Task Force Commanders;
(2) OTC to his own Task Group (or Unit) Commanders (if he commands a Task Force);
(3) OTC broadcast (F Method) circuit. (May be assumed continuously effective during Tactical Problems and only on schedule or via radio guardship (i.e., relay) during Strategical Problems);
(4) Special circuits as desired by OTC; e.g., air task groups, etc.;
(5) Each Task Force Commander to his Task Group (or Unit) Commanders.

Each circuit list will be headed as follows: (shown in "—");
"List ___" (A,B,C, etc., without repeating).
Maximum daylight distance required in miles "—— miles".
Whether key or voice by indicating "voice" when desired.
Circuit designation - ("A combination of two letters representing the List and Band Letters.") (If more than one parallel circuit is being provided for simultaneous transmissions each should be indicated with its distance in miles, if different.)

EXAMPLES: "List A | or | "List F voice | or | "List B
AQ 3000 miles" | | FN 100 miles" | | BH 800 miles
R/T 40 miles | | | | BL 800 miles" (Two simultaneous circuits)

The circuit designation will be entered in the "Circuit space" on the message form.
Sets marked "To aircraft" will be reserved for that purpose.
Avoid repetition of bands as far as practicable.

Fig. 48
Frequency plans

Gunfire

Naval War College war-game gunfire rules were designed to replicate what might be expected to occur in actual practice under average conditions. In many instances they were arbitrary assumptions, based on either average or expected conditions for purposes of simplicity in scoring gunfire or on a "reasonable" estimate of conditions for which there was no factual or singular mathematical basis. In view of this, while an endeavor was made to afford a comparative basis, there was no assurance that the Rules were necessarily "factual." (Page f-1.)

The first area covered, Fire Control Conditions, was divided into Gunlaying by the Director and the Pointer; Spotting by Top Spot (the spotter in the top of the mast), Plane Spot, Radar Spot, or Local Control; and Method, which was Direct, Indirect, Barrage, Bombardment, or Radar Gunfire. In Gunlaying, Director Fire could be used—subject to loss through damage—of main, secondary, and antiaircraft batteries on major combatant vessels (except submarines) and on auxiliary ships that were equipped with gun directors. Pointer Fire had to be used by ships that did not have gun directors or had lost their director systems because of damage. The Top Spot method could be used by all types of ships except by submarines and ships that had lost their Top Spot through damage. The Plane Spot could also be used for the main battery of a ship that was equipped with directors unless the plane was lost by damage. Also, a student who proposed to employ Plane Spot had to ensure that there was a definite assignment of a spotting plane for the ship and that communication provisions had been made.

In addition, at least one plane allotted to that ship alone had to be in the spotting position; the plane had to be able to distinguish the ship's target and the splashes from the ship's guns and to determine the approximate line of fire of the vessel for which it was spotting. A plane was considered to be in spotting position when two miles nearer the target ship than the recognition range given in the Visibility Tables, at an altitude neither less than one thousand feet nor less than 4 percent of the range. Local Control had to be used by submarines, by other types of ships that had lost their planes, or by ships that had lost their radar tops, radars, or low position spots through damage. Local Control also had to be used by other types of ships that fired guns in excess of those that could be controlled from the prescribed control stations. Finally, Radar Spot could be used for the control of main and secondary batteries on all major combatant ships except for submarines, but Radar Spot was also obviously subject to loss through damage. (Pages f-1 to f-2.)

In terms of method, the Direct Method entailed a ship firing on a selected target that was visible. The Indirect Method involved firing on a selected target that was not visible to the firing ship but was visible to a detached observer with which the firing ship had suitable communications. In addition, shore batteries could use the Indirect Method of fire provided that a Plane Spot was available and the target was visible from an established observation station. In the Barrage Method, a ship fired with a fixed gun range in the direction of a point of aim visible from the firing ship through which the target would pass. The purpose was to confuse enemy fire control. Barrage fire could be used either with Pointer or Director Gunlaying. The Bombardment Method was one in which a ship fired into a specified target such that ships using Director Fire had a predetermined compass bearing and gun range. All director ships could use the Bombardment Method, provided they had auxiliary points of aim or radar. For bombardments conducted by the Indirect Method, shore fire control observers or air spotters would be necessary. In the Radar Method, major combatants equipped with fire control radar were considered to have fire control sufficiently accurate to warrant opening fire at the extreme ranges of their guns. Ships equipped with fire control radar, itself subject to loss through damage, were considered to deliver fire effect equivalent to Plane Spotting. Thus, sixteen-inch guns had a range of thirty thousand yards; fourteen-inch and twelve-inch guns had a range of twenty-six thousand yards; eight-inch guns had a range of twenty-three thousand yards; six-inch guns had a range of twenty thousand yards; and five-inch guns had a range of fifteen thousand yards. (Pages f-2 to f-3.)

In terms of measurement of gunfire, range, target angle, and bearing for ship or shore, gunfire normally would be measured from the ship's position at the end of a Maneuver Move, except in special cases to be determined by the Director. For purposes of scoring gunfire, all moves, of whatever actual length, were divided into three-minute increments, unless smaller increments were necessary for scoring extremely short-range actions. Range was the distance in thousands of yards from a firing ship or shore battery to its target. As an example, actual ranges of zero to 1,500 yards were considered to be one thousand yards, actual ranges 1,500 to 2,500 yards as two thousand, and so forth. The maximum range imposed on each type of gun by its mounting was as given in the manual *Fleets* for ship's guns. For shore batteries, maximum ranges were established as in figure 49.

Target angle, for purposes of scoring gunfire, was measured as the angle between the line of fire and the keel line of the target ship. With a target broadside, for instance, target angle was ninety degrees. In order to determine whether or not hits were penetrative against vertical armor, target angle between zero and ninety degrees was recorded to the nearest multiple of forty-five degrees. Penetration of deck armor, however, was considered independent of target angle. Bearing was measured in relative degrees in order to determine the portion of the battery that would bear on the target. Special cases of gunfire measurement included fire of or on destroyers being measured from the center of the destroyer division or a half-flotilla of ships. When fire was masked by a smoke screen or by the interposition of ships, it was to be measured proportionally to the time it had not been masked. Finally, measuring fire by bombardment meant that target areas had to be designated by the firing ship's commander in one or more hundred-yard squares, each square measuring one hundred yards on a side. If the target was within one or more of the designated squares, it was subject to damage. (Pages f-3 to f-4.)

There were also basic conditions for determining Fire Effect. In one situation, a ship would be undamaged, have no list, and be in normal trim. In another situation, fire was not masked, the ship was firing by Direct or Radar Method, and the range had been established. Further, a target could remain the same as during the preceding move, the fire of the ship might be not divided among targets, and only one ship might be firing at one target. These conditions prevailed if the ship was under Normal fire, was not under Enfilade Fire, was on a steady course, and did not change course or speed. The conditions would continue to prevail if the rate of change of range was small; yaw, roll, and pitch were not excessive; visibility was good; and there was no sun glare, silhouette, smoke, spray, or gas interference. Fire Effect for these basic conditions was contained in the Fire Effect Tables and Fire Effect Diagrams and was expressed in terms of equivalent fourteen-inch penetrative hits per three-minute move. (Pages f-4 to f-5.)

Caliber	Range (Yards)
16"	30,000
14" & 12"	26,000
8"	23,000
6"	20,000
5"	15,000

Fig. 49
Maximum gun ranges

The Fire Effect Tables were designed to set forth the expected average fire effect for one gun of each type used afloat at ranges within the practical limits of that gun against each class of target. Percentage bonuses were provided for deck penetration, as well as for deck and side penetration. Rates of fire for guns were tabulated in terms of rounds per gun per three-minute move. To use the tables, it was necessary to consult *Fleets* for the types and numbers of firing ship guns, ammunition allowance, size and class of the target, thickness of target armor, and target life. Fire Effect Diagrams were also compiled from data contained in *Fleets* and in the Fire Effect Tables in order to expedite rapid scoring of gunfire during the Maneuvers by specific guns on specific targets. Other data contained in *Fleets* and the Fire Effect Tables were graphically incorporated into the Fire Effect Diagrams. For instance, the Fire Effect of auxiliaries and armed merchantmen—except for fleet auxiliaries equipped with main battery directors—was to be 50 percent of that shown in the Fire Effect Tables, since it was assumed that fire control systems had been installed in such ships with "relative" inefficiency compared to combatants. No distinction was to be made in the Fire Effect Tables or the Fire Effect Diagrams between Director and Pointer Fire, but Director Fire was to be less penalized under certain firing conditions. (Pages f-5 to f-6.)

Other Fire Effect conditions were shown in the Fire Effect Tables and Fire Effect Diagrams, subject to an increase or decrease by increments of a tenth in three successive corrections. Correction I was to determine the remaining Fire Effect after damage had been received; Correction II was to determine conditions affecting one's own rapidity of fire; and Correction III was to determine the conditions affecting one's own accuracy of fire. As an example, Correction I determined that Above Water Damage reduced the potential Fire Effect by an amount equal to one-tenth the life of the firing vessel, to the nearest tenth. This damage was to be determined independently of and in addition to other reductions, including the shock effect of a torpedo or mine hit or of heavy bomb hits resulting in Under Water Damage equivalent to at least 1.8 fourteen-inch hits. This damage would render the ship incapable of using its weapons for three minutes thereafter. Masked fire would also reduce the Fire Effect of ships during a move by a number of tenths in proportion to the time masked. The Director would decide on the size of the reduction, but Correction II came in when ships attempted to fire over ships of their own size or larger that were a thousand yards or less from them.

Correction II also came into play when ships attempted to fire over ships that were seven hundred yards or less from them or over physical obstructions where the angle of elevation would not carry the projectiles over the obstruction. Correction II additionally came into effect when a ship was firing by Direct Method and the ship's fire was halted by the interposition of an effective smoke screen. When more than four ships were proceeding in a line of bearing of no more than fifteen degrees, the fire would be masked if the ship opened fire or continued firing on the third move after it had first maneuvered into the line of bearing. The situation did not hold, however, if it was evident that errors in station keeping could not mask

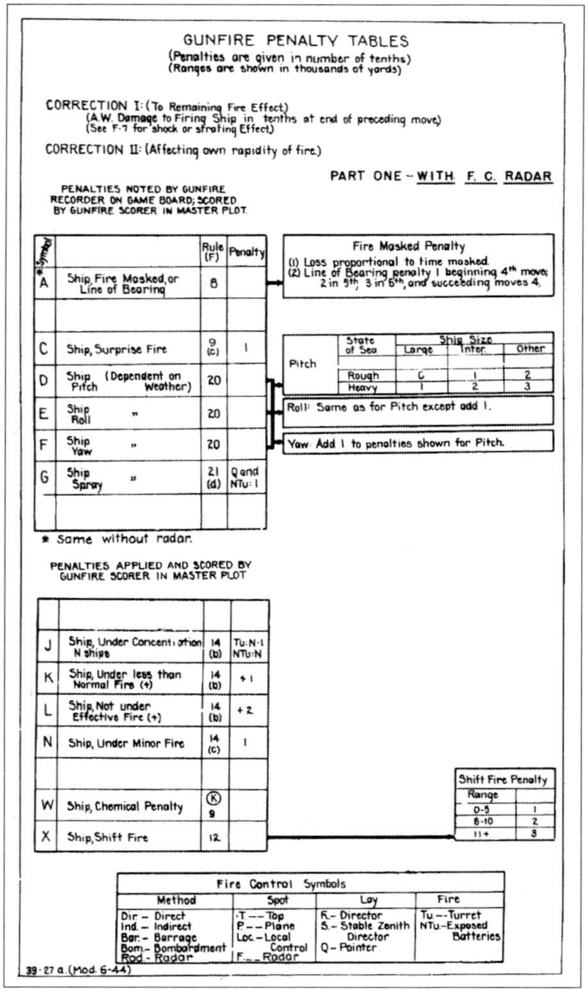

GUNFIRE PENALTY TABLES
(Penalties are given in number of tenths)
(Ranges are shown in thousands of yards)

CORRECTION I: (To Remaining Fire Effect)
(A.W. Damage to Firing Ship in tenths at end of preceding move)
(See F-7 for shock or strafing Effect)

CORRECTION II: (Affecting own rapidity of fire.)

PART ONE – WITH F. C. RADAR

PENALTIES NOTED BY GUNFIRE RECORDER ON GAME BOARD; SCORED BY GUNFIRE SCORER IN MASTER PLOT

*Symbol		Rule (F)	Penalty
A	Ship, Fire Masked, or Line of Bearing	8	
C	Ship, Surprise Fire	9 (c)	1
D	Ship Pitch (Dependent on Weather)	20	
E	Ship Roll "	20	
F	Ship Yaw "	20	
G	Ship Spray "	21 (d)	Q and NTu: 1

* Same without radar.

Fire Masked Penalty
(1) Loss proportional to time masked.
(2) Line of Bearing penalty 1 beginning 4th move; 2 in 5th, 3 in 6th, and succeeding moves 4.

	State of Sea	Ship Size		
		Large	Inter.	Other
Pitch	Rough	C	1	2
	Heavy	1	2	3

Roll: Same as for Pitch except add 1.

Yaw: Add 1 to penalties shown for Pitch.

PENALTIES APPLIED AND SCORED BY GUNFIRE SCORER IN MASTER PLOT

J	Ship, Under Concentration N ships	14 (b)	Tu: N-1 NTu: N
K	Ship, Under less than Normal Fire (+)	14 (b)	+1
L	Ship, Not under Effective Fire (+)	14 (b)	+2
N	Ship, Under Minor Fire	14 (c)	1
W	Ship, Chemical Penalty	(K) 9	
X	Ship, Shift Fire	12	

Shift Fire Penalty	
Range	
0-5	1
6-10	2
11+	3

Fire Control Symbols			
Method	Spot	Lay	Fire
Dir. – Direct Ind. – Indirect Bar. – Barrage Bom. – Bombardment Rad. – Radar	T — Top P — Plane Loc. – Local Control F — Radar	R. – Director S. – Stable Zenith Director Q – Pointer	Tu — Turret NTu – Exposed Batteries

39-27 a. (Mod. 6-44)

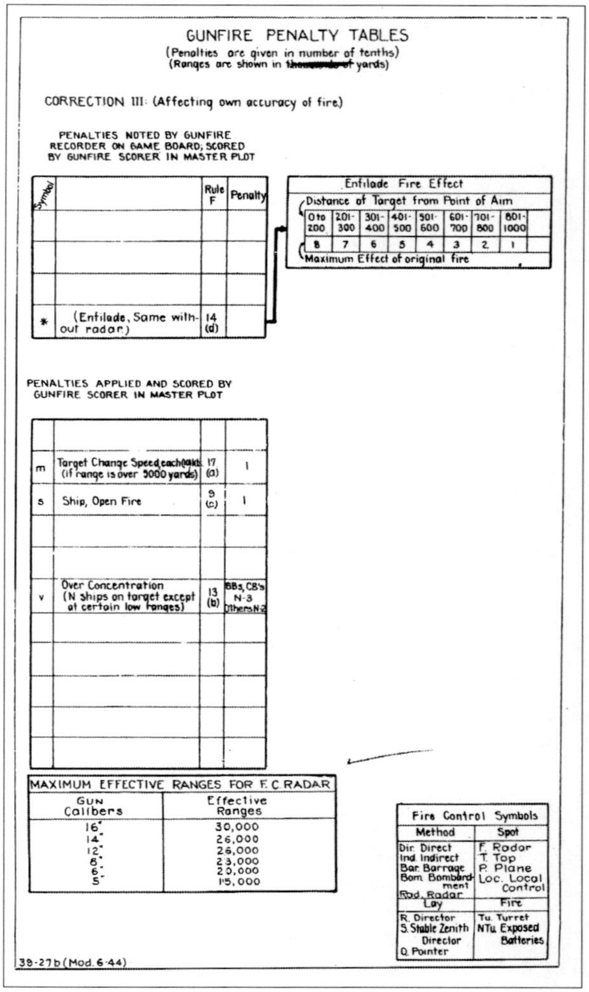

GUNFIRE PENALTY TABLES
(Penalties are given in number of tenths)
(Ranges are shown in ~~thousands of~~ yards)

CORRECTION III: (Affecting own accuracy of fire.)

PENALTIES NOTED BY GUNFIRE RECORDER ON GAME BOARD; SCORED BY GUNFIRE SCORER IN MASTER PLOT

Symbol		Rule F	Penalty
*	(Enfilade, Same without radar.)	14 (d)	

Enfilade Fire Effect							
Distance of Target from Point of Aim							
0 to 200	201-300	301-400	401-500	501-600	601-700	701-800	801-1000
8	7	6	5	4	3	2	1
Maximum Effect of original fire							

PENALTIES APPLIED AND SCORED BY GUNFIRE SCORER IN MASTER PLOT

m	Target Change Speed each 10 kts. (if range is over 5000 yards)	17 (a)	1
s	Ship, Open Fire	9 (c)	1
v	Over Concentration (N ships on target except at certain low ranges)	13 (b)	BBs, CB's N-3 Others N-2

MAXIMUM EFFECTIVE RANGES FOR F. C. RADAR	
Gun Calibers	Effective Ranges
16"	30,000
14"	26,000
12"	26,000
8"	23,000
6"	20,000
5"	15,000

Fire Control Symbols	
Method	Spot
Dir. Direct Ind. Indirect Bar. Barrage Bom. Bombardment Rad. Radar	F. Radar T. Top P. Plane Loc. Local Control
Lay	Fire
R. Director S. Stable Zenith Director Q. Pointer	Tu. Turret NTu. Exposed Batteries

39-27b (Mod. 6-44)

Fig. 50
Gunfire Penalty Tables, parts 1 and 2

the fire. The penalty in these cases would be one-tenth for ships whose fire was masked during the third move after the formation was first maneuvered into the line of bearing; two-tenths for ships maneuvering in the fourth move; three-tenths for ships maneuvering in the fifth move; and four-tenths for ships maneuvering in the sixth and seventh moves. (Pages f-6 to f-8.)

Range for the establishment of fire could be obtained if it was within the maximum range of the battery as shown in *Fleets,* the Fire Effect Tables, and the Fire Effect Diagrams. In addition, there were various stipulations for the Direct, Indirect, Barrage, and Bombardment Methods. In employing any of these methods, the student was to submit a "feasible" plan to the Director. During the move in which gunfire was opened in establishing the range for ships not equipped with fire control radar, Fire Effect was reduced in tenths for ranging before the target was straddled. In addition to and independent of these reductions, when fire was opened by Direct Method on a target that had just appeared in view of the firing ship, the time

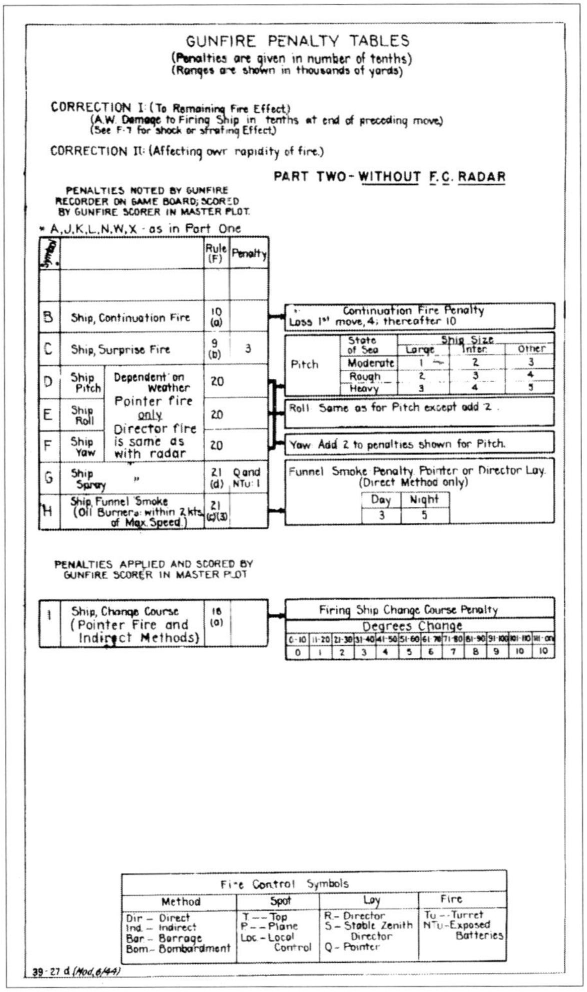

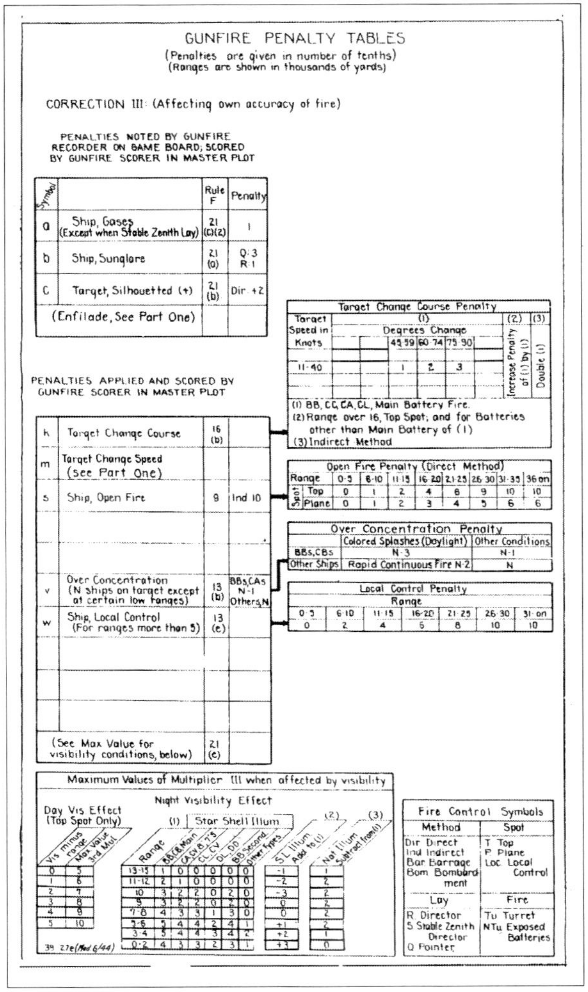

Fig. 50
Gunfire Penalty Tables, parts 3 and 4

required to get the battery in train and to set up the fire control problem would reduce the Fire Effect for that move by three-tenths. For radar gunfire control, Fire Effect was reduced at all ranges by a tenth for ships using radar fire control during moves in which gunfire was opened to establish the range. (Pages f-8 to f-9.)

For reduction of Fire Effect when continuing to fire on the same target, reduction of four-tenths was made for the Direct, Barrage, and Bombardment Methods during the first Maneuver Move after the target was lost from view, and then entirely in succeeding moves, unless the target reappeared or fire was shifted to the Indirect or Radar Methods. For indirect fire, with the range established by direct fire, the fire could be continued by the Indirect Method at any time, subject to the rules noted above. In reestablishing fire by the Indirect Method, ships that had ceased fire for one or more moves had to, upon reopening fire, reestablish the range. Fire Effect for the move in which fire was reestablished by the Indirect Method was to be reduced by five-tenths. When using the Direct or Radar Methods, ships that had ceased fire for one or more moves also had to reestablish the range upon reopening fire on the same or a new target. They were also subject to an "open fire" penalty, unless there was a new line of fire within fifteen degrees of that previously used; there was a new range within four thousand yards of that previously used; and there had been no more than fifteen minutes elapsed since the previous effective fire. Ships were also excused from an open-fire penalty if the range was not greater than twenty-five for battleships or cruisers, fifteen for large or intermediate-sized ships, or ten for small fighting ships. Additionally, ships were excused from that penalty if they had been subject to reduced visibility in the last firing Maneuver Move or upon reopening fire. (Pages f-9 to f-10.)

(b) During the move in which gunfire is opened in establishing the range for ships not equipped with fire control radar, fire effect is reduced in tenths for ranging before the target is straddled, as follows:

(1) For direct method: (Corr. III).

	Range	0-5	6-10	11-15	16-20	21-25	26-30	31 plus
	Top	0	1	2	4	6	9	10
Spot	Plane	0	1	2	3	4	5	6

In addition to, and independently of the foregoing reduction, when fire is opened by direct method on a target which has just appeared in view of the firing ship, the time required to get the battery on in train and to set up the fire control problem will reduce the fire effect for that move, <u>for surprise fire</u>, three-tenths. (Corr. II)

(2) For indirect method: 10. (Corr. III)

(3) For barrage method: same as for <u>direct</u>.

(4) For bombardment method: 0, except <u>for surprise fire</u> penalty as described above under direct method.

Fig. 51
Fire effect ranging reductions

Ships using the Indirect Method could not shift the fire of a battery more often than once in four moves. When fire by any method was shifted to a new target, a penalty was to be applied to the Fire Effect for that move of three-tenths at ranges of and above eleven thousand yards; two-tenths at ranges from six to ten thousand yards; and one-tenth at ranges from zero to five thousand yards. There would be no penalty, however, if the new target was adjacent to the old target. Also, for Maneuver Board purposes, two ships were considered adjacent targets when they were not more than 1,500 yards apart, were in the same formation, or had no intervening ships between them and the firing ship. Shore targets were considered adjacent if not more than a thousand yards apart. Ships would be penalized for shifting fire if the difference in range from the old to the new target was over four thousand yards; the change in target bearing was more than fifteen degrees; and the new target was on a materially different course or speed from the old target. (Page f-11.)

Concerning the distribution of gunfire, a ship's battery that divided fire between two targets had to have each set of guns separately controlled. Divided fire by the Bombardment Method was not subject to penalty, but other types of divided fire were. The fire of main batteries of capital ships and cruisers could be divided, with forward and after turrets firing either on the same side or on opposite sides. Again, however, the turret groups had to be under the direction of two separate primary control stations. The Fire Effect of these guns, moreover, would be reduced for conditions of Local Control if the number of guns firing was in excess of the number that could actually be controlled from their primary control stations. Still, if there was divided fire on an adjacent target there would not be any loss in Fire Effect, since the conditions of spotting for each group of guns governed the resulting fire. Fire from an after group of main-battery guns on ships other than battleships and cruisers, however, would be reduced for conditions of Local Control if the ships divided their fire. Further, an overconcentration of fire (too many shells fired on the target; a waste of ammunition) might result if batteries from two or more ships firing on the same target had spotting difficulties. In this situation, the ships in question could be subject to a penalty unless the fire method was Bombardment. Additionally, for Maneuver Board scoring methods, a division of destroyers was considered as a firing unit for that type of ship. A destroyer squadron leader was considered a separate firing ship. No ship, however, was to be penalized for overconcentration during daylight if the fire was by the Direct Method and if the gun firing had sufficient range. (Page f-11.)

The Fire Effect of each ship of any type armed with batteries greater than eight inches firing on the same target at ranges in excess of those allowed would be reduced by up to three-tenths when using fire control radar; during daylight when using colored splashes to mark the fire; and under all other conditions. The Fire Effect of a ship of any type armed with guns eight inches or less that were firing on the same target in ranges in excess of those allowed would also be reduced up to two-tenths when using fire control radar or during daylight when using rapid, continuous fire in accordance with stipulations in the manual *U.S. Fleet Tactical Orders and Doctrine.* Under all other conditions, these ships would have their Fire Effect reduced by computing overconcentration penalties in which the penalty was equal to the number of ships firing on the same target. If a capital ship and one ship of any other type were concentrating on the same target, neither ship would be penalized. Also, Fire Effect would be reduced in tenths when ships were firing using the Indirect Method with Local Control or when damaged. If the range was zero to five, the penalty was zero, but if the range was six to ten, the penalty was two. If the range was eleven to fifteen, the penalty was four, and if the range was sixteen to twenty, the penalty was six. Finally, if the range was twenty-one to twenty-five, the penalty was eight; and if the range was twenty-six or above, the penalty was ten. (Pages f-12 to f-15.)

For the effect of enemy fire, the volume and accuracy of a ship's fire were to vary according to the intensity of enemy fire, the types of ships firing and under fire, the number of guns per salvo, the caliber of guns involved, and the frequency of the striking salvo. For purposes of scoring gunfire on the Maneuver Board, a ship was considered to be under one of four degrees of fire intensity. A destroyer division or half-flotilla was considered again as a target unit or ship, while a destroyer or flotilla squadron leader was again a separate target. A ship was considered under Normal Fire when under the fire of the main battery—as shown in *Fleets*—of a ship of its own general type, provided that the main battery or batteries firing on the ship were at least two-thirds and less than four-thirds the number of main-battery guns with which the target ship was capable of returning fire. An exception here would be if a ten- or twelve-inch-gun battleship was under fire from a six-inch-gun cruiser, which would be considered Normal Fire. A ship was considered under a Concentrated Fire when under the fire of the main batteries of two or more enemy ships of its own general type, provided the guns firing on the ship were at least four-thirds the number of main battery guns that the target ship was capable of returning fire with. The exception here would be if the same ten- or twelve-inch-gun battleship were now under fire from two six-inch-gun cruisers. When under the Concentrated Fire of a given number of ships, the Fire Effect of the target ship was to be decreased in various formulas for its turret and exposed batteries. (Pages f-14 to f-16.)

A ship was considered under Less Than Normal Fire when under the effective fire of the main batteries of ships of its own general type and the number of guns firing on the target ship was less than two-thirds the number of main battery guns that the target ship could fire back. The exception was now a six-inch cruiser, which was considered to deliver Less Than Normal Fire if not firing more than four of its guns. The Fire Effect of a ship under Less Than Normal Fire would be increased by one-tenth. A ship was considered under Ineffective Fire when the number of enemy main battery guns firing at it was two or fewer. The Fire Effect of a ship not under Effective Fire was to be increased by two-tenths. For cases not covered by these contingencies, decisions would be made by the Director either in advance or when the situation arose. In these decisions, "due consideration" was to be given to the practical limitations of Fire Control and Fire Distribution. (Page f-16.)

A capital ship under fire of a number of guns that were appreciably smaller in caliber than those of its own main battery was to have its Fire Effect decreased one-tenth for being under "Minor Fire," provided that the volume of Minor Fire was at least that of eight guns of 7.5 to ten inches, fourteen five-to-six-inch guns, or twenty guns under five-inch caliber. When a ship other than the target ship was under fire by any method except Bombardment, it was liable to damage by enfilade. To be subject to damage by Enfilade Fire, a ship had to be within one hundred yards of the line of fire and within a thousand yards of the target. Change of course obviously impacted Fire Effect as well. A change of course by a firing ship reduced its Fire Effect for that move, in tenths, in various ways according to figure 52, especially if using pointers for aiming or using the Indirect Method. Tables in that figure also come into play for a target ship that changed course during a move, such that the Fire Effect of the battleships and cruisers firing on the target ship also saw their Fire Effect reduced in tenths, depending on whether the fire was by Direct or Barrage Methods and whether Plane Spotting was being used. (Pages f-16 to f-17.)

These penalties increased by a tenth if the fire of the main batteries from the battleships and cruisers was by Direct or Barrage Method, as well as with a Top Spot, at ranges over sixteen thousand yards. That penalty would also be awarded

(1) When using pointer lay:

	Degrees Change						
	0-10	11-20	21-30	31-40	41-50	51-60	etc.
Penalty:	0	1	2	3	4	5	etc.

(Corr. II).

(2) When using indirect method:

Same penalties as for pointer lay. (Corr. III)

(1) Fire of main battery of BBs, CCs, CAs or CLs by direct or barrage methods, with plane spot:

Target	Degrees Change		
speed	45-59	60-74	75-90
11-40	1	2	3

Fig. 52
Fire Effect from change of course

if the fire was by the same methods in the battleships and cruisers but by other than their main-battery guns. For fire by the Indirect Method, these penalties were doubled. There were no penalties, however, if the fire was by Bombardment or Radar Methods. If the target ship changed speed during a move, the Fire Effect of the ships firing on it would also be reduced by a tenth for each ten-knot change in speed, provided that the range to the target was greater than five thousand yards. If the change of course, speed, or range took place at night or in low visibility, a ship firing at close range might experience a very rapid change of relative bearing and range to the target due to the relative courses and speeds of the firing and target ships. This might seriously interfere with the rapidity and accuracy of the fire and illumination, and it might make it impossible for the firing ship to pick up and follow the target with its battery. Taking into consideration all of these factors, the Director of the Maneuver could assign penalties to reduce Fire Effect. (Pages f-18 to f-19.)

Sea conditions were also taken into consideration. When the character and direction of the sea were such as to cause a firing ship to pitch, roll, or yaw more than usual, the effect on gunlaying and ammunition handling might slow the fire. For Maneuver Board purposes, penalties to a ship's Fire Effect were assigned for Moderate, Rough, and Heavy Seas. More specifically, penalties were assigned for pitch when the sea was from a direction twenty degrees or less from ahead; for roll, when the sea was from a direction between fifteen degrees forward of the beam and sixty degrees abaft the beam; and for yaw, when the sea was from a direction thirty degrees or less from astern. Fire Effect for conditions of pitch, roll, or yaw was reduced in tenths when using any method other than Pointer Fire. Similarly, sun glare had an impact. When a ship was firing into the glare of the sun, the accuracy of its fire was reduced when using the Direct Method or the Indirect Method with a Plane Spot. The sun-glare penalty was operative if the sky was clear during the two hours after sunrise and the two hours preceding sunset when the target was bearing within fifteen degrees of the sun. The sun-glare penalty reduced Fire Effect by a tenth when using the director, by three-tenths when using the pointer, and by three-tenths when using the Indirect Method with a Plane Spot. (Pages f-20 to f-21.)

A target ship was judged to be silhouetted against the horizon when bearing within thirty degrees of the sun during the twilight period if the sky was clear, as well as during the sun-glare period if there was a haze or a cloudy sky. The degree of silhouetting was to be decided by the Director. If the target ship was silhouetted, a firing ship using Direct Method against the target ship would not be penalized for reduced visibility but would have its Fire Effect increased by two-tenths. The firing ship was penalized in its Fire Effect, however, when using the Direct Method if—owing to the relation between its course and speed, on the one hand, and the direction and force of the wind, on the other—gun gases, funnel gases, or funnel smoke from the firing ship or other nearby ships drifted across the line of fire. The gas penalty, effective when gases but no funnel smoke intervened, was a Fire Effect reduction of one-tenth. The smoke penalty—effective when funnel smoke intervened at speeds within two knots of maximum speed for oil-burning ships, and at all speeds for coal-burning ships—was to be a reduction of Fire Effect by three-quarters during the day and by half at night. When sea and wind conditions relative to the course and speed of a firing ship caused sea spray over gun positions, the Fire Effect of the ship might also be reduced. For Maneuver Board purposes, the spray penalty was operative when the wind had a force of 4 or more and its true direction was from fifteen degrees forward of the beam to dead ahead on the engaged side. For ships using Director Fire and the Direct Method, the spray penalty was not effective with regard to the fire of either turrets or enclosed mounts. With this one exception, the spray penalty also reduced Fire Effect by a tenth. (Pages f-21 to f-22.)

Pitch:

State of Sea	Ship size: Large	Inter.	Other
Moderate	1	2	3
Rough	2	3	4
Heavy	3	4	5

Roll: Same as for Pitch, except add 2 if using other than director lay.

Yaw: Add 2 to penalties shown for Pitch. (Corr. II).

Pitch:

State of sea	Ship Size: Large	Inter.	Other
Rough	0	1	2
Heavy	1	2	3

Roll: Same as for Pitch except add 1.

Yaw: Add 1 to penalties shown for Pitch.

Fig. 53
Fire Effect penalties for sea conditions

Visibility conditions "better than normal" did not increase Fire Effect and a ship could not fire by the Direct Method at a range exceeding the range of visibility. With Direct Method by Top Spot during daylight and at ranges greater than five thousand yards when the gun range was within four thousand yards of the range of visibility, the maximum value in tenths was given for computation. With Plane Spot, no penalty was assigned for limited visibility, nor was there any with Radar Spot. With the Direct Method during night, Fire Effect was curtailed on account of difficulties in spotting and control. The outside range limits of visibility for gunfire purposes at night were given in Section D of the Maneuver Rules. Within these limits, Night Fire Effect would be computed in the same way as for daylight, with maximum values in tenths as shown in figure 55. (Pages f-22 to f-23.)

As to methods of scoring gunfire between ships, except by the Bombardment Method, corrections to Fire Effect were converted in various scenarios as a first step in obtaining the Fire Effect inflicted. Correction I saw Above Water Damage in tenths on the firing ship at the end of the preceding move, plus any damage inflicted from shock effect that was applicable to the succeeding move. Corrections II and

Visibility minus gun range	Max. Value Multiplier III.
0	5
1	6
2	7
3	8
4	9
5	10

Fig. 54
Daylight Fire Effect multiplier

III were each the sums, in tenths, of the penalties or bonuses incurred affecting, respectively, the rapidity and accuracy of fire. Multipliers for Corrections I, II, and III were equal to ten-tenth minus each Correction total, and the Final Multiplier was the product of multipliers for Corrections I, II, and III and of the number of guns firing as used in Multiplier IV. The Fire Effect inflicted was, with the exception of Enfilade Fire, the product of the Final Multiplier and the Fire Effect taken from the Fire Effect Tables and the Fire Effect Diagrams. Fire Effect inflicted by enfilade was added to that which might be inflicted by other means and was scored such that the Fire Effect inflicted was the product of the Final Multiplier, as determined for the firing ship on the target, the Fire Effect as taken from the Fire Effect Tables and Fire Effect Diagrams for the size of the ship under fire, and a multiplier based on the distance of the enfiladed ship from the target vessel. From zero to two hundred yards the multiplier in tenths was eight, going down to one as the increments of distance went up to a thousand yards. The scoring of gunfire by the Bombardment Method against an anchored ship would depend on whether the target was within the hundred-yard square designated earlier in Section F of the Maneuver Rules. If it was, Corrections I and II would be determined in the same manner as for other methods of fire between ships, except that Correction II penalties or bonuses would be based on figures for the Direct Method with Plane Spot. There was no Correction III in this case, and the Multipliers for Corrections I and II were determined in the same manner as for other methods of fire between ships. The Final Multiplier was the product of Multipliers for Corrections I and II and of the actual number of guns firing. (Pages f-23 to f-26.)

Range	With Star Shell or Flare Illumination (A)						With Search-light Illumination Only Or in Combination With Star Shell (or flares) Add to (A)	With no Artificial Illumination. Subtract from(A)
	BB,CC Main	CA,CV 8" 7"5	CL CV	DL DD	BB Sec	Other		
13-15	1	0	0	0	0	0	-1	1
11-12	2	1	0	0	0	0	-2	2
10	3	2	2	0	2	0	-3	2
9	3	2	2	0	2	0	0	2
7-8	4	3	3	1	3	0	0	2
5-6	5	4	4	2	4	1	plus 1	2
3-4	5	4	4	3	4	2	" 2	1
0-2	4	3	3	2	3	1	" 3	0

Fig. 55
Nighttime Fire Effect ranges

In addition, the Fire Effect for the type of gun bearing was taken from the Fire Effect Tables. This Fire Effect was reduced in proportion to the total target areas occupied and the total areas designated. It was further reduced by half. Fire Effect inflicted would then be the product of the foregoing resultant Fire Effect and Final Multiplier. Students were referred to the manual *Landing Operations Doctrine, U.S. Navy* for the Fire Effect of naval guns against shore targets. The fire of shore batteries against ships, however, was covered in game documents and was to be scored in the same manner as for fire between ships, except that when using the Direct Method or the Indirect Method with direct terrestrial observation, the normal Fire Effect for Plane Spot given in the Fire Effect Tables would be multiplied so as to allow for the increased accuracy of shore batteries. As noted in the Conduct of the Maneuvers, damage to ships was classified as Above Water Damage (due to hits from gunfire or bombing) or Under Water Damage (due to hits by torpedoes, mines, bombs, ramming, or grounding). Only Above Water Damage inflicted a permanent reduction in gunfire effectiveness, as far as determining the Fire Effect of the firing ship was concerned. (Pages f-25 to f-27.)

Range	Multiplier	Range	Multiplier	Range	Multiplier
0-5	1.0	12	1.7	19	3.4
6	1.1	13	1.8	20	4.0
7	1.2	14	1.9	21	4.5
8	1.3	15	2.0	22	5.0
9	1.4	16	2.2	23	5.5
10	1.5	17	2.5	24 plus	6.0

Fig. 56
Fire Effect of shore batteries on ships

The Maneuver Rules described how damage was to be assigned in terms of equivalent fourteen-inch penetrative hits against the life of the ship, converted into "percentage of life" for determining not only the ship's remaining Fire Effect but also the specific handicap to its battle efficiency. In this context, a destroyer division or half-flotilla was again considered as a target group or ship. In addition, there were provisions so that Under Water Damage inflicted by any one bomb or torpedo would not be scored against more than one destroyer of a group. Under Water Damage to the ship that was hit, however, reduced its original maximum speed in proportion to the percentage of damage received, expressed in one-percentage-point increments. Such reduction was to be assigned in addition to any Above Water Damage that might result in speed losses. If the Above Water Damage was 30 percent, any ship, other than a capital ship, was to lose 10 percent of its original maximum speed, its radar was to be considered destroyed, and the time it needed to code and decode messages was doubled. If 50 percent Above Water Damage was inflicted, a ship could no longer use Director Gunlaying unless a capital ship. The ship would also not be able to use a Plane Spot and would lose all of its secondary battery and antiaircraft capability. All planes on its deck would be lost, its catapults would be inoperable, and it could no longer fly planes on or off. In addition, if a capital ship, it would lose 20 percent of its original maximum speed. All other ships would lose an additional 10 percent of their original maximum speed; all ships would lose half of their deck tubes and torpedoes on each side of the ship. A ship thus damaged would lose any minelaying capability it might have had, would no longer (if a submarine) be able to submerge, and could no longer transmit messages by radio on frequencies below 2,000 KCS. Finally, gunnery and signal searchlights would no longer be operational. (Pages f-27 to f-28.)

If a ship received 70 percent Above Water Damage, it could no longer use its Gunlaying Director if a capital ship. If other than a capital ship, it now had to use Local Control for its guns. All deck tubes and torpedoes would now be lost, as would an additional 30 percent of its original maximum speed. Radio messages could no longer be transmitted, and if the ship was a surface minelayer with at least 50 percent of its mines onboard, it would be ruled to have blown up. With 80 percent Total Damage a ship could not make more than five knots, could not fire its guns or torpedoes, and could no longer transmit messages even by flag hoist. By the time 90 percent Total Damage had been inflicted, the ship was considered dead in the water and sinking. (Pages f-29 to f-30.)

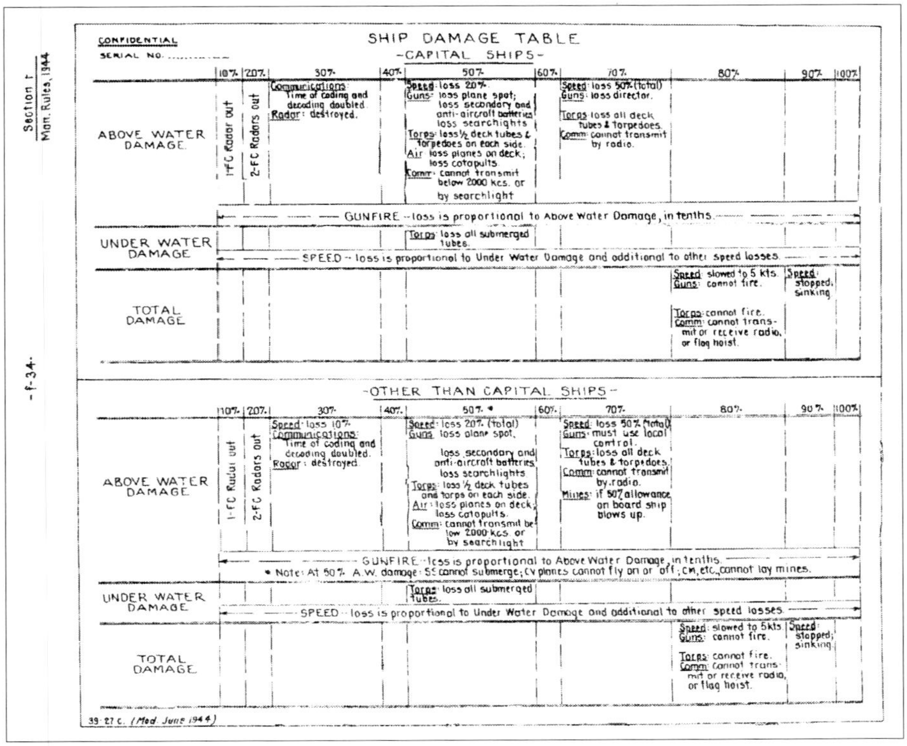
Section I
Man. Rules, 1944

-f-34-

CONFIDENTIAL
SERIAL NO. ....

SHIP DAMAGE TABLE
-CAPITAL SHIPS-

	10%	20%	30%	40%	50%	60%	70%	80%	90%	100%
ABOVE WATER DAMAGE	1-FC Radar out	2-FC Radars out	Communications: Time of coding and decoding doubled. Radar: destroyed.		Speed: loss 20%. Guns: loss plane spot; loss secondary and anti-aircraft batteries loss searchights. Torps: loss ½ deck tubes & torpedoes on each side. Air: loss planes on deck; loss catapults. Comm: cannot transmit below 2000 kcs. or by searchlight		Speed: loss 50% (total) Guns: loss director. Torps: loss all deck tubes & torpedoes. Comm: cannot transmit by radio.			
	GUNFIRE – loss is proportional to Above Water Damage, in tenths.									
UNDER WATER DAMAGE					Torps: loss all submerged tubes.					
	SPEED – loss is proportional to Under Water Damage and additional to other speed losses.									
TOTAL DAMAGE								Speed: slowed to 5 kts. Guns: cannot fire. Torps: cannot fire. Comm: cannot transmit or receive radio, or flag hoist.	Speed: stopped, sinking	

-OTHER THAN CAPITAL SHIPS-

	10%	20%	30%	40%	50% *	60%	70%	80%	90%	100%
ABOVE WATER DAMAGE	1-FC Radar out	2-FC Radars out	Speed: loss 10%. Communications: Time of coding and decoding doubled. Radar: destroyed.		Speed: loss 20% (total) Guns: loss plane spot, loss secondary and anti-aircraft batteries loss searchlights. Torps: loss ½ deck tubes and torps on each side. Air: loss planes on deck; loss catapults. Comm: cannot transmit below 2000 kcs. or by searchlight		Speed: loss 50% (total) Guns: must use local control. Torps: loss all deck tubes & torpedoes. Comm: cannot transmit by radio. Mines: if 50% allowance on board ship blows up.			
	GUNFIRE – loss is proportional to Above Water Damage, in tenths. * Note: At 50% A.W. damage: SS cannot submerge; CV planes cannot fly on or off; CM, etc., cannot lay mines.									
UNDER WATER DAMAGE					Torps: loss all submerged tubes.					
	SPEED – loss is proportional to Under Water Damage and additional to other speed losses.									
TOTAL DAMAGE								Speed: slowed to 5kts. Guns: cannot fire. Torps: cannot fire. Comm: cannot transmit or receive radio, or flag hoist.	Speed: stopped; sinking.	

39-27 c. (Mod. June 1944)

Fig. 57
Ship Damage Table

Torpedo Fire

As with gunfire, the Maneuver Rules had more specific procedures for torpedo fire than did the Conduct of the Maneuver, as well as "special rules" for the use of torpedoes by submarines and surface ships. For instance, for prewar modernized destroyers (ODDs), complete data in regard to adjustments for torpedo fire were to be submitted to the Torpedo Umpire at the beginning of a Board Maneuver: the type of torpedo fire to be employed, such as curved fire or straight; the spread setting between torpedoes; and the depth setting. Since the older destroyers did not have outside setting devices on their tubes, torpedo settings could not be changed during an action. For other types of surface ships, it was assumed that settings could be changed, at least during periods of preparation. Torpedo Fire Blank Forms included the type of torpedo to be fired and—in the case of variable-speed Type E or Type G torpedoes—the speed setting to be used. Torpedo armament for ships was given as in *Fleets.* Submerged tubes for submarines were taken to be aligned forward and aft; in other ships, tubes were assumed to be on the beam, firing over the side. (Pages g-1 to g-2.)

Submerged tubes could not be trained in azimuth, broadside above-water tubes could be brought to bear within thirty degrees, and centerline tubes could be brought to bear within forty-five degrees of fore-and-aft, unless otherwise noted in *Fleets.* Additionally, the Director could reduce these arcs of fire, especially in the case of ships with quadruple or quintuple tubes. Every torpedo was also to be fitted with an angle-fire device so that it could be fired at a ninety-degree angle or less with the direction of the tube. Torpedoes, however, could not be fired from submerged tubes if the ship's speed was in excess of sixteen knots. Each tube was to be considered as loaded at the beginning of the Maneuver, and reloads were regarded as equally distributed among tubes at the beginning of torpedo fire.

Type	Size	Speed	Hit Value	Range
A	21"	26	2.4	17,000
B	21"	29	3.0	15,000
C	21"	27	2.4	13,500
E	18"	30	1.8	6,000
E-1	18"	33	1.8	3,000
E-2	23"	30	2.4	4,500
E-3	23"	33.5	2.4	6,000
F	21"	36	3.0	3,500
F-1	21"	34	2.4	5,500
F-2	18"	35	1.8	3,500
G (X	21"	27	3.0	15,000
G (Y	21"	34	3.0	10,000
G (Z	21"	46	3.0	6,000
H (Y	21"	32.5	3.0	9,000
H (Z	21"	46	3.0	4,500

Fig. 58
Torpedo characteristics

Torpedoes that had been loaded into deck tubes, however, could not be withdrawn during an action, and they could not be shifted between submerged and deck tubes or between forward and after tubes while under way. Torpedoes could be shifted between starboard and port submerged tubes, subject to certain restrictions. For instance, a ship with broadside submerged tubes could fire torpedoes from each submerged tube at the rate of one torpedo per move until the torpedoes on that side had been exhausted. A torpedo on the other side of the ship could then be shifted, but two moves would be required to carry that out. The first of these shifted torpedoes could then be fired in the third move, and others in alternate moves thereafter. A ship with deck tubes and spare torpedoes could reload only when its deck was not being made unsteady by the action of the sea or rudder. The minimum time for torpedo reloads was fifteen minutes. (Pages g-2 to g-3.)

A submarine could fire all its loaded torpedoes in a three-minute move. A torpedo tube in any torpedo room could then be reloaded whenever it was empty,

TORPEDO FIRE BLANK

8/14/41 **Exercise 2**
Desdiv 15

Use separate blank for each salvo. Prepare in Duplicate.

Fired at 0701 (See Rule G-8) During Move 1
(Scratch one)

Firing Ships	Type torpedo and Speed	What Tubes fired NOTE A	Number torpedoes fired	Point of aim NOTE B	Sight angle. Base torpedo course. (Scratch one) NOTE C	Spread (degrees)	Depth (feet)	For Use of Umpire: Damage at beginning of move	Not under fire. Under Fire. Behind Smoke or Exceptional Conditions	Number of torpedoes that run	Firing Range	Track Angle	Results Obtained
Desdiv 15	**G-3 50 Kts**	**All**	**9**	**----**	**BTC 257°** **(TA 55° SP 22 Kts SA 21° Brg 236 BTC 257°)**	**8°**	**22**						

Signature ______________

NOTE A Enter tubes fired as "Stbd", "Port", "Ford", "Aft", "All", "Deck" or "Submerged" as may be appropriate after giving due consideration to the position and course of the firing vessel at the time of firing.
NOTE B Not required if base torpedo course plan of firing is used.
NOTE C Scratch either "Sight angle" or "Base torpedo course". Make entry in this column in degrees from 0 to 360.
The sight angle should be the angle you desire used between the line of sight and the track you desire your base torpedo to run on.

Fig. 59
Torpedo Fire Blank Form

under certain restrictions. Essentially, two torpedo tubes could be reloaded in two three-minute moves, and four tubes could be reloaded in three moves. Before a torpedo could be fired from a submerged submarine, the periscope had to be exposed for at least ten seconds. A submarine, however, could fire entirely by sound, subject to the limitations of accuracy given in Section D of the Maneuver Rules.

Single ships firing on their own initiative could fire during or at the end of a three-minute move, handing in the Torpedo Fire Blank Form after the move had been made but before the next move had been called for. For all other torpedo fire, the form had to be handed in before the firing move was made on the board, and the form could not specify torpedo fire earlier than the end of the first minute of movement by the target formation subsequent to turning in of the form. Such fire was considered completed at the Maneuver Time designated on the form—provided it conformed to the above restriction—unless the Torpedo Umpire was informed to the contrary immediately after this time was reached. If torpedo fire was held up in this manner and the settings were not to be changed, it could be completed at any subsequent Maneuver Time by informing the Torpedo Umpire. If firing settings were changed for any reason, a modified form had to be handed in, with the same conditions regarding actual movement of the target formation as for the original form. (Pages g-3 to g-4.)

At the time of firing, units had to be "suitably" deployed and, if firing torpedoes with curved fire ahead, had to be sufficiently slowed. Students obtained data for torpedo fire from their tactical plots, but the Maneuver Staff furnished bearings and ranges when requested, if the information would be available under actual conditions. In addition, the condition of the firing ship that was going to fire torpedoes was considered to apply at the actual time of firing, meaning that a requested firing in a move would not happen if the damage assessed at the end of a move made it impossible. A ship receiving 50 percent or more Above Water Damage was to lose 50 percent of its deck torpedo tubes on each side, and a ship receiving 70 percent or more Above Water Damage was to lose all of its deck tubes. A ship taking 80 percent or more Above Water Damage could no longer fire any torpedoes, and a ship receiving 50 percent or more Under Water Damage was to lose all of its submerged tubes. Additionally, all torpedoes of a salvo fired from a division of ships in formation were considered to have been fired from one point, the center of the division. The exception was that at ranges inside five thousand yards, the commander of the firing ships could elect to have the torpedoes plotted as coming from an individual ship. Of the total number of torpedoes intended to be fired from a surface ship not under Effective Fire, only 75 percent—due to wartime conditions—were to be considered as running effectively and functioning properly. When surface ships were under Effective Fire or fired from behind smoke screens on signals from an exposed leader, only 50 percent of the torpedoes would "run."

Under "exceptional" conditions—such as a ship being under unusually heavy fire or in a melee, or making frequent and significant changes of course while under fire—the Director could decrease the number of torpedoes that would run. In the case of submarines, 80 percent of total torpedoes were considered to run. In Moderate to Rough Seas, however, torpedo depth settings of less than fifteen feet could result in torpedoes functioning improperly. (Pages g-4 to g-5.)

Torpedo fire was also to be plotted on the Master Plot, on which was drawn the firing points and the paths of running torpedoes. Arcs showed the length of the torpedo runs during each three-minute move; in a salvo of two effective torpedoes from a surface ship, their right and left dispersion limits were shown. When the number of effective torpedoes was three or more, one torpedo was considered to run on the right limit of dispersion, one on the left, and the remainder equally spread between these two. Under conditions in which the discharge or tracks of the torpedoes could be seen, the Director would give "appropriate" information to the student or students concerned. A student receiving this information could also take appropriate action at that time, even modifying a move previously made. Unless otherwise prevented by the existing conditions, the student could start messages to maneuver his own ships or to warn other ships. This student could also, in an emergency, order his ships to maneuver independently at the initiative of their commanding officers. The tracks and positions of ships endangered by torpedoes were added to the Master Plot. From the plots of the torpedoes and the intended and unintended targets in "torpedo water" whose draft exposed them to danger, the Director was to determine the number and allocation of torpedo hits, using various methods. First, he would consider each torpedo and ship individually, studying their simultaneous positions, and determine if a hit had been achieved. In the case of a formation of ships endangered by a salvo of torpedoes, the number of torpedo hits was to equal the quotient obtained when *T* (the total length of the targets projected onto a line at a right angle to the main torpedo track) was divided by *D* (the average distance between effective torpedoes as measured at the time of passing the target information). (Pages g-5 to g-6.)

When ships were in approximate column formation—a line of bearing of 180 degrees—the problem was simplified by using the actual instead of projected length of the formation through which the torpedoes would pass, as well as the actual rather than projected total length of ship targets. The number of hits thus determined could be reduced by the Director when the paths of the torpedoes were at least twice as far apart as the width of the target ships. Hits could also be reduced in number if the ships had sufficient warning of the approach of the torpedoes or had straightened out on a course parallel to the tracks of the torpedoes. The number of hits could additionally be reduced if the ships that were being fired at could see these tracks, were maneuvering independently to avoid them, and had enough speed and maneuverability to keep clear. Under the conditions favorable for dodging torpedoes, the chances of avoiding them were better by running away from the torpedoes than turning to meet them.

The effects of torpedo hits were calculated on the basis of an equal number of penetrative hits from fourteen-inch shells. For instance, a hit by one Type B, F, G, or H torpedo equaled three fourteen-inch penetrative hits; a hit by a Type A, C, E-2, E-3, or F-1 torpedo equaled 2.4 fourteen-inch hits; and a hit by a Type E, E-1, or F-2 torpedo each equaled 1.8 fourteen-inch penetrative hits. The effect of additional torpedoes that hit within fifteen minutes of a preceding torpedo hit would be increased by a third for the first additional hit and by two-thirds for any successive ones. Torpedo damage was added to damage from other causes and was communicated to the Damage Recorder. Independently of permanently damaging effects, the shock of a torpedo hit would render a ship incapable of using its weapons, flying planes on or off, or maneuvering for a period of three minutes. (Pages g-6 to g-7.)

The outcome of a ship being hit by torpedoes in the region of the propeller shaft or the rudder was to be judged by the Director in accordance with the table in figure 61, which was based on the angle that the track of the torpedoes made with the keel of the ship. This measurement was made from ahead to the right through 360 degrees. Aside from the speed loss inflicted, a torpedo hit on a propeller shaft on

TABLE FROM WHICH D (of Rule G-224(b)(2) Maneuver Rules) MAY BE OBTAINED IN COMPUTING TORPEDO HITS.

Confidential.

TENTHS OF DEGREES BETWEEN ADJACENT EFFECTIVE TORPEDOES.

RANGE	.5	.6	.8	1.0	1.2	1.4	1.5	1.6	1.8	2.0	2.2	2.4	2.5	2.6	2.8	3.0	3.25	3.5	3.75	4.0	4.5	5.0	6.0
1	9	10	14	17	21	24	26	28	31	35	38	42	44	45	49	52	57	61	65	70	78	87	105
2	17	21	28	35	42	49	52	66	63	70	77	84	67	91	98	105	113	122	131	140	157	174	209
3	26	31	42	52	63	73	79	84	94	105	115	126	131	136	147	157	170	183	196	209	235	261	314
4	35	42	56	70	84	98	105	112	126	140	154	168	174	161	195	209	227	244	262	279	314	349	418
5	44	52	70	87	105	122	131	140	157	175	192	209	218	227	244	262	283	305	327	349	392	436	523
6	52	63	84	106	126	147	157	168	188	209	230	251	262	272	293	314	340	366	392	419	471	523	627
7	61	73	98	122	147	171	183	195	220	244	268	293	305	318	342	366	397	427	456	488	549	610	732
8	70	84	112	140	168	195	209	223	251	279	307	335	349	363	391	419	454	488	523	558	628	697	836
9	79	94	126	157	188	220	235	251	283	314	346	377	393	408	440	471	510	549	589	628	706	784	941
10	87	105	140	175	209	244	262	279	314	349	384	419	436	454	489	523	567	611	654	698	785	872	1045
11	96	115	154	192	230	269	288	307	346	384	422	461	480	499	537	576	624	672	719	767	863	959	1150
12	105	126	168	209	251	293	314	335	377	419	461	503	523	544	586	628	680	733	784	837	942	1046	1284
13	113	136	181	227	272	318	340	363	408	454	499	544	567	590	635	680	737	794	850	907	1020	1133	1359
14	122	147	195	244	293	342	367	391	440	489	537	586	611	635	684	733	794	855	916	977	1098	1220	1463
15	131	157	209	262	314	366	393	419	471	523	576	628	654	680	733	785	850	916	981	1046	1177	1307	1568
16	140	168	223	279	335	391	419	447	503	558	614	670	698	726	782	837	907	977	1046	1116	1255	1395	1672
17	148	178	237	297	356	415	445	475	534	593	652	712	741	771	830	890	964	1038	1112	1186	1334	1482	1777

Enter table with range of target formation from firing point and with angle between adjacent effective torpedoes of a spread to obtain the distance in yards at target between adjacent effective torpedoes.

42-13. The total projected length of target units in the formation divided by the above figure gives number of torpedo hits obtained.

Fig. 60
Torpedo hits computation table

Angle		Propeller shaft	Rudder
0	360	0	0
45	315	1/10	0
90	270	1/5	1/20
135	225	1/2	1/10
180	180	4/5	1/5

Fig. 61
Chance of hits on rudder and propeller

one side of the ship with two or more screws would reduce the ship's maximum speed to three-quarters of whatever would otherwise have remained to it. A torpedo hit on a ship with just one propeller, or hits in the propeller-shaft area on both sides of a ship with two or more screws, would render that ship incapable of self-propulsion. A torpedo hit in the rudder area of a ship would render that ship incapable of steering by use of its rudder. Finally, a torpedo hit could cause a ship to assume a list, which would also affect its ability to use its weapons, fly planes on or off, or maneuver. The amount, duration, and effect of such a temporary list would be assessed by the Director. (Pages g-7 to g-8.)

V The Maneuver Rules Concluded, June 1945

The Maneuver Rules on mine employment were classified by type and use. There were Anchored, Drifting, and Ground Mines, designated *A, D,* and *G,* respectively. Mines were also classified as to kind and by the methods by which they were fired. These were Contact, Influence (acoustic, magnetic, etc.), Antenna, or Observation Mines, designated *C, I, N,* and *O,* respectively.

Mines

Accordingly, mines were designated as to type and kind by a two-letter symbol in which the first letter indicated type and the second indicated kind. Mines carried by submarines were suited for discharge either from special mine trunks or from torpedo tubes. In designating these special types of mines by symbol, students added the letter *M* for "mine trunk" or *T* for "torpedo tube" to the two letters above. Mines carried by aircraft were designated *P.* Mines were not convertible as to type or kind.*

Anchored Mines could be Contact, Influence, Antenna, or Observation, and they could be laid in any depth of water up to 250 fathoms. These mines could also be set to ride at any depth below the surface at slack, low water. Influence Mines had a limit of two hundred feet, the optimum distance being thirty to a hundred feet below the hulls of their intended targets. In a current, the submergence depth of an Anchored Mine was considered to be increased by an amount of "dip" calculated as, in feet, 4 percent of the length of the mine's anchor cable times the strength of the current in knots. In a current more than four knots or in rough seas, Anchored Mines dragged and became inoperative. The Director of the Maneuver would judge the conditions under which all or any part of a minefield dragged and became inoperative. (Pages h-1 to h-2.)

Drifting Contact Mines were considered to float on the surface, and Drifting Antenna Mines were assumed to remain suspended at fixed depths, each mine being supported by a small surface float, to which it was connected by antenna wire.

---

* "Maneuver Rules," June 1945, p. h-1, NHC, folder 2525-E, box 124, RG 4, NWC. Subsequent page-number references in this chapter are to this source.

Unless otherwise stated, the length of the suspending antenna wire was thirty feet. Ground Mines lay on the bottom and could be either Observation or Influence Mines. Contact Mines in general were exploded by actual contact of the mine case with a passing ship. Influence Mines in particular had internal firing mechanisms actuated by external influences, such as distortion of the earth's magnetic field by the residual magnetism of the target ship or its acoustical influence. These mines were capable of selective delay and could fire after a set number of actuations by successive target ships, which for the purpose of the Maneuvers was set at a maximum of twenty-five feet. The maximum depth of these mines was two hundred feet, with optimum depth at thirty to a hundred feet; the degree of protection a ship had against mines of this sort would be decided by the Director. (Page h-2.)

An Observation, or Controlled, Mine was exploded by closing an electrical firing circuit when it was determined by direct observation that the target vessel was in its "proper" position. Antenna Mines were exploded by the contact of metal with either the mine case, the antenna wire, or the antenna float. Anchored Antenna Mines had antenna wires leading upward and suspended from a small float, as noted; a section of the anchor line leading downward also acted as an antenna. Unless otherwise stated by the student, the length of the upper antenna wire was taken as ten feet less than the intended slack-low-water submergence depth of the mine. This depth was not to exceed seventy feet; the length of the downward antenna section of the anchor line was taken as one hundred feet. Mines were also designed to become inoperative in particular circumstances: Anchored Mines would become inoperative if they came adrift from their anchor gear, and Drifting Mines became inert a certain period after they had been laid. Influence Mines were not so designed, but the life of a Drifting Mine, in accordance with the Hague Convention, was to be one hour (the Director could give such a mine longer life for purposes of the Maneuver). (Pages h-3 to h-4.)

Moreover, some proportion of initially defective mines was to be expected, and some proportion was expected to become defective after being laid. With Antenna Mines, some proportion were expected to explode prematurely, while they were being laid. These proportions varied greatly in actual practice, but for Manuever purposes averages were assumed. Five percent of Contact Mines, 10 percent of Antenna Mines, and 15 percent of Influence Mines would be initially defective. Premature explosions would not take place with Contact Mines, but 10 percent each of the Antenna and Influence Mines would be lost in this manner. Two percent of Contact Mines already in place, 5 percent of Antenna Mines, and 10 percent of Influence Mines were to become inoperative per month. The ratio of effectiveness of a group of mines was the quotient of the total number originally laid minus the number that became inoperative, ineffective, or prematurely exploded, divided by the number originally laid. Unless otherwise stated, for game purposes the mine equipment was as given in *Fleets,* and if not otherwise announced, students having mine ships could select the type of mine each ship would carry, informing the Director before the Maneuver began. (Page h-4.)

Dummy mines—pieces of wood shaped and painted to resemble drifting surface mines or antenna floats—were included in the Exercises. Surface ships carrying mines could carry a number of dummy mines equal to the true mine complement, as allowed by the Director; students employing them informed the Director before the Maneuver began. Contact Mines floating on the surface could be seen from ships up to five hundred yards away, under ideal conditions—a glassy sea, good height of eye, and good light. Under any other conditions, these mines could not be seen from ships proceeding at moderate speeds at sufficient distances to avoid them. Aircraft would be able to see Contact Floating Mines within a horizontal radius of five hundred yards under ideal conditions—a crew flying at not over one thousand feet and concentrating on the surface to look specifically for mines, torpedoes, or submarines. Dummy mines on the surface had the same visibility as real ones and could not be distinguished from real ones at ranges greater than fifty yards. Antenna floats on the surface were also visible under ideal conditions, but only at fifty yards, or by aircraft flying at altitudes of two hundred feet or less. Surface ships proceeding at any speed would not be able to see them in time to avoid them. Mines or floats below the surface were considered invisible to surface ships but discernible by aircraft under favorable conditions. (Pages h-4 to h-5.)

Minelaying by surface ships was to be visible under ideal conditions at ranges up to seven thousand yards for small or high-speed minelayers and planes, and up to twenty thousand yards for large, slow minelaying ships. Under less favorable conditions, these ranges were reduced. Premature explosions during minelaying could be visible at greater distances, but their cause might not be evident. Mine explosions, however, were to be considered audible from surface ships and submarines for a distance of ten miles. Mines could be laid by surface ships at intervals of three seconds or more and at any speed up to thirty knots. Submarines, either surfaced or submerged, could discharge mines at intervals of not less than one minute between mines carried at the same end of the boat but not in excess of four knots when discharged from forward. In the case of tube mines, submarines could load only one per tube and would then have to reload, a task taking three minutes per tube. Submarines were to have no less than forty feet of water under the keel when laying mines, and there were time intervals between mines being laid at various speeds (see figure 62). The explosion of a mine would also destroy the effectiveness of mines closer to it than two hundred feet. (Pages h-5 to h-6.)

Prior to beginning a Chart or Board Maneuver, each student having mines on board his ship was to furnish the Mine Umpire a memorandum (updated as necessary before mines were sown) showing the types of mines, the number on each vessel, and the length of antenna setting for Antenna Mines, and the number of skipped activations for delayed-action Influence Mines. In a Chart Maneuver, when mines were laid, the student handling the minelaying ships was also to furnish the Director a Mine Laying Blank, showing for each minelaying operation the date, hour, and locations when minelaying started and stopped; which ships carried it out; the formation and maneuvers used; the number of mines laid by each ship and the total number laid; types and kinds of mines laid; the number of lines of mines, with their intervals, spacing, and depth settings; and any details about the mines that differed from standard characteristics. During the times that mines were intended to be effective, the student would mark on his chart and tracings the area and location of the minefield and submit this information to the Director. In a Board Maneuver, a student laying mines would state on the Move Blank the ships concerned and then supply the necessary details on the Mine Laying Blank, again to be submitted to the Director. In either case, the Director plotted the mine area on the Master Plot and decided the ratio of effectiveness of the mines. (Pages h-6 to h-8.)

Speed of layers	Time intervals between mines for different distances apart			
(a) knots	(b) 600 ft.	(c) 500 ft.	(d) 400 ft.	(e) 300 ft.
10	36	30	24	18
11	33	28	22	17
12	30	25	20	15
13	28	23	18	14
14	25	21	17	13
15	24	20	16	12
16	23	19	15	11
17	20	17	14	10
20	18	15	12	9
22	16	14	11	8
25	14	12	10	7
28	13	11	9	7
30	12	10	8	6

Fig. 62
Minelaying time intervals and speeds

It was assumed that paravanes would be carried by all combatant ships of more than three thousand tons displacement, by Navy auxiliary ships of more than five thousand tons, and by merchant ships as might be permitted by the Director. To stream or take in paravanes, a ship had to run for six minutes at a speed of not more than twelve knots. With paravanes out, a ship could not make more than twenty-eight knots without losing them. For a ship proceeding on a straight course, paravanes afforded "practically" complete protection against Anchored Contact Mines to ships immediately astern and on the beams. Mines of that type that came within the spread of a paravane would have their anchor lines cut, causing them to float to the surface. If a ship changed course in a minefield, however, a paravane could not afford complete protection against swinging its afterbody into a mine. Paravanes, moreover, were no protection against Drifting Surface Contact Mines, Influence Mines, or Antenna Mines, whether drifting or anchored. Contact of a paravane with a mine antenna would cause the mine to explode and, whatever else it might do, destroy that paravane. If a ship entered a mined area, the Director would have its track plotted on the Master Plot and determine the number of mine hits, if any. The degree to which a ship was threatened by a particular line of mines it crossed would depend on the depth of its underwater hull. In addition, the mine threat would depend on the water level at which or through which the mines were operative, especially their depths of submergence at the time and the vertical lengths of their antennas. The threat to a particular ship would further be determined by whether the ship had a paravane out, whether the paravanes protected against the type of mine encountered, and whether the ship was properly degaussed for the locality. (Pages h-8 to h-9.)

If a single ship on a straight course crossed a single line of mines, the chance of its hitting a mine was computed by a method involving the distance between the mines in feet, the mine-effectiveness ratio, the angle between the course of the ship and the line of mines, the effective width of the ship in feet, and the effective distance between mines. A similar formula was used to figure the chance of a ship hitting a mine if it passed through successive lines of mines. Probabilities could be figured separately if the characteristics of the lines of mines differed. If the ship was changing course while passing through a line of mines, its chance of hitting one was increased because its hull presented an increased effective width while turning. The Director could decide that special sea or local conditions also altered the calculated chances of a ship missing a line of mines. If a number of ships crossed a mine area or a ship crossed a mined area repeatedly, the total number of mine hits to be expected for the total number of ship traverses involved was the sum of the chances of hitting on each traverse, with the exceptions noted above. A ship that crossed a mine area directly in the wake of another ship not smaller than itself had a negligible chance of taking a hit. When a ship did strike a mine, other ships nearby would see that the ship had been mined or torpedoed, and the Director would inform students accordingly. He might also permit them to modify the moves of such ships. (Pages h-9 to h-10.)

A mine hit inflicted Under Water Damage equal in amount to three fourteen-inch penetrative hits; second and succeeding hits in the same minefield equaled four fourteen-inch penetrative hits. This amount could be increased to no more than five and six fourteen-inch penetrative hits, respectively, in the cases of Ground Influence Mines, at the discretion of the Director. In terms of permanent effect on the life and capabilities of a ship, mine damage was added to that from other causes and was to be communicated to the Damage Recorder. Aside from permanent damage, a mine hit, like a torpedo hit, inflicted shock damage that rendered the ship incapable for three minutes of using its weapons, operating aircraft, or maneuvering. The mine hit could also cause a temporary list, which would impede those functions; the Director would judge the amount, duration, and effect of such a list. Finally, a surface minelaying ship would lose mines in proportion to the Above Water Damage received. If the ship took 50 percent Above Water Damage, it could no longer lay mines. If the ship had at least a 50 percent mine allowance on board and suffered 70 percent Above Water Damage, it would blow up. (Pages h-10 to h-11.)

Submarines

The Submarine Section of the Maneuver Rules asserted that a submarine under way was considered to be either in Surface Condition or Submerged. While on the surface, a submarine was always ready for quick dives. Changing from Surface to Submerged or back again took certain amounts of time. If on the surface and under way with all engines, a submarine needed one minute to disappear; it would also take one minute to go from Submerged to Surface Condition. The speed of the submarine changing from one condition to the other could not exceed its maximum submerged speed. The time during which a submarine was submerged was taken to be from the instant it began to submerge until it began to return to the surface.

Students operating submarines had to recall such preparations as securing the radio set while flooding and preparing to dive. When submerged at greater than periscope depth, submarines could neither transmit nor receive radio messages. *Fleets* was the source for submerged speeds and radii for one hour of operation; endurance at lower speeds could be obtained from the curve in figure 63. Time and fuel required for charging batteries was to be computed by the formulas in figure 64. It was assumed that submarines and submarine minelayers had four main engines and one auxiliary charging engine. The formulas also took into account when the submarine was taking current from a submarine tender or shore base. The formulas additionally accounted for the submarine not being under way but charging batteries with the submarine's engines. Finally, the formulas took into account the submarine being under way on the surface and charging the submarine's batteries with the submarine's engines. (Pages i-1 to i-4.)

Fleets also provided a table for computing fuel expenditures in tons per hour at designated speeds. While under way on the surface and charging batteries, the limiting speeds that the submarine could make were 80 percent of maximum with two main engines, 90 percent on one main engine, and all of maximum speed while charging on the auxiliary engine. For prewar modernized submarines and submarine minelayers, the formula for figuring the fuel that was expended for recharging batteries was given in the Time and Fuel Table. This formula assumed the boat had two main engines and was taking current from a tender or shore base or had stopped on the surface to use both engines for charging. If the maximum speed was less than fifteen knots, the fuel expenditure at maximum speed was to be used; the Fuel Table called for fuel expenditure as if the boat was making fifteen knots. Using one engine to charge batteries while running on the surface with the other engine would increase these fuel expenditures by 25 percent. Total fuel expenditures for both engines were to be computed for a speed of three-halves that of actual operation, so the maximum speed that could actually be made was two-thirds of that which could have been made if the boat was not charging. If the charge was being taken from a tender, fuel expenditure as computed above would be charged to the

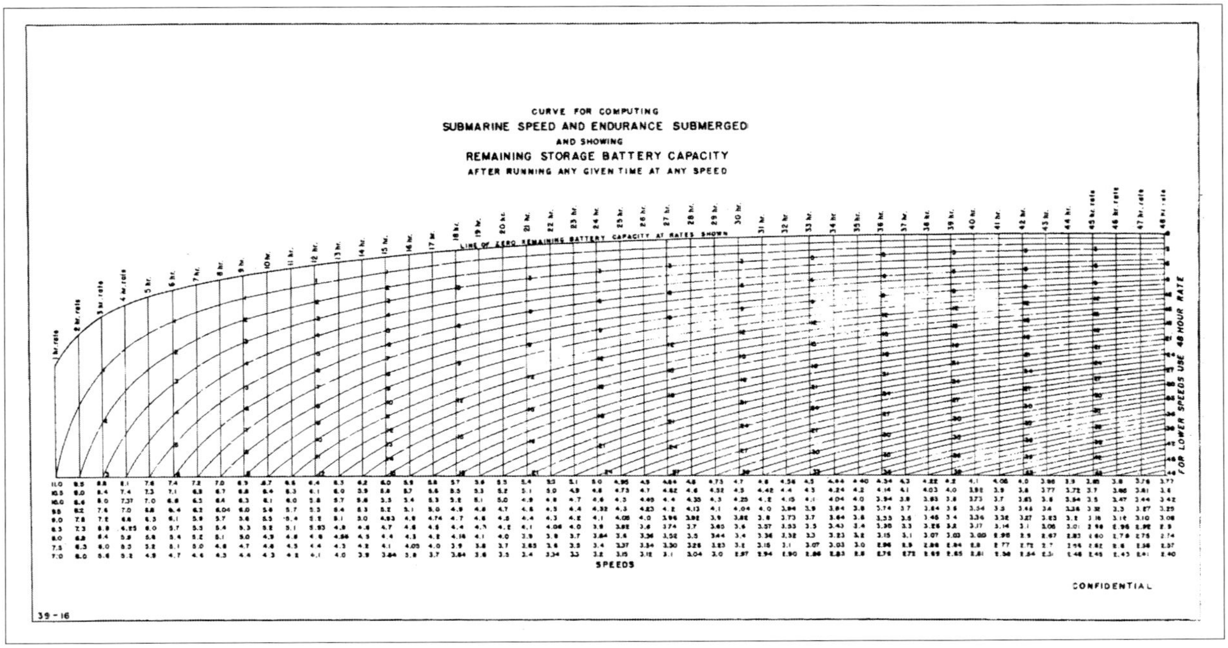

Fig. 63
Curve and notes for computing submarine submerged speeds

tender, not the submarine. If the tender used coal as fuel, the figures above in fuel oil would be converted to coal on the basis of one hundred tons of oil to 150 tons of coal. (Pages i-4 to i-6.)

The Rules dealt with a wide range of other details that could be critical in tactical situations. Submarines were assumed to be fitted with Recognition Signal Ejectors to release colored smoke bombs when submerged; a submerged submarine could be identified one minute after firing the Recognition Signal, at a distance of four miles. A surface ship making eleven to fifteen knots could lay depth charges at a rate of five per minute; at sixteen to twenty knots the rate was ten per minute, and at twenty-one knots or more, fifteen. If a surface ship sighted a periscope that disappeared as the ship approached, the ship could lay depth charges and the Director would assess damage based on the track and depth of the submarine; the tracks of attacking ships; the number, times, and positions of depth charges dropped; and their depth settings. Damage assessed could include the disabling of the submarine for a number of hours, rupture of the outer hull, damage to the rudder or propeller (or both), or total destruction.

In consideration of the limitations imposed by fuel, fresh provisions, supplies, habitability, materiel, and reduction in the efficiency of personnel, submarines could operate away from tenders or bases for only so long. Submarines and submarine minelayers could operate on their own for raiding purposes when the submarine was required to conduct only intermittent submerged operations, but for no more than fifteen of a total of seventy-five days. High speeds would "considerably reduce" this figure. (Page i-6.)

On patrol operations in which the submarine was to remain submerged during most of daylight hours and be on station for thirty days—plus time required for proceeding to station and returning to base—total time under way was not to exceed sixty days. The length of time out also depended on the speeds used; fuel was the limitation. Prewar submarines and submarine minelayers could participate in raiding operations for not more than ten days of a twenty-five-day cruise, or the limit of their fuel capacities if less than twenty-five days. These figures also assumed only intermittent submergence during these operations. On patrol operations, when the submarine would need to be submerged most daylight hours, ten days on station plus the time for transiting to station and back was also not to exceed twenty-five days, less if the boat's fuel capacity was even more limited. The Director might also reduce the length of these operations when weather or other unfavorable circumstances warranted. Additionally, upkeep and rest periods had to be allowed for submarine crews after the newer boats had been out twenty-one days and the older ones ten days. An operating period that was less than the maximum was to be followed by a rest period reduced in the same proportion. Moreover, a navy-yard overhaul period of sixty days was required after five maximum-duration operating periods. Finally, submarines could launch mines from their torpedo tubes, but the torpedo allowance given in *Fleets* was reduced by one torpedo for each pair of mines carried. Two mines could be stowed in a torpedo tube, but only one when the submarine was actually laying mines. Two mines, however, could be stowed in each torpedo storage rack. (Page i-7.)

Engines in use	Time in hours for charging under given engine combinations the following % of charge				Fuel expended same as for SS Making
	100%	90%	75%	37½%	
(1) 2M	2.0	1.7	1.0	0.6	15 knots
then 1M	2.6	2.6	2.5	1.0	11 knots
then A	2.0	2.0	2.0	1.6	10 knots
Total time of charge	6.6	6.3	5.5	3.2	
(2) 1M	6.0	4.8	4.0	1.2	11 knots
then A	2.0	2.0	2.0	2.0	10 knots
Total time of charge	8.0	6.8	6.0	3.2	
(3) A	12.0	11.5	9.0	4.0	10 knots

Engines in use	Time in hours for charging under given engine combinations the following % of charge				Fuel expended same as for SS Making
	100%	90%	75%	37½%	
(1) 2M	2.0	1.7	1.0	0.6	15 knots
then 1M	2.6	2.6	2.5	1.0	11 knots
then A	2.0	2.0	2.0	1.6	10 knots
Total time of charge	6.6	6.3	5.5	3.2	
(2) 1M	6.0	4.8	4.0	1.2	11 knots
then A	2.0	2.0	2.0	2.0	10 knots
Total time of charge	8.0	6.8	6.0	3.2	
(3) A	12.0	11.5	9.0	4.0	10 knots

% of charge	Time in hours	Fuel expended same as for SS making
100%	8.0	15 knots for 2 hours, then 8 knots for 6 hours
90%	6.0	15 knots for 2 hours, then 8 knots for 4 hours
75%	4.0	15 knots for 2 hours, then 8 knots for 2 hours
37½%	2.0	15 knots for 1 hour, then 8 knots for 1 hour

Fig. 64
Time and fuel for charging submarine batteries

Aircraft

The Maneuver Rules for Aircraft were adopted to give students a "comparative basis" for the employment of aircraft in strategic and tactical maneuvers. These conventions were based on what the College thought "average conditions"; the Director modified their application whenever it was necessary for realism. During bad weather or other hazardous flying conditions, for instance, the Director could rule that aircraft were unable to take to the air at all or not without "unusually" high percentages of losses. (Page j-1.)

"Airdromes" (as the Rules called them) used for the regular operation of heavily loaded land planes were to have at least three runways, fewer if the prevailing wind was dependable or a large area that could include runways. The minimum size of each runway was to be five thousand feet by three hundred feet, clear of obstacles at either end by at least a thousand feet. A runway was to be level across its short dimension and "approximately" level along the other. There were to be no soft spots, sand, soft earth, mud, or marsh; tree clumps on the long sides were considered valuable as cover. The length of time needed to prepare each runway by an engineer or labor battalion with appropriate machinery was three days in flat, open terrain requiring no cuts or fills. Two weeks, however, would be required for the same type of ground that needed significant clearing; ten days for moderately rolling (but open) terrain with some cuts and fills; and three or four weeks for moderately rolling terrain that needed heavy clearing. (Pages j-1 to j-2.)

All flights of individual aircraft were to conform to the pamphlet *Airplane Characteristics*, which listed the tactical qualities of aircraft then in use by the U.S. Navy. When airplanes of different types were proceeding in company, the Director would rule as to speed and fuel consumption. Airplanes were assigned to squadrons or other organizations on board ships or at shore stations, but students could change these assignments by appropriate orders; airplanes that were to be available would be listed in the Statement of the Problem. All squadrons were considered as having spare, disassembled aircraft equal in number to 20 percent of the operating allowance, unless otherwise specified in the Statement of the Problem. During a Maneuver, these spares could replace crashed airplanes after sufficient time was allotted for their assembly, and they could be flown if flight personnel were available. Seaplane tenders, in addition to supporting naval patrol planes, might also be assigned to service other types of aircraft, in which case the tender could transport on board all the observation scout, scout observation, or fighter planes so assigned, or 20 percent of the scout bombers and torpedo bombers assigned. No more than five naval patrol planes, however, could be transported by one airplane tender. (Page j-2.)

Airplanes were considered able to rendezvous without difficulty over a visible object at a time and altitude specified, but the absence of visible references meant that they had to use intermediate-frequency radio to coordinate if starting from points separated by more than seven times the limits of aircraft-sighting visibility. Amphibian airplanes could operate either as seaplanes or land planes. Airplanes taking off or landing at night—either from a ship or an airfield—had to use lights, which in turn were visible at a distance of five miles if natural obstructions did not interfere. Land planes landing in the water, seaplanes setting down on land, and airplanes shot down by antiaircraft fire or in aerial combat were to be considered "wholly" crashed, not available for future employment. Damaged but flyable airplanes, though, could be used for gunnery spotting or the discharge of chemical agents or parachute flares. Small planes searching singly were somewhat more subject to navigational errors and communication failures than would be several such planes in company. Their ability to keep a good lookout and to sight objects within the assumed range of visibility was reduced, and the effectiveness of search by single, small planes was accordingly considered no more than 75 percent of that of a search by two or more similar planes in company. (Page j-2.)

There were special rules for carrier-based aircraft. Since space on carriers was limited, students had at all times to be prepared to inform the Director as to the location of all airplanes, the order in which they were spotted on the hangar and flight decks, whether or not their wings were folded, and which were disassembled. Students were also to keep records of the fuel in these airplanes and make proper time allowances for movement between the hangar and flight decks and for warming up, taking off, landing, reservicing, and checking. One torpedo bomber was considered to occupy as much space as two fighters, and three fighters occupied as much space as two scout planes, two dive-bombers, or two torpedo bombers with wings folded. Airplanes could be landed in the usual manner carrying bombs or torpedoes. All carriers were assumed to have two elevators between the hangar and flight decks; either could carry one plane of any type, or two fighters. A round trip between the hangar and flight decks required one minute. All airplanes were assumed to be either fully assembled or fully disassembled, and all carrier planes to have wings that could fold (in ten seconds). (Page j-4.)

Given a complete allowance of planes, a carrier could either launch or land two-thirds, but not simultaneously. The other third was to be in the hangar during launching or landing operations. Planes could be launched by catapult—regardless of the number on deck—at the rate of one plane per minute per catapult, provided that the catapult or catapults were clear. For battle carriers (CVBs), however, planes could be launched and landed simultaneously, with not more than a quarter of the carrier's allowance of planes on deck. On all carriers, no more than half the allowance of planes could be placed in the hangar at the same time. Also, if the carrier's course was more than twenty degrees off the true wind, five minutes were needed to head into the wind and then begin landing planes or launching them. Planes could be launched by deck catapult with no change of course, but only if the apparent wind was forward of the beam. To launch planes without catapults, the apparent wind had to be within ten degrees of the ship's course and the wind over the deck had to be no less than thirty knots for fleet carriers and light carriers (CVLs). For escort carriers (CVEs), the wind over the deck had to be no less than twenty-five knots. To land planes, the apparent wind had to be within five degrees of the ship's course, and the wind over the deck had to be no less than twenty-five knots for fleet carriers and CVLs, twenty knots for CVEs. During Daylight and under Normal Conditions, planes on deck could be launched at intervals of fifteen seconds and landed every thirty seconds. During Rough Seas and at Night, these intervals were doubled. Airplanes could be launched simultaneously from deck and hangar catapults. Aircraft could also be respotted on deck in the time required for refueling and rearming. Unscheduled launches could start five minutes after the decision was made to conduct them, provided the planes were on deck ready for takeoff and were not to go beyond a hundred miles from the carrier. For longer flights, preparations would require fifteen minutes before the start of the launch; if the planes took off at once, planning was to require fifteen minutes with the planes circling overhead. A carrier that had received 50 percent Total Damage or had its flight deck destroyed could no longer launch or land airplanes. (Pages j-4 to j-6.)

Special rules also existed also for seaplanes. They needed at least five knots of apparent wind down a catapult to be launched. Each catapult could launch a plane every five minutes; a submarine could launch its plane by floating it off ten minutes after surfacing or recover it in the same amount of time. After stopping, a ship could hoist one airplane out of the water on each side every five minutes if the Wind Force was not above 2. If above Force 2, a ship could hoist one airplane on the lee side every five minutes. After stopping, or coming to a course within thirty degrees of the wind and slowing to ten knots, a ship could hoist one plane on each side every ten minutes if the wind was, again, not above Force 2; above that, a ship under the same conditions could hoist in one airplane on the lee side every ten minutes. Other time allowances applied to assembling and disassembling the aircraft of all types. Squadrons were only to have one assembly crew for every six aircraft in their allowance. When a ship's allowance was less than six, it was considered to have one complete assembly crew attached to it, provided that the ship normally carried airplanes at all. One assembly crew was to be able to assemble or disassemble one fighter, observation scout, or scout observation plane in eight hours, one torpedo bomber or scout bomber in twenty-four hours, or one naval patrol plane in forty-eight hours. The assembly of spare airplanes could not be carried out by squadron personnel while they were checking squadron airplanes. (Pages j-7 to j-9.)

Aircraft, having completed one flight, could not make another flight until sufficient time had been allowed for refueling, rearming, or checking, time that might be varied by the Director to simulate exceptional circumstances. Regular reservicing times indicated in figure 65 included the times necessary for takeoff and landing, except in the case of aircraft carriers. Refueling seaplanes from tenders or at temporary ramps required fifteen minutes for each plane, regardless of the number of planes. Battleships, cruisers, and submarines required thirty minutes to refuel and rearm one or more small seaplanes in their allowance after these had been hoisted onboard. At night, the above times increased by 50 percent; in the open sea and on smooth water, ships with gasoline could refuel one patrol plane in one hour. Under any other conditions in open sea, they could not refuel the planes.

Rearming with bombs, machine-gun ammunition, or torpedoes was normally carried out in the same time as refueling but could take longer. For seaplane tenders

AIRPLANE CLASS	LANDPLANES AT FLYING FIELD	SEAPLANES AT PERMANENT RAMP	SEAPLANES AT EACH TENDER (AV)
1 to 20 VOS,VF,VSO	20 minutes	1 hour	1 hour
Each additional 20	Add 25 minutes	Add 5 minutes for each plane.	Add 1 hour
1 to 12 VSB,VTB,VP or VB (Army)	30 minutes	2 hours, 30 minutes.	1 hour
Each additional 12	Add 25 minutes.		

Fig. 65
Times required for refueling

and shore-based squadrons, it was assumed that it took the same time to load either one large bomb or one torpedo, regardless of size, and that each squadron had on hand half as many bomb trucks or floats as its allowance of airplanes. In short, a crew at an airfield could rearm a land plane with a bomb or torpedo in five minutes, while it would take a crew fifteen minutes for a seaplane at a permanent ramp. At a temporary ramp or a tender, a crew would need thirty minutes to load a bomb or torpedo. For carrier planes, landing time was thirty seconds per plane; refueling and rearming with machine-gun ammunition or bombs would take twelve minutes per squadron, regardless of the number of planes. To reload with torpedoes would take five minutes per plane, and takeoff time was fifteen seconds per plane. During Rough Seas or under Night Conditions, these times doubled. Airplanes also had to be laid up for maintenance checks after operating in the air for thirty hours. It would take a squadron with all of its personnel available six hours at a regular base to check six airplanes and ten hours to check the entire squadron of eighteen planes. Airplanes operating from other than regular bases would need twice this time. (Pages j-9 to j-10a.)

Pilot allowances were such that each battleship or cruiser had 150 percent of the numbers of planes in its operating allowance. Each carrier squadron was to have 125 percent more, including fighter, observation scout, scout bomber, and scout observation planes. Each Army dive-bomber, Navy torpedo bomber, and naval patrol plane squadron was to have 225 percent of the number of planes. Planes allowed to seaplane tenders or other auxiliaries were for "administrative use" only. Multiple airplanes making flights longer than eight hours had to carry two pilots; pilot fatigue rules for fighters, observation scout planes, scout observation planes, scout bombers, and torpedo bombers limited flights to twelve hours in a day. Flights between nine and twelve hours in one day had to be followed by a twelve-hour rest period; no pilots flying these types of aircraft were permitted to fly more than fifty hours in one week. Army dive-bomber and Navy patrol plane pilots could fly longer than twenty hours in any one day, but those flights had to be succeeded by a rest period of twenty-four hours. Flights between nine and twenty hours in one day had to be followed by a rest period of twelve hours, and no Army dive-bomber pilot or Navy patrol plane pilot was to fly more than seventy hours in one week. Pilots in squadrons or of airplanes that were not in the air during one or more days, however, could relieve pilots in other squadrons or planes of the same general type for that period. When pilots flew under unusually hazardous conditions or had been repeatedly exposed to heavy antiaircraft fire or aerial combat, the Director could rule some of them temporarily out of action because of excessive fatigue or minor wounds. (Page j-11.)

Aircraft operational losses could occur from engine failure, landing and takeoff crashes, collisions, damage suffered while being hoisted in and out of ship, crashes from poor weather or high winds, and forced landings due to fuel exhaustion (normally caused by faulty navigation or failure to rendezvous with a parent ship). During the Maneuver, a percentage of loss was to be applied to all flights, and whenever the percentage of loss for a squadron added up to 100 percent, one airplane of that squadron was removed permanently from the Maneuver. Penalties involving temporary delays due to minor damage could be assigned by the Director. As shown in the table in figure 66, conditions at takeoff, landing, and en route were to govern application of the loss penalty. In a sheltered harbor, Smooth Seas were always considered to exist, regardless of the Wind Force. Airplanes could be damaged (but pilots not lost) by the blast of shipboard guns. The Director would assign this damage by taking into consideration the caliber of the guns and their positions with respect to the airplanes. Finally, a fuel reserve of 20 percent was allotted for formation keeping, making rendezvous, climbing, and possible adverse weather or navigational conditions. Planes whose flights were expected to use more than 80 percent of their fuel were to be assessed a crash probability of 20 percent. If they were to consume more than 90 percent of their fuel, their crash probability would be 50 percent. (Pages j-12 to j-13.)

Speed and navigation were governed by the air speeds given in *Airplane Characteristics*. Ground speed was obtained by adding or subtracting the full force of the wind to or from the Air Speed on any course that had a wind component with or against the airplane's course. Ground speed was to be taken into account to the nearest five knots, and the time of flight was to be taken to the nearest five minutes. A resultant of 2.5 or more was to be counted as five minutes or five knots, as the case might be. Airplanes were considered to make good the courses they steered, with lateral drift being disregarded; the Director could judge the navigation of airplanes to be erroneous if need be for game purposes. On partly cloudy days, airplanes flying above the lower level of clouds were to have a 50 percent chance of seeing or being seen, in accordance with the visibility tables in the Maneuver Rules. If planes attempted to bomb from above the lower cloud level, they would be able to drop their bombs only within a half-hour after the time the student had set for the bombing and in whichever ten-minute interval the Director determined by a throw of dice. (Pages j-13 to j-14.)

(b) PERCENTAGE OF LOSS DURING OPERATION

TYPE OF AIRPLANE	CLASS OF AIRPLANE	SMOOTH SEA	MODERATE SEA	ROUGH SEA
Seaplane	VF, VOS, VSO	0.5%	1% to 50%	20% to 100%
do	VTB, VSB	0.5%	10% to 50%	100%
do	VP	0.2%	1%	25%
do	Any class, wind above force 6 -		50%	100%
Add for catapulting, any class		0.0%	0.0%	1%
Carrier Plane, any class		0.5%	0.7%	1%

Landplanes operating from shore have a crash percentage of 0.1% per flight.

FOG take-offs of ship based planes have a crash percentage of 50%. Fog landing of seaplanes have a crash percentage of 50% in smooth seas; all other fog landings, 100%.
NIGHT SEAPLANE LANDINGS IN STRANGE HARBORS have a crash percentage of 5%. Night seaplane landings in the open sea have a crash percentage triple that for day landings.

Fig. 66
Percentage of loss during operations

Aircraft could be seen in the air from other aircraft at the horizontal distances given in figure 67—with these distances in miles—provided that the difference of altitude between the observer and the plane observed was not over two thousand feet. For greater distances of altitude, the figures in Columns 1 and 2 were decreased by half a mile for each additional thousand feet. The figures given were for Normal Visibility; for High Visibility one mile was to be added to the values, and for Low Visibility, one-third subtracted. Airplanes could be seen by an observer on a surface ship or on land one mile farther than the distances in Column 1, and airplanes could be seen by submarine periscopes at the same distance as by observers on ships, but the chance of seeing in each all-around observation was one in twenty at maximum range, one in sixteen at three-quarters of maximum range, one in ten at half the maximum range, and one in six at one-quarter maximum range. Rigid airships could be seen from aircraft under the various conditions of visibility from one-third farther away than Large Ships could be seen. Nonrigid airships could be seen a third again the distance from which Intermediate-sized Ships could be seen by aircraft. While the sun was visible, the figures noted above would be reduced "considerably" if the observer was looking in its general direction. Each of these cases, however, would be decided by the Director. (Pages j-14 to j-15.)

Sound detection of planes was computed in various ways. Sound detectors on shore could locate approaching aircraft according to the table in figure 68, taking into account ground speed. Distances of detection by radar were governed by the Maneuver Rules. Aerial photographs could be taken by planes at any altitude up to the plane's maximum, provided that previous arrangements had been made and included on the Aircraft Flight Form. Under Normal Visibility Conditions, such photographs were to be considered as a record of the presence of ships of Intermediate or Large size to a maximum distance of twenty miles. These ships could be recognized as to type from a maximum distance of fifteen miles. For Small Ships, the corresponding figure was twelve miles. Photographs from handheld cameras would be considered developed and printed thirty minutes after the plane returned to base, while photographs from mapping cameras required three hours for developing and printing after the plane returned to base. (Pages j-15 to j-17.)

(1) Normal visibility:

Class seen	Number in group	Col. 1 Same Altitude or higher than observer	Col. 2 Lower than observer
VP, VTB, VB (Army)	1	5	3
VP, VTB, VB (Army)	2-6	7	4
VP, VTB, VB (Army)	7 plus	8	5
VF, VSO, VOS,VSB	1	4	2
VF, VSO, VOS,VSB	2-6	6	3
VF, VSO, VOS,VSB	7 plus	7	4

Fig. 67
Distances for sighting aircraft

Under favorable conditions, aircraft could sight submarines if the submarine was submerged to a depth of one hundred feet or less. In addition, aircraft could sight mines submerged to a depth of ten feet. These conditions would be affected by the depth and clarity of the water, the nature of the bottom, the condition of the surface, light effects, altitude, and the relative position of the observer. The Director

No. planes in flight	Located accurately yards	Located approximately yards.
1	11,000	20,000
3	12,000	22,000
7	13,000	25,000
13	14,000	27,000
18	15,000	30,000

3611-1925
19/Jun/44el - j-16 - Section J

Ground winds decrease the ranges as follows:

Wind Beaufort	Decrease range percent of above.
1	0
2	10
3	20
4	30
5	40, etc.

Fig. 68
Sound detection distances of airplanes, with ground speed

would decide each case after considering the attendant conditions. Related to this, five-hundred-pound and thousand-pound bombs were to have delayed-action fuses and so would be effective against submarines to a depth of forty feet. Depth charges could also be dropped from planes listed as having that capability in the Statement of the Problem. When used against a submerged submarine, a depth charge had the same percentage of hitting as would a bomb on a small target. One full hit by a depth charge would be enough to sink a submarine. Aircraft would also be able to see the wakes of torpedoes near their tracks and if not otherwise engaged; the Director would rule whether or not an aircraft could see torpedo tracks. Electrically propelled torpedoes, however, could not be seen from aircraft at all. Much of this assessment would depend on the roll of a dice, which would determine which torpedoes would become apparent during a three-minute "torpedo Move" or the time that torpedoes were in the water. (Page j-17.)

Aircraft could lay smoke screens by means of smoke boxes dropped on the ground or into the water or by discharging a chemical carried in tanks in lieu of bombs. Smoke screens were effective for eighteen minutes and drifted with the wind. Smoke boxes could be carried in lieu of hundred-pound bombs and in the same numbers. They could be dropped from any altitude, but to be effective a screen had to be laid using at least six boxes not over two hundred feet apart. Heavy smoke curtains were laid by naval patrol bombers, scout bombers, or torpedo bombers, light ones by scout observation, observation scout, or utility planes. Screens laid by airplane smoke tanks, in accordance with those in *Airplane Characteristics,* were laid 50–150 feet above the surface. It was assumed that the lower edge would drop to the surface and the upper edge would rise to a hundred feet. One large tank would be able to lay an effective curtain about two thousand yards in length, and a small tank an effective curtain of about six hundred yards. It was estimated that 25 percent of aircraft smoke screens would be improperly placed owing to hostile opposition or incorrect estimates of position, as much as five thousand yards out of position or too high to be effective. The amount and direction of smoke-screen displacement would be decided by the Director. (Page j-18.)

When aerial combat occurred, the Director would assign losses after consideration of conditions, making allowance for altitudes, formations, and positions. Each aerial engagement expended a quarter of available machine-gun ammunition. Losses from antiaircraft fire were given according to "Conventions for Scoring

Anti-Aircraft Gunfire." The Director could modify the application of these Conventions when unusual conditions existed. Airplanes of a bombing or torpedo unit suffering antiaircraft fire losses in a move were to be considered as having done so prior to dropping their ordnance, and the number of effective bombs or torpedoes was reduced accordingly. Airplane losses upon retirement would also be decided by the Director. As to the Conventions themselves, antiaircraft guns with greater than two-inch bores were considered effective when target airplanes were within fourteen thousand yards, but against dive-bombers and fighters only when they were strafing. Antiaircraft machine gun fire of 40 mm or 1.1-inch caliber was effective within a range of three thousand yards, 20 mm and below within 1,500 yards. (Pages j-18 to j-19.)

"Attackers" refer to those planes making the combat attack, and "Defenders" refers to those receiving the combat attack.

ATTACKERS	DEFENDERS	ATTACKERS' LOSS	DEFENDERS' LOSS
n VF	m VF	m x .1	n x .15
n VF	m VOS(s)	m x .2	n x .3
n VF	m VSB, VTB	m x .2	n x .2
n VF	m VP	m x .3	n x .3

Fig. 69
Aerial combat losses

In computing the effectiveness of antiaircraft fire, a battery of four antiaircraft guns of over 4.5 inches was taken as a unit of antiaircraft fire. The effectiveness of batteries of guns of other calibers was computed according to the table in figure 70. The effectiveness of the antiaircraft fire of each unit could be reduced by certain unfavorable conditions, expressed in cumulative fractional reductions. The percentage of planes of any attack group that were shot down was based on the same table, which indicated the estimated loss percentages sustained by bombers attacking from an altitude of three thousand feet but being fired at only by antiaircraft guns. For other conditions of attack, the basic percentages were to be lessened by multiplying them by the reduction factors in the table. Thus, to figure losses from antiaircraft fire and find the number of airplanes that remained effective for discharging bombs or torpedoes, the student entered the number of airplanes in each attack group, determined the number of units firing against each attack group, and then obtained the effectiveness of antiaircraft fire by the reductions shown. After this, the student had to figure the basic percentage of loss for each attack group and carry out multiplications to obtain final airplane losses in each group. (Pages j-20 to j-22.)

The effects of aerial bombing first considered the bombs supplied to aircraft, which were 1,600, 1,000, 500, or 100 pounds in size. Bombing methods included Sight or Horizontal Bombing, Dive Bombing, and Skip Bombing. Planes dropping bombs from altitudes less than those indicated in figure 70 would be considered destroyed by blast effect and the fragments of their own bombs.

Specifically, a plane dropping 100-, 325-, or 500-pound bombs with impact fuses of any type had to drop from a minimum altitude of six hundred feet, while 650-, 1,000-, and 1,600-pound bombs with fuses set for delays of a tenth of a second or less had to be dropped from at least eight hundred feet. A two-thousand-pound bomb with the same type of fuse as a thousand-pound bomb had to be

TABLE "B". AA FIRE EFFECTIVENESS OF SHIPS

IN TERMS OF UNITS OF FIRE.

Multiply the total number of firing guns by the following

Constants:-

Kind of AA fire.	Caliber of Guns Firing		
Heavy AA Guns.	2".01 to 3".9	3".91 to 4".5	over 4".5
	.15	.20	.25
AA Machine Guns (20 mm = .8") (40 mm = 1".6)	Under 0".50	0".50 to 1".0	1".01 to 2".0
	.05	.10	.15

NOTE: Not more than 8 units of heavy AA guns can fire effectively against one airplane group.

TABLE "C". LOSS OF EFFECTIVENESS OF GUNFIRE OF AA UNITS

Condition	Fractional Reduction
Firing ship under effective gunfire	0.20
Firing ship firing main or secondary battery	0.20
Firing ship strafed by at least 6 small planes just before attack	0.50
AA battery permanently damaged by previous strafing	0.30
Firing ship has 50% AW damage	1.0
Firing ship in rough seas	0.20

NOTE: If more than one unfavorable condition exists, the losses of effectiveness are cumulative.

TABLE "D". BASIC PERCENTAGES OF AIRPLANE LOSSES BY AA FIRE

NOTE: After computation of airplane losses due to aerial combat and AA fire, the Director will rule whether an airplane group will be able to complete its attack.

Number of Planes	Number of Firing Units in Terms of BBs.							
	1	2	3	4	5	6	7	8
5	60	80	100	100	100	100	100	100
10	50	70	90	100	100	100	100	100
15	40	60	70	80	80	80	90	90
20	30	50	60	60	70	70	80	80
30	20	40	50	50	60	60	70	70
40	17	30	40	40	50	50	60	60
60	12	20	25	30	40	40	50	50
80	10	15	20	25	30	35	40	45
100	8	12	17	22	25	30	35	40

TABLE "E". REDUCTION FACTORS FOR DIFFERENT CONDITIONS OF ATTACK

Condition	Multiplier for Table "D"
Sight bombing, 3000' and up	1.0
Sight bombing, 7000' and up	0.6
Sight bombing, 11000' and up	0.2
Dive bombing approach (Use only AA gun units)	0.1*
Dive bombing dive (Use only AA machine gun units)	0.5*
Strafing (Use only AA machine gun units)	0.4
Torpedo or skip bombing attack, no smoke	0.6
Torpedo or skip bombing attack, with smoke	0.3
Night bombing, planes illuminated by searchlight, or day bombing from behind clouds, reduce losses by 50%.	
Night bombing, planes not illuminated, AA fire is ineffective, unless firing ship is radar-equipped, in which case fire effect is same as for day AA fire.	

NOTE: The Director may increase or decrease the effect of the AA fire by as much as 30 percent to take care of special or unusual conditions.

* Compute these items separately, and then Add them.

Fig. 70
Tables for AA gunfire effectiveness

dropped from a thousand feet. Fighters, scout bombers, scout observation planes, and observation scout planes could also attack ships by dropping small bombs and strafing with machine guns, but torpedo bombers, scout observation planes, observation scout planes, and naval patrol planes were not capable of dive-bombing. Under battle conditions, with adequate visibility, planes meeting no resistance and catching a target on a steady course would be awarded a number of direct hits for a given number of bombs based on analyses of data available to the umpires for bombs dropped on different-sized targets. Factors that could reduce the number of bomb hits obtained would be attacks on the strike group by other planes, antiaircraft fire, and maneuvering by the target during the approach. Attacks by hostile planes during the approach, it was assumed, would cease when the bombers came under effective antiaircraft fire. If a bombing unit came under effective antiaircraft fire before releasing its bombs, the number of planes shot down would be determined by the rules outlined above; only the remaining planes would be dropping bombs. (Pages j-23 to j-24.)

The effectiveness of the antiaircraft batteries of a ship subjected to strafing by three or more fighter planes just prior to a bombing or torpedo attack would be reduced by 50 percent during that attack. Only carrier planes and some land-based planes—including Army types—were considered "strafing planes" in Naval War College Maneuvers. In each strafing attack, airplanes were deemed to have dropped all of their bombs and expended a fourth of their gun ammunition. Owing to the variety of conditions under which the attacks could occur, no rules were laid down to cover permanent damage; the Director decided the amount of any increase or decrease in the effectiveness of the attack. Reduced visibility would not impact bombing effectiveness as long as the bomber could actually see the target. In the case of night bombing, the Director was to decide the amount of damage. Normally, planes bombing at night in good visibility conditions or on a target illuminated by flares or searchlights would have an accuracy rate 50 percent of that of a daylight strike. (Pages j-25 to j-26.)

The percentage of hits was also determined by the standard deck areas of different-sized targets. In addition to Above Water direct bomb hits, hits alongside the ship—within forty feet—would be judged to have been made. For each Above Water bomb hit, damage in fourteen-inch penetrative hits would be allowed both above and below the waterline, the latter depending on the average value of one Under Water bomb explosion within forty feet and the ratio of probability of an Above Water hit for targets of each size and class. Fractional hits would also be computed and scored this way. More exactly, the student would obtain the number of aircraft lost in combat prior to the bombing, determine losses from antiaircraft fire, figure the percentage of direct bomb hits, multiply this percentage by the "effectiveness factor" against maneuvering targets, multiply this figure by the number of bombs remaining, and multiply the number of direct hits by the values in the table in figure 71 to get Above Water, Under Water, and Total Damage, expressed in fourteen-inch penetrative hits. (Pages j-25 to j-26.)

Bomb damage to an aircraft carrier's flight deck and airplanes was figured by assuming that a 1,600-, 1,000-, or 500-pound bomb hit would temporarily destroy half the flight deck and one of the carrier's elevators. Moreover, such a hit over the hangar would destroy half the airplanes and the spare planes in the hangar. The time required to repair such damage can be seen in figure 72. When either half of the flight deck had been destroyed and before temporary repairs had been completed, planes would be able to fly on or off after half an hour had elapsed in "double normal" interval, but only half of the carrier's allowance of planes could now be parked on deck. After temporary repairs had been completed, planes could fly off and on in normal time, and parking restrictions would be removed. Also, each large Above Water bomb hit would destroy 20 percent of the assembled aircraft on board at the time of the bombing, and each hundred-pound bomb hit would destroy 5 percent of the planes on deck. (Pages j-26 to j-27.)

TABLE "F". PERCENTAGE OF DIRECT BOMB HITS, ALL CONDITIONS OF BOMBING

Target size	Dive Bombing	Sight Bombing 11000'	Sight Bombing 7000'	Sight Bombing 3000'
Large	16	8	15	30
Intermediate	9	5	9	20
Small	6	3	5	14
DD and SS	3	1	3	9

TABLE "G". EFFECTIVENESS FACTOR AGAINST TARGET IF IT MANEUVERS

Speed of Target	Single Ship				Ship in Formation			
	Dive Bombing	Sight Bombing 11000'	7000'	3000'	Dive Bombing	Sight Bombing 11000'	7000'	3000'
0 to 12 kts.	1.0	1.0	1.0	1.0	1.0	1.0	1.0	1.0
12 to 25 kts.	0.7	0.5	0.6	0.7	0.8	0.6	0.7	0.8
Over 25 kts.	0.5	0.3	0.4	0.5	0.6	0.4	0.5	0.6

TABLE "H". DAMAGE CAUSED BY ONE DIRECT HIT IN 14" PENETRATIVE HITS

	BB	CV	CA	CL	DD-SS	AUX. TYPE
100-POUND BOMB						
Above Water	0.15	0.15	0.15	0.25	0.25	0.25
Under Water	0.00	0.00	0.00	0.00	0.00	0.00
Total	0.15	0.15	0.15	0.25	0.25	0.25
500-POUND BOMB						
Above Water	1.00	1.25	1.25	1.75	1.75	1.75
Under Water	0.25	0.25	0.75	1.00	1.50	0.75
Total	1.25	1.50	2.00	2.75	3.25	2.50
1000-POUND BOMB						
Above Water	2.00	2.50	2.50	3.50	3.50	3.50
Under Water	0.75	0.75	1.50	2.50	3.50	1.50
Total	2.75	3.25	4.00	6.00	7.00	5.00
1600-POUND BOMB						
Above Water	3.00	3.75	3.75	5.25	5.25	5.25
Under Water	1.25	1.25	2.25	4.00	5.50	2.25
Total	4.25	5.00	6.00	9.25	10.75	7.50

Fig. 71
Tables for aerial bomb damage

Formations of torpedo bombers that were attacked in the air or came under effective antiaircraft fire would lose a proportion of their planes prior to dropping their torpedoes; torpedoes of lost planes were not counted as effective. Aerial torpedoes would be counted as fired from a single point with a spread and on courses designated by the player, who was also to indicate the relation of this point to the target; the Director would judge the position of this point to be in error by a

TIME REQUIRED TO REPAIR FLIGHT DECK

NOTE: These times are applicable only to CVs, not to CVLs or CVEs. If statement of problem specified that special repair equipment is not available, the time of repair by ship is doubled.

Damage	Temporary Repairs by Ship	Permanent Repairs by Base
Less than 2 500# hits	4 hours	10 days
Less than 2 1000# hits	6 hours	10 days
2 to 4 large bomb hits	36 hours	15 days
More than 4 large bomb hits	Cannot be repaired on board	30 days
5 or less small bomb hits	½ hour delay	None
6 to 10 small bomb hits	4 hours	2 days
11 to 20 small bomb hits	8 hours	5 days
More than 20 small bomb hits	Cannot be repaired on board	7 days

Fig. 72
Time required to repair flight deck bomb damage

maximum of 1,500 yards. Of the effective torpedoes—that is, those arriving at the firing point—60 percent would be assumed to function properly throughout their run, unless curved fire was used, in which case only 40 percent would function. Hits would be determined by plotting the movement of the torpedoes and their targets. (Page j-28.)

Airships had to operate from hangars and mooring masts at land bases or from mooring masts on ships that were specially fitted with them—if the latter did not experience "undue" motion. Airships could make and maintain physical contact with aircraft carriers to refuel, rearm, or transfer personnel, but also only in waters that did not cause much motion to the ship. Carriers, moreover, at such times could launch but not land planes. The airship could leave or enter a hangar only when the wind was Force 5 or less, provided the direction of the wind was within forty-five degrees of the longitudinal axis of the hangar. It could enter or leave the hangar when the wind was less than Force 4 and only if the direction was forty-five degrees of the hangar's longitudinal axis. An airship could moor to a mooring mast with wind from any direction as long as the wind did not exceed Force 8; it could land on the ground in winds up to Force 6, provided personnel and facilities were available. The airship could stay in the air in winds of any force, but dangerous sectors of circular storms had to be avoided. Rigid airships could carry, launch, and recover up to five suitably fitted airplanes. The airship's speed had to be at least fifty knots to recover the planes, and it had to be at an altitude above a thousand feet to launch them. (Pages j-28 to j-29.)

At the completion of a flight, an airship would not be considered available for a subsequent flight until an interval had elapsed sufficient for servicing it—a minimum of two hours and a maximum of six hours, depending on the duration of the previous flight. The Director would judge the time required for flights under special conditions. Airships filled with helium had ranges of ten, nine, and four thousand yards at speeds of forty, fifty, and seventy knots, respectively, if rigid, and eight and five hundred yards at these speeds if nonrigid. A helium-filled rigid airship could also remain in the air and operate for a limited time with 30 percent of its gas capacity lost to gunfire. A loss of 40 percent, however, would bring it down as a "wreck." (Page j-29.)

Chemical Warfare

The effect that could be obtained from chemicals depended on weather conditions, with strong winds and wet weather decreasing effectiveness. Cold weather or cloudy days would retard the evaporation of liquids and thereby increase their persistence, while hot, sunny weather would increase the rate of dissipation. Chemical effectiveness would also depend on the type of ship against which the chemicals were used. New combatant ships could be expected to be "fairly well protected," but it was thought that the cost of necessary alterations in older ships and auxiliaries was "almost prohibitive." The impact of chemical weapons would also depend on whether or not they were carried into the ship or expended against the outside "shell." Nonpersistent chemicals would dissipate almost immediately from weather decks, but if gases were carried into the interior of the ship by shell or bomb, an "extremely powerful" concentration could be built up, and was thought to be able to penetrate gas masks. Persistent gases, like mustard gas, however, could be cleared from a ship, provided decontamination agents were available. The condition of the ship at the time of the attack was another factor. If surprise was achieved and the ship was caught with its ventilation systems running and hatches open, the effect would obviously be much greater than at General Quarters with all openings shut. The degree of "gas discipline," training, and experience in the crews mattered, as did the quantity and quality of available protective closing and decontamination agents. Given these factors, the efficiency of gas munitions would vary, just as gunfire effects varied with range, visibility, and the nature of the target. Since it was impossible to evaluate all of these factors and the attempt would make the Maneuver Rules too complicated, certain "arbitrary" rules were given by the Director, which he could modify as he thought suited the situation. (Pages k-1 to k-2.)

Chemical munitions could be used in Naval War College Maneuvers, unless they were specifically prohibited by the Staff. However, the Director had to be notified prior to the beginning of the Maneuver and before the Student Commander-in-Chief authorized the use of chemical munitions among his forces. Each Student Commander at the start of the Maneuver was to specify, using *Fleets* as a guide, what percentage of shells and aircraft bombs would be allowed to be loaded with chemical agents or displaced by spray tanks. The kind of chemical agent also had to be specified; these might include mustard gas, lewisite, phosgene, tear gas, tear gas solution, white phosphorus, and chloropicrin. Tear gas solution, mustard gas, lewisite, phosgene, and white phosphorus could be delivered by aircraft bombs in all sizes but were most effective in small bombs. These bombs were to have no effect on materiel, just on personnel. White phosphorus, moreover, would create a smoke screen around a ship such that nothing would be visible to leeward for 1.5 minutes; a thirty-pound bomb would contaminate an area of 1,250 square yards.

Mustard gas, lewisite, tear gas solution, and chloropicrin could be sprayed from aircraft in either of two methods. First was a Small Drop Method for use by small planes at low altitudes against marching troops or landing craft, though it could be used against ships as well. The Large Drop Method was suitable for use by large planes and was more accurate. Tanks used for chemical spray were of two sizes as well, fifteen gallons for the Small Drop Method and fifty gallons for the Large

Drop Method. Aircraft could use these tanks as replacements for their allowance of smoke tanks as indicated in *Airplane Characteristics.* Tanks used for the Small Drop Method discharged their contents in ten seconds and Large Drop tanks in thirty seconds; in both cases, the length of the spray was determined by the speed of the airplane. (Pages k-2 to k-3.)

When chemicals were employed to contaminate an area—such as a landing beach or camp—the effective width of the spray was as shown in figure 73. The contaminated area would be more heavily contaminated, and the agent was known to persist longer, along the upwind edge of the shower, both effects dropping to zero along the downwind edge, though vapor made the area to leeward of such sprays dangerous for about a thousand yards. For Naval War College Maneuver purposes, these two factors were considered to cancel each other out; the area designated in the table would be considered contaminated for various periods of time figured by formulas outline below. It was not considered practical to decontaminate areas that had been contaminated by airplane spray, and penalties were applied to them. In addition, all sprays—no matter the type of tank used or the altitude at which released—would be considered for Naval War College purposes effective and equally potent for the length of time the chemical would persist. (Pages k-3 to k-4.)

Altitude	Wind across line of flight			
	0 kts.	10	20	Over 20
Atomization Method				
150	10 yds.	30 yds.	90 yds.	Ineffective
500	20 "	150 "	300 "	"
1000	100 "	350 "	700 "	"
Over 1000	Ineffective			
Pressure Method				
150	10 yds.	30 yds.	50 yds.	Ineffective
500	20 "	50 "	80 "	"
1000	30 "	60 "	100 "	"
2000	50 "	100 "	180 "	"
Over 2000	Ineffective			

Fig. 73
Effective width of chemical spray

Mustard gas, lewisite, phosgene, and tear gas could also be dropped by "tin can" bombs, ordinary one-gallon cans, such as might have originally contained lubricating oil, thrown from airplanes or set like land mines. They could also be secured above water to stakes driven a short distance from shore; in this case, the mustard was caused to float by adding fuel oil. The cans were exploded, contaminating 250 square yards. Mustard gas, tear gas solution, lewisite, white phosphorus, and chloropicrin could also be delivered by artillery, except for star shells. These munitions were contained in special thin-wall shells for use against shore objectives only. It was to be foreseen, however, that the enemy might do the unexpected and fire them from destroyers against armored warships. Accordingly, the gas shells could replace ammunition allowances, weight for weight, and the effect would be determined by the Director. A 75 mm shell was judged to contaminate an area of 150 square yards, while a 155 mm shell did the same to an area of 1,900 square yards. (Pages k-4 to k-5.)

Student Commanders were to specify on Aircraft Flight Forms when chemical bombs or spray tanks were to be carried by planes, as well as the kind of agent to be loaded. They were also to specify on their Move Blanks when tear-gas shells or thin-wall shells were being fired. To be effective, shells, bombs, and spray had actually to hit the target; therefore, penalties were possible. Shells partially loaded with tear gas would hit in accordance with the Gunfire Rules. When these hits accumulated such that the damage reached a percentage equal to a sixth the life of the ship, penalties for wearing masks would be applied. Shells and bombs loaded only with chemical agents would also hit in accordance with the Gunfire and Aircraft Bombing Rules but would have no damaging effect on materiel. Chemical bombs were ruled to explode on the upper deck without penetrating. One hit by a bomb of any size containing tear gas solution, mustard gas, or lewisite would force all topside personnel leeward of the contaminated area to don masks for certain periods of time; the Director would decide by throwing dice the location of the hit. One white phosphorus shell or bomb of any size would create a "smoke ribbon," the size of which would be calculated by combining wind velocity with an emission of 1.5 minutes. The ribbon would drift to leeward and would be opaque for two moves; it was plotted in the same way as a smoke screen. Ships and shore stations hit by chemical sprays were penalized in various ways. (Pages k-5 to k-6.)

Chemicals sprayed by aircraft were considered ineffective if dropped from over one thousand feet on a maneuvering target or from two thousand feet on a shore target, and in any case if the apparent wind over the target was more than twenty knots. Subject to these limitations, one chemical spray—which was counted in the Maneuvers as one or two tanks' worth—from each group of planes remaining after losses due to antiaircraft fire would hit if certain minimum percentages of chemical spray hits could be made by the planes after they reached their spraying positions. If the agents hit, the effect on the target during the contamination period would be determined in several ways.

First, it was assumed (in an acknowledged Naval War College artificiality) that all of the chemical agents would give personnel at the target sufficient warning to don masks and protective clothing. The College did, however, impose certain penalties on a fleet unit's fighting efficiency after chemical attack. Gunfire—whether main, secondary, or antiaircraft—would be reduced by a second correction of two-tenths during the first hour after the target was hit by gas, then four-tenths until the contamination was removed. The maximum speed remaining to a ship was reduced by three-tenths on a coal-burning ship during the first hour and by six-tenths until the contamination was removed; oil-burning ships, however, were not to be reduced in speed. Coding and decoding times for all messages sent and received by ships would be doubled during the first hour, then quadrupled, again until the contamination was

Rule K-16. Probable Casualties on Shore

Situation		% Casualties of men on or immediately down wind from area gassed. No protection available	Gas Mask worn	Gas mask and Protective clothes worn.
Column marching on road or in an area when actually sprayed with persistent vesicants from planes		80	50	20
Column marching on hard surface road previously neutralized with persistent vesicants. No dust.	On area 30 min.	50	10	5
	On area 10 min. or less	5	2	0
Troops occupying area previously neutralized by persistent vesicants (vapor effect only)		80	40	10
Troops attacked by Livens projectors with phosgene type agents		80	20	20
Troop marching over fields or through brush neutralized by persistent vesicant chemical agents. Area 400 yards across		50	25	10
Troops in position attacked by phosgene type agents in gas shell	Wooded	80	10	10
	Open	40	10	10
Troops in position attacked by persistent vesicant type agents in gas shell. Area evacuated at once.	Wooded	50	25	20
	Open	25	20	10

Fig. 74
Probable casualties on shore

Number of planes necessary to hit a target, using "Pressure" Method.

Target	Altitude	Apparent wind over target: 0 kts.	10 kts.	20 kts.
Maneuvering ship	150	1*	1	2
	500	2	2	3
	1000	4	6	9
Fixed target, of same size as a large ship	150	1*	1*	1
	500	1	1	2
	1000	2	3	5
	2000	4	6	9

Note: Using "Atomization" Method, the number of planes required is double that shown above except for the numbers marked (*), and except that the atomization method is ineffective from an altitude of above 1000 feet.

Fig. 75
Number of planes necessary to hit a target

removed. Until contamination was removed, times for plane handling on all ships affected would be doubled, as would all activities performed on land in areas that had been contaminated. (Pages k-7 to k-8.)

Ships were considered decontaminated when masks and protective clothing were no longer necessary. It was assumed that if an effective decontamination agent was available, it would take one hour to decontaminate from mustard gas, longer if no agent was handy. Lewisite would take a half-hour to decontaminate and phosgene one hour. If tear gas solution was used as a partial filler in shells seven inches or larger—the hits being equivalent to a sixth of the life of the ship—decontamination would take fifteen hours, but only one hour if the solution was in the form of a spray. On shore, if a target was sprayed by a plane with any of the above-mentioned agents, removal by a contamination agent was not considered "practicable." Tear gas solution and phosgene would evaporate in two hours, but mustard gas and lewisite would persist longer, possibly for days. Lewisite, however, would evaporate with rainfall; it was presumed that one man shoveling chloride of lime for an hour could neutralize eighteen square yards contaminated with tear gas or phosgene. A tank truck spraying chloride of lime could decontaminate nine thousand square yards per hour in terrain suitable for this type of operation. Lewisite in shells, bombs, or tin cans could be neutralized in the same manner, with water as the decontaminating agent; the agent would be removed ten minutes after rainfall. Personnel could not remain masked against these chemical agents longer than twelve hours, and at least four hours had to elapse before masks could again be worn. (Pages k-8 to k-9.)

Temperature	Ship 100°	Ship 80°	Ship 60°	Shore 100°	Shore 80°	Shore 60°
*Plane spray (Atomization)	2 hr.	4 hr.	7 hr.	4 hr.	6 hr.	10 hr.
*Plane spray (Pressure)	10 "	13 "	25 "	15 "	20 "	36 "
Tin can bomb or land mine	3 days	4 d.	7 d.	5 d.	7 d.	12 d.
30 lb. bomb	2 "	3 "	4 d.	4 "	5 "	7 "
100 lb. bomb	4 "	6 "	8 "	8 "	12 "	15 "
6 inch shell	1 "	1 "	2 "	2 "	4 "	5 "

* It is not practicable to decontaminate land areas hit by a plane spray.

Fig. 76
Mustard gas concentration periods without decontamination

When gunfire between two forces had subjected most ships on both sides to chemical agents, the Director could order that individual records of chemical penalties be discontinued and that all ships on both sides be considered subjected to penalties as long as they were "heavily" engaged. He could also assess that chemical agents on both sides canceled each other out and direct that chemical penalties be discontinued while both sides were heavily engaged. Otherwise, and unless decontamination was carried out, mustard gas was counted as persisting in dangerous concentrations for periods indicated in figure 76. Blue battleships, fleet carriers, and new light cruisers were considered to have effective decontaminating agents for mustard gas at the outbreak of a war; whether or not that was true for other Blue ships would be decided by the Director. After six months of conflict, however, all Blue ships would be considered as having the agents. Whether or not ships of other navies would be similarly equipped was to be decided by the Director. (Pages k-9 to k-10.)

Smoke Screens

Blue destroyers more modern than the prewar, modernized DD 347 class; destroyer leaders (DLs); and *Brooklyn*-class light cruisers were considered to carry a battery of cylinders filled with sulfur trioxide. Each cylinder contained sufficient chemicals to smoke for a total of twenty-one minutes. In order to produce an opaque screen, it would be necessary for these ships to use at least one cylinder if the apparent wind was from zero to ten knots, three cylinders if from ten to twenty knots, and five if twenty to thirty knots. The number of cylinders went up proportionately with ten-knot increases in apparent wind velocity. These cylinders could be distributed among any number of ships in a formation up to a division. A "Standard Screen" was opaque for thirty minutes; a "Double Standard Screen" used twice the number of cylinders and was opaque for forty-five minutes. These screens were to be plotted in the same manner as stack-exhaust smoke screens, except that they could be "turned on and off" at any time and were considered immediately effective regardless of the weather conditions. In other words, the quality of a screen depended on the number of cylinders used and not on its length. (Pages k-10 to k-11.)

The Double Standard Screen could be laid as a protection against airplanes, in which case it was termed a "Blanket Screen." This type of screen would provide some degree of protection against bombing and gunfire by Plane Spot. To be effective, it had to be laid down by at least four ships which could not be more than 150 yards apart for fifteen minutes; the vessel being screened had to maneuver so that it took advantage of the screen; and the course of the formation had to be crosswind or downwind. Under these conditions, a ship could remain in the screen for up to one hour, during which time it could engage surface targets by indirect fire but could not launch or land airplanes. For a period of five minutes within each fifteen minutes of screen—as determined by a throw of the dice—the ship was subject to aerial observation and bombing but could use its antiaircraft guns. (Pages k-11 to k-12.)

Aircraft carrying sulfur trioxide smoke tanks could lay a blanket screen or cloud that could provide protection against bombing and the fall of observation shot. This screen could also provide protection against ground observation of aerial movements. To be effective, it had to be laid at or below 1,500 feet of altitude. The effective width of the cloud was five hundred yards for each pair of planes laying it, and the cloud was not to become effective sooner than three minutes after the time the planes began to make smoke. The effective length of the cloud produced by each plane was six hundred yards for a small tank or two thousand yards for a large one; the cloud, which would drift with the wind, would be effective for thirty minutes. The Director could allow the cloud to be laid as far as two thousand yards out of position. No fewer than four planes had to make smoke simultaneously to screen a fixed objective; no fewer than nine had to make smoke simultaneously to screen a ship under way. The ship could maneuver to attempt to remain under its screen. The protection afforded a ship that maneuvered to remain under a cloud was the same as under a Blanket Screen. (Page k-12.)

Conclusion

The Maneuver Rules represented an attempt to simulate most situations that might be encountered in the conduct of Naval War College Exercises and Maneuvers. While not all actual conditions could be provided for, the College obviously strove conscientiously to think about as many contingencies as possible in as great detail as possible and integrate the resulting procedures into games that gave the student officers a feel for at least some of the realities of modern naval warfare. The following chapters will illustrate how the Maneuver Rules were put to use in the Exercises and Maneuvers and why the games were carried out in the manner just described.

VI *Operations Problem 4*

Night Search and Attack, September 1945

In September 1945, the Command and Staff Class of December 1945 undertook a Night Search and Attack war game as Operations Problem 4. Conducted between 18 and 22 September, the game was outlined to the students by Commodore Bowdey. Bowdey told the students that Operations Problem 4 was a Chart and Maneuver Board Exercise that was to be played according to Naval War College Maneuver Rules and the Conduct of the Maneuver. Ships were to be set up on the Maneuver Board in initial dispositions and locations; they could be changed by the Officers in Tactical Command and Student Commanders after the Maneuver started. In order to simulate night conditions, Student Commanders would keep charts of movements by their own units as well as those of opposing forces in the rooms assigned to them in Pringle Hall. Supervision of planned action would be effected according to information received from the Master Plot. Students assigned to Blue and Orange forces would not be allowed to visit the Maneuver Room or any room not assigned to them. Contact information would be received by telephone, written messages, or flimsies. These flimsies, as well as the Move and Gunfire Sheets, would be filled out by students and transmitted to the Master Plot via pneumatic tubes.[1]

After the Student Commanders had received the allotted information, they would have only five minutes in which to make their decisions, which then had to be transmitted in appropriate form to the Master Plot. No communication between Student Commanders was to take place other than through "regular" channels. Dispatches, for instance, were to be sent according to the Maneuver Rules, very-high-frequency (VHF) range would be limited to visual range, and Talk Between Ships (TBS) communication would be on forms provided. Radio, radar countermeasure, and detection would be employed as prescribed in tactical manuals and the Maneuver Rules, subject to additional capabilities established by the Director of the Maneuver. The students were provided with a summary of rules for night visibility, a visual representation of the Maneuver Rules, and a list of questions that could assist Student Commanders in coming to decisions. A Critique, he said, would also be held on the Maneuver Board, during which ships would be moved by the Combat Information Center (CIC) Evaluators as the narrative progressed.[2]

Director of the Maneuver (Capt. Samuel Hurt, Instructor, Department of Tactics, Naval War College)

Instructions

Contact Officers were to report to the Maneuver Board as soon as the Maneuver began; other students were to report to their assigned rooms. Forces were set up on the Maneuver Board prior to the first contact. A geographical plot was made of all subsequent movements in accordance with the Move Sheets of the various Student Commanders. Each Contact Officer was also to see that forces assigned to him were moved according to the Student Commanders' directions. "Care must be taken," the instructions declared, "to *disclose only that information the Commanders would receive under the prevailing conditions,* by radio, radar, sound or within range of visibility." Contacts and information were given to Unit Commanders in such a manner as to reproduce the difficulties and uncertainties that prevailed in night formation: "*Too little,* rather than too much information is preferable." Since numerous Unit Commanders would be present in the various rooms, it was necessary to communicate information in low tones of voice so as not to spread information to "unauthorized personnel," which "spoils the realism of the problem."[3]

Authorized information about the various night conditions, including that revealed by star shells and searchlights, was in accordance with the Rules and as discussed in chapter 3. As with star shell target information, that from searchlights would be "vague." When any ship was sighted by an enemy, the Director or an Assistant Director would take charge of the contact and the information to be furnished to the Student Commander, unless there were directions otherwise.[4]

The Contact Officers were then given instructions about using Information Blanks to sketch out the information a Unit Commander was allowed to know. They were to time with their own watches the five-minute intervals allotted for Unit Commanders to make their decisions. Once a Unit Commander made a decision, the unit was to be moved on the Maneuver Board at the same course and speed until new orders were given to the Contact Officer. Star shell illumination was simulated with paper models that had times written on them indicating when the shell went off and when its illumination would burn out. Simulated paper searchlight beams were also to be employed in the same way and left on the Maneuver Board as long as the searchlight was to be simulated. Contact information given by the Radar Umpire would give IFF, range, and bearing of all objects within the limits of those instruments. Radar information would be listed on a blackboard by the Radar Umpires for each period of the Exercise, and Contact Officers would take the radar data to the Student Commanders for action. Finally, as far as conditions permitted, Contact Officers were to keep the Director informed about information given to the Unit Commanders and the decisions the Student Commanders made. Contact Officers were also to obtain advance torpedo information from the Student Commanders before the Maneuver began.[5]

The Blue Statement

The Blue Statement of the Problem was put forth on 18 September. The Exercise was primarily being conducted to afford students the opportunity to make quick decisions without the benefit of time to prepare a detailed Solution to the Problem. The specific Problem was fleet defense against a light surface force attack at night in low visibility conditions. It was set in the southern Marshall Islands area, to familiarize students with that region. The Blue Situation entailed Blue's being at war with Orange, the Blue 3rd Fleet to the north of Jaluit and covering a Blue Expeditionary Force that was to seize the island. The Blue 3rd Fleet then moved about two hundred miles southeast of Jaluit; the Blue 3rd Fleet Commander (COM3FLEET) was the Officer in Tactical Command, and he knew the Orange Main Body to be in the area. He surmised that if the Blue Main Body arrived at a position of about fifteen miles to the east of Northeast Pass by 0500 the next morning, it would be able to interpose itself between the Orange Main Body and the Blue Expeditionary Force. At 1600 on the present day, a Blue plane from a cruiser in the Main Body had reported an Orange force consisting of destroyers and possibly cruisers about 150 miles from the Blue position. This flight also investigated Mili Atoll and thought it saw combatant units at anchor, though the pilot could not discern ship types. Heavy rain covered the area, so the pilot had to return before further observation could take place. The Blue Officer in Tactical Command determined that there was not enough time to take offensive action before dark and that the Orange destroyers that had already been discovered would probably deliver a night attack. He therefore disposed his force against this attack, and by 2000 the Blue Main Body was in a Night Cruising Disposition on course 310 degrees at a speed of fourteen knots. It therefore had about 150 miles to transit by 0500 (requiring a speed of advance of just over sixteen knots). Blue's battleships were limited to eighteen knots in battle-line formation.[6]

In the Exercise, the Blue Officer in Tactical Command was to furnish the Director with the course and speed of the Blue disposition after 2000 as well as any changes desired in the Cruising Disposition. Any changes in course and speed were to be undertaken to effect a "timely" arrival at the selected destination. Students were issued flimsies showing both Blue and Orange units within radar range. Each Commander would maintain communications and running plots that would help him visualize changing situations and take appropriate action. Further, students were referred to the Naval War College publication *Cruisers and Destroyers in General Action*, especially the information and instructions in its appendix III. The weather was included in the preparatory information; the wind was from the northwest, the sea was smooth with frequent rain squalls, and visibility was normal except in the squalls themselves. Sunrise was at 0600, sunset was at 1800, and evening twilight

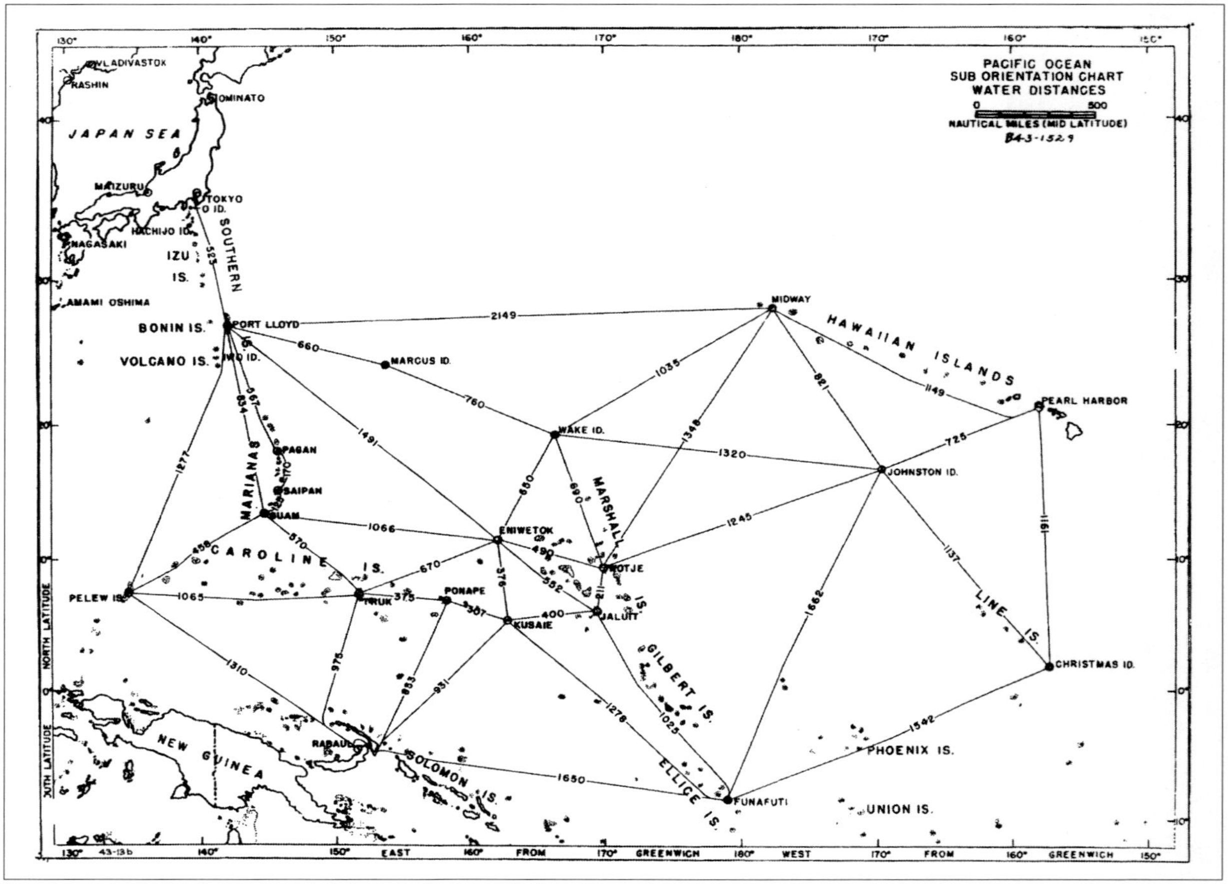

Map 1
Pacific Ocean Sub-Orientation Chart: Water Distances

was half an hour before dark. There was no moon, aircraft would not be employed, and all ships were equipped with radar. The Blue Officer in Tactical Command was to furnish the Director with a copy of his Radar Plan if it was different than indicated in his cruising diagram. Students serving as Unit Commanders also had to publish a summary of the doctrine they were employing so that Subordinate Commanders would know what doctrine to follow during the exercise.[7]

The Blue 3rd Fleet, for purposes of this Exercise, was organized into a battle line of Battleship Divisions (BATDIVs) 2, 3, and 4, and nine battlewagons under Admiral BA. Two fleet carriers were under the command of Rear Admiral BC, while a Sonar Screen consisting of Destroyer Division 50—six destroyers—was under Captain B-10. A Radar Screen under Captain B-11 consisted of Destroyer Squadron 1, broken down into Destroyer Divisions 1 and 2 and including six destroyers and a destroyer leader. Sector 1 Force, under Rear Admiral BD, consisted of Cruiser Division (CRUDIV) 9 of four light cruisers; CRUDIV 4, of four heavy cruisers; CRUDIV 3, another four light cruisers; and Destroyer Squadron 5 of nine destroyers, organized into Destroyer Divisions 9 and 10. The Sector 2 Force, under Rear

Admiral BA, Commander, Blue Battleline (Cdr. James Shaw, U.S. Navy, Command and Staff Class of December 1945)

Rear Admiral BC, Commander, Blue Carriers (Capt. Donald Hilton, U.S. Navy, Command and Staff Class of December 1945)

Captain B-10, Commander, Blue Sonar Screen (Col. Jack Saunders, U.S. Army, Command and Staff Class of December 1945)

Captain B-11, Commander, Blue Radar Screen and Destroyer Squadron 1 (Lt. Cdr. Richard Summers, U.S. Navy, Command and Staff Class of December 1945)

Rear Admiral BD, Commander, Blue Sector 1 Force and Cruiser Division 4 (Lt. Cdr. Valentine Holzapfel, U.S. Navy, Command and Staff Class of December 1945)

Rear Admiral BE, Commander, Blue Sector 2 Force (Capt. Franklin Lane, U.S. Navy, Command and Staff Class of December 1945)

Rear Admiral BF, Commander, Blue Sector 3 Force and Cruiser Division 5 (Lt. Cdr. Talbot Harper, U.S. Navy, Command and Staff Class of December 1945)

Admiral BE, consisted of Destroyer Squadron 4, another nine destroyers, organized into Destroyer Divisions 7 and 8; Destroyer Squadron 6, nine destroyers organized into Destroyer Divisions 11 and 12; and two light cruisers. The Sector 3 Force, under Rear Admiral BF, consisted of Cruiser Division 5, four heavy cruisers; CRUDIV 6, another four heavy cruisers; and Destroyer Squadron 2, another nine destroyers, organized into Destroyer Divisions 3 and 4.[8]

The Orange Statement

The Orange Statement outlined the same teaching objectives for the Operations Problem. The Orange Situation had the Orange 1st Fleet at sea and endeavoring to reduce the strength of the Blue Fleet through attrition attacks as Blue advanced into the Central Pacific. Orange still held Jaluit and Mili but expected a Blue attack

Rear Admiral OB, Commander, Orange Attack Force, Task Force 52, and Destroyer Flotilla 2 (Capt. Wells Thompson, U.S. Navy, Command and Staff Class of December 1945)

to seize Jaluit as soon as it consolidated its position in the Gilbert Islands. Orange aerial scouts furnished the Orange 1st Fleet Commander, Admiral OA, with information on a Blue force consisting of nine battleships, two fleet carriers, fifteen to twenty cruisers, and thirty to forty destroyers then located two hundred miles southeast of Jaluit. These scouts also reported a force of transports and cargo vessels escorted by combatant ships to the south of the Blue Force. From this information, Admiral OA deduced that the northern force was the Blue Main Body covering a Blue Expeditionary Force headed to Jaluit. He therefore decided to proceed with his Main Body toward Jaluit to oppose the probable operations, specifically to reduce Blue battleship strength with a supported destroyer night attack in anticipation of a fleet action. In accordance with his decision, Admiral OA directed Rear Admiral OB, Commander, Destroyer Flotilla 2, who had been reinforced by Cruiser Divisions 6 and 7, to attack the Blue Main Body for the purpose of reducing its battleship strength. Admiral OB's force was designated the Orange Attack Force, or Task Force (TF) 52.[9]

Admiral OA also decided to reinforce Task Force 52 with Battleship Division 1, which had just completed refueling at Mili and which he did not think had yet been sighted by Blue scout planes because of frequent rain squalls at the atoll. Battleship Division 1, three battleships capable of high speed, joined Task Force 52 at 1930 hours that day, where they were alerted to join up at night at high speed per previous information about the operation. Battleship Division 1 was placed in the stern of the Orange formation; it was capable of twenty-five knots. Subsequently, Rear Admiral OB put out a dispatch to the Commanders of Task Groups (TGs) 52.1, 52.2, and 52.3, giving the strength and location of the Blue Force and an assumed base course for the Blue Force. Admiral OB ordered that his force would conduct night search and support attacks on the enemy battleships and carriers in order to prevent the Blue occupation of Jaluit.[10]

Admiral OB's forces specifically consisted of TG 52.1, under the command of Captain O-1 and consisting of sixteen destroyers (Destroyer Divisions 2, 15, 16, and 24, organized into Destroyer Squadron 4) with a light cruiser as flagship; the Task Group was to locate and develop the Blue disposition with a radar-guided torpedo attack when directed. Additional forces included TG 52.2, the three battleships organized into Battleship Division 1, and the eight heavy cruisers of CRUDIVs 6 and 7, all under Rear Admiral OD. This force was to support TGs 52.1 and 52.3. Task Group 52.3, under Captain O-5, consisted of a light cruiser flagship and twelve destroyers of Destroyer Divisions 8, 9, and 18, these units together constituting Destroyer Squadron 2. Captain O-5's force was to destroy enemy battleships and carriers by torpedo attack and coordinate the torpedo attack of Task Group 52.1. TG 52.1 was further to scout, steaming in division units in column at a speed of twenty-five knots, with a distance of seven miles between the Guides of the various

Captain O-1, Commander, Orange Task Group 52.1 and Destroyer Squadron 4 (Lt. Cdr. Lodwick Alford, U.S. Navy, Command and Staff Class of December 1945)

Rear Admiral OD, Commander, Orange Task Group 52.2 (Col. Herbert Ehrgott, U.S. Army, Command and Staff Class of December 1945)

Captain O-5, Commander, Orange Task Group 52.3 and Destroyer Squadron 2 (Col. Richard Hutchinson, U.S. Army, Command and Staff Class of December 1945)

divisions. The plan designated which commanders were in authority; Orange Student Commanders were instructed to plot 2000 (game time) positions for all units of Task Force 52. Unit Commanders were also to carry out the Search Plan as laid out and plot those movements as well. The same weather data as was given to the Blue Force Commander was furnished to Orange; the Orange Officer in Tactical Command was ordered to inform the Director of the Radar Plan to be used if it changed after 2000. The Orange Officer in Tactical Command was further to publish his doctrine for his Subordinate Commanders and additional orders as he saw fit, as long as they did not conflict with the directives about maintaining the Plotting Sheets.[11]

The History of the Maneuver

This Exercise's History of the Maneuver, dated 21 September, placed the Blue formation initially two hundred miles southeast of Jaluit in its Night Cruising Disposition. This entailed a circular formation, with Battleship Divisions 2, 3, and 4 in division columns on a bearing of 090. The two fleet carriers were in column about two miles astern, with their support ships in circular spacing, in this order: CRUDIVs 4 and 9, along with light cruiser CL-43; Destroyer Squadron 5; CRUDIV 3; Destroyer Squadron 4; two modernized prewar light cruisers (OCLs), OCL-7 and OCL-8; Destroyer Squadron 6; CRUDIV 6; Destroyer Squadron 2; and CRUDIV 5. The Sonar Screen of six destroyers was in the Van, and seven additional destroyers acted as radar pickets about sixteen miles out. Orange was in a Scouting Disposition on course 225 at a speed of twenty-five knots. Its Main Body, Task Force 52, was in its approach position; Destroyer Divisions 2, 15, 16, and 24, as well as light cruiser CL-25, were 7.5 miles from the Orange center bearing 180 degrees, with additional destroyers on a Scouting Line bearing 090 degrees and on the same course and speed as the Main Body. The Orange Officer in Tactical Command was aware of the presence and location of the Blue Force; for his part, the Blue Officer in Tactical Command surmised that the Orange Force would undertake a night surface attack with destroyers and possibly cruisers. It did not appear, however, that the Blue Officer in Tactical Command was aware at this point of Orange's reinforcement of three battlewagons from Battleship Division 1.[12]

In Move 1 of the Maneuver, Blue maintained its course and speed but changed its fleet axis thirty degrees at 2000. A minute later, Cruiser Divisions 3, 4, 5, and 9 were ordered to take station on Circle 5, on current bearings. In the next minute, Commander, Battleship Division 2 (COMBATDIV 2) was made Commander, Battleline, by Admiral BA, the Blue OTC. At 2003, Destroyer Division 50 was ordered to take position on Circle 5, on the same bearing. Orange continued course and speed, except for Cruiser Division 7, which was ordered to take station one mile ahead of the center of the Scouting Line. In addition, Orange Cruiser Division 6 was ordered to divide into two sections and maintain present formation on the destroyers of Task Group 52.3, with two cruisers on each bow. At 2005, Orange Task Force 52 was ordered to maintain boiler power for maximum speed; at 2028, mutual radar contact was made between the opposing forces. Radar contact with Orange picket units was made at thirty-eight thousand yards by Blue Destroyer Divisions 15 and 16, light cruiser CL-25, and attached destroyers DD-353 and DD-355 in Sector 1. At this time, the Orange Officer in Tactical Command ordered Task Group 52.1 onto a parallel course with Blue while developing the contacts, but these orders were considered conflicting and so were canceled at the beginning of Move 2.[13]

In Move 2, all forces maintained course and speed; all of Orange's Scouting Line ships had the Blue Radar Screen on their scopes, while Blue's Radar Screen in Blue Sector 1 and destroyer DD-348 in the Sonar Screen had radar contact on the Orange Scouting Line. At 2030, the Orange Commander of Destroyer Division (COMDESDIV) 16 informed Commander, Task Force 52 that he was directing COMDESDIV 24 to close with Task Force 52 in order to avoid separation while developing the contact—this in answer to an inquiry as to why Destroyer Division 24 was turning right, toward the Orange Scouting Line. At the beginning of Move 3, a rain squall separated the two forces; both continued course, speed, and formation. The exception was the Blue Sonar Screen, which was ordered to form column astern of the Division Commander, the latter being directed to maintain course and speed.[14]

The entire Orange Scouting Line was now within radar range of various units of the Blue Radar Screen, the Sonar Screen, and Cruiser Division 4. The Orange Scouting Line had various units of the Blue Radar and Sonar Screens on its radars but was experiencing jamming in the vicinity of Blue Sector 1 Radar Screen. All units with radar contact had the rain squall on their scopes. (One of the umpires noted at this point in the presentation of the History of the Maneuver that students in general were having problems evaluating radar countermeasures, such as jamming, probably due to lack of experience as radar operators or evaluators. These countermeasures had been introduced, however, to remind the students that they could become a reality at any time and they had to be on the alert for it.) At 2037, Commander, 3rd Fleet directed the Blue Force to prepare to repel the Orange Force with gunfire on the closest targets when they were within range. In addition, the destroyers were to fire half-salvo torpedoes at the Orange heavy units at their discretion, and the Sonar and Radar Screens were to concentrate. There was to be no illumination or radar jamming. Also at this time, Orange Destroyer Division 16 came left to a southerly course to investigate numerous radar contacts.[15]

By the beginning of Move 4, the rain squall was beginning to dissipate to the southwest, and the Blue battleships were ordered to form column on the Center Division, in the order Battleship Divisions 2, 3, and 4. At the same time, Blue fleet carriers CV-3 and CV-9 were ordered to take position on Circle 3 while maintaining their present bearing astern of the formation, though this move was very difficult to achieve at night and was potentially "embarrassing" for Destroyer Squadrons 4 and 6, through which the carriers had to pass. Meanwhile, the Blue Radar Screen was commencing its concentration in accordance with COM3FLEET's orders, with one division on the port beam and one on the starboard quarter. Blue Cruiser Division 6 and Destroyer Squadron 2 were ordered to increase speed to twenty-eight knots, maintaining a course of 310 so as to gain position on Cruiser Division 5 and make a coordinated attack with that unit. The rest of the Blue formation remained as before, as did the Orange formation except for Cruiser Division 7, which was coming up to take position one mile ahead of the Scouting Line to support it. By this time, Blue had the Orange Scouting Line and Cruiser Division 7 on its radar scopes, and the Orange Scouting Line had radar contact on Blue Cruiser Divisions 4, 5, and 9, as well as Destroyer Squadrons 5 and 50. (The Staff found it unclear why Blue was withdrawing his Radar Screen so soon and why the port beam and starboard quarter were being reinforced when Blue was threatened on its port bow.)[16]

By Move 5 at 2048, the weather was clear, and the Orange Scouting Line was ordered to take station to the west, fifteen miles ahead of the Blue Main Body. Cruiser Division 7 was sent to cover the destroyers, while the remainder of the Orange Force stayed in the same formation. Also at this time, Blue Destroyer Squadron 5 was deploying into a line of division Guides in order to unmask its torpedo tubes, and Blue Cruiser Division 4 opened fire on Orange Destroyer Division 16 at a range of twenty-eight thousand yards. Orange Destroyer Division 16 changed course to the right at 2051, the end of the move, under fire from Blue CRUDIV 4. At the beginning of Move 6, the two formations were in the same positions, with the exception of Blue Destroyer Division 50, which was ordered to take station on Circle 7 on the same bearing so as to unmask Blue Cruiser Division 9's guns. Orange Cruiser Division 7 was instructed to have its first section support the left end of the Scouting Line and its second section support the right. Blue Destroyer Division 50 at this time was beginning to obtain the Orange Main Body on its radar scopes, but the Orange Scouting Line only had the outer screen of Blue forces in Sector 1 on radar. Blue Cruiser Division 4 continued to fire on Orange Destroyer Division 16; Blue Cruiser Division 9 and light cruiser CL-43 opened fire on the same target. Orange Destroyer Division 16 sustained 10 percent damage from this fire.[17]

Move 7 saw the formations still in essentially the same positions. Blue Cruiser Division 6 was ordered to fire "window" (small strips of reflective material, producing a large, diffuse false contact) shells for radar deception on a true bearing of 310 at a maximum range of seventeen thousand yards. Blue Cruiser Division 9, with light cruiser CL-43, continued to fire on Orange Destroyer Division 16, as did Blue Destroyer Division 50 and Blue destroyers DD-348 and DD-365. Blue Cruiser Division 4 then shifted its fire from Orange Destroyer Division 16 to Destroyer Division 24. At the same time, Orange Cruiser Division 7 opened fire on Blue Destroyer Division 50 and destroyers DD-348 and DD-365, as well as on Blue heavy cruisers CA-40 and CA-42. Orange heavy cruisers CA-43 and CA-44 also opened up on Blue Cruiser Division 9 and light cruisers CL-43, CL-47, and CL-48. By the end of this move, however, Orange Destroyer Division 16 was 50 percent damaged, while Blue Destroyer Division 50 and Blue destroyers DD-348 and DD-365 had sustained 20 percent damage. Again, an umpire interceded here during the post-Maneuver presentation to point out that Orange Destroyer Division 16 had been under effective medium-caliber fire by two groups of Blue ships but that this unit by 2100 was in an "ideal" position to fire half of its torpedoes and perhaps a full salvo at Blue Cruiser Division 4. "The division commander does not appear to realize that part of the mission of the scouting line is to destroy enemy ships when these ships are in a favorable position relative to their own main body as well as to penetrate the screen to the enemy main body." The umpire asked why the division commander had assumed that all units of "ball carriers" on any team had to be blocking backs or tacklers; "He should realize that someone has to take out the screen." Vice Admiral Pye added that since Move 5, Blue's battleships had been strung out in a long column, a very poor night formation. He suggested that the best formation was probably division columns, with division Guides on a line of bearing normal to the bearing of the Orange Main Body.[18]

In Move 8, Blue Cruiser Divisions 3 and 4 and Destroyer Squadron 5 turned to the right on course 021, maintaining their prior speed. This turn was taken to avoid torpedoes that it was assumed Orange Destroyer Divisions 16 and 24 had fired; those two units had turned radically to the right, and the Commander of Blue Sector 1 forces surmised that this turn had been taken after launching torpedoes.

Orange Cruiser Division 6 and Destroyer Squadron 2 were now ordered to concentrate, with the destroyer divisions in column in numerical order and the cruisers to the south in covering position. They were also ordered to increase speed to thirty knots and change course to 254. This move would take the formation ahead of the Blue Force and allow them to attack from ahead. Orange Destroyer Division 16 then fired all thirty-two of its torpedoes, of which sixteen ran, while the destroyer unit itself was still under effective fire. The umpire assumed the target was Blue Cruiser Division 4, though Orange Commander, Destroyer Division 16 violated doctrine by setting his torpedo depths at twenty-two feet instead of six. The latter depth was supposed to be maintained unless one was certain he was firing at capital ships.[19]

In the meantime, window fired by Orange Cruiser Division 6 and Destroyer Squadron 2 was effective approximately eight miles ahead of the Blue formation. Orange Destroyer Division 16 fired star shells at the Blue formation, but they fell short and failed to illuminate any targets. Blue Cruiser Division 9 and light cruiser CL-43 were still firing at Orange Cruiser Division 7 and Blue Cruiser Division 4 was still firing on Orange Destroyer Division 24, while Blue Destroyer Division 50 and destroyers DD-348 and DD-365 were still firing on Orange Cruiser Division 7. Blue Cruiser Division 5, meanwhile, was firing at Orange Destroyer Squadron 2, Destroyer Division 15, and light cruiser CL-25. Orange heavy cruisers CA-32, CA-40, CA-42, and CA-43 by this time had taken Blue Cruiser Division 4, heavy cruiser CA-31, light cruiser CL-47, Destroyer Division 50, and destroyers DD-348, DD-353, DD-355, DD-360, and DD-365 under fire. (One of the umpires wanted to know why the four ships of Orange Cruiser Division 7 had selected such widely separated targets for their gunfire. He was also interested in knowing why Blue Destroyer Squadrons 4 and 6—along with light cruisers OCL-7 and OCL-8 in Blue Sector 2—did not appear to be taking any action, either to repel other Orange attacks or make attacks themselves.) Move 8 ended with Orange heavy cruiser CA-43 having sustained 40 percent damage; heavy cruisers CA-32 and CA-42 and light cruiser CL-25 all having sustained 10 percent damage; and all of Destroyer Division 24 sustaining 10 percent damage. Blue Destroyer Division 50 had also taken 40 percent damage, while Blue light cruiser CL-47 and Destroyer Division 2, along with an attached destroyer, had each taken 10 percent damage.[20]

By the beginning of Move 9, at 2112, the window fired by Blue had disappeared, and the Orange Officer in Tactical Command ordered all of his units to attack. All Blue forces were also ordered to turn to course 021, except Destroyer Division 50, which was to retire to Circle 3 of the formation on bearing 320 after firing its torpedoes. Blue Cruiser Division 4 was also ordered to return to its base course. While this was going on, Orange Cruiser Division 6 and CL-25 were firing window, and Orange Destroyer Division 24 was firing star shells at fifteen thousand yards. Their illumination showed a column of six destroyers on a northerly course—Blue

Destroyer Division 50 and its two attached destroyers. Each ship of Orange Cruiser Division 7 now fired six torpedoes in the general direction of the Blue Main Body, though these were probably actually directed at Cruiser Division 9 and Destroyer Division 50. Orange Destroyer Division 24 also fired torpedoes, thirty-two in all, with the Blue battleships as the apparent targets. Blue Destroyer Division 50 then fired forty torpedoes on base course 315, the apparent target being Orange Cruiser Division 7 and the Orange Scouting Line. In addition, Blue had been listening in on the Orange Talk Between Ships and sent a false message on the Orange frequency that one of the destroyer units was to turn on its searchlights and open fire. (One of the umpires immediately asked, during game play, what Orange would do to nullify this Blue action.) By this time, Orange Destroyer Division 24 had fired on Blue Destroyer Division 50, Orange Destroyer Division 16 had taken Blue Cruiser Division 9 under fire, and Orange Destroyer Division 7 was firing on Blue heavy cruiser CA-28, light cruiser CL-48, and Destroyer Division 50. In addition, Orange Cruiser Division 6 and light cruiser CL-25 were firing window, while Blue Cruiser Division 4 was firing at Orange Destroyer Division 24.[21]

At this time, Blue Cruiser Division 9 was firing on Orange heavy cruisers CA-32 and CA-42; Cruiser Division 5 was shooting at Orange heavy cruisers CA-32, CA-40, and CA-43, light cruiser CL-25, and Destroyer Division 15; Blue Cruiser Division 6 fired on this same concentration; and Blue Destroyer Division 50 fired at Orange heavy cruisers CA-32 and CA-43, which were also under fire from Blue Cruiser Division 5. (It was unclear to one of the umpires during the postgame presentation why Blue had ceased firing on Orange Destroyer Division 16. The cease-fire allowed that unit to keep Blue Cruiser Division 9 under Enfilade Fire, even though badly damaged.) At the end of Move 9, Orange had lost heavy cruiser CA-43, heavy cruiser CA-32 was dead in the water and sinking, light cruiser CL-25 had 30 percent damage, and heavy cruiser CA-40 had 20 percent damage. Blue Destroyer Division 50 had taken 60 percent damage, light cruiser CL-48 had sustained 70 percent damage, and heavy cruiser CA-28 had received 20 percent damage. (It was pointed out in the discussion that a torpedo spread fired by Orange Cruiser Division 7 to the east had placed an "effective" barrier south of the Orange Main Body that was to cause "considerable" embarrassment for Orange maneuvers later on and result in Orange sinking one of its own destroyers.)[22]

The beginning of Move 10 saw the Orange Force deploying to attack, with the exception of Battleship Division 1, Destroyer Squadron 2, and Destroyer Divisions 16 and 24. The Blue Main Body remained on its same course, and both sides had full pictures of the other from radar. Orange Destroyer Divisions 15 and 24 continued firing on Blue Destroyer Division 50; Orange Battleship Division 1 opened fire on that unit also, and Orange Destroyer Division 16 shifted its fire to the Blue unit as well. Orange Battleship Division 1 also opened fire on Blue light cruisers

CL-47 and CL-48, and Orange light cruiser CL-25 shifted fire to Blue light cruiser CL-48 after ceasing to fire window. Orange Cruiser Division 6 additionally ceased firing window to shift to Blue Cruiser Division 5. Orange heavy cruiser CA-40 also fired on Blue Destroyer Division 3, while Orange heavy cruiser CA-42 and Orange Destroyer Division 2 fired on Blue destroyer DD-417. Blue Cruiser Division 4 was firing on Orange Destroyer Division 16, and Blue Destroyer Squadron 2 fired on Orange Destroyer Division 15, while Blue Cruiser Division 6 was firing on Orange Destroyer Divisions 2 and 15. At the same time, Blue Cruiser Division 9 was firing on Orange Destroyer Divisions 2 and 15, as well as on Orange heavy cruiser CA-32. Meanwhile, Blue Cruiser Division 5 was taking Orange heavy cruisers CA-32, CA-36, and CA-39, and light cruiser CL-25 under fire, along with Orange's Destroyer Division 15.[23]

About this time, destroyers DD-109 and DD-110 from Blue Destroyer Division 15 fired a half-salvo of torpedoes on a base course of 090, with Orange Cruiser Division 9 as the obvious target. The torpedoes were fired from DD-109 and DD-110 rather than the first two destroyers of the formation because the former ships had torpedoes with forty-five-knot speeds but short range, while the latter had longer-range but slower torpedoes. By the end of this move, Orange Destroyer Division 16 was 70 percent damaged, Orange had lost heavy cruiser CA-32, and Destroyer Divisions 2 and 15 were each 30 percent damaged. Blue had seen its entire Cruiser Division 5 and heavy cruisers CA-24, CA-25, and CA-27 take 10 percent damage, heavy cruiser CA-28 take 40 percent damage, Destroyer Division 3 as a whole sustain 10 percent damage, the entire Destroyer Division 50 and destroyer DD-417 take 90 percent damage, light cruiser CL-47 take 40 percent damage, and light cruiser CL-48 sink. (It was unclear at the end of this move why the Orange battleships had turned to the west when a southeasterly course would have allowed them to close on the Blue battleships. Up to this time, Blue had had no indication of the presence of Orange's three battlewagons. The commenting umpire also thought that Orange Cruiser Division 6 and Destroyer Squadron 2 had engaged in "unusual and difficult" maneuvers. He was additionally unclear on why Orange Destroyer Divisions 16 and 24 had retired when Destroyer Division 24 was still relatively undamaged and had earlier received orders to attack.)[24]

As Move 11 began, Blue and Orange were listening on each other's Talk Between Ships circuits and attempting deceptions, but without marked success. The Blue Force then turned to 340 in order to carry out its original mission and allow itself "sea room" to the north in case it had to take evasive action later. The Orange Force by this time had entirely deployed for attack, with the exception of Destroyer Division 24, which was retiring. (Destroyer Division 24's actions were unexplained.) Orange Destroyer Division 8 and Destroyer Squadron 2's flagship, light cruiser CL-27, now turned to 090, with Destroyer Divisions 9 and 18 following suit but remaining on the port quarter of the leading formation. This course, however, would take CL-27 and Destroyer Division 8 into the wakes of their own torpedoes if it was maintained or into collisions with their own ships on their port quarter if they turned to the left. (Questions therefore arose whether these ships were aware of the torpedo tracks and as to what their tactical intentions were.) Meanwhile, Orange Destroyer Squadron 4's flagship, light cruiser CL-25, and Orange Destroyer Divisions 2 and 15 penetrated the outer Blue screen on 090 and fired fifty-two torpedoes. But CL-25 and all of Destroyer Division 2 were sunk. (The umpire later commented that the Commander of Orange Destroyer Squadron 4 should have taken into account that the Blue battleships would have to turn back to their original course soon if they were to carry out their missions. Orange Destroyer Squadron 4's Commander, in firing long-range torpedoes, should have chosen a base course somewhat more in the direction of the Blue battleships.)[25]

In this move, Blue Cruiser Division 4 opened gunfire on Orange Destroyer Divisions 16 and 24, while Blue Cruiser Division 9 fired on CL-27. Cruiser Division 9's secondary batteries were firing on Orange Destroyer Division 15, as were Blue Cruiser Divisions 5 and 6. The secondary batteries of those Blue units were also shooting at CL-25, and Blue Destroyer Division 50 was firing on Orange Destroyer Squadron 4. In addition, Blue Destroyer Division 3 was firing on CL-25, and Blue Destroyer Division 4 was firing on Orange Destroyer Division 2. At the same time, Orange Destroyer Division 24 was firing on targets of opportunity, Orange Destroyer Division 16 was firing on a unit of Blue Cruiser Division 4, Orange light cruiser CL-25 was firing on Blue light cruiser CL-43, and Orange Destroyer Division 2 was firing on Blue Destroyer Division 4. Additionally, Orange Destroyer Division 15 was firing on Blue Destroyer Division 50, while Orange Cruiser Division 6 and heavy cruisers CA-40 and CA-42 were firing at Blue Cruiser Division 5. In addition, Orange battleship BB-11 was firing on light cruiser CL-46 from Blue Cruiser Division 9; Orange battleship BB-12 was firing on Blue light cruiser CL-47; and Orange battleship BB-13 was firing on Blue Destroyer Division 50.[26]

The result of all of this was that by the end of the move every unit of Blue Destroyer Division 50 had been sunk, destroyers DD-348 and DD-365 from Blue Destroyer Divisions 1 and 2 had sustained 80 percent damage, Blue heavy cruisers CA-25 and CA-27 had each taken 30 percent damage, heavy cruiser CA-28 had sustained 60 percent damage, heavy cruiser CA-24 had taken 40 percent, heavy cruiser CA-32 had taken 10 percent, as had Blue light cruiser CL-43, and light cruiser CL-47 had had 50 percent damage inflicted on it, as had Blue Destroyer Division 4 as a whole. Orange came away with its entire Destroyer Division 15 damaged 60 percent; all of Destroyer Division 2 and Destroyer Squadron 4's flagship, light cruiser CL-25, sunk; Destroyer Division 24 40 percent damaged; Destroyer Division 16 80 percent damaged; and Destroyer Squadron 2's light-cruiser flagship, CL-27, sustaining 70 percent damage. As the move ended, Orange Destroyer Division 24 was firing star shells from a range of fifteen thousand yards, as it would do for three minutes.[27]

At the beginning of Move 12, the Blue Force changed base course to 280, all units having evaluated that the movements of Orange Destroyer Divisions 2 and 15 as well as Orange light cruiser CL-25 indicated torpedo launches at the Blue Main Body. Orange continued its previous movements as command of the Orange Force passed to Commander, Cruiser Division (COMCRUDIV) 6, who was also Commander, Task Group (CTG) 52.2. Blue Destroyer Squadron 5 at this time was ordered to fire window in the direction of the Orange Main Body, while Orange Destroyer Division 15 fired star shells with a range of three thousand yards and illuminated Blue Cruiser Division 5 on its northerly course. Orange Destroyer Division 24, having obtained illumination of Blue Cruiser Division 4 during the previous move, fired thirty-two torpedoes on base course 185 and continued to retire to the northeast. Blue destroyers DD-353, DD-355, and DD-360, all from Blue Destroyer Squadron 1, now fired twenty-four torpedoes on base course 345. (Given the depth setting of six feet, it was unclear what the target was, unless Orange Destroyer Division 24.) It was at this time that Orange lost destroyer DD-118, sunk by its own torpedoes fired in Move 9.[28]

In gunfire exchanges in this move, Orange Cruiser Division 6 fired on Blue light cruisers CL-43 and CL-49, of Blue Cruiser Division 9, as well as on Blue Destroyer Division 4. Orange Cruiser Division 7 heavy cruisers CA-40 and CA-42 fired on light cruisers CL-46 and CL-47, of Blue Cruiser Division 9. Orange battleship BB-11 fired on CL-47, and Orange battleship BB-12 fired on Blue heavy cruiser CA-31 with its main batteries and at CL-46 with its secondary batteries. At the same time, Orange battleship BB-13 fired on Blue heavy cruiser CA-32 with it main batteries and on Blue light cruiser CL-49 with its secondary batteries. Blue Cruiser Division 4, for its part, fired on Orange Destroyer Division 16; Blue Cruiser Division 9 fired on Orange light cruiser CL-22; and heavy cruisers CA-40 and CA-42 of Orange Cruiser Division 7 accidentally fired on Orange Destroyer Division 15. Blue Cruiser Division 5 also fired on Orange Destroyer Division 15, while Blue Destroyer Division 2 fired on Orange Destroyer Division 16, and Blue Destroyer Division 3 fired on Orange Destroyer Division 15, as did Blue Destroyer Division 4.[29]

At the end of this move, Orange had lost all of Destroyer Divisions 15 and 16 and the light cruiser CL-47; heavy cruisers CA-40 and CA-39 had suffered 20 percent and 30 percent damage, respectively. In addition, Destroyer Division 24 took 40 percent damage, and Orange Destroyer Division 8, as a whole, 10 percent. In the entire Maneuver, Orange had two heavy cruisers, two light cruisers, and three whole destroyer divisions sunk. Blue in this move lost light cruiser CL-47, sunk; light cruiser CL-46 suffered 60 percent damage and CL-43 and CL-49 50 percent damage; heavy cruisers CA-31 and CA-32 received 50 and 60 percent damage, respectively; and Destroyer Division 4 took 90 percent. Total Blue losses by sinking included light cruisers CL-47 and CL-48 and all of Destroyer Division 50. Damaged Blue units also included heavy cruiser CA-28 with 70 percent damage; CA-32, 60 percent; CA-24 and CA-31, 50 percent each; and CA-25 and CA-27, 30 percent each. In addition, Blue light cruiser CL-46 took 60 percent damage; light cruiser CL-43 took 50 percent; destroyers DD-348, DD-365, and DD-417 took 80 percent; Destroyer Division 4 as a whole received 90 percent damage; and Destroyer Divisions 2 and 3 and destroyer DD-360 each had 10 percent damage.[30]

The Critique

A few days later, Commodore Bowdey spoke about the communications aspect of Operations Problem 4. Bowdey started off by stating that no matter how well planned a military operation was, unforeseen developments would arise that would require "modification" of action. This, he asserted, could be accomplished only through coordination, whereby the command received information on the changing situation and conveyed orders to meet it. Communications were the key to handling a changing situation, and in a night action the problem was greatest. He reminded the students about previous lectures illustrating how disastrous the night battles around Guadalcanal had been because of poor communications. In Operations Problem 4, communications had been effective, but not to the highest degree. Orange had cleared seventy-five messages during the Maneuver and Blue fifty-six, but the Blue and Orange Officers in Tactical Command had been frustrated by material failures and had been in the dark at times about enemy dispositions and even their own. "It can be understood how disturbing this would be isolated on board a ship."[31]

Bowdey also spoke to communications security, in particular to interception. He considered call signs, codes, and code words "helpful" in preventing the enemy from understanding one's plans by listening in on radio circuits. Blue, for instance, had successfully used a code to conceal course and speed changes, but both sides had been able, by radio interception, to figure out the other's types of ships, organization, and numbers when plain language calls like "COMDESDIV 16" were used. Simple coded call signs, Bowdey thought, could prevent this, and the same kind of communications security could be applied to such things as the location of enemy units at various points in the battle when the enemy does not attend to the security of its communications. Also in terms of communications security, Bowdey pointed out how both sides had attempted to deceive the other but without success since each side used authenticator systems. Blue's system, however, was more secure than Orange's, and both sides' had achieved temporary confusion and slowing of the opponent's circuits. This temporary disruption of enemy communications Bowdey found worth the effort.[32]

Bowdey next focused on the fact that in a night action, radio traffic had to be handled primarily through voice circuits, because of the need for visual silence (that is, flashing light gave away location, and flag hoists, etc., could not be seen). Speed was vital at night to ensure passage of all traffic. The way to ensure this speed, he thought, was to use proper procedures, such as call signs, which not only helped with security but were faster and easier to say over the radio than full unit designations. Both Blue and Orange had these signs but sometimes in the heat of battle had failed to use them. Message texts had to be short and succinct. Flag hoists, he argued, saved time, but neither side had used them. Also, students did not understand how to use radio protocols such as "Roger," "Over," and "Out." Some spoke too rapidly and indistinctly on many messages, requiring repetition of the message in part or whole. He saw a further problem during the Maneuver in that ten minutes had been allowed to clear communications for a six-minute move; on several occasions not all of the messages on hand had been cleared, because of the slowness of deliveries. Some orders were therefore delayed, resulting in confusion. "At sea only six minutes would be available to clear the orders for six minutes of operations." To Bowdey, half of the messages that originated in Operations Problem 4 could have been cleared in actual operations at sea. He ended by asking for comments on improving the "effectiveness and realism of communications in these problems."[33]

Conclusion

Operations Problem 4 was a night surface action fought in support of an amphibious assault, in the case of Blue, and in defense of an island position, in the case of Orange. Although Blue had two fleet carriers, they were not employed. That could be explained by the fact that Operations Problem 4 was a night action, though the Navy did have at least a very limited night carrier capability by the latter part of the Pacific War.[34] Actions in support of amphibious landings in areas such as Micronesia or in defense of those positions make sense in certain contexts as war games for an immediate postwar Naval War College class. In addition, given the battles in the Solomon Islands in 1942–43, surface warfare cannot have been far from the minds of many American naval officers coming back from the Pacific theater.[35] Moreover, these officers must have understood that most of the battles in the Pacific War had featured surface engagements rather than carrier battles.[36] Finally, the United States was still significantly focused on the battle fleet as the centerpiece of its strategy as late as 1940–41 and only refocused on carriers then because of the absence of fortified bases in the western Pacific, the damage to the battle line at Pearl Harbor, and the lack of steel for armor plating for battleships and battle cruisers after December 1941.[37]

Fighting Orange in Micronesia at night in a small-scale version of the long-looked-for "decisive battle" was also sensible in several potential scenarios if the

United States found itself fighting a foe in that region again. The Joint Chiefs of Staff (JCS) for a time entertained the idea that in case of a disaster that drove the United States from East Asia and the western Pacific and threw it back on its postwar positions in Micronesia or even farther east, a defense-in-depth would be necessary in the Pacific Basin. U.S. forces would use southern and eastern Pacific bases to fight their way once again through to the western Pacific and East Asia.[38] The facts that the Soviet Union then had so little naval power and that the only other potential navy in the region was Great Britain's suggested that the Naval War College staff saw a resurgent Japanese navy as a future foe or thought the Soviet Union might build and deploy a fleet similar to prewar Japan's.[39]

In addition, Operations Problem 4 offered a variant of the Solomon Islands battles in which the United States had fared so poorly in late 1942.[40] The surface-warfare significance of Operations Problem 4 lies also in wartime Pacific Fleet doctrine and operational practice. Pacific Fleet commanders had continued to see a need for major heavy surface forces to defeat the Combined Fleet in detail; Admiral Spruance, as wartime commander of the Fifth Fleet, and Admiral William Halsey, Commander of the Third Fleet, had never thought they had enough carriers to defeat the Japanese fleet entirely, or at least not in all possible scenarios. With the concurrence of Admiral Chester Nimitz, wartime commander in chief of the U.S. Pacific Fleet, both Spruance and Halsey prepared numerous operations orders and battle plans in which heavy surface forces were still vital to amphibious operations, night operations, and action in weather that precluded carrier operations. Finally, as has been pointed out in chapter 1, the purpose of the Exercises and Operations Problems was not to re-create the past or even predict the future, but to expose student officers to decision-making learning opportunities.[41]

Adm. William Halsey, Commander, wartime Fifth Fleet

NOTES

1 Command and Staff Class of December 1945, "Operations Problem 4: Instructions and Procedures," 17 September 1945, p. 1, NHC, folder 2533, box 125, RG 4, Naval Historical Collection, NWC.

2 Ibid., p. 2.

3 Command and Staff Class of December 1945, "Operation Problem Y," Annex A, "Instructions for contact officers," 30 March 1945, p. 1, NHC, folder 2533, box 125, RG 4, NWC. The date on this document is puzzling, since the document is addressed to the Class of December 1945. Since many of the war games were repeated from class to class, it is probable that this was a recycled instruction sheet from a previous war game and that the old date was inadvertently left on.

4 Ibid., pp. 1–2.

5 Ibid., pp. 2–3. For the methodology of war gaming at NWC prior to World War II, see Ronald Spector, *Professors of War: The Naval War College and the Development of the Naval Profession* (Newport, R.I.: NWCP, 1977), pp. 71–87; and Michael Vlahos, *The Blue Sword: The Naval War College and the American Mission, 1919–1941* (Newport, R.I.: NWCP, 1980), pp. 133–56.

6 Command and Staff Class of December 1945, "Operations Problem 4: The Blue Statement, Section

A," 18 September 1945, pp. 1–2, NHC, folder 2533, box 125, RG 4, NWC.

7 Ibid., pp. 2–3.

8 Ibid., p. 4.

9 Command and Staff Class of December 1945, "Operations Problem 4: The Orange Statement, Section B," 18 September 1945, pp. 1–2, NHC, folder 2533, box 125, RG 4, NWC.

10 Ibid., pp. 2–4.

11 Ibid., pp. 4–5. For interwar American surface-ship doctrine, see Thomas Hone and Trent Hone, *Battle Line: The United States Navy, 1919–1939* (Annapolis, Md.: Naval Institute Press, 2006), pp. 68–89; Craig Felker, *Testing American Sea Power: U.S. Navy Strategic Exercises, 1923–1940* (College Station: Texas A&M Univ. Press, 2007), pp. 110–13; and William McBride, *Technological Change and the United States Navy, 1865–1945* (Baltimore: Johns Hopkins Univ. Press, 2000), pp. 111–202.

12 Command and Staff Class of December 1945, "Operations Problem 4: History of Maneuver," 21 September 1945, pp. 1–2, NHC, folder 2533-C, box 125, RG 4, NWC.

13 Ibid., p. 3.

14 Ibid., p. 4.

15 Ibid., p. 5.

16 Ibid., pp. 6–7.

17 Ibid., pp. 7–8.

18 Ibid., pp. 8–9.

19 Ibid., p. 10.

20 Ibid., pp. 10–11.

21 Ibid., p. 12. There was apparently at least one example during a Pacific War surface battle of successful radio deception by the United States against Japan; see John Lorrelli, *The Battle of the Komandorski Islands, March 1943* (Annapolis, Md.: Naval Institute Press, 1984), p. 139.

22 Command and Staff Class of December 1945, "Operations Problem 4: History of Maneuver," p. 13.

23 Ibid., p. 14.

24 Ibid., pp. 14–15.

25 Ibid., pp. 16–17.

26 Ibid., p. 17.

27 Ibid.

28 Ibid., p. 17a.

29 Ibid., pp. 17a–18.

30 Ibid., p. 18.

31 "Memorandum for Command and Staff Class of December 1945, Operations Problem 4: Communications," 26 September 1945, p. 1, NHC, folder 2533, box 125, RG 4, NWC. For an example of poor communications early in the Pacific War, see Dennis Warner and Peggy Warner, *Disaster in the Pacific: New Light on the Battle of Savo Island* (Annapolis, Md.: Naval Institute Press, 1992), pp. 215–59.

32 "Memorandum for Command and Staff Class of December 1945, Operations Problem 4," pp. 1–2.

33 Ibid., p. 2.

34 Clark Reynolds, *The Fast Carriers: The Forging of an Air Navy* (Annapolis, Md.: Naval Institute Press, 1968), pp. 228–29.

35 Warner and Warner, *Disaster in the Pacific;* Russell Crenshaw, *The Battle of Tassafaronga* (Baltimore: Nautical and Aviation, 1995).

36 Vincent O'Hara, *The U.S. Navy against the Axis: Surface Combat, 1941–1945* (Annapolis, Md.: Naval Institute Press, 2007), pp. xiii–xvi; James Levy, "Race for the Decisive Weapon: British, American, and Japanese Carrier Fleets, 1942–1943," *NWCR* 58, no. 1 (Winter 2005), pp. 137–50.

37 For the argument that the absence of fortified bases in the western Pacific caused the Navy to reformulate its strategy, place the battle line as a strategic reserve, and deploy battleships only after enemy air strength had been destroyed by naval aviation and the enemy battle line had been reduced by submarine and light-force attacks, see John T. Kuehn, *Agents of Innovation: The General Board and the Design of the Fleet That Defeated the Japanese Navy* (Annapolis, Md.: Naval Institute Press, 2008), pp. 162–79. For the assertion about the damage to the battle line and the lack of steel for armor plating, see Joel Davidson, *The Unsinkable Fleet: The Politics of U.S. Navy Expansion in World War II* (Annapolis, Md.: Naval Institute Press, 1996), pp. 11–12, 14–16, 19–21, 23–24, 32, 34, 60, 96–97, and then 34–37, 44, 76, 95, 97, 108, 110, 131–32, 135, 143–44, 148–54, 157, 163–64, 169–71, 175, and 177.

38 For the idea of a resurgent Japan seen by the JCS as a potential future foe in 1945–46, see Hal Friedman, *Creating an American Lake: United States Imperialism and Strategic Security in the Pacific Basin, 1945–1947* (Westport, Conn.: Greenwood, 2001), pp. 1–62; and *Governing the American Lake: The U.S. Defense and Administration of the Pacific, 1945–1947* (East Lansing: Michigan State Univ. Press, 2007), pp. 83–115. In fact, James Levy makes the important observation that there were only five carrier-to-carrier battles in the Pacific War and that fifteen of the seventeen engagements in the Solomon Islands were between surface forces; Levy, "Race for the Decisive Weapon," pp. 137–50.

39 See Friedman, *Creating an American Lake,* pp. 1–62; Friedman, *Governing the American Lake,* pp. 83–115; Eric Morris, *The Russian Navy: Myth & Reality* (London: Hamish Hamilton, 1977), p. 25; and Jurgen Rohwer and Mikhail Monakov, *Stalin's Ocean-Going Fleet: Soviet Naval Strategy and Shipbuilding Programmes, 1935–1953* (London: Frank Cass, 2001), pp. 178–220.

40 Thomas Mahnken, "Asymmetric Warfare at Sea: The Naval Battles off Guadalcanal, 1942–1943," *NWCR* 64, no. 1 (Winter 2011), pp. 95–121.

41 See Friedman, *Creating an American Lake,* pp. 1–62; Friedman, *Governing the American Lake,* pp. 83–115; Trent Hone, "U.S. Navy Surface Battle Doctrine and Victory in the Pacific," *NWCR* 62, no. 1 (Winter 2009), pp. 67–105; and Hone, e-mails to author, 28 February, 2 March, 3 March, and 14 March 2011.

VII Operations Problem 5

The Blue Statement, October 1945

Operations Problem 5 was an exercise in estimating a "strategical" situation that flowed from the plan of an immediate superior, preparing the subordinate plan, and supervising the planned action. There was also an objective of familiarizing students with the capabilities and limitations of ships and planes of the Blue and Orange fleets as well as with the western Pacific, the southwest Pacific, and the seas around Malaysia. In the Blue General Situation, war continued between the United Nations (UN, founded by, primarily, the wartime Allies in 1945) and the Axis powers. Purple (the Soviet Union), though at war with Black (Germany), was faithfully observing a nonaggression pact with Orange. The war in Europe continued to demand practically all of Red's (Britain's) strength and resources. Therefore, the war in the Pacific had developed into a war principally between Orange and Blue, with Yellow (China), Brown (the Netherlands), and Red assisting Blue as circumstances permitted. Orange thus far had lost control of most of New Guinea, Timor, Java, Borneo, Celebes, Halmahera, the Marianas, and the Palaus but still retained Sumatra, Banka, Billiton, Singapore, Malaya, Thailand, Burma, Indochina, and the Philippines. Ceram, Boeroe, the Aroes, and the McCluer Gulf had been neutralized and isolated by Blue, but Orange forces were still present in those areas. Orange had very important lines of transportation and supply to protect, by which raw materials ran from Burma through the Straits of Malacca and from Sumatra, Malaya, Banka, and Billiton up the western part of the South China Sea, along the coast of Indochina, and then northward to the Orange homeland. In effect, Orange was known to be willing to accept significant risks in order to maintain these routes.[1]

It was estimated that between five and ten Orange ships traveled this route per week carrying rice, grains, rubber, tin, sugar, oil, hemp, gold, chrome, and tobacco. Average net tonnage was about 3,500, and 75 percent of the ships proceeded at an average speed of ten knots. Blue sources also indicated that most of Orange's battleships and carriers were based at Amami Oshima, just north of Okinawa, and that most of its older heavy cruisers, many of its light cruisers, destroyers, and patrol vessels (PEs), and a few escort carriers and submarines were being used to protect supplies in the Bay of Bengal, the Straits of Malacca, and the South China Sea. Blue submarines reported that Orange merchant ships proceeded in these areas in convoys. It was additionally known that Orange had a number of Main and Subsidiary Bases distributed throughout the areas of Orange control, with Main Fleet Bases at Manila and Camrahn (as spelled in Problem documents) Bay; Subsidiary Fleet Bases at Singapore and Tutu Bay; Main Air Bases at Manila and Camrahn Bay; and Subsidiary Air Bases at Rangoon, Bangkok, Medan, Singapore, Banka, Puerto Princessa, Davao, Palembang, and Tutu Bay. All Orange bases were known to be strongly fortified, mined, "adequately" garrisoned, and to possess air contingents of one or more fighter squadrons at each base. Previous air raids on these targets had been costly to Blue and "unproductive" of "lasting" damage. Therefore, even temporary seizure by United Nations forces was considered beyond present capabilities. Blue understood that the bulk of Orange's Army Air Forces was regularly stationed in Manchuria, the Philippines, Thailand, Indochina, Malaya, and Burma. One or two patrol squadrons, or more, were also regularly based at Bangkok and Rangoon, and several others were stationed in the Philippines.[2]

The major portion of the Blue Pacific Fleet was based at Eniwetok; Saipan, Guam, and the Palaus were available for "large detachments." Blue submarines of the Central Pacific Force continued to operate off Orange's home ports. The United Nations' Main Bases in the southwest Pacific were Manus, Darwin, Batavia, and Surabaya. Each of these bases included a land-plane base, a seaplane base, a submarine base, and a destroyer base. Each also had a small dry dock that was capable of taking ships up to ten thousand tons' displacement and several marine railways for hauling destroyers and "small-sized" vessels out of the water. Kaoe Bay was being developed as a base and was already available as a fleet anchorage, but with services available only from afloat. Each base possessed a supply depot, a fuel depot with diesel oil and aviation gasoline, and one or more "large" hospitals. Repair facilities were "extensive," and spare parts were "readily available" or could be manufactured within "reasonable" time limits. Each base additionally offered "ample" office space with "comprehensive" communication equipment. Each was protected by Blue, Red, and Brown troops. However, these troops were neither available nor suitable for any other duty. Another negative factor was that the UN bases were crowded; ships that were not required to be at these bases had to go elsewhere. Also, at Batavia's harbor, five miles east of the city, only a limited number of large vessels could moor at docks to buoys inside the basin. Other vessels had to anchor off shore in exposed waters.[3]

The Blue Navy had also established Secondary Bases at Timor and at Saleh Bay. These bases each contained a small land and seaplane base capable of servicing fleet- and light-carrier air groups. They also had docks and cranes for servicing destroyers and seaplane tenders; larger ships had to depend on tender services for repairs. The bases additionally had ramps and anchorage spaces for patrol planes, as well as two fifty-thousand-gallon aviation gasoline tanks at each base from which gasoline tankers (AOGs) could replenish their cargoes. Ships, however, had to be refueled from fleet oilers (AOs). Still, the anchorages had been well surveyed, and each base had twenty-five berths for large ships, with mooring buoys ready. No dry docks or marine railways, however, were available. The importance of the bases was reflected by the fact that each was defended by a Blue Marine Defense Battalion, reinforced with antiaircraft units. Each Blue base had twelve fighter planes for local defense, was "well equipped" with radar of the latest type, had "well trained" aircraft warning services, and was ready to conduct fighter direction. "Adequate" antiaircraft batteries were so placed as to give good protection to all of these base facilities, but Blue had captured Borneo, Halmahera, and Celebes so recently that there had been no opportunity to restore the refineries and airfields destroyed by Orange. As concerned potential future bases, Tawi Tawi had been reconnoitered, and no enemy forces, defenses, or artificial harbor obstructions existed there. Balikpapan had a "serviceable" airstrip capable of handling two squadrons of planes, but fuel and service facilities would have to be provided. UN naval forces in the southwest Pacific at this time consisted of Blue forces only. These forces were under Vice Admiral BC, a subordinate of Admiral BN, Commander-in-Chief of the Pacific Fleet and Commander-in-Chief of the Pacific Ocean Areas (CINCPOA). The Blue naval component of United Nations Forces, Southwest Pacific was organized as the Blue 7th Fleet.[4]

Vice Admiral BC, Commander, 7th Fleet (Col. Leland Kuhre, U.S. Army, Command and Staff Class of December 1945)

Admiral BN had decided that the time had come to begin operations for extending Blue control westward to the China coast, with the immediate objective of capturing Formosa and the ultimate objective of forcing the Orange Fleet into a decisive action. The Combined Chiefs of Staff (CCS) had approved this plan. To assist BN, they had decreed that General BX, Commander, Southwest Pacific (COMSOWESTPAC), would command all ground, air, and naval forces south of the equator and east of 132 degrees east longitude. Red Admiral LMB, Commander of United Nations forces in Southeast Asia, would command all ground, air, and naval forces in India, Burma, the Bay of Bengal, and the Straits of Malacca west of 100 degrees east longitude, as well as those in the Indian Ocean west of 95 degrees east. Vice Admiral BC would remain directly subordinate to Admiral BN but would command all ground, air, and naval forces west of 132 degrees east longitude and east of the eastern boundary of the area commanded by Admiral LMB, as well as those forces between latitude 20 degrees north and 20 degrees south. Vice Admiral BC was subsequently directed by a Letter of Instruction from Admiral BN to plan a "systematic" campaign for the destruction of Orange ships, port facilities, and other shore installations in his area. Ultimately, as noted above, the purpose of the operation was to create "a diversion of Orange naval strength (especially battleships [*sic*] strength) from the north, in order to present an opportunity for the Blue Pacific Battle Fleet to bring the Orange Fleet (or important elements of it) to decisive action."[5]

Admiral BN further directed that these operations commence on the first of the next month and that they be prosecuted with "vigor" from the outset. BN informed BC that at such time as it appeared that important Orange Battle Fleet strength was being drawn northward, and Orange capital ships were being sent southward, to oppose the new Blue threat to Orange sea-lanes, the Blue Pacific Fleet would proceed westward to prevent it. Vice Admiral BC was further informed by Admiral BN that the Blue 7th Fleet would be reinforced by a detachment, designated Task Force 18, from the Blue Pacific Fleet and that BC was free to use this force as he saw fit. Task Force 18 would arrive in Kaoe Bay on the 25th of the month so that it could be 95 percent fueled prior to the commencement of operations from its own oilers, which were then to return to the Central Pacific. The Letter of Instruction also instituted an "orderly" flow of aircraft to the Blue 7th Fleet in weekly groups as replacements for operational losses, with each weekly flight to number about 10 percent of each type of plane in the 7th Fleet. The first flight of these replacement aircraft would arrive at Saleh Bay on the 30th, but 7th Fleet air strength could not be increased after that owing to a lack of facilities in the 7th Fleet area. Finally, Admiral BN notified Vice Admiral BC that General BX and Admiral LMB were undertaking similar but independent operations in their theaters of responsibility.[6]

Rear Admiral BC, Commander, Blue Carriers (Capt. Donald Hilton, U.S. Navy, Command and Staff Class of December 1945)

It was the Blue 7th Fleet that was to draw the Orange forces to where they could be decisively attacked by the Blue Pacific Fleet. The Blue 7th Fleet was a substantial force. First listed in its Type Organization was Cruisers, 7th Fleet, consisting of six modernized prewar light cruisers, organized into Cruiser Divisions 13 and 14 and commanded by Rear Admiral B-15 at Surabaya. Next, Destroyers, 7th Fleet, commanded by Rear Admiral B-14 at Saleh Bay and Koepang, consisted of four destroyer tenders (ADs), ten destroyer leaders, and eight destroyers, organized into Destroyer Squadrons 4 and 21. Aircraft, 7th Fleet was commanded by Rear Admiral B-3 at Surabaya and Batavia; it had ten seaplane tenders and eight squadrons of PBY Catalina, PBM Mariner, and PB4Y Liberator naval patrol bombers (VPBs); these units were constituted into Fleet Air Wings (FAIRWINGs) 7 and 10 and were all equipped with the latest bombing and search radars. Mine Craft, 7th Fleet was commanded by Captain B-31 and consisted of thirteen destroyer minesweepers (DMSs) organized into Mine Squadron 2 (MINRON 2), based at Surabaya and Batavia. Also at those two locations, as well as Saleh Bay, was Service Force, 7th Fleet, under Rear Admiral B-17, twelve destroyer escorts (DEs) organized into Escort Squadron (CORTRON) 16, as well as seven fleet oilers, two gasoline tankers, two ammunition ships (AEs), two provision store ships (AFs), and two cargo

Rear Admiral B-14, Commander, Destroyers, 7th Fleet, and Rear Admiral B-17, Commander, Service Force, 7th Fleet and Task Force 79, Service and Escort Force (Col. Ralph Zwicker, U.S. Army, Command and Staff Class of December 1945)

Rear Admiral B-3, Commander, Aircraft, 7th Fleet and Task Force 75, Air Force (Col. William Woodward, U.S. Army, Command and Staff Class of December 1945)

Rear Admiral B-4, Commander, Cruiser Division 20, 7th Fleet, and Rear Admiral B-6, Commander, Carrier Division 16, 7th Fleet and Task Force 78, Striking Force (Col. Glenn Nye, U.S. Army, Command and Staff Class of December 1945)

Rear Admiral B-2, Commander, Cruiser Division 6, 7th Fleet (Col. Jack Saunders, U.S. Army, Command and Staff Class of December 1945)

ships (AKs). Operations Problem documents gave cargo-fuel and aviation-gasoline capacities for each type of ship that could carry these products, pointing out that while the gasoline tankers could discharge to barges and seaplane tenders, they could not refuel planes directly.[7]

Task Force 18, from the Pacific Fleet, was to be at Halmahera in preparation for operations. It consisted of the three battle cruisers of Cruiser Division 20, under Rear Admiral B-4; four heavy cruisers under Rear Admiral B-2, commanding Cruiser Division 6; and eleven light cruisers organized into Cruiser Divisions 11, 12, and 15, commanded by Rear Admirals B-11, B-7, and B-12, respectively. In addition, there were two fleet carriers and three light carriers organized into Carrier Divisions (CARDIVs) 5 and 11, respectively, commanded by Rear Admirals B-6 and B-9, respectively. These carriers employed a total of 144 Hellcat fighters, seventy-four Helldiver scout bombers, and sixty-three Avenger torpedo bombers. Moreover, each carrier division was to have a relief air group ready for service at Saleh Bay, in addition to the weekly replacement aircraft that were to be expected. Task Force 18 had the destroyer leader and eight destroyers of Destroyer Squadron 5, under Captain B-9; the destroyer leader and eight destroyers of Destroyer Squadron 6, under Captain B-23; and nine destroyer leaders commanded by Captain B-24. Blue 7th Fleet submarines operated under a separate plan that was coordinated with those of other Blue Fleet submarines. Specifically, the Blue Fleet had two boats observing Camrahn Bay and another two off Manila, as well as one each off Davao Gulf and Tutu Bay. Another nine submarines were dispatched in solitary patrols at various sectors in the operating area.[8]

Weather was planned for in some detail. The operational period was in summer, the season of the southwest monsoon, entailing a significant amount of rainy and cloudy weather, hard squalls, and heavy gusts of wind and rain. Weather in the Philippines would be "exceedingly" wet—an important factor in ship and aircraft operations. Typhoons, normally originating south and southeast of the Marianas and Carolines, would be "relatively frequent," and local storms in the Philippines known as "collas" could interfere with aircraft operations for as long as ten days at a time. Except during typhoons and local storms, however, it was assumed that the South China Sea, the Sulu Sea, and the northern part of the Celebes Sea would be favorable for surface, submarine, and aircraft operations. Weather in the southern Celebes Sea and around the Palaus was to be generally favorable for the operation of all craft—except during daily rain squalls, which were of "considerable" extent. The Java and Arafura Seas would also have good weather as far south as Australia, but with significant haze and "rough" seas in the frequent squalls south of Java. The Bay of Bengal would have monsoons that brought clouds, rain, and occasional cyclones. In areas of heavy rain, squalls, or typhoons, aircraft operation would be subject to "frequent" interruption. In general, long flights were considered hazardous, air navigation was difficult and likely to be inaccurate, and pilots had to be skilled in "blind" flying (i.e., on instruments). Discharges of static electricity were known to interfere with aircraft radio, and the wet atmosphere would generally increase the difficulty of airplane maintenance. In all, "unfavorable weather conditions may be expected to have a marked effect upon naval and military operations . . . between June and October." A weather map accompanied the analysis, as did tables for sunrise, sunset, moonrise, and moonset.[9]

Requirements of the Student Officer

Operations Problem 5 was divided into three phases. In Phase 1, student officers each produced a Solution of the Basic Problem, which would ultimately lead to a Basic Operation Plan for the force as a whole. Student officers would in this phase receive assistance from Staff Solutions furnished by the Maneuver Staff. Annexes for the Operation Plan were to be prepared to cover logistics, such as fuel for surface vessels, gasoline for aircraft, and aircraft replacements. Additionally, a Communication Plan had to be prepared, as did any other plans deemed necessary by the student officers for the "proper" supervision of the planned action.

Phase 2 was to cover the transition from Solutions of the Problem to the major subdivisions of the Basic Plan that would be played; Phase 3 was comprised of the Supervision of the Planned Action in a Chart Maneuver, followed by a Critique of all the phases. Student officers could be assigned as Commanders of major subdivisions of the fleet or as members of a Commander's Staff. They could also be assigned to assist the Maneuver Staff in the Master Plot. The Exercise was held between 24 September and 17 October (that is, actual, not game, dates), with Phase 1

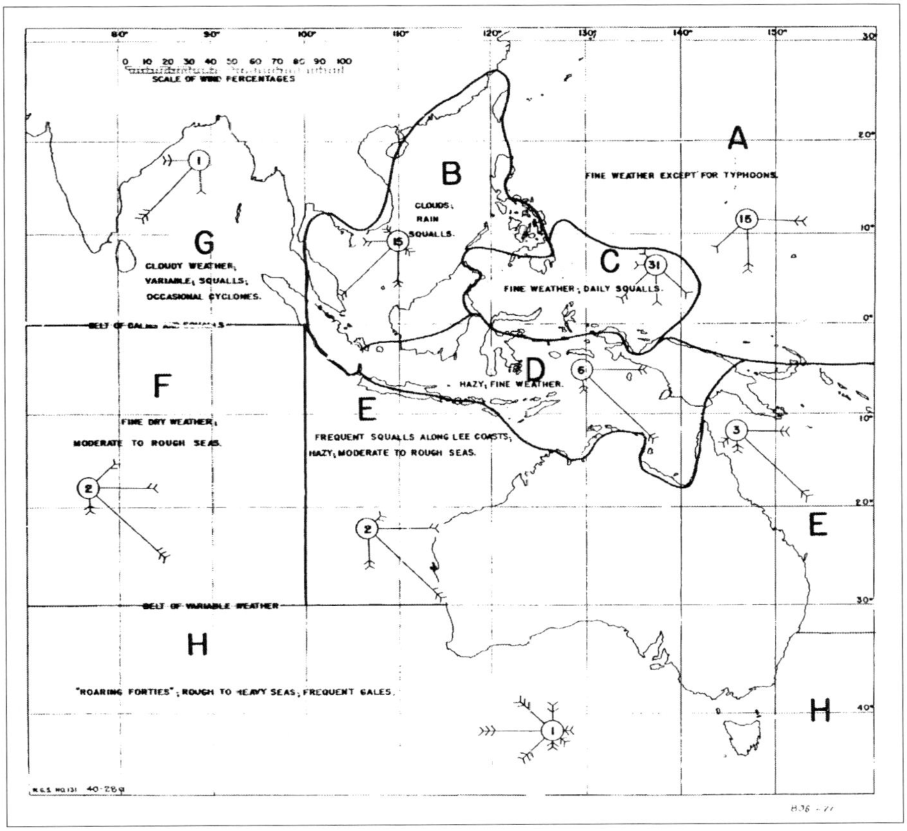

Map 2
Weather map, southwest Pacific

between 24 and 26 September, during which time student officers would draw up Blue and Orange Statements and Solutions of the Problem, Summaries and Conclusions of Relative Fighting Strengths, and, for each side, the Operation Plan itself. Phase 2 was to take place between 26 September and 4 October and would entail copying the Operation Plan and ensuring that each student officer who was a Subordinate Commander in the Plan was personally capable of solving entirely the problem confronting him. The various plans were to be presented after completion, finalized, and distributed; then preparations were to be made for the Chart Maneuver. Phase 3 would consist of the Chart Maneuver itself, which would run from 5 to 15 October and be followed on 16–17 October by discussions and Critiques.[10]

The Blue Solution

In the Estimate of the Situation, Vice Admiral BC recalled the General Situation, with a particular emphasis on the desire of Admiral BN, CINCPOA, to force a decisive action on the Orange Fleet in order to expand Blue control westward to the China coast. He also reiterated Orange air and naval strengths in the area, which Blue Commanders would control the various forces in the coming operation, and where those forces were stationed. He also noted that Blue patrol planes could operate freely throughout the mid-Pacific islands and that the General Plan for extending Blue control to the China coast was by seizing Formosa and, again, by forcing a decisive action. He thought, however, his own specific mission—that of destroying Orange ships, port facilities, and shore installations in the 7th Fleet area—was "so restrictive as to constitute a predetermined course of action." It would be possible to draw the Orange battleships north for the desired decisive action in several different ways, but that had to be done in a way that forced Orange to send naval units to the theater of operations to protect shipping and facilities. Vice Admiral BC thought he could do this with extensive mining or air raids, but neither mining nor air raids by themselves would force Orange to send capital ships to the area. BC assumed that "only by the use of ships can I force Orange to do that."[11]

In the next step of the formal process—the Completion of the Plan—Vice Admiral BC reiterated that destroying Orange ships, port facilities, and shore installations was merely the means by which to get Orange capital ships to divert to the north; he understood also that his mission was not just to destroy Orange naval strength but to do so in such a way as to "wreak havoc," to such an extent that Orange would have to withdraw its major naval strength northward. Mines, submarines, and aircraft could accomplish part of this mission, but such platforms, as he had assessed previously, would bring Orange responses only in the form of small forces and aircraft. "Ships in strength will have to be used in the effort to force Orange to reply in kind; and I have that strength, particularly in my large cruisers and carriers." Given this situation, Vice Admiral BC did not consider that a distinct, separate preliminary phase would be necessary. He did not think his information on Orange was as complete as it should be, but his general knowledge of the key trade routes and focal points was a sufficient basis for planning the initiation of the campaign. Along these lines, he wanted to avoid stepping up air patrols and searches until about D+6 Day (D-Day being the date on which strikes on harbors began, as explained below), in order to achieve some degree of surprise. These searches, once commenced, would be coordinated with ship movements, and he assumed he would receive considerable information on Orange naval activity in the northern sector of the South China Sea from Blue submarines once those units were in their patrol areas.[12]

The following times of sunrise and sunset, moonrise and moonset are L.C.T. on the equator at 120° East Longitude (zone minus eight time). This data is for use in this problem only. The times of sunrise and sunset vary little from year to year but the time of moonrise and moonset is subject to radical change.

The times given herewith may be used throughout the area, although there may be appreciable variation as the latitude and longitude changes.

July	Sunrise	Sunset	Moonrise	Moonset	Remarks
					25 June Full Moon
1	0600	1807	1515	0230	
2	0600	1807	1559	0326	Approximate Brightness of
3	0600	1808	1643	0422	the Moon:
4	0600	1808	1728	0520	
5	0600	1808	1816	0618	Full (Standard) 1
6	0601	1808	1907	0718	2 Days from full 2/3
7	0601	1808	2001	0819	3 " " " 1/2
8	0601	1809	2058	0920	4 " " " 1/3
9	0601	1809	2156	1019	5 " " " 1/4
10	0601	1809	2254	1115	6 " " " 1/5
11	0602	1809	2350	1206	7-10 " " 1/10
12	0602	1809	----	1253	
13	0602	1809	0043	1335	
14	0602	1810	0133	1413	
15	0602	1810	0222	1451	
16	0602	1810	0308	1527	
17	0602	1810	0354	1602	
18	0602	1810	0441	1639	
19	0602	1810	0529	1719	
20	0602	1810	0618	1801	
21	0603	1810	0710	1847	
22	0603	1810	0804	1937	
23	0603	1810	0859	2031	
24	0603	1810	0963	2129	
25	0603	1810	1047	2227	Full Moon
26	0603	1810	1138	2326	
27	0603	1810	1227	----	
28	0603	1810	1313	0025	
29	0603	1809	1357	0122	
30	0603	1809	1441	0218	
31	0603	1809	1526	0315	
Aug.					
1	0603	1809	1612	0413	
2	0603	1808	1701	0511	
3	0603	1808	1753	0610	
4	0603	1808	1848	0710	
5	0602	1808	1945	0809	
6	0602	1808	2042	0905	
7	0602	1807	2138	0957	
8	0602	1807	2232	1045	
9	0602	1807	2324	1129	
10	0602	1807	----	1201	
11	0602	1807	0014	1248	
12	0602	1806	0101	1324	
13	0602	1806	0148	1400	
14	0601	1806	0234	1436	
15	0601	1806	0321	1514	
16	0601	1806	0410	1554	
					23 August Full Moon.

Fig. 77
Appendix 2, sun and moon data in the theater of operations

In the next step in the Solution procedure, Vice Admiral BC reviewed Orange's dependence on raw materials from Southeast Asia, its pattern of shipping from that region to the Orange homeland, the Orange southern sector of operations, around Singapore, and the more northerly sector focused on Thailand, Indochina, and the Philippines. He thought that his own forces were advantageously placed for operations against these Orange shipping concentrations. The problem was that he

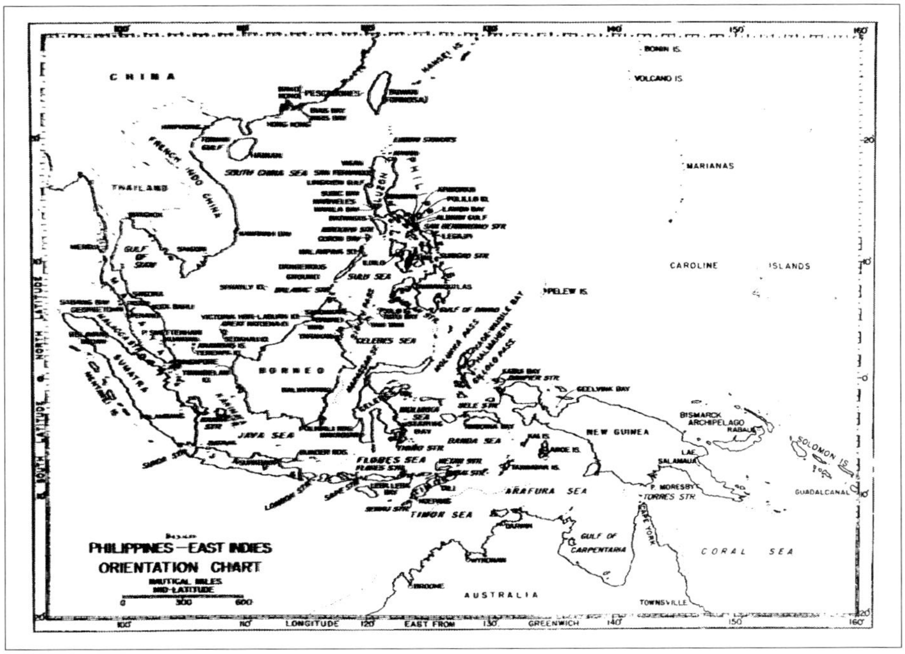

Map 3
Philippines and East Indies Orientation Chart

was not certain about the location of Orange forces, other than the fleet at Amami Oshima. He thought that the Orange naval base nearest his own base was Singapore, six hundred miles away by water from Batavia and five hundred miles by air, and he did not think that Orange would have put major fleet units at a Subsidiary Base like Singapore. He guessed, in fact, that Orange had its major fleet units near where he was going to start the campaign. Orange would then be able to dispatch reinforcements quickly, but since his mission was to destroy major portions of the Orange Fleet, he did not see this proximity as a disadvantage to him.[13]

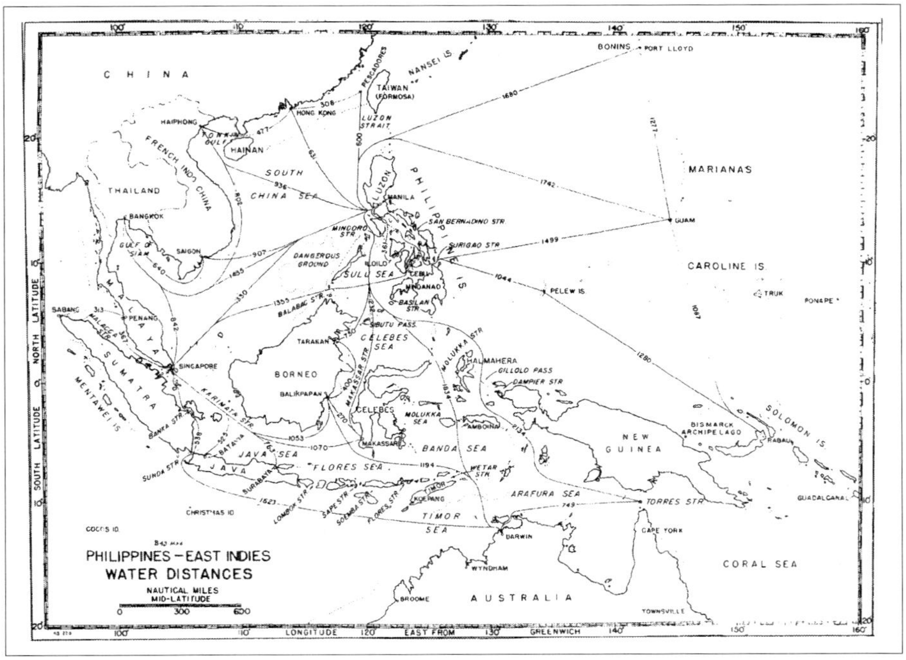

Map 4
Philippines and East Indies Water Distances

Admiral BC further reasoned that since Orange's conduct of the war was based on its ability to obtain raw materials from its colonies and conquered countries, its ability to wage war would become questionable if it were denied them. Given the direct bearing his mission had on the success of the Blue-Red-Yellow alliance, Vice Admiral BC assumed that Orange would "take the bait" with its heavy fleet units. BC did not assume that its recent setbacks had lowered

Orange morale but felt that perhaps they had even given it a greater determination to win. Cutting this last major line of transportation and supply, however, might be Blue's best way to lower Orange morale seriously. He was taking all safeguards to ensure that information did not leak to Orange, especially given the "well known propensity of the Orange people as spies."[14]

Vice Admiral BC now compared Blue and Orange combatant ships that might encounter each other in the coming campaign, The Orange *Kongo*-class battleships, with an average age of twelve years, had twenty-seven-knot speeds and main armaments of eight fourteen-inch guns with a range of thirty-two thousand yards. BC thought these ships would be a "fair match" for his battle cruisers, but he saw the Orange ships as inferior to his in some ways, mainly in terms of speed and average age. He went on to the Orange *Nagato*-class battleships, with speeds of only twenty-three knots and ages of nearly nineteen years on average but main batteries of eight sixteen-inch guns with ranges of thirty-six thousand yards. The *Ise*-class ships had eight fourteen-inch guns with ranges of thirty-two thousand yards and speeds of twenty-three knots. Ships of the *Yamato* class had more speed, were newer, and carried, he thought, sixteen-inch guns that had ranges of forty thousand yards. Thus, he did not think that Blue battle cruisers were a match for Orange battleships, except perhaps the *Ise*- and *Kongo*-class ships, which his battle cruisers exceeded in speed and range. None of the Orange battleships, however, could match the speed of his battle cruisers.[15]

As regarded heavy cruisers, Vice Admiral BC pointed out that Orange heavy cruisers, with few exceptions, had more main-battery guns per ship than Blue heavy cruisers and had an advantage in range, if a very small one. Orange heavy cruisers also had a small advantage, he thought, in speed. A significant advantage for the Orange ships, however, was that they carried twenty-four to thirty-six torpedoes and could lay mines. Blue heavy cruisers had no torpedoes or mines, though their antiaircraft batteries were equal to Orange's. "On the whole, the CAs of the two navies appear to be well-matched, except in torpedoes." In terms of light cruisers, Orange light cruisers were of about half the displacement of Blue light cruisers, except for the lighter Blue *San Diego* and *Oakland* classes. The ages of these Orange ships were comparatively high, indicating to BC that they must be "tough" ships. Nevertheless, their gun batteries were considerably weaker than those of their Blue counterparts, except the *Agano* class, which compared favorably to the Blue *Cleveland* class. Still, Orange's modernized prewar light cruisers appeared to be weaker than Blue's. Again, however, Orange light cruisers carried torpedoes, whereas only the *San Diego* and *Oakland* classes, as well as the prewar ships, did so in the Blue fleet. In spite of the torpedo advantage, Vice Admiral BC saw Blue as having a "decided" advantage in terms of light cruisers.[16]

BLUE

Class	No.	Speed Max.	Radii	Life	Batteries	Range	Torpedoes Tubes	Mines/DC Planes
					Large Cruisers			
CB-1	3	33	15-12000	10.7	9-12"/50 12-5"/38h 44-40 mmh 40-20 mmh	36 17	- -	0/0 6VSO
					Heavy Cruisers			
CA-32	4	32.6	29-2420 12-8600	4.6	9-8"/55 8-5"/25h 16-40 mmh 12-20 mmh	30 14	- -	0/0 4VSO
					Light Cruisers			
CL-43	3	32.5	29-2830 12-9160(2) 9800(1)	4.5	15-6"/47 8-5"/25h 16-40 mmh 20-20 mmh	24 14	- -	0/0 4VSO
CL-53	2	32.5	29-3110 15-7590	3.3	16-5"/38h 8-40 mmh 13-20 mmh	17	12/8	0/16 0
CL-95	2	32.5	29-3110 15-7590	3.3	12-5"/38h 16-40 mmh 14-20 mmh	17	12/8	0/16 0
CL-55	4	32.5	29-3630 10-11290	4.6	12-6"/47 12-5"/38h 12-40 mmh 21-20 mm	24 17	- -	0/0 4VSO
OCL-8	6	32.5	29-2000 10-8900 (Ave.)	3.7	10-6"/53 6-3"/50h 6-40 mmh 8-20 mmh	20 10	10/6	0/0 2VSO
					Destroyers			
DL-357	3	34.5	31-1440 15-5580	1.5	5-5"/38h 8-40 mmh 6-20 mmh	17	8/8	0/27
DL-445	18	38	15-6500	1.5	5-5"/38h 6-40mmh 10-20mmh	17	10/10	0/46-300# 22-600#
DD-364	16	35	32-1380 15-6160	1.4	4-5"/38h 4-40 mmh 5-20 mmh	17	12/12	0/20
DD-380	4	36	32-1500 15-4815	1.4	4-5"/38h 8-20 mmh	17	16/16	0/27
DD-386	4	35	32-1380 15-6160	1.4	4-5"/38h 2-40mmh 6-20mmh	17	16/16	0/20
DE-739	6	24	12-6000	1.3	3-3"/50h 2-40 mm 4-20 mm	10	3/3	0/75 300#
DE-217	6	24	12-6000	1.3	3-3"/50h 44-1"/MGh 4-20 mm	10	3/3	0/75 300#
					Minesweepers			
DMS-1	13	31.1 to 32.5	27-1160 to 1770 15-3500 to 4600	1.2 1.3	4-3"/50h 5-20mm h	10	---	---
					Aircraft Carriers		**Planes**	
CV-12	2	33	30-6440 15-17200	10.6	12-5"/38h 44-40mm h 46-20mm h	17	36 VF 37 VSB 18 VTB	
CVL-22	3	32.5	30-3400(e) 15-1100	4.6	26-40mm h 16-20mm h	—	24 VF 9 VTB	

Class	No.	Speed	Radii	Life	Auxiliaries	Rations	Stores	Torps	Cargo Fuel
AD-3	1	16	10-6760	3.1	4-5"/38h 16-1".1 16-20mm	13DD-90 das.	9 mos.	216	1532
AD-9	1	13	10-10400	2.5	4-5"/50h 8-1".1 16-20mm	13DD-45 das.	9 mos.	12	- -
AD-14	1	16.5	16.5-10000	3.4	4-5"/38h 8-40mm 12-20mm	18DD-90	6 mos.	105	
AD-16	1	16.5	10-20000	3.2	4-5"/38h 8-1".1 4-20mm	18DD-90 das. (Assumed same as AD-14)	6 mos.	105	
AE-3	2	16	10-19500	2.4	1-5"/51 4-3"/50h 8-1".1 8-20mm	Cargo - 1930 T. ammunition 160 T. meat			
AF-10	2	15.5	10-13000	2.1	1-5"/51 4-3"/50h	Cargo-845 T. refrig. meat 1300 T. vegetables 1875 T. dry prov.			
AK-16	2	11.5	10-9760	2.0	1-5"/38h 4-3"/50h 8-20mm	Cargo space - 288,000 cu. ft.			

Destroyers (Cont)

Class	No.	Speed Max.	Radii	Life	Batteries	Range	Torp/Tubes		Mines/DC
						Cargo Oil			Cargo Gasoline
AO-1	1								
AO-1	1	14	10-5610	2.3	2-5"/51 4-3"/50h 8-20mm	7854			536
AO-2	1	14	--	2.3	1-5"/38h 4-3"/50h 8-1".1 h 8-20mm h	7854			617
AO-36	5	18	--	2.3	1-5"/38h 4-3"/50h 4-40 mm h 12-20 mm h	16,000(assumed)			1200
AOG-1	2	14	14-6000	1.5	4-3"/50h 12-20 mm h				100 (assumed)
						Rations	Stores	Torps	Cargo Oil
AV-4	1	18	16.5-10000 12-12000	3.2	4-5"/38h 22-20 mm.h 8-40 mm h	24VP-60 das.	3 mos.	-	- 917 T. (Gas)
AV-8	2	16.5	16.5-10000	2.3	1-5"/38h 4-3"/50h 20 m.g.h				575 T. (Gas)
AVD-1	6	25	22-2360 12-3800	1.3	2-3"/50h 7-20 mm h				30,000
AVP-6	1	14	12-5150 10-6200	1.1	2-3"/50h 4-20 mm h	An AVP can act as a			
AVP-11	4	18.5	15-6000	1.4	4-5"/38h 4-40 mm h 6-20 mm h	temporary base for 12 VPB for 1 week.			

Fig. 78
Blue major combatant characteristics

Moving on to destroyers, BC noted that the Orange destroyer leaders of the *Takanami* and *Terutsuki* classes were a bit larger and more heavily gunned than Blue's *Fletcher* class, while the Orange destroyers of the *Fubuki* class compared "favorably" with the *Fletchers*, though of slightly less displacement. Orange's 1,500-ton destroyers appeared to have some advantages—especially in gun power and torpedo fire—over the Blue classes, whereas the Orange destroyers of 1,400-ton displacement were inferior to his own destroyers. He also thought that his destroyer escorts were "suitable" and effective as antisubmarine screens and had "some" value as antiaircraft screens. Looking at aircraft carriers, Vice Admiral BC found the Orange *Shokaku*-class fleet carriers to compare favorably with the Blue *Essex* class; other Orange fleet, light, and escort carriers of varying displacements had numbers of planes ranging from fourteen to seventy-two. All of the Orange carriers except for the *Shokaku* class were slower than the Blue fleet and light carriers. Admiral BC stated that there was no definite information about the type or location of Orange aircraft, so he could only examine aircraft generally by type, in order to determine "maximum economical" performance. After referring to tables (figures 79 and 80) of the range, air speed, and load capacity of both Blue and Orange ship-based and land-based fighters, dive-bombers, torpedo planes, and scout planes (VSs), he pointed out that all Blue planes were equipped with IFF. He assumed that all Orange planes—which included Zeke fighter planes, Kate torpedo planes, Betty land-based bombers, Pete ship-based observation scout planes, and Emily land-based scout planes—were similarly equipped, and he also assumed that all Orange patrol planes had the latest bombing and search radars, since Blue's did as well.

Type	Load Condition	Air Spd.	Range (Econ) 40%	Range (Econ) 80%
Land-based				
VPB or	Patrol-Max Fuel	120	1490	2980
VTB	Bomber-16-100#	115	1324	2648
"EMILY -22"	or 2 Torpedoes	115	1324	2648
VB	Patrol	135	1142	2284
"BETTY-22"	Bomber-4-500#	135	980	1960
	or 1 Torpedo	135	980	1960
Carrier or land-based.				
VF	Fighter Nor.Fuel	130	392	782
"ZEKE-52"	Fighter Max.Fuel	125	568	1136
VTB	Scout	115	648	1296
"KATE-12"	Torpedo	115	424	848
Other ship-based				
VOS	Scout	115	226	452
"PETE-11"	A/S Patrol-2-325 #DC	115	208	416

Fig. 79
Orange combat aircraft characteristics

Other Aspects of the Blue Plan: Geography and Harbors

Vice Admiral BC summarized again the location of his Main Bases, their facilities for such things such as ship repair, and their security from Orange attacks. The logistical capabilities of his Main Bases and of the supply ships attached to his fleet, combined, made his logistical situation "sufficient" for his purposes and his mission. He similarly assessed Orange bases to be well defended and to possess "adequate" logistics, since Orange had occupied the areas for so long.[17] He then examined at length the characteristics of his theater of operations. Looking at its geographic boundaries—which ranged from Thailand, the Malayan Peninsula, and Sumatra, to the Netherlands East Indies and the Philippines—BC thought that the existence of many large and small islands, many passages and small seas, and the large amount of shoal water to the south might affect to a "considerable" degree his "freedom of action." The principal entrances to the area were the Malacca and Sunda Straits, from the west and southwest; the Bali, Lombok, Alas, Flores, and Ombai Straits, from the south; and the Dampier, San Bernardino, and Formosa Straits, Gilolo and Molucca Passages, and Balintang and Bashi Channels from the east and north. Bali Strait was narrow with strong currents, while Alas was shallow and adequate only for sailing craft. Important passages in the area were the

4948-4348
19/Mar/45 lm

Type of Plane	Load Condition	Economical Speed: Air Speed	Economical Speed: 40% Miles	Economical Speed: 40% Hr.	Economical Speed: 80% Miles	Economical Speed: 80% Hr.	Economical Speed: Fuel Used GPH	Cruising Speed: Air Speed	Cruising Speed: 40% Miles	Cruising Speed: 40% Hr.	Cruising Speed: 80% Miles	Cruising Speed: 80% Hr.	Cruising Speed: Fuel Used GPH	Units	No. of Planes	Location
Hellcat	Fighter-Max Fuel	145	592	4.1	1184	8.2	40	215	412	1.9	824	3.8	85	VF-12	36	CV-12
(VF)														VF-13	36	CV-13
F6F-5	Fighter-1-1000#	145	486	3.4	972	6.7	50	200	360	1.8	720	3.6	90	VF-22	24	CVL-22
	Fighter-Norm.Fuel 6 Rockets	140	364	2.6	728	5.2	40	215	248	1.2	496	2.3	85	VF-23	24	CVL-23
														VF-26	24	CVL-26
														VF-5R	36	SA EH BAY
														VF-11R	24	SALEH BAY
Hell diver	Scout-Max Fuel	130	644	4.9	1288	9.8	45	165	560	3.4	1120	6.8	65	VS-12	37	CV-12
(VSB)	Bomber-2-1000#	135	468	3.4	936	6.9	50	165	440	2.7	880	5.4	65	VS-13	37	CV-13
SB2C-4	Patrol-2-325# DC	125	500	4.0	1000	8.0	45	160	400	2.5	800	5.0	70	VS-5R	37	SALEH BAY
Avenger	Scout-Max Fuel	145	694	4.8	1388	9.6	50	160	520	3.2	1040	6.5	75	VT-12	18	CV-12
(VTB)	Bomber-1-1000#8R.	135	350	2.6	700	5.2	50	160	288	1.8	576	3.6	75	VT-13	18	CV-13
TBM-3	Bomber-1-1000# 2-Wingtanks	135	546	4.0	1092	8.0	55	160	480	3.0	960	6.0	70	VT-22	9	CVL-22
														VT-23	9	CVL-23
	Torpedo-1 Mk XIII	130	524	4.0	1048	8.0	55	160	428	2.7	856	5.4	80	VT-26	9	CVL-26
														VT-5R	18	SALEH BAY
														VT-11R	9	SALEH BAY
Seahawk	Scout-Max Fuel	110	378	3.4	756	6.8	25	165	288	1.8	576	3.5	50	Misc.	63	CB,CA,CL,
(VSO)	Patrol-2-325# DC	110	228	2.0	456	4.0	25	165	148	0.9	296	1.8	55			and Misc.
SC-1																
Catalina	Patrol-Max Fuel	100	884	8.8	1768	17.7	75	115	584	5.1	1168	10.2	115	VP-101	12	SURABAYA
(VPB/VPT)	Bomber-4-1000# or 4-650# DC	95	440	4.6	880	9.2	70	110	330	3.0	660	6.0	110	VP-102	12	"
PBY-5A														VP- 63	12	"
	Patrol-4-325# DC	100	816	8.3	1632	16.6	80	115	584	5.1	1168	10.2	115	VP- 72	12	"
	Torpedo-2 Mk XIII	95	522	5.5	1044	11.0	70	110	418	3.8	836	7.6	105	VP- 73	12	"
														VP- 91	12	"
Mariner	Patrol-4-650#DC	115	926	8.0	1852	16.1	115	130	824	6.3	1648	12.6	145	VP-201	12	SURABAYA
(VPB/VPT)	Bomber-8-1000#	120	644	5.4	1288	10.7	130	135	588	4.4	1176	8.7	160	VP-203	12	"
PBM-3D	Bomber-8-1600#	120	348	2.9	696	5.8	135	130	312	2.4	624	4.8	165			
	Torpedo-2 Mk XIII	115	814	7.0	1628	14.1	130	130	720	5.5	1440	11.0	165			

AIRCRAFT BLUE

CONFIDENTIAL

Fig. 80
Blue combat aircraft characteristics

Banka, Gaspar, and Karimata Straits between the South China and Java Seas; the Makassar Strait, which connected the Celebes and Java Seas; the Sibutu Passage and Basilan Strait between the Celebes and Sulu Seas; the Mindoro and Balabac Straits, connecting the Sulu and South China Seas; and the Palawan Passage, which lay between Palawan Island and the South China Sea. It appeared to Admiral BC that all of these areas could be mined except for Makassar Strait and Palawan Passage, whose waters were too deep. Karimata Strait offered, to him, the most direct route to the principal focal point of Orange transportation and supply activities in the South China Sea but he also thought the route via the Sulu Archipelago and Balabac Strait "worthy" of consideration.[18]

The Karimata Strait was the largest of the three passages that joined the South China and Java Seas, and the only one through which a large body of ships could move in company. The passage was 115 miles wide, though islands and reefs reduced that width to about forty-five miles at some points. Because of these reefs, large vessels had to keep to a passage about seventeen miles wide between Ontario Reef and Seroetoe Island, but the passage was about sixteen fathoms deep, and had no other navigational hazards or strong tidal currents. The entire area, however, could be mined, as could the South China Sea below 10 degrees north latitude. The entire Java Sea could be mined, as could the Timor and Arafura Seas. Philippine waters too could be mined, except for the Sulu Sea. The remaining waters of the theater were too deep for mining except near coastlines. BC also pointed out that while the shoal water in the lower part of the South China Sea increased the feasibility of mining, so also did it impede submarine operations: the shallow water and coral bottom made it easier to detect submarines, especially from aircraft. The topography of the region too was going to impact operations, especially in the air. The land was mountainous along the entire west side of Sumatra, through central Java and the islands to the east of Java, and throughout Borneo, Celebes, and the Philippines. During the monsoon season, the tops of these mountains were covered with clouds, making aviation operations quite difficult. As concerned harbors, however, Sandakan Harbor on the northeast coast of Borneo had a basin anchorage of more than three miles in diameter, with depths up to sixteen fathoms. The harbor itself extended to the south and west about fifteen miles, and BC thought that a seaplane anchorage could be established there.[19]

In addition, Brunei Bay, on the north coast of Borneo, was twenty-one miles wide, with depths of eleven to twenty fathoms; seaplane anchorage would be possible near the city of Brunei. Victoria Harbor, in Labuan Island in the entrance to Brunei Bay, was small but had a good anchorage of five fathoms' depth, and Kuching in Sarawak also had anchorage adequate for seaplanes. Tarakan, 150 miles south of Sandakan, had a protected harbor of eight square miles, with a maximum depth of thirty-eight fathoms, and Balikpapan, 270 miles south of Tarakan, was an important oil port. It had a good anchorage in a harbor six square miles in area and with depths of three to twelve fathoms, but also a bar that limited the drafts of ships that could enter. There was an airstrip at Balikpapan that could handle two squadrons of planes, but Admiral BC knew he would have to provide services to it. Batavia, on the island of Java itself, was a Blue Main Base. Ships could anchor outside its breakwater in from four to nine fathoms of water or moor inside. Large ships had to anchor outside in exposed conditions, but there were "extensive" aviation facilities for both land planes and seaplanes, as well as docking facilities for ships up to ten thousand tons; general repairs of any nature could be undertaken. Surabaya was another Blue Main Base. Its harbor was not deep enough for battleships and fleet carriers—which had to use Bunder Roads, fifty-three miles to the east off Madura Island—but it afforded a good anchorage for seaplanes, and it could handle land planes as well. Bunder Roads had an anchorage that could accommodate a limited number of large ships, but it could be mined and be sealed off by nets. While it was exposed to monsoons from the south and east, it was protected from the west.[20]

Saleh Bay in Sumbawa extended in an east-southeasterly direction about forty-five miles and had an average width of about ten miles. Fleet- and light-carrier aircraft could be serviced at this Blue Secondary Base; Kaoe Bay in Halmahera had a large protected anchorage but did not have shore base facilities usable for the current operation. Admiral BC next looked at Koepang Bay, on the southwestern end of Timor Island. This bay was open to southwest monsoons, but in other seasons it offered a protected anchorage with between four and twenty-five fathoms of water

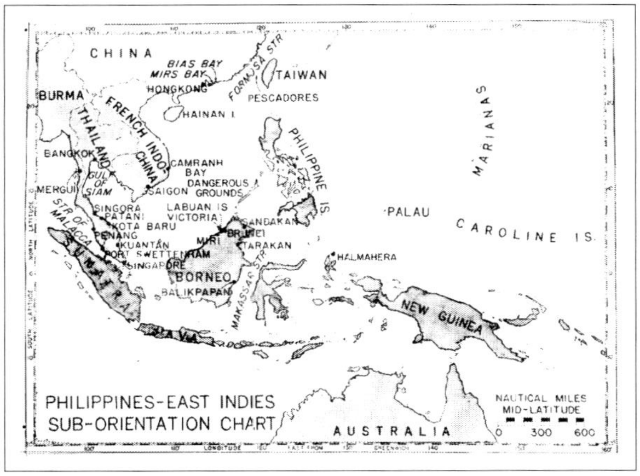

Map 5
Philippines–East Indies Sub-Orientation Chart

in an area of about twenty square miles. A more protected anchorage was located just to the south in Semaoe Strait, which was also a Blue Secondary Base. This anchorage had two entrances and depths of from eighteen to fifty-eight fathoms, but a three-knot current ran through it. It had, however, sites and facilities for land and seaplanes, and fleet- and light-carrier aircraft could be serviced. Darwin, a Blue Main Base, had a harbor with more than fifteen square miles of anchorage area and depths up to eighteen fathoms. The port was open to northwest monsoons but was protected from winds from other directions. Tidal currents there were 2.5 to five knots and tidal ranges were from sixteen to twenty-two feet, but the port had two large military airfields with full facilities, as well as a seaplane base. Finally, Manus, another Blue Main Base, had a harbor capable of accommodating the entire fleet and facilities that were sufficient for all of BC's needs for both ships and aircraft.

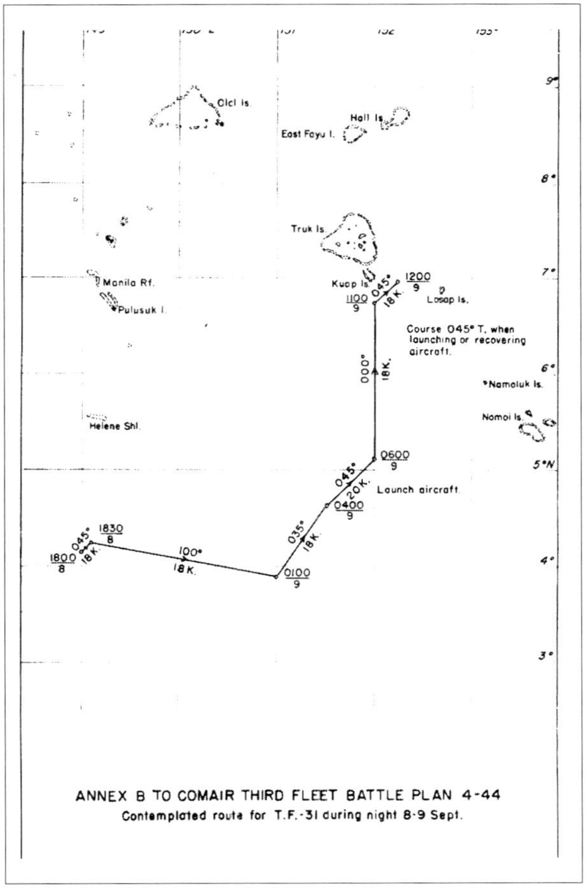

Map 6
Annex B to COMAIR 3rd Fleet Battle Plan

As noted earlier, this period of operations was to coincide with the monsoon season in the South China Sea area. BC acknowledged the need for his pilots to be able to handle difficult air navigation, hazardous long flights, and instrument flying and that aircraft maintenance would be more difficult in the generally wet atmosphere.[21]

Vice Admiral BC further noted that his bases were located on a general east–west line on the Malay Barrier, with Batavia about 475 air miles and 630 sea miles from Singapore, the nearest Orange base. Bases had not been established yet at Blue forward positions at Halmahera, Borneo, and Celebes. Balikpapan represented an advance of some five hundred miles toward Orange-held Tutu Bay and could be used as a temporary fleet-supplied air base for staging aircraft, but Tawi Tawi was unoccupied. BC felt it necessary to "knock out" the Orange airfield and facilities at Tutu Bay on Jolo Island before his fleet could venture into the Sulu Sea. BC thought that Orange ships would be relatively safe until they reached Indochina, which was flanked by

Blue-controlled Borneo and Java. Moreover, most of his lines of communication were in waters already controlled by Blue, though Orange flanked the southern water route to West Borneo by way of Sumatra, Singapore, and Indochina. The northern route of West Borneo was flanked by Sulu and Palawan. BC ended this part of the Staff Solution by looking again at the Orange cargoes being carried through the South China Sea, the average speed of the ships, their average tonnage, and the operational and logistical capabilities of the Blue and Orange bases, especially in terms of ship berthing and repair, fighter defense, mine defenses, and up-to-date communications.[22]

Blue Staff Solution: Strengths and Weaknesses

Admiral BC counted as among his strengths the opportunity to take the initiative, provided that Orange did not become aware of Blue's plans. He thought he also had "relatively" stronger forces than Orange, numerous harbors and anchorages from which to operate submarines and patrol aircraft, and a good flanking position on Orange's principal trading route from Borneo. He further counted as a strength the fact that his surface-ship bases were well located, well protected, and "adequate." Finally, he thought that his mobile bases for logistics were superior to Orange's. However, he felt that he was generally unfamiliar with the area of operations, lacked surface-ship reinforcements, and did not have information he might "reasonably" expect on the status of Orange's forces. He next noted Orange's ability to cover the area of operations with shore-based aircraft. In addition, Admiral BC did not have Army aircraft to assist him in this operation, and his surface ships could reach Orange lines of supply only by first passing through restricted waters. Orange strengths included what BC thought were better information services about the theater than Blue had. Moreover, Orange's main fleet was closer to the theater of operations, Orange could mine the area, and all Orange cruisers, destroyers, and submarines could carry both torpedoes and mines. Orange also had protected ports along its trade routes, long experience with operations in the area, and interior lines of transportation and supply. Orange was further aided by the fact that shallow waters in the lower part of the South China Sea would make it easier to find Blue submarines with aircraft. These strengths were offset by weaknesses that included an inability to divert shipping from its present trade route, the need to disperse forces in order to protect shipping, weak naval forces in the area at the moment, and a great dependence on raw materials supplied from Southeast Asia and Sumatra.[23]

In general, BC saw that he had a "fast, hard-driving" force that was "normally" well suited to conducting warfare on trade routes. Unless Orange correctly interpreted the purpose of Blue's ships concentrating in the Netherlands East Indies and Halmahera, BC thought it was within his power to take the initiative. He knew he needed this initiative, since his forces had to pass through restricted waters before

Orange could block his passage. Since Orange had army and other shore-based aircraft at his disposal (while Blue had only naval patrol planes and a dozen fighter planes at each base), BC knew that Orange could stage aircraft from the north as soon as it learned that Blue had ships in the area. Admiral BC thought that Orange only had a few surface ships and carriers to oppose his force but that the enemy could concentrate a large number of ships in the South China Sea. He did not, however, think that Orange would send forces down from Amami Oshima until Blue cut the Orange lines of transportation and supply; he could handle the Orange forces he faced until and unless that happened. If Orange did deploy most or all of those forces, however, Blue long-range patrol aircraft, BC assumed, would report this before the Blue 7th Fleet carrier force, known as the Striking Force, entered the South China Sea—and the Blue 7th Fleet would have accomplished its mission. Along these lines, he was going to have the Karimata and Balabac Straits patrolled while operations were being conducted in the South China Sea in order to ensure for his force a safe retirement route from the area if that became necessary. He reiterated that he had the strategic initiative and that his objectives were Orange ships, port facilities, and other shore installations, in that order.[24]

Tasks for Subdivisions of the Blue Force

BC now began analyzing the objectives of his force and how he planned to subdivide his forces so as to accomplish them. His first main purpose was to locate and destroy enemy shipping in the South China Sea. For this, he had three battle cruisers, four heavy and seven light cruisers, four antiaircraft cruisers (CLAAs), thirty-six destroyer leaders and destroyers, two fleet and three light carriers, and nine destroyer-minesweepers. He thought that his force was strong enough to handle anything that Orange could muster, except for battleships. However, his force could outrun the Orange battleships should they appear, and his force should be able to get assistance from shore-based forces when it was in the South China Sea. The problem, as noted above, was entering the South China Sea. The first steps were therefore to have his destroyer-minesweepers conduct exercises to ensure they could sweep Karimata and Balabac Straits of Orange mines and to make sure his air forces could neutralize Orange airfields in the region before his ships attempted passage. His second task, for which he had fighter squadrons available, was to make daily searches of the Sulu Sea and the southern sector of the South China Sea so that Orange airfields and air strength could be identified in preparation for neutralization. He also wanted these fighter patrols to determine Orange surface ship strength in the area. His third task followed from the second—to obtain photographic reconnaissance on D–8, D–3, and D–1 of Orange airfields, oil refineries, mines, smelters, and harbors at Tutu Bay, Davao, Singapore, Banka, Puerto Princessa, and Palembang. This photo reconnaissance by the fighter squadrons should supply him the information he needed about Orange port facilities and shore installations.[25]

His next task was to ensure that his Striking Force and his naval patrol plane squadrons were available to destroy targets at the Orange bases designated by him. These operations were to commence on D-Day, followed by antisubmarine patrols off Batavia, Surabaya, and Saleh Bay by naval patrol plane squadrons. Next, his Service Force was to provide support and escorts for the operation in the form of six prewar modernized light cruisers, nine destroyer leaders and destroyers, four destroyer-minesweepers, twelve destroyer escorts, four destroyer tenders, six fleet oilers, two ammunition ships, two provision ships, two cargo ships, and two gasoline tankers. The destroyer-minesweepers were to be made available to the Striking Force if necessary to augment the sweeping of Karimata and Balabac Straits. The Service Force had to be provided air cover, which was to come from island Base Commanders as the Service Force entered and left the respective bases, and by Commander, Striking Force when the Striking Force was within range of the operating area. The Base Relief Carrier Air Group was to supply the fleet carrier air groups with sixty fighters, thirty-seven scout bombers, and twenty-seven torpedo bombers at Saleh Bay and, at the same base, Replacement Aircraft was to ensure a complement of twenty fighters, ten naval patrol planes, eleven dive-bombers, and nine torpedo bombers. These aircraft would be supplied and supported by Commander, Aircraft in readiness for use on immediate notice by Commander, Striking Force. Finally, BC reiterated the need to conduct minesweeping operations in the shallow waters ahead of the Striking Force.[26]

Analysis of Enemy Capabilities

Given the defensive nature of Orange's objective to maintain its lines of transportation and supply, BC inferred that Orange would have to escort its shipping in convoys, cover that shipping against Blue attacks, patrol the lines of transportation and supply from Malaya–Sumatra to the homeland, and defend its port facilities and shore installations. BC also assumed, once the operation began, that Orange would attempt "nuisance" raids by aircraft on whatever Blue forces it could find, especially at Batavia, since it was so close to Orange territory. Orange capabilities to escort its convoys, he assumed, would be augmented as soon as Orange figured out the Blue objective of destroying shipping. The Orange escorts were primarily for protection from submarine and air attack, but were also there to defend against surface strikes. From the information he had on Orange forces, Admiral BC thought that the enemy could provide "fairly adequate" protection of its convoys from submarine attack, "some but not adequate" attention from air attack, and "little" protection from surface attack. BC surmised that Orange would establish covering forces in probable areas of Blue attack whenever important convoys were in motion. Because of the limited number of approaches that Blue could use—and since Orange lines of transportation and supply were flanked by Orange-controlled territory with hydrographic and topographic features favoring the defense—

Orange could cover its convoy movements and could probably mine the approaches to the theater and conduct air strikes on Blue bases. However, BC did not think that Orange could generally oppose his forces, at least not with surface ships. To do that, Orange would have to bring its capital ships into the area, the drawing out of which was BC's primary goal.[27]

Admiral BC asserted that his forces could enter the South China Sea through the Balabac and Karimata Straits, since Borneo and Karimata Island were under his control. The Karimata Strait could be mined, but he had thirteen destroyer-minesweepers. Moreover, by bombing Banka prior to making the transit and by making the transit at night, he thought his forces could get through, though they could not achieve surprise. Once in the lower part of the South China Sea, he assumed, his forces would be capable of "coping" successfully with any enemy forces that were now there, except perhaps land-based aircraft. He did not currently know what this strength consisted of but was hopeful that air reconnaissance would provide that information prior to the commencement of operations—"Such reconnaissance must be made before I despatch [*sic*] any of my forces into either area." His current information indicated that Orange had little air strength in the area, but he was going to avoid the Balabac Strait and the Sulu Archipelago, since these routes were very restrictive of movement and Orange had air bases in the Philippines, Palawan, and Tutu Bay.[28]

Admiral BC saw Orange capability for patrolling its lines of transportation and supply as consisting of air patrols of the sea-lanes in the theater as well as surface or submarine patrols of entrances to the theater and of focal points. He saw these Orange patrols as primarily meant to provide information on his forces and did not think that his forces could prevent that. He did think, however, that his forces could prevent Orange patrols from inflicting any damage. Orange's ability to protect its port facilities and other shore installations had to first be countered by Blue reconnaissance of those targets and determination of Orange strength, especially airpower. Once in the area, BC surmised, he would strike Orange shore installations with carrier aircraft rather than surface ships, because of the difficulty of movement in the area given its features and the threat of mines. He thought that preliminary strikes on Orange shore installations could be carried out by his naval patrol bombers, which had a "considerable" bomb load and radar for night strikes. His carrier bombers could carry, altogether, 135 tons of bombs for 375 miles and still return to their ships. Of course, some of the fighters were needed for protection of the carriers and the strike groups, but they could carry thousand-pound bombs if necessary; also, he had his relief air group at Saleh Bay. Unless Orange surface ship capability proved "vastly" superior to what BC thought it was, he did not think that Orange would employ surface ships to attack Blue bases in the Netherlands East Indies. He thought that Orange submarines might attempt a blockade of his major ports in Makassar Strait and Molucca Passage and that there would be Orange air strikes, but to oppose these threats, he had, respectively, light forces for antisubmarine patrols and a "considerable" air force. "The most that Orange could accomplish by this capability would be of a harassing nature and could by no means prevent me from accomplishing my mission."[29]

Tests of Proposed Tasks

Vice Admiral BC characterized Task Force 18, from the Central Pacific—plus his additional destroyers and destroyer-minesweepers—as representing "considerable" power. All of this power would be available to him once these units were in the South China Sea. He knew that Orange had its heaviest shipping between Camrahn Bay and Singapore. He did not think it would be difficult to find this shipping, given the number of aircraft he had. The destruction of Orange shipping would depend on the size of the convoys, the strength of the escort, and the distance from his task force. Still, he thought, he had sufficient force unless Orange brought in his battleships, in which case Blue would have to retire. For his pre-operations aerial reconnaissance, he would need twenty-four naval patrol planes to make daily searches eight hundred miles out from Batavia. Twelve fighter planes, he thought, would be sufficient to make daily searches from the Sulu Sea to Balikpapan; he wanted a tender deployed to Balikpapan to service the seaplanes. If the searches could be completed with fewer patrol planes, he wanted the remainder used as naval patrol bombers. He was concerned about seaplane losses—though there had not been heavy losses of seaplanes in previous operations—since he could not give them fighter cover. He saw D–6 and D–3 as the days for aerial reconnaissance of Orange shore installations and port facilities, reserving D–1 for searches, so as to lessen the load on his naval patrol plane squadrons. Moreover, Task Force 18 could furnish fighter cover for the photographic missions on D–6 to Tutu Bay and Davao, since it would be at Kaoe Bay by that time. He also saw it feasible on D–2 for TF 18 to furnish fighter cover for the reconnaissance of Puerto Princessa. A light carrier could furnish this fighter cover as well as employ its own photographic reconnaissance aircraft. Naval patrol planes with fighter cover could additionally conduct missions to Singapore, Banka, and Palembang. He saw these tasks as divided into two phases. The first would consist of photographic reconnaissance on D–6, D–3, and D–2 of airfields, oil refineries, mines, smelters, and harbors at Tutu Bay, Puerto Princessa, and Davao. Subsequent missions would take place on the same days to targets at Singapore, Banka, and Palembang. In the first set of missions, Commander, Striking Force would furnish the aircraft, while Commander, Aircraft would furnish the units for the second phase.[30]

In order to destroy the Orange targets at bases designated by BC commencing on D-Day, Commander, Air Forces would furnish forty-eight naval patrol planes to Commander, Striking Force, who would prioritize the targets. Commander, Air

Forces would furnish another sixty naval patrol bombers to neutralize Orange airfields and installations in the area. These operations would allow Commander, Striking Force to take his units into the South China Sea. Since this involved aircraft from different commands, BC would coordinate their operations, but the attacks would not be made until the photographic reconnaissance missions had been carried out. He had an additional twelve naval patrol planes to establish and maintain an antisubmarine patrol off Batavia, Surabaya, and Saleh Bay. This patrol force was also supplied by Commander, Air Forces. Air-base fighters, however, were not available for this duty, nor were replacement or relief carrier aircraft. Establishing this barrier, therefore, would have to be left to the Base Commanders.[31]

The task of providing services and escorts in support of operations was mainly logistical. Blue destroyer leaders and destroyers were capable of furnishing "strong" antiaircraft and antisubmarine escort, while Commander, Striking Force could provide air cover for refueling operations under way; local defense forces would do the same in Blue-held areas. Commander, Striking Force would provide replacement and relief aircraft—specifically, sixty fighters, thirty-seven scout bombers, and twenty-seven torpedo bombers as the Base Relief Carrier Air Group and another twenty-one fighters, ten patrol planes, seven scout bombers, and six torpedo bombers as replacement aircraft. All shore-based aircraft, as well as all carrier aircraft when ashore, would be under the authority of Commander, Aircraft. The Relief Carrier Air Group and replacement aircraft would be maintained in a state of readiness to meet the requirements of the carriers, but in an emergency, Commander, Air Forces could employ them with "due regards" for Commander, Strike Force's needs. BC's high-speed destroyer-minesweepers were also available to sweep the narrow straits and shallow waters in advance of the Striking Force as it moved into the South China Sea. In addition, there were to be continuous air patrols of Karimata Strait, since Orange would probably have mined it and would probably operate submarines there to prevent Blue egress from the South China Sea. Commander, Air Forces would be responsible for ensuring that the Blue Striking Force could leave the South China Sea if necessary. Finally, Commander, Striking Force was to avoid an engagement with superior enemy surface forces.[32]

"Freedom of Action" Problems and Their Solutions

To carry out these operations, Vice Admiral BC decided to divide his fleet into three task forces. With Blue's submarines operating under separate command, he was going to create Task Force 75, Air Force, under Rear Admiral B-3; Task Force 78, Striking Force, under Rear Admiral B-6; and Task Force 79, Service and Escort Force, under Rear Admiral B-17. He was sending an immediate dispatch with this new administrative organization. Reinforcements for the Blue 7th Fleet would be fueled and ready to sail from Kaoe Bay by 25 June. He developed a Communication Plan, in accordance with real-world Pacific Command (PACOM) procedures, that emphasized immediate and accurate transmission and receipt of contact reports as essential to the success of the campaign. All contacts were to be reported immediately to the Officer in Tactical Command of the task force the contacting ship or aircraft reported to or the TF operating in the area. If the contact was made within fifty miles of the task force, abbreviated plain language with authenticators was to be used. Beyond that range, the contact report was to be encrypted—an encryption that was to be repeated and authenticated with each transmission. Radio Manus was to place these contact reports on fleet broadcast schedules. To ensure rapid receipt of aircraft contact reports within Task Force 75, TF 78 was to maintain a watch on the TF 75 aircraft frequency. Task Force Commanders and others responsible for communications security were reminded that they would be operating in enemy waters, that Orange was known to be vigilant and "efficient" in the use of radio intelligence, and that success of Blue's operations depended to a great extent on the secrecy of the location of Blue forces. Therefore, the use of radio and radar had to be restricted to a minimum consonant with the particular military situation at hand. Officers in Tactical Command were to prescribe conditions of silence, and authenticators for aircraft and voice transmissions were to be taken from tables provided. In addition, jamming of Orange radio circuits would be undertaken only upon orders from Commander, 7th Fleet, except when in actual contact with the enemy.[33]

Authenticators were to be used with all plain-language reports and any radio transmission in which authenticity might be questioned. If authentication was requested by a receiving station, it was to be transmitted with Date-Time Group as well as an authenticator group. Authenticators themselves were to be built around the words "Determination," "Reinforcement," "Strategic," "Unhappily," "Consequence," "Rationalize," "Justifiable," "Intermediate," "Disadvantage," and "Procrastinator." The word for the particular day of the month was to be obtained from the above list by dropping the first digit from the date numeral when the date was the 10th or later. Two letters were then chosen from the selected word, any letter plus the next letter following, between the 1st and 10th of the month, inclusive; any letter with the second letter following, from the 11th to the 20th, inclusive; and any letter with the second letter preceding from the 21st to the last day of the month, inclusive. The third letter of the authenticator group was any one of the first three letters or figures of the first group in the body of the transmission.

With the exception of Destroyer Squadron 12, all vessels were expected to be at ports at which they could be fully fueled, supplied, and provisioned by 1 July, before proceeding to carry out the operation. BC reviewed the cargo capacity, in terms of oil and gas in tons, of his seven fleet oilers and two gasoline tankers. He also noted that fuel oil, including diesel fuel, was "amply" available at Darwin, Batavia, Manus,

Ship	Cargo oil (tons)	Cargo gasoline (tons)
AO-1	7854	536
AO-2	7854	617
AO-36,37,38,39,40	16,000(each)	1200(each)
AOG-1,2	----------	1000(each)
Total	79,708	7953

Fig. 81
Tankers available to Blue 7th Fleet

Type	Fuel Capacity	Consumes 24 hrs/ 15 kts	24 hrs/ 15 kts plus 20%	No. of ships	Total Capacity	Total Con. 24 hrs/15 kts plus 20%	%24 hrs. Consump. to tot.capacity
CB-1	4000	120	144.0	3	12000	432	3.6
CA-32	2100	96	115.2	4	8400	461	5.5
CL-53,95	1353	64.8	77.8	4	5412	311	5.7
CL-55	2275	81.6	97.9	7	15925	685	4.3
DL-357	637	40.8	49.0	2	1274	98	7.7
DL-445	491	28.8	34.6	18	8840	622	7.0
DD-364	510	28.8	34.6	16	8160	554	6.7
DMS-1	300	28.8	34.6	9	2700	311	10.7
CV-12	6915	144	172.8	2	13830	345	2.5
CVL-22	2640	86.4	103.7	2	5280	207	3.9
DE-217	240	18 (e)	21.6	12	2880	259	9.0
					84701	4284	5.1

Fig. 82
Fuel capacity and consumption of major Blue 7th Fleet combatant ships

and Surabaya. No fuel oil was available at Saleh Bay or Koepang, except for two fifty-thousand-gallon tanks for aviation gasoline at each of those Secondary Bases. He did not think that exact requirements for fuel oil for surface craft could be determined at this time, but estimates of probable consumption could be made on the basis of the performances of these ships. Seventy percent of the total capacity of these ships was just over fifty-nine thousand tons; he would not want any ship to fall below that percentage, so an average of twenty-five thousand tons of fuel oil would be needed every six days. Two large fleet oilers would be required to supply this amount. Seven fleet oilers were available, but a shuttle service between Batavia or some other source of supply and the Striking Force had to be established. Interim fueling of destroyers and destroyer-minesweepers would also be required, but this could be provided by the heavy ships. Surface vessels from TFs 75 and 79 would be, for the most part, at or near bases where fuel was available. Ships and auxiliaries of the task forces operating at sea or from advanced bases or anchorages where oil was not available, however, would need to request logistical services from Commander, Task Force (CTF) 79.[34]

Oilers were to obtain fuel from Batavia, Surabaya, Manus, or Darwin. CTF 79 would schedule regular refueling for all the task forces, arranging the first refueling before the operation commenced. Since rendezvous for subsequent refuelings were dependent on operations and movements of the task forces, these were to be arranged as operations progressed; the details were left to the Task Force Commanders. These arrangements could include informing the oilers week by week as to the next rendezvous, relaying via a friendly plane, sending a destroyer to the tankers to transmit the details, or flying a plane to the nearest base. Requests for fuel were to be made directly to CTF 79; Vice Admiral BC was to be an information addressee.[35]

Ship	Capacity (tons)	Planes	1 Complete Fueling (Tons)
CV-12, 13	917	36 VF, 37 VB, 18 VT	152
CVL-22,23,26	300	27 VF, 9 VT	50
AV-4	917)	24-PBY-5A	
AV-8,9	575)		60 (per squadron)
AVD-1,4,5,7,11,12	100)	24-PBM-3D	100 (per squadron)
AVP-6	100)	48-PB4Y	
AVP-11,12,21,24	250)		
Base	**Capacity**	**Planes**	**1 Complete Fueling**
Main	Unlimited	See above for PBY-5A, PBM-3D, and PB4Y	
Secondary			
SALEH BAY	none	36 VF, 37 VB, 18 VT	152) minimum
KOEPANG	none	27 VF, 9 VT	50) minimum

Fig. 83
Gasoline requirements for aircraft

Gasoline for the aircraft embarked on the carriers and seaplane tenders was the next subject. Admiral BC noted that the fleet and light carriers carried enough gasoline to refuel their embarked squadrons completely six times: the total capacity for these ships was 2,734 tons, and the total gasoline-carrying capacity for the large oilers was six thousand tons. As the carriers would probably never reach the absolute bottom of their gasoline tanks, he was confident that the seven fleet oilers would be able to provide the gasoline needed by the carriers for a considerable period. However, so little gasoline was available at Saleh Bay for the planes there that it was doubtful whether its depot could supply gasoline to planes other than those temporarily based at its airfield. For that reason, Admiral BC did not think it was practical to base patrol planes there, except for operations from tenders. In fact, he surmised that supplying Saleh Bay, Koepang, and temporary advanced bases with gasoline would take the entire services of the two gasoline tankers. Fleet carriers, light carriers, and all types of aircraft tenders were to fuel to capacity at Main Bases whenever practical. Replenishment at sea of the aircraft carriers would be by the fleet oilers, while aircraft based ashore would obtain fuel from stocks there. Aircraft based at Saleh Bay and Koepang would obtain fuel from tenders when they could, thereby preserving shore stocks for emergency use. Aircraft based at newly established advanced bases would obtain gasoline from tenders as well, and the tenders would be replenished by the gasoline tankers upon request of CTF 75.[36]

Admiral BC expected Orange action to hamper or prevent logistical support in terms of fuel supplies to surface craft, fuel supplies to aircraft, and the replacement of aircraft. He thought that Orange actions would take the forms of interception and destruction of oilers and gasoline tankers; destruction of shore fuel-oil and gasoline storage tanks; and the interception and destruction of replacement aircraft in flight or on the ground. He felt that offensive action by his Striking Force was the best way to prevent these Orange actions; the means and facilities at his disposal left him few alternatives for logistics. Provision had been made for adequate air cover for auxiliaries, except for the Service Force—he hoped that offensive employment of the carrier force would reduce the Orange threat to the Service Force.

A concern was that the intelligence information on Orange upon which the initial Blue air strikes were to be based had not yet been obtained. His intelligence thus far consisted of "generalities," nothing of a definite nature. He had, as noted above, provided for photographic flights late in June and he expected information from coast watchers and submarines, but he needed to know the location and size of the Orange bases at Tutu Bay, Davao, and Puerto Princessa, as well as the availability of aircraft and the location of antiaircraft guns at those bases. He also needed equivalent information on Banka, Singapore, Camrahn Bay, and Palembang. He additionally wanted to know the location of tin mines and smelters at Banka and Singapore; the location of oil fields and refineries at Palembang; confirmation of the presence and types of enemy merchant ships and combatants at Manila, Camrahn Bay, and Singapore; the movement of enemy shipping and combatants through the South China Sea; and estimates of the numbers and types of Orange aircraft at Manila and Camrahn Bay.[37]

Admiral BC divided responsibility for this intelligence between the Intelligence Section of his own staff, the intelligence staffs of Task Forces 75 and 78, and the submarines. Radio reports on these matters were to conform to his Communication Plan, and information was to be disseminated "by the manner most clearly indicated to ensure efficient accomplishment of the operation." Admiral BC also intended

Task Force 75, Air Force	Rear Admiral B-3
AV-9	1 AV
Fairwing 7	1 AV, 3 AVP, 4 AVD, 48 VP
Fairwing 10	1 AV 48 VP
Relief Carrier Air Group	
Replacement Carrier Aircraft	
Desdiv 8	4 DD
Task Force 78, Striking Force	Rear Admiral B-6
Crudiv 20	3 BB
Crudiv 6	4 CA
Crudiv 12	4 CL
Crudiv 15	3 CL
Crudiv 11	4 CL AA
Cardiv 5	2 CV
Cardiv 11	3 CVL
Desrons 5, 6	2 DL, 16 DD
Desrons 21, 47	18 DL
Minron 2, less Mindiv 6	9 DMS
Task Force 79, Service and Escort Force, Rear Admiral B-17	
AD-3, 9, 14, 16	4 AD
Crudivs 13, 14	6 OCL
Desron 4, less Desdiv 8	1 DL, 4 DD
Cortron 16	12 DE
Mindiv 6	4 DMS
AO-1, 2	2 AO
AO-36 to 40	5 AO
AOG-1, 2	2 AOG
AE-3, 4	2 AE
AF-10, 11	2 AF
AK-16, 17	2 AK

Fig. 84
Task organization and apportionment of command

to construct an Intelligence Plan, designated the "Commander, 7th Fleet Intelligence Plan," as an annex to the Operation Plan; this would lay out all current information on Orange. The Intelligence Plan was also to be updated as information became available. BC intended that geographical positions be referred to by the Joint Army-Navy Grid System and that every measure necessary be taken to keep knowledge of the operation from the enemy. He assumed that Orange patrols would report his movements and that the Orange Commander would take note of the increased air activity commencing on D–6. He thought, however, that if he could prevent Orange from noticing anything suspicious until then, Blue would have won the element of surprise. "By maintaining the top secret status of this operation, I should be able to effect surprise in initial attacks on all objectives." Thereafter, subsequent attacks had to be made in the most rapid possible succession in order to exploit the initial advantage.[38]

In summary, Vice Admiral BC repeated the General Situation, the composition and basing of his forces, the location of Orange forces, and the plan of drawing Orange capital ships out for a fight by attacking Orange combatants and merchant ships in the South China Sea and seizing Formosa. More specifically, Task Force 75 was to commence on D–1 making daily searches of the Sulu Sea and the southern sector of the South China Sea, as well as photographic reconnaissance of the facilities and forces at Singapore, Banka, and Palembang.

D-Day was to see the commencement of operations to destroy designated targets at these locations, with priorities set by Commander, Striking Force. Task Force 75 was then to maintain continuous air patrols of Karimata Strait, while Blue forces operated in the South China Sea and reported on any Orange submarine or mining activities. TF 78 was to locate Orange shipping in the South China Sea and commence destroying it on D-Day, while avoiding engagement with superior enemy surface forces. TF 75 was also to make photographic reconnaissance on D–2, D–3, and D–6 of Orange facilities and forces at Tutu Bay, Puerto Princessa, and Davao; conduct additional air strikes as might be required on D-Day; furnish air cover to the Service Force when it was operating in the theater of operations; and conduct minesweeping operations in the narrow straits and shallow waters in advance of the Striking Force. TF 79, of course, was to provide service and escort in support of these operations.[39]

Blue Staff Solution: Final Arrangements

In the next formal step of the Solution process, Vice Admiral BC reiterated the contents of his planning comments: Task Organization, the General Situation, the composition of his and the Orange forces, and the tasks he was assigning to his Air Force, Striking Force, and Service and Escort Force. He also re-summarized his Communication Plan; his Authenticator List and Instructions; his Logistics Plan; his Intelligence Plan; the Information on Orange Shipping, Lines of Transportation and Supply D; Bases Located in the 7th Fleet Operating Area; and the Distribution of Enemy Aircraft. There were also some temporary, pre-operation command changes. First, Task Force 18, upon arrival at Kaoe Bay, was to temporarily dissolve. In addition, Cruiser Divisions 6, 11, 12, 13, 14, 15, and 20 would be placed under command of Rear Admiral B-4 as Commander, Cruisers, 7th Fleet, while Rear Admiral B-6 would take command of Carrier Divisions 5 and 11 as Commander, Carrier Divisions, 7th Fleet. Destroyer Squadrons 5, 6, and 47 would report to Commander, Destroyers, 7th Fleet for duty. Additionally, there was to be a pre-operation command conference at Surabaya.[40]

FORM S-5
1944

MESSAGE

FROM: Commander Seventh Fleet (TITLE) (SHIP) (STUDENT) (ROOM NO.)

TO:

*	Title	Student	Room No.	TOD
Z	CTF 18			
Q	Seventh Fleet			
Q	CINCPAC			

"Q" - for Information. "Z" - for Action.

TO BE FILLED IN BY UMPIRE: Words / Min. / Time; Coding; Call and Head; Text; Decoding; Delivery; Each Relay; TOTALS: No relay / 1 relay / 2 relays

SENDER INDICATE BY CHECK: Plain Lang. — Urgent or Abb. Proced. — Radio (Key) XX — R XF G T Radio (Voice) — Exec. at; Sig. Book, Con. Code, or Air Code — Priority — Bridge Circuit; Code (Spd.) (See Table E-4-II) — Routine XX — Visual — Flag / Other — Dist.; Deferred — Sound; Code XX — Acknowledge — Wire — Carrier — Circuit

Additional delay of ...... min. in delivery to: Reason: Actual Time of release by Umpire: UMPIRES FILE No.

HEADING — — — — 200100 (DATE AND TIME OF ORIGIN GROUP) — — — —

TEXT: On arrival KAOE BAY task force 18 temporarily dissolved X Rear Admiral B-4 assume duties as commander cruisers Seventh Fleet comprising crudivs six eleven twelve thirteen fourteen fifteen and twenty X Rear Admiral B-6 assume duties as commander carrier divisions Seventh Fleet comprising cardivs five and eleven X Desrons five six and forty seven report to commander destroyers Seventh Fleet for duty.

DO NOT USE BACK OF THIS BLANK FOR ADDITIONAL TEXT

Fig. 85
Message, COM7FLEET

FORM S-5
1944

MESSAGE

FROM: Commander Seventh Fleet (TITLE) (SHIP) (STUDENT) (ROOM NO.)

TO:

*	Title	Student	Room No.	TOD
Z	Comcardivs			
Z	Comcrudivs			
Z	Comair			
Z	Comserfor			
Z	Comdes			

"Q" - for Information. "Z" - for Action.

TO BE FILLED IN BY UMPIRE: Words / Min. / Time; Coding; Call and Head; Text; Decoding; Delivery; Each Relay; TOTALS: No relay / 1 relay / 2 relays

SENDER INDICATE BY CHECK: Plain Lang. — Urgent or Abb. Proced. — Radio (Key) XX — R XF G T Radio (Voice) — Exec. at; Sig. Book, Con. Code, or Air Code — Priority — Bridge Circuit; Code (Spd.) (See Table E-4-II) — Routine XX — Visual — Flag / Other — Dist.; Deferred — Sound; Code XX — Acknowledge — Wire — Carrier — Circuit

Additional delay of ...... min. in delivery to: Reason: Actual Time of release by Umpire: UMPIRES FILE No.

HEADING — — — — 200101 (DATE AND TIME OF ORIGIN GROUP) — — — —

TEXT: Addressees report to me at Headquarters SOERABAYA at 1300 21 June for conference X Air travel directed.

To be transmitted in a crypto-channel held only by the adees.

Fig. 86
Message, COM7FLEET

This summary was followed by recent developments: radio dispatches from CTF 78 to Commander, 7th Fleet reported that photographic reconnaissance by CTF 78 on D–6 had not found any Orange airstrips or aircraft at Tutu Bay, while only a few fighters were sighted on the ground or in the air at Puerto Princessa. Task Force 78 had, however, located oil and gasoline storage tanks near the wharf there and two large oil or gasoline tanks at Davao. TF 78 pilots had also sighted three Orange fighter planes on the ground at an airfield ten miles west of that city and they deduced from general activity that Orange was using Manila as a Main Base. At the same time, Commander, Task Force 75 reported to COM7FLEET and CTF 78 that Palembang refineries were in full operation and oil storage tanks were located in the immediate vicinity of the refineries. No other targets existed there. Moreover, Banka had an airstrip and gasoline storage but no other facilities of "much value." At Singapore, Orange was using the airfield on the north end of the island as a fighter base, with six planes on the ground and one airborne. Oil tanks were also located in the harbor area near the docks and two tanks near the northern airfield were probably gasoline storage tanks. Several merchant ships were also seen in the harbor by each flight of Task Force 75 planes.[41]

The results of the photographic reconnaissance were then followed by last-minute command arrangements by Subordinate Commanders. For instance, Commander, Task Force 78 directed Commander, Destroyer Squadron (COMDESRON) 21 and Commander, Mine Division (COMMINDIV) 4 to have all of the destroyers of that squadron and two of the destroyer-minesweepers from the division join his force by 1500 on D+3, while the rest of MINRON 2 was to proceed to Kaoe Bay to report to him no later than 28 June. In terms of refueling, Commander, Task Force 79 told the Commanders of the 7th Fleet, Task Force 78, Mine Division 6, and Task Group 78.2 that the first refueling would take place at their request; that Mine Division 6, plus oilers and escorts, would join Task Group 78.2 at a place and time designated; and that Mine Division 6 would be made available for sweeping operations ahead of the Task Group all the way to the refueling rendezvous. CTF 78 replied that the first refueling would take place at Kuching Bay on D+5 at 1800. Mine Division 6, with oilers and escorts, was requested to join TG 78.2 at 1500 on D+3. The minesweepers were requested for operations up to the refueling rendezvous, but CTF 78 stated that they would be released instead to return to their base with the oilers and escorts.[42]

Conclusion

This Operations Problem entailed a strategic scenario that might very well have come to fruition if the United States had not decided to invade the Philippines but had instead seized Formosa as a way of interdicting Japanese lines of supply and communication to Southeast Asia. As mentioned above, Operations Problems did not necessarily reflect actual wartime events, but in this case there was a very clear use of combined-arms naval forces to reconnoiter and strike Orange land-based forces and facilities as a way of forcing the enemy fleet to sortie. Vice Admiral BC's plan entailed scenarios for the use of land-based naval aviation assets, carrier forces, and surface ships to engage and destroy significant portions of the Orange Fleet, while submarines performed a scouting function. About the only element of late–Pacific War U.S. Navy doctrine left out was Blue submarine interdiction of Orange supply lines. The objective of this Operations Plan, therefore, was exactly that of the U.S. Pacific Fleet throughout the second half of the Pacific War, and the planning by Naval War College students to carry out the campaign almost precisely mirrored that of the wartime Third and Fifth Fleets as well. This close approximation of the Operations Problems to both Pacific Fleet and Imperial Japanese Navy (IJN) doctrine and practice will be seen in additional Exercises and Operations Problems as well.[43]

NOTES

1 Command and Staff Class of December 1945, "Operations Problem 5: Blue Statement and Staff Solution," 6 September 1945, p. 1, NHC, folder 2536, box 125, RG 4, NWC.

2 Ibid., pp. 2–3.

3 Ibid., p. 3.

4 Ibid., pp. 3–4.

5 Ibid., pp. 4–5.

6 Ibid., pp. 5–6.

7 Ibid., pp. 7–8, 9.

8 Ibid., pp. 8–9.

9 Command and Staff Class of December 1945, "Operations Problem 5: Blue Statement and Staff Solution," app. 1, "Weather during July and August in the Theatre of Operation," p. 10.

10 Command and Staff Class of December 1945, "Operations Problem 5: Blue Statement and Staff Solution," app. 3, "Required of the Student Officer," pp. 12–15.

11 Command and Staff Class of December 1945, "Operations Problem 5: The First Step—the Estimate of the Situation," 6 September 1945, pp. 17–20.

12 Command and Staff Class of December 1945, "Operations Problem 5: The Second Step—the Completion of the Plan, Section One," 7 September 1945, p. 21.

13 Ibid., pp. 22–23.

14 Ibid., pp. 23–24.

15 Ibid., pp. 24–28.

16 Ibid., p. 28.

17 Command and Staff Class of December 1945, "Operations Problem 5: Effectiveness of Logistics Support," 7 September 1945, p. 31.

18 Command and Staff Class of December 1945, "Operations Problem 5: Characteristics of the Theater of Operations," pp. 32–33.

19 Ibid., pp. 33–34.

20 Ibid., pp. 34–35.

21 Ibid., pp. 35–37.

22 Ibid., pp. 37–43.

23 Command and Staff Class of December 1945, "Operations Problem 5: Blue Staff Solution, Step 2, Section Two," 7 September 1945, pp. 45–46.

24 Ibid., pp. 46–47.

25 Ibid., pp. 48–49.

26 Ibid., p. 49.

27 Ibid., pp. 50–51.

28 Ibid., pp. 51–52.

29 Ibid., pp. 52–53.

30 Ibid., pp. 54–55.

31 Ibid., pp. 55–56.

32 Ibid., pp. 56–58.

33 Command and Staff Class of December 1945, "Operations Problem 5: Blue Statement and Staff Solution, Step II—Sections III and IV," 7 September 1945, pp. 59–60.

34 Ibid., pp. 62–64.

35 Ibid., p. 64.

36 Ibid., pp. 65–66.

37 Ibid., pp. 66–67.

38 Ibid., pp. 67–68.

39 Ibid., pp. 69–72.

40 Command and Staff Class of December 1945, "Operations Problem 5: Blue Statement and Staff Solution, Step III," 8 September 1945, pp. 73–88.

41 Command and Staff Class of December 1945, "Operations Problem 5: Blue—Section A," 26 September 1945, n.p.

42 Command and Staff Class of December 1945, "Operations Problem 5: Blue—Section A," 2 October 1945, pp. 9–12.

43 For the decision to strike the Philippines as opposed to Formosa, see Ronald Spector, *Eagle against the Sun: The American War with Japan* (New York: Free Press, 1985), pp. 417–44. For the theoretical basis of the Operations Problems, see Francis McHugh, *A Manual of War Games* (Newport, R.I.: Naval War College, 1960), pp. 1–3; McHugh, *Fundamentals of War Gaming* (Newport, R.I.: Naval War College, 1966); McHugh, "War Gaming and the Navy Electronic Warfare Simulator," *NWCR* 19, no. 7 (March 1967), pp. 22–32; and McHugh, "Eighty Years of War Gaming," *NWCR* 21, no. 7 (March 1969), pp. 88–90. For the continued use of modified interwar doctrine in the wartime Pacific Fleet, see Ivan Musicant, *Battleship at War: The Epic Story of the USS* Washington (New York: Harcourt Brace Jovanovich, 1986), pp. 257–59, 295–303, 318; Trent Hone, "U.S. Navy Surface Battle Doctrine and Victory in the Pacific," *NWCR* 62, no. 1 (Winter 2009), pp. 67–105; and Hone, e-mails to author, 28 February, 2 March, 3 March, and 14 March 2011. For U.S. Navy submarine doctrine during the Pacific War, see Clay Blair, *Silent Victory: The U.S. Submarine War against Japan* (Annapolis, Md.: Naval Institute Press, 1975), pp. 557–873.

SAIPAN

VIII *Operations Problem 5*

The Orange Statement, October 1945

In its own Statement of the Problem, Orange summarized its strategic position in the western Pacific, with particular attention to its losses to Blue. Orange's Statement also focused on its one remaining major supply line between Southeast Asia and the Orange homeland. Most of Orange's remaining battleship and carrier strength was now at Amami Oshima, and its older heavy cruisers, many of its light cruisers, destroyers, and patrol vessels, and a few escort carriers and submarines were being employed for the protection of trade in the Bay of Bengal, the Straits of Malacca, and the South China Sea. Orange-controlled areas south of its 3rd Naval District had been organized as the zone of the Southwest Area Fleet, with its headquarters at Manila. Its Subsidiary Fleet Base was at Singapore, and its Main Air Bases were at Camrahn Bay and Manila. As Blue surmised, Subsidiary Air Bases were at Rangoon, Banka, Bangkok, Puerto Princessa, Singapore, Medan, Davao, Malampaya Sound, and Dumanquilas. Orange's Main Fleet Bases had repair and supply facilities for all types of ships and planes, and they were heavily defended by land, air, and local naval forces. In addition, Subsidiary Fleet Bases contained "nominal" repair and supply facilities, as well as storage for ten thousand tons of fuel oil and four thousand tons of diesel fuel. Singapore was defended as if it were a Main Fleet Base, and it could handle vessels of any size for docking, but not for heavy or prolonged repairs. Main Air Bases contained all the repair and supply facilities usually found at a first-class land-plane and seaplane base, while Subsidiary Air Bases could undertake minor repairs, replacement of parts, and replenishment of fuel and ammunition. Each Subsidiary Air Base had two tanks of aviation gasoline, each of 150,000 gallons capacity, but these had to be refilled from fleet oilers or gasoline tankers, using a fifty-thousand-gallon barge available at each base. These bases also had accommodations for one spare air group, though no groups had been placed there. Each base additionally had good anchorages for seaplanes and "satisfactory" landing strips for land planes. All bases, the Orange Solution stated, regardless of type, were "well defended" by troops, antiaircraft batteries, and a squadron of twelve fighter planes under Orange Army command.[1]

Orange acknowledged the advance of Blue across the Pacific and its control of island chains such as the Solomons and those of Micronesia; Orange understood that though Blue was not in great strength in these forward positions, it could prevent Orange from using any of them. Orange surmised that the major portion of Blue's Pacific Fleet was based at Wotje, with detachments at Eniwetok, Truk, Ponape, and Jaluit. Other points—Guam, Saipan, Halmahera, and the Palaus—having been recently captured, could probably not be used by Blue for some time. Orange also noted that the Blue Pacific Fleet had recently been reported as having deployed several of its new battle cruisers. "Reliable" sources had also reported that Blue had Main Fleet Bases at Darwin, Surabaya, and Shark Bay; Subsidiary Fleet Bases at Batavia and Koepang; and advanced bases at Borneo, Celebes, Timor, and New Guinea. Blue Main Air Bases were believed to be at Batavia, Surabaya, Koepang, and Darwin, and Subsidiary Air Bases at Sumbawa, Wetar Pass, Wyndham, and Cape York. These locations, however, had not been confirmed. Orange thought that other Blue Subsidiary Air Bases might exist. Orange also assumed that all of these bases were strongly protected and garrisoned by Blue, Red, and Brown units, with aircraft of the "usual" types.[2]

Vice Admiral OD, Commander, Task Force 2 and Southern Control Force (Col. Herbvert Ehrgott, U.S. Army, Command and Staff Class of December 1945)

Rear Admiral O-1, Commander, Battleship Division 1 and Task Group 2.1, and Rear Admiral O-3, Commander, Carrier Division 3 and Task Unit 2.1.1 (Col. George Moseley, Jr., U.S. Army, Command and Staff Class of December 1945)

Orange intelligence indicated that a Blue offensive was imminent in the western Pacific. An offensive was all the more likely since powerful naval reinforcements, including the battle cruisers, had been dispatched to Blue naval forces operating in Australian waters. Blue heavy cruisers and destroyers had also recently been reported entering Kaoe Bay, and other forces had been seen in the Flores Sea area. "It is possible that such increased activity may take the form of heavy attacks on vital Orange southern maritime communications." Noting that all Orange forces south of Manila were subject to sporadic air attacks by United Nations forces and that this knowledge was common to Orange flag officers, the Commander-in-Chief of the Orange Combined Fleet decided to counter the increasing Blue threat to his vital southern maritime supply route by combining all ships and units operating in the southern theater into a single force, designated Task Force 2, the Southern Control Force. Commanded by Vice Admiral OD, Task Force 2 was notified of the recent movement of Blue naval forces in the southwest Pacific and their possible intentions, and given the task of protecting Orange maritime transportation and supply lines in the area. Vice Admiral OD was also to be prepared to concentrate his 1st Mobile Fleet in Philippine waters if the Blue Pacific Fleet initiated a major effort in that direction. The control of Orange merchant shipping was to be effected by Commander, Southwest Area Fleet, who, at Manila, was now placed under OD's command. Vice Admiral OD was also informed that replacement fuel, aviation gasoline, and aircraft for Task Force 2 were available at Manila and Camrahn Bay.[3]

The Southern Control Force consisted of two battleships (Battleship Division 1, under Rear Admiral O-1) and Carrier Divisions 1 and 2, under Rear Admirals O-3 and O-2, respectively, four fleet carriers in all. There were also significant cruiser forces. Eight heavy cruisers were organized into Cruiser Divisions 4 and 5, under Rear Admirals O-4 and O-5, respectively, while thirteen light cruisers constituted Cruiser Divisions 10, 16, 17, and 21, under Rear Admirals O-11, O-9, O-10, and O-15, respectively. Battleship Division 1 and Carrier Division 1 were based at Amami; CARDIV 2 was at Camrahn Bay. Cruiser Divisions 4 and 10 were located at Amami and Camrahn, respectively, while CRUDIV 5 was at Rangoon, CRUDIV 17 was at Tutu Bay, and CRUDIVs 16 and 21 were at Manila. Destroyers, commanded by Rear Admiral O-20, were also numerous. Seventeen destroyers—Destroyer Squadron (DESRON) 10, commanded by Captain O-25—were based at Amami and Manila while DESRON 2, consisting of another seventeen destroyers under Captain O-21, was at Singapore and Port Lloyd. The thirteen destroyers of DESRON 7, under Captain O-30, were at Camrahn Bay. Orange had eighteen submarines operating in the theater under Commander, Submarines. Six of these were in reserve at Singapore, two were on patrol in the Makassar Strait, one was on patrol in the Molucca Passage, and another was north of Talaud Island. There was also an Orange submarine south of Talaud, one in the Ceram Sea, two in the Java Sea, two in the Gaspar Strait, two in the Karimata Strait, and two in the Sulu Sea near Balabac Strait.[4]

Admiral OD also had twenty-four patrol vessels, eight each in Patrol Divisions (PATDIVs) 30 (Commander O-48, at Manila), 31 (Captain O-31, at Singapore), and 32 (Commander O-27, Camrahn Bay). OD's train consisted of two repair ships (ARs) at Manila, under Commander O-41; four fleet oilers, each with a cargo of seven thousand tons of fuel oil and one thousand tons of diesel or aviation gasoline,

Rear Admiral O-5, Commander, Cruiser Division 5 and Task Unit 2.1.2, and Commander O-41, Commander, Task Unit 2.1.4 (Col. Richard Hutchinson, U.S. Army, Command and Staff Class of December 1945)

Rear Admiral O-9, Commander, Cruiser Division 16 and Task Unit 2.3.1, Escort Group No. 1 (Lt. Col. John Greco, U.S. Army, Command and Staff Class of December 1945)

Captain O-30, Commander, Destroyer Squadron 7 and Task Unit 2.3.2, Escort Group No. 2, TF 2 (Capt. Wells Thompson, U.S. Navy, Command and Staff Class of December 1945)

under Commander O-75, also at Manila; and two aviation gasoline tankers, each with a cargo of six thousand tons of aviation gasoline, under Commander O-90, at Manila. OD's Fleet Air Wing 1 was under Rear Admiral O-22, with a tender at Tutu Bay and five naval patrol plane squadrons, each with twelve Emily seaplanes, at Tutu, Malampaya Sound, Dumanquilas, and Camrahn Bay. Orange carrier aircraft consisted of twenty-four Zeke fighters, twenty-four Judy scout bombers, and twelve Myrt torpedo planes on board each of the four fleet carriers. Replacement aircraft to fill the present allowances were available at Manila and Camrahn Bay, with up to two complete carrier air groups at Camrahn and 100 percent replacements for the naval patrol planes at Manila. These could be delivered upon request by the Task Force Commander.[5]

Rear Admiral O-22, Commander, Fleet Air Wing 1 and Task Group 2.2, TF 2, and Commander O-55, Commander, Task Unit 2.2.1, Search Unit 1 (Lt. Cdr. Jack Gibson, U.S. Navy, Command and Staff Class of December 1945)

The main convoy route, which was under the control of Commander, Southwest Area Fleet, originated at Singapore. Vessels from Burma, Palembang, Banka, and Billiton departed these ports unescorted and arrived at Singapore twenty-four hours prior to the scheduled departure of a convoy. Vessels from Bangkok and Saigon followed an inshore route unescorted and joined northbound convoys at latitude 12 degrees north, longitude 111 degrees east (Point K). At latitude 15 degrees north, longitude 113 degrees east (Point J), convoys from the homeland met and exchanged escorts; the escort from Singapore returned with the southbound convoy. From Point K vessels bound for Saigon and Bangkok proceeded independently to their destinations. At Singapore, vessels bound for Billiton, Banka, Palembang, and Rangoon also proceeded independently. A ten-knot convoy and a sixteen-knot convoy would depart Singapore each week, normally keeping five miles to the right of their prescribed tracks. Shipping from the southern Philippines proceeded independently to Manila. There, convoys formed when required and were escorted to Point J by light antisubmarine warfare (ASW) units. "Past experience indicates that about one convoy every two weeks will be formed." As in the Blue Statement, data on weather as well as sun and moon data were provided. Requirements for the Orange student officers, which mirrored those of the Blue, were also outlined in similar detail.[6]

Operations Problem 5: The Orange Solution

In the first step of the Orange Solution, Vice Admiral OD stated that Blue's rapid expansion of control into the western Pacific—its recent deployment of large numbers of major Blue fleet units, their exact strength and locations still unknown—was "a source of worry to the Emperor." Admiral OD noted that the forces under his command were in various ports from Amami to Port Lloyd in the north down to Rangoon in the south; none of them would be ready for sea until between 27 and 30 June. He also predicted that the 1st Mobile Fleet would operate north of the Philippines and would therefore be of no assistance to him in carrying out his mission unless the Blue Pacific Fleet made a major effort in the direction of the Philippines.

Assuming instead that Blue was going to raid and disrupt Orange supply lines in the area between latitudes 15 north and 5 south and longitudes 90 and 130 east, OD took his mission to be the protection of Orange supply and transportation lines in that area, though he found the word "protect" in his orders a bit lacking in specificity as to the exact actions he was to take. The fact that the Commander-in-Chief of the Combined Fleet ordered him to cooperate with the 1st Mobile Fleet, however, also indicated that the Blue offensive might, in part, be intended to draw major elements of the Orange Fleet into southern waters. Whether Blue's objective was diversion of the Orange Fleet or disruption of raw materials to the Orange homeland, Admiral OD assessed that the initial Blue action would be raiding Orange's southern supply lines. OD also noted the location of his and Blue's bases as well as the distances between the various Orange and Blue bases. Citing his own weakness in having no aircraft except for carrier aircraft and naval patrol planes, he pointed out that Blue had advanced bases in Borneo—as well as bases at Surabaya, Batavia, and Saleh Bay—from which Blue would be capable of strikes on Orange shipping as well as port facilities. Still, he thought he had sufficient bases from which to protect shipping in the area in spite of the fact that Blue had enough bases for submarines, surface ships, and aircraft to make his task a "difficult" one.[7]

Admiral OD then reviewed the raw materials that Orange transported along these routes, emphasizing that they were so vital that the Orange High Command was willing to accept "grave" risks to protect the shipping lanes; any "reduction in the flow of essential materials to the homeland will probably have an adverse effect on the morale of the Orange people." His knowledge of Blue's force composition was inexact, too "meager" to make a detailed analysis; he hoped to receive information from spies and "friendly native" sources. He knew that Blue had many destroyers, several cruisers, and three "large" warships in the southwest Pacific, and he assumed that Blue naval forces in the southwest Pacific consisted of other cruisers, destroyers, submarines, and auxiliaries. He additionally thought that a Blue deployment of battleships was a real possibility, and he assumed that Blue's surface ship operations would be supported by both land-based and carrier-based aircraft. As for his aerial strength, he discounted the sixty-nine Pete scout planes on his battleships and cruisers as offensive weapons. His lack of land-based strike aircraft was a "severe handicap," especially since he thought Blue had "large numbers of shore-based aircraft for both scouting and attacks on shipping." Still, he took his bases to be well defended, and he saw his personnel as superior to Blue's, though for practical purposes he counted this factor as equal.[8]

Coast-watcher reports had observed three Blue warships that were thought to be larger than battleships; Vice Admiral OD did not know whether he was facing battleships or battle cruisers. His own two battleships had speeds of twenty-eight knots and main batteries of nine sixteen-inch guns and twelve 6.1-inch guns, with ranges of forty and twenty-seven thousand yards, respectively; Admiral OD thought his two battleships were equal in strength to the suspected three Blue battle cruisers. He additionally thought that against an equal number of Blue battleships, his ships were equivalent in age and firepower to anything more modern than the Blue BB-45 class and superior to any ships earlier than this. Moreover, he took his battlewagons to be superior in speed to any Blue counterparts except the BB-61 class. His four aircraft carriers possessed speeds of twenty-seven knots, sixteen five-inch guns with ranges in excess of fifteen thousand yards, and sixty aircraft each. Corresponding Blue carriers carried 50 percent more planes of each type, but any one of his fleet carriers was equal in number of planes to two of Blue's light carriers. In all other areas, Admiral OD assumed Blue carriers to have "marked" advantages. Concerning cruiser forces, his eight heavy cruisers and thirteen light cruisers had speeds of thirty to thirty-three knots and carried between eight and thirty-two torpedo tubes; two classes of light cruisers could deploy up to a hundred mines. His heavy cruisers each had ten eight-inch guns, with ranges in excess of thirty thousand yards, while his light cruisers had batteries of six or seven 5.5-inch or 6.1-inch guns, with ranges of twenty to twenty-seven thousand yards. He understood his heavy cruisers to be similar to Blue's but superior in that his could carry torpedoes. Against modern Blue light cruisers, however, his heavy cruisers had little advantage at short ranges. He also knew Blue to have a large number of ten-thousand-ton light cruisers, any one of which was superior in tonnage, age, and firepower to his ships. "Any two of my CLs of the 5.5" (that is, 5.5-inch-gun) class are about equal in fighting strength to one Blue CL." Still, all of his light cruisers also carried torpedoes. Additionally, while not currently carrying mines, the cruisers could be loaded with those weapons at Singapore, Camrahn Bay, or Manila.[9]

Admiral OD's forty-six destroyers each had speeds between thirty-four and thirty-six knots and carried six 5.1-inch guns and between eighteen and twenty-seven torpedo tubes. He thought he had a six-to-four superiority in guns over Blue's 1,500-ton destroyers. He did, however, surmise that Blue had a large number of 2,100-ton destroyers, with some in the southwest Pacific, specifically at Kaoe Bay. Admiral OD did not have anything of equivalent strength. He did not have direct command of the eighteen submarines on patrol in the area but he had "reasonable" confidence that he could expect assistance from them. However, he did not assume that the submarines could cover the entire potential area of Blue deployment, nor did he assume that the submarine reports would be 100 percent accurate. His twenty-four submarine chasers (PCs and SCs) had maximum speeds of twenty knots, had good depth-charge and sonar equipment, and were "well adapted" to antisubmarine operations. As for aircraft, and in light of wartime experience, Admiral OD admitted that "enemy aircraft are somewhat superior to our own in protection, fire

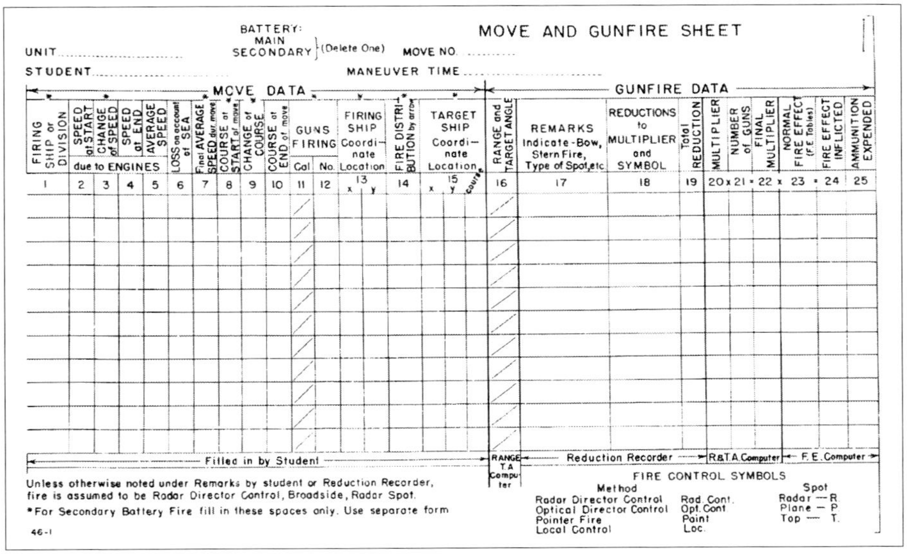
UNIT .... BATTERY: MAIN / SECONDARY (Delete One) MOVE AND GUNFIRE SHEET MOVE NO. ....

STUDENT .... MANEUVER TIME ....

MOVE DATA — GUNFIRE DATA

FIRING SHIP or DIVISION	SPEED at START	CHANGE of SPEED	SPEED at END	AVERAGE SPEED	LOSS on account of SEA	Final AVERAGE SPEED dur. move	COURSE at START of move	CHANGE of COURSE	COURSE at END of move	GUNS FIRING Cal.	GUNS FIRING No.	FIRING SHIP Coordinate Location	FIRE DISTRIBUTION by arrow	TARGET SHIP Coordinate Location	RANGE and TARGET ANGLE	REMARKS Indicate-Bow, Stern Fire, Type of Spot, etc.	REDUCTIONS to MULTIPLIER and SYMBOL	Total REDUCTION	MULTIPLIER	NUMBER of GUNS	FINAL MULTIPLIER	NORMAL FIRE EFFECT (F.E. Tables)	FIRE EFFECT INFLICTED	AMMUNITION EXPENDED
1	2	3	4	5	6	7	8	9	10	11	12	13 x y	14	15 x y course	16	17	18	19	20 x	21 =	22 x	23 =	24	25

SPEED: due to ENGINES

Filled in by Student — RANGE T.A. Computer — Reduction Recorder — R&T.A. Computer — F.E. Computer

Unless otherwise noted under Remarks by student or Reduction Recorder, fire is assumed to be Radar Director Control; Broadside, Radar Spot.

*For Secondary Battery Fire fill in these spaces only. Use separate form

46-1

FIRE CONTROL SYMBOLS

Method		Spot	
Radar Director Control	Rad. Cont.	Radar	R
Optical Director Control	Opt. Cont.	Plane	P
Pointer Fire	Point	Top	T.
Local Control	Loc.		

Fig. 87
Orange BATDIV 1 Move Number 1 Move and Gunfire Sheet

power, and weight of bomb load carried."[10]

Vice Admiral OD thought he had an adequate number of fleet and air bases for his surface, air, and submarine forces, along with ample supplies of fuel oil and aviation gasoline at Manila and Camrahn Bay. However, fuel supplies at his Subsidiary Fleet Bases were low, and unloading facilities at his Subsidiary Air Bases for aviation gasoline were poor. He thought the fuel situation could be "handled" but that the loss of any oiler or storage facilities would have "serious" consequences. Replacement aircraft, he surmised, were available in adequate numbers as were dry-docking and ship-repair facilities. He did not believe he had sufficient strength to interrupt Blue logistical facilities or supply lines in any way.[11]

Admiral OD also reviewed the hydrography of the operational area, noting the numerous shoals that would restrict lines of transportation and that the southern part of the South China Sea and its southern approaches were well suited to mining. He knew that the Banka, Gaspar, and Karimata Straits were the only routes for Blue access to the South China Sea from the Java Sea, that the Karimata was the most suitable of these, and that he had control of the aids of navigation in the more dangerous areas. East of Java, Blue had numerous "excellent" passages to the south and numerous bays in the Netherlands East Indies for rendezvous and fueling. However, he did not see topography impacting operations—other than through weather. Aside from the climate generally, which he found hot, humid, and "disagreeable," he expected rain in the South China Sea and heavy clouds, causing low visibility. In the Java, Flores, Molucca, and Banda Seas, however, there should be excellent flying weather. He saw the same in the Celebes Sea and the southern Sulu Sea, with some rain squalls and strong winds. Above latitude 10 degrees north, he assumed there would be occasional typhoons and rain squalls. Admiral OD thought that poor flying weather would hinder his air operations more than Blue's, given the weather in the respective basing areas. He argued, however, that low visibility in the South China Sea should afford his shipping some protection.[12]

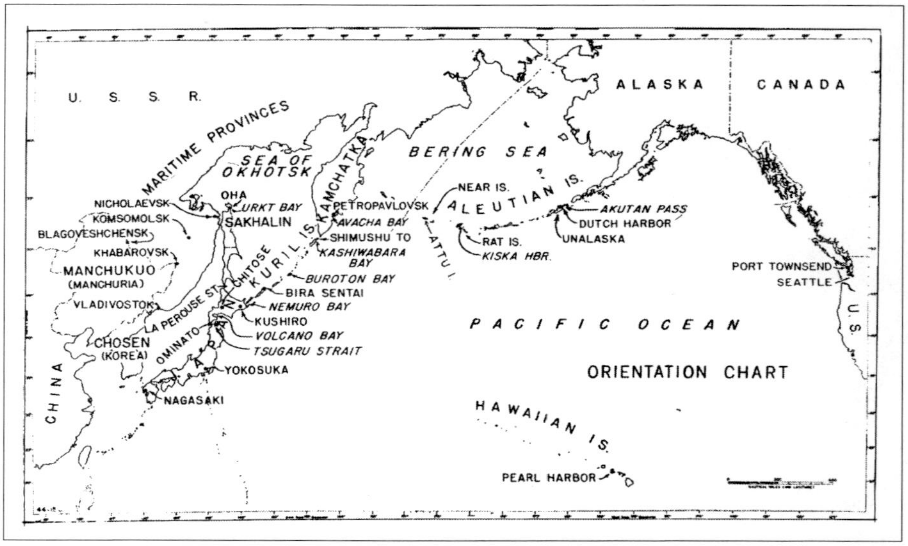

Map 7
Pacific Ocean Orientation Chart

OD noted how easy it would be for Blue to raid Orange shipping routes. Blue itself had longer supply lines—they stretched back to Hawaii and Australia—but these had been shortened by the loss of Orange-controlled islands in the Central Pacific and OD did not have the strength to disrupt them, adding to the difficulty of his task. Moreover, although his logistical situation was adequate, fuel, aviation gas, and aircraft spares had to be transported from the Main Fleet Bases at Manila and Camrahn Bay to all of the other bases. Also, as noted earlier, his Subsidiary Fleet Base at Singapore could not carry out heavy or prolonged repairs. Vice Admiral OD did note, however, that all of his bases were linked by radio and that his communications facilities at Manila and Camrahn Bay were "excellent"; he assumed that Blue communications capabilities—while he had no information on them—were "adequate."[13]

Admiral OD next compared and contrasted the strengths and weaknesses of his forces and Blue's. In his favor, he had a series of fleet and air bases that flanked most of the transportation and supply routes he needed to protect. In addition, many of the narrow passages that Blue had to use to enter the South China Sea could be mined, and OD had seven light cruisers that could carry a hundred mines each. His forces were more familiar than Blue with the passages and the aids to navigation, his cruisers were armed with torpedoes, and he had initial control over the theater of operations. His weaknesses included a long supply line to protect against all forms of attack, a lack of alternative routes for his shipping, and his need to concentrate with the 1st Mobile Fleet. He additionally lacked shore-based bombers, and part of his line of transportation and supply was flanked by enemy-held territory. Finally, Palembang, Singapore, and the Straits of Malacca were all near probable enemy air bases.[14]

Blue probably had overall superiority in naval strength. OD further thought that the Blue battle cruisers were well adapted to commerce raiding and that the Blue Commander had the advantage of the initiative. Admiral OD further surmised that Blue had an advantage in light cruisers, shore-based aircraft, and ability to effect strong local concentrations of naval and air strength. He additionally thought that Blue radar was superior to his. However, the weather in the area, OD guessed, would hamper Blue air operations and the shallowness of the South China Sea would make submarine operations difficult. In addition, Blue had long lines of supply and transportation from his bases and large areas to search. Overall, OD was doubtful to some extent about the effectiveness of protection he could afford Orange lines of supply and transportation in the Bay of Bengal. "No form of protection which I may be able to give to shipping or shipping facilities will guarantee absolute security. Some losses must be expected." Nevertheless, he thought, there was a "reasonable" chance that he could prevent "critical" losses to shipping and supplies unless Blue raids were in "overwhelming" strength. In any case, he was not willing to question the feasibility or acceptability of the mission, since the raw materials were so vital to the Orange war effort and the Commander-in-Chief of the Combined Fleet was willing to accept "grave risks" to Orange forces in order to carry it out.[15]

In analyzing his mission, Vice Admiral OD assumed that Blue's targets were not only Orange and neutral shipping but also Orange port facilities and concentrations of materials. OD further assumed that Blue could take action against these objectives unopposed, and then move on to Orange forces protecting the shipping convoys and the Orange bases whence those forces came. Since Blue's bases were so far from the area of operations, OD accepted the Commander-in-Chief's belief that Blue would conduct raiding operations only as a way of disrupting Orange supply lines. Therefore, OD thought he could limit his planning to countering Blue raiding forces.

He additionally thought that Blue was capable of operating long-range shore-based or tender-based aircraft throughout the theater and would probably use carrier aircraft for scouting. In the South China Sea, however, air operations would be "greatly hampered" by poor flying weather. Blue could deploy submarines at "focal points" along the routes to gain information; OD took it for granted that Blue knew the location of Orange bases and their defensive strength from previous observations. He also took it for granted that Orange shipping ports were known to Blue. In looking at raiding capabilities per se, OD foresaw "strong" Blue raiding forces composed of carriers, cruisers, and destroyers, as well as, probably, battleships or the new battle cruisers. He also expected raids by land-based aircraft and submarines. He additionally assumed that the waters off his shipping ports and bases, as well as the focal points along his lines of shipping, would be mined by surface vessels, submarines, and aircraft. It would also be consistent with previous operations if Blue's surface raiders were covered by a strong force and if its primary target was Orange shipping in the South China Sea itself.[16]

OD stated that all of his southern ports were within striking distance of Blue shore-based or carrier aircraft. His principal source of oil and oil products was Palembang on Sumatra, which was "relatively" close to "probable" Blue bases for air operations. Admiral OD therefore expected "particularly heavy" Blue air raids on shipping in this area, as well as on oil refineries and storage facilities. Moreover, he saw Blue attempting to interrupt Orange supply lines further by striking at Orange fleet and air-base facilities to hamper OD's protective efforts. Locating Orange forces that were protecting that shipping and striking Orange port facilities and concentrations of raw materials might be just as effective as capturing or destroying Orange shipping. Accordingly, Vice Admiral OD thought that a Blue strike on the Orange fleet and air bases was the least likely possibility, but he could not afford to ignore it.[17]

Admiral OD did not see his options limited by the fact that Blue naval forces in the southwest Pacific had been reinforced, but the Commander-in-Chief's assumption that Blue raiding operations would begin soon meant that OD did not have to worry about Blue attempts to occupy Orange territory. Therefore, his potential courses of action were to use his forces to escort and cover Orange shipping and to destroy any force threatening lines of supply and transportation. He saw escorting convoys of merchant shipping as not feasible. "To escort each convoy with an escort capable of defending shipping against all comers would unnecessarily tie up large numbers of combatant ships." To give each convoy a small antisubmarine escort within what he considered the limits of "feasible acceptance" was a possibility, but this would leave convoys open to attack by long-range, land-based bombers, which Blue had in abundance. Adding light cruisers to any "tentative" escort group would relieve the antiaircraft situation to some extent, but he considered the antiaircraft power of his light cruisers very weak. The protection afforded by forming all combatant units into a covering group that could be interposed between convoys or that could operate between the transportation lines and any opposition was what was desired. Moreover, this covering force would represent an economy of force, though only against surface raiding forces. Such a force would afford no protection from air or submarine threats. Another option, that of destroying any Blue force that appeared, had the merit of taking the offensive but it also required a far greater force than Vice Admiral OD had, especially given the numerous directions Blue could come from. He therefore determined that a combination of escorting and covering convoys or shipping would afford protection against submarine attacks, some degree of protection from air attack, and some degree of protection from surface raiders as well.[18]

Given the importance of the mission to the Orange war effort, his orders from the Commander-in-Chief of the Orange Combined Fleet, and the heavy concentration of Orange shipping in the South China Sea, Admiral OD was going to concentrate his forces to cover the Karimata, Banka, Gaspar, and Balabac Straits. If the weather assisted him and he could maintain a "well-balanced" force in the South China Sea, he thought he had a "reasonable" chance to deny Blue surface forces entry to that area. The requirement that his force was to concentrate with the 1st Mobile Fleet in Philippine waters, however, precluded the possibility of operating anything other than small escort vessels and aircraft in the Indian Ocean or the Straits of Malacca. He also pointed out that shipping was proceeding at the rate of two convoys each week, both to and from Singapore, with each convoy consisting of thirty slow ships and ten fast ones. Admiral OD had no details about the escorts being employed by Commander, Southwest Area Fleet, but he knew those forces to be light. Since he could not provide strong carrier and light force escorts to each convoy, he was going to concentrate his forces in the areas where shipping was heaviest; rely for the protection of his bases, port facilities, and raw material concentrations on base-defense forces and army fighter planes; and assist those forces with his carrier fighters when he could. OD assumed that if he could prevent a strong Blue surface raiding force from operating in the South China Sea, he could protect his bases and ports from a "serious" surface-ship bombardment. He did not, however, think that his fighters could destroy all of Blue's "numerous" naval patrol planes; in general, he was "faced with a task which requires me to do the best I can with the forces available even if the costs are excessive. If I fail to prevent excessive losses in shipping and materials, the consequences will be serious."[19]

Vice Admiral OD perceived that his mission would entail operations over a considerable length of time and in separated areas, involving simultaneous actions relating to protecting two or more physical objectives—Orange shipping, bases, and ports. He assessed that his best bet was to prevent Blue forces from even entering the South China Sea. However, many of his units would not be ready in time; his next option was to maintain a strong force in the South China Sea to prevent Blue from carrying out its mission. Since he was not sure of the strength of Blue forces, he thought it questionable to divide his forces, which consisted of two battleships, four fleet carriers, eight heavy and nine light cruisers, and twenty-six destroyers. With this force, he expected to be able to meet a Blue force of three battleships or battle cruisers, fleet carriers, and supporting ships. If he could confine the operations of his heavy combatant units to near the Philippines, his fleet units could concentrate there. The low visibility from the monsoon season in the South China Sea would afford Orange shipping some protection from submarines and aircraft. Because Blue submarines would find the shallow waters south of Singapore and in the Straits of Malacca hazardous, his ships would also have some natural protection in those regions.[20]

He decided to provide the heaviest escorts—four light cruisers, twenty destroyers, and twenty-four patrol craft—to the ships of his Service Force and the ships engaged in transporting fuel, maintaining existing convoy schedules. When Blue became more active, these schedules might have to change; Singapore would no longer offer a shipping concentration point when Blue raids became frequent.

Any air cover for ships would have to be provided by the Army; Admiral OD needed any early information on Blue movements that could be obtained. To that end, he required about thirty naval patrol planes to conduct daily searches of the various seas and straits, but only had about half that number of planes immediately available. Commander, Search Group was therefore ordered to relocate any squadrons necessary for this task. Commander, Southwest Area Fleet would handle all routing of shipping, under OD's direction; that officer would also deal with fuel and gasoline support. Remaining logistical support would be the responsibility of Fleet and Subsidiary Base Commands from which Admiral OD's Southern Control Force was to operate. OD considered these latter facilities "adequate." OD thought that if he could maintain local superiority in the South China Sea, he could accomplish his mission; maintaining local superiority, however, was the primary problem. His forces were scattered from Amami to Rangoon and had to be concentrated, at Camrahn Bay. This concentration was not considered particularly hazardous; transit at high speeds by units such as Cruiser Division 5 at Rangoon would reduce the danger from Blue submarines. However, OD did not want these units engaging superior Blue forces until they were concentrated.[21]

Admiral OD thought he initially had enough planes for his daily air search plan and was even willing to accept severe losses, because of the importance of early detection of the Blue Force. He was concerned, however, about weather conditions in the South China Sea and the availability of aircraft over a long period. If Blue's location was determined early on, areas to be searched could be reduced. His four light cruisers, twenty destroyers, and twenty-four patrol craft assigned as escort forces where Blue submarine and air attacks were most likely would also be under Commander, Southwest Area Fleet, since that officer was responsible for convoy routing. He noted the necessity of "excellent" communications facilities and close liaison with that commander, Vice Admiral OF. OD thought he might have to increase the number of escorts directly from his attack group if shipping losses became "unacceptable," once the strength of the Blue raiding force was known. He also had complete confidence in Rear Admiral O-1, Commander, Battleship Division 1, who would command the major combatant group, designated Task Group 2.1. This force would consist of the battleships, carriers, cruisers, and destroyers noted above, as well as the two repair ships and two fleet oilers. Rear Admiral O-22 would command Task Group 2.2, with its seaplane tender and sixty naval patrol planes. Vice Admiral OF would command Task Group 2.3, with forces noted above, as well as local defense forces, two fleet oilers, and two gasoline tankers. Admiral OD would establish his own headquarters ashore at Manila.[22]

OD issued a separate Communications Plan, but his General Instructions included details along that line. The Communications Instructions were to take effect at 0600 on 27 June. Units of Task Group 2.2 were to broadcast contact reports addressed to Commander, Task Force 2, and units of all forces were to guard that circuit. Units of Task Groups 2.1 and 2.3 would broadcast contact reports to their own Task Group Commanders, who would relay them to Commander, Task Force 2. TG 2.1 and TG 2.3 would also be responsible for guarding their own circuits, specifically Fox Circuit (Radio Manila). Task Group and Task Unit (TU) Commanders could assign high-frequency (HF) voice circuits as desired, and zonal time was to be used. Since his operations might continue for months, OD focused on the use of the Fleet Base at Camrahn Bay for logistics, with secondary use of the Subsidiary Base at Singapore. The logistical functions that Vice Admiral OF would have to perform would be minor and that assignment of the oilers and gasoline tankers would alleviate any foreseeable logistical problems for Commander, Southwest Area Fleet. The defense of fuel facilities at his bases by the Army was adequate. Therefore, he did not worry about fuel support as long as Blue could not destroy his fuel facilities at these bases. As regarded intelligence, OD ordered that radio direction finders and radar installations be used for gaining information and to supplement air search. Information could be obtained from neutral ships. Finally, OD laid out a Shipping Control Plan for OF, entailing convoys proceeding from Indochina, Thailand, Malaya, Sumatra, Burma, and the Philippines. There would be ten convoys from each of the first four areas in each direction each week, five from Burma, and fifteen from the Philippines.[23]

	List A-RO-4000 CTF-2 FOX	List B-SR-4000 Task Group Comdrs.	List C-RQ-4000 Task Unit Comdrs.	List D-PN-4000 Task Units	List F-QV-3000 K 500 V Patrol Aircraft	List G-QU-500 K 100 V Carrier Aircraft	List H-ZY 4000 Shore Circuit
C.T.F. Fox Radio MANILA	TR	TR	TR	R	R		TR
Task Gr. Comdrs.	R	TR	TR	R	R		
Task Unit Comdrs.	R		TR	TR	R		
Task Units	R			TR	R		
Patrol Aircraft					TR		
Carrier Aircraft						TR	
Carriers						TR	
Shore Bases	TR			TR			TR

Fig. 88
Task Force 2 Radio Frequency Plan No. 1

Task Organization Operational Plans

Subordinate units then began devising Operations Plans of their own. Rear Admiral O-1, commander of Task Group 2.1 (Battleship Division 1), outlined his forces for his Operation Plan No. 1-45. Task Unit 2.1.1, under the command of Rear Admiral O-3, consisted of Battleship Division 1's two battleships, Carrier Division 1's two fleet carriers, Cruiser Division 4's four heavy cruisers, CRUDIV 10's three light cruisers, and Destroyer Squadron 10—less Destroyer Divisions (DESDIVs) 16 and 17—with nine destroyers. Task Unit 2.1.2, under Rear Admiral O-5, consisted of four heavy cruisers organized into Cruiser Division 5, one fleet carrier, three light cruisers of CRUDIV 17, the nine destroyers of DESDIVs 16 and 17, and one fleet oiler. Rear Admiral O-16 commanded TU 2.1.3, which consisted of three light cruisers (Cruiser Division 21), one fleet carrier, eight destroyers from DESRON 2—less DESDIVs 15 and 29—and one fleet oiler. Commander O-41 led Task Unit 2.1.4, which comprised two repair ships and a fleet oiler.

Task Group 2.1	Present Location	Readiness for Sea
Batdiv 1	AMAMI	0600, 27 June
Cardiv 1	AMAMI	0600, 27 June
Cardiv 2	KAMRANH BAY	0600, 29 June
Crudiv 4	KAMRANH BAY	0600, 29 June
Crudiv 5	RANGOOD	0600, 28 June
Crudiv 10	AMAMI	0600, 27 June
Crudiv 17	TUTU BAY	0600, 27 June
Crudiv 21	MANILA	0600, 28 June
Desdivs 4, 10	AMAMI	0600, 27 June
Desdivs 16, 17	MANILA	0600, 27 June
Desdivs 11, 24	PORT LLOYD	0600, 27 June
AR-2, 3	MANILA	0600, 27 June
AO-2, 4	MANILA	0600, 27 June
Task Group 2.2		
AV-1	TUTU BAY	0600, 27 June
VP-2	MALAMPAYA	0600, 27 June
VP-3	DUMANQUILAS	0600, 27 June
VP-4, 6	TUTU BAY	0600, 27 June
VP-5	KAMRANH BAY	0600, 27 June
Task Group 2.3		
Crudiv 16	MANILA	0600, 27 June
Desron 7	KAMRANH BAY	0600, 29 June
Desdivs 15, 29	SINGAPORE	0600, 30 June
Patdiv 30	MANILA	0600, 28 June
Patdiv 31	SINGAPORE	0600, 28 June
Patdiv 32	KAMRANH BAY	0600, 28 June
AO-5, 6	MANILA	0600, 27 June
AOG-1, 2	MANILA	0600, 27 June

OD,
Vice Admiral,
Commander Task Force Two
and Southern Control Force.

AUTHENICATED:

ABC,
Commander,
Flag Secretary.

Fig. 89
Location and dates of readiness for sea of units of Task Force 2

In the SOUTH CHINA SEA this is the period of the southwest monsoon, which works up in June, blows hard during July and August, and diminishes in September. Much rainy and cloudy weather occurs during these months. Hard squalls are common, blowing especially out of the GULF OF SIAM, characterized by clouds rising from the horizon in the form of an arch, and accompanied by heavy gusts of wind and rain.

In the PHILIPPINES, including the SULU SEA and the northern part of the CELEBES SEA, the rainy season brings exceedingly wet weather from June to November. Frequent storms occur during this season, and the weather is likely to be an important factor in ship and aircraft operations. Typhoons are relatively frequent above latitude 9 degrees north. They generally originate to the south and southeast of the MARIANAS and CAROLINE ISLANDS, and pass westward or west northwestward through the archipelago, or curve gradually to the north before reaching the islands. The local storms of the PHILIPPINES during the rainy season are known as "collas". They generally occur in the afternoon or night, with thunder and lightning, rain in torrents and violent winds, but are usually of short duration. They may, however, interfere with aircraft operations for as long as ten days at a time.

Except during the typhoons and local storms, the SOUTH CHINA SEA, the SULU SEA, and the north part of the CELEBES SEA will be favorable for surface, submarine and aircraft operations.

Weather in the south part of the CELEBES SEA as far east as the PELEWS will be generally good, and favorable for operation of all types of craft, with, however, daily rain squalls of considerable extent.

The southeast trades prevail in the JAVA and ARAFURA SEA and off the east and west coasts of AUSTRALIA as far south as latitude 30° south. The trades bring fine weather but with considerable haze in the JAVA and ARAFURA SEAS. In the waters south of JAVA the seas are rough and squalls make up frequently along the lee shores as far west as longitude 100° east. West of that meridian fine dry weather prevails with rough seas.

In the BAY OF BENGAL, the southwest monsoon blows continuously force 3 to 9, bringing rain and cloudy weather, and cyclones may be expected to occur occasionally in the extreme north portions.

In general, in the areas subject to heavy rain, squalls, and typhoons, airplane operation is subject to frequent interruption. Long flights are hazardous, air navigation is likely to be difficult and inaccurate, and pilots must be skilled in blind flying. Static electric discharges interfere with the use of aircraft radio. The wet atmosphere generally increases the difficulty of airplane upkeep.

These unfavorable weather conditions may be expected to have a marked effect upon naval and military operations in Areas A, B, G and H between June and October.

A weather map is appended. The wind roses show the general character of the winds within the areas in the same manner as on the monthly pilot charts.

Fig. 90
Weather forecast, 0600 30 June to 0600 1 July

Admiral O-1, after outlining the General Situation, the information he had on Blue forces, and the assumptions about Blue objectives, stated that his group's task was to destroy enemy naval forces in the South China Sea south of 15 degrees north latitude in order to protect Orange maritime transportation and supply between Rangoon, the South China Sea ports, and the Orange homeland.[24]

Rear Admiral O-16, Commander, Task Unit 2.1.3 (Col. Ralph Butchers, U.S. Army, Command and Staff Class of December 1945)

Task Unit 2.1.1 was to support TUs 2.1.2 and 2.1.3 in destroying major Blue naval forces that succeeded in penetrating and operating unopposed to the north, west, and northwest of Bintan Island–Borneo–Palawan. Task Unit 2.1.2 was to destroy Blue naval forces or ships attempting to operate north of the line Bintan Island–Singkawang Bay. In the event that superior Blue forces tried to engage this group, Task Unit 2.1.2 was to fall back on TU 2.1.1. Task Unit 2.1.3 was to destroy raiding Blue forces or ships that attempted to operate west of the line running from the south tip of Palawan to the north tip of Borneo. In the event of encountering superior Blue forces, Task Unit 2.1.3 was to fall back on TU 2.1.1. Task Unit 2.1.4 was to establish itself at Camrahn Bay and effect repairs to ships putting in to that base. All Task Unit Commanders were ordered to be prepared to steam to Philippine waters on short notice from Commander, Task Group 2.1.[25]

Of Task Group 2.1's components, Battleship Division 1, Carrier Division 1, Cruiser Division 10, and Destroyer Divisions 4 and 10 were at Amami as of 0600 on 27 June; CARDIV 2 and CRUDIV 4 were both at Camrahn Bay as of 0600 on the 29th. Cruiser Division 5 was at Rangoon by 0600 on 28 June, CRUDIV 17 was at Tutu Bay by 0600 on 27 June, and CRUDIV 21 was at Manila by 0600 on 28 June. By 0600 on 27 June, Manila also held Destroyer Divisions 16 and 17, along with the two repair ships and the two fleet oilers; DESDIVs 11 and 24 were on station at Port Lloyd as of the same time and date. O-1 outlined the roles, forces, and area of operations for Task Groups 2.2 and 2.3, as well as the stationing of the eighteen submarines under Commander, Submarines that had been ordered to provide assistance. Ships in need of repair were to head to Camrahn Bay, but repairs were not to begin without his "express direction"; vessels too large to be repaired at Camrahn Bay would be sent to Manila or Singapore. Destroyer Divisions 11, 16, 17, and 24 were all to top off fuel at Manila, while Cruiser Division 5 did the same at Singapore. After concentration, refueling would be done at sea; Commander, Task Unit 2.1.4 was to send a fleet oiler to Point K on 2 and 19 July to refuel TU 2.1.1. A fleet oiler would be sent to Point K on 9 and 26 July to refuel TU 2.1.3. CTU 2.1.2 was to provide escorts from Manila to the concentration point and thereafter would deploy the oiler at his own discretion but would not permit it to steam south or east of Point K. Two complete carrier air groups were available at Camrahn Bay; Task Unit Commanders were to inform CTG 2.1 of losses above "normal" attrition so that "timely" requests for replacements could be made to the task force commander.[26]

Admiral O-1 issued an amplified Radio Frequency Plan as well as concentration points for Task Units 2.1.1, 2.1.2, and 2.1.3. Since all units at Amami were part of Task Unit 2.1.1, they would simply proceed to their concentration point when and as directed by Commander, Task Group 2.1. At Manila, Commander, Cruiser Division 21 was to command all ships of the Task Group from the time of receipt of Plan No. 1-45 until these ships had all been released to their assigned task units. As soon as Destroyer Divisions 16 and 17 had topped off, Commander, CRUDIV 21 was to assume "proper" protective disposition, especially with reference to the safety of the fleet oilers and repair ships, and steam for concentration with Task Unit 2.1.2. At Camrahn Bay, fleet carrier CV-8 would await instructions and escort from CTU 2.1.3, while fleet carrier CV-9 would concentrate with TU 2.1.2. Cruiser Division 4 would sortie from Camrahn Bay so as to arrive at Task Unit 2.1.1's concentration point at the prescribed time. Cruiser Division 5 and Destroyer Squadron 2, at Rangoon and Port Lloyd, respectively, would proceed to their task unit concentrations at maximum "safe" speeds, topping off fuel en route. Cruiser Division 17 at Tutu Bay was to proceed to Task Unit 2.1.2's concentration point, and all TUs were to complete concentration by 1200 on 1 July.[27]

At the same time, Rear Admiral O-22, Commander of Task Group 2.2 (Fleet Air Wing 1), devised Operation Plan No. 7-45, which divided his forces into four components. Task Unit 2.2.1 (Search Unit 1), under Commander O-55, consisted of forty-nine naval patrol planes organized into Naval Patrol Plane Squadrons 3, 4, and 6, with elements from Naval Patrol Plane Squadrons 2 and 5 attached. Task Unit 2.2.2, or Search Unit 2, was under Commander O-32 and consisted of another six naval patrol planes from Naval Patrol Plane Squadron 2, while TU 2.2.3, Search Unit 3, under Commander O-56, had an additional five naval patrol planes from Naval Patrol Plane Squadron 5. Commander O-89 commanded the Mobile Base Unit, which was the seaplane tender. Rear Admiral O-22 also advised his command of the assumption—given probable Blue objectives—that Banka and Singapore would most likely be under air attack so often that Orange patrol planes could no longer be based there for refueling, rearming, and emergency repairs. The specific task at hand in this context was to conduct searches of the sea north of the Sumatra–Java–Timor line to latitude 10 degrees north, between longitudes 132 degree east and the Asian mainland. Task Group 2.2 was to locate and report enemy contacts and then attack when directed.[28]

Commander O-32, Commander, Task Unit 2.2.2, Search Unit 2, and Commander O-89, Commander, Task Unit 2.2.4 and Mobile Base Unit (Cdr. John Hatch, U.S. Navy, Command and Staff Class of December 1945)

Search Unit 1 was to conduct continuous barrier-patrol searches between Davao or Tutu Bay and Banka. Search Unit 2 was to conduct continuous single-plane patrols over the Balabac Strait and to the east toward the western entrances to Surigao

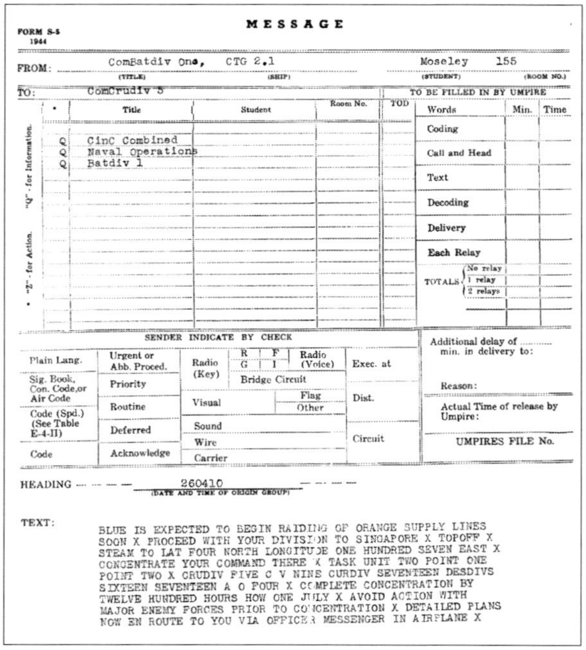

FORM S-5
1944

MESSAGE

FROM: ComBatdiv One, CTG 2.1 (TITLE) (SHIP) Moseley (STUDENT) 155 (ROOM NO.)

TO: ComCrudiv 5

•	Title	Student	Room No.	TOD
Q	CinC Combined			
Q	Naval Operations			
Q	Batdiv 1			

"Q" - for Information. "Z" - for Action.

TO BE FILLED IN BY UMPIRE

	Min.	Time
Words		
Coding		
Call and Head		
Text		
Decoding		
Delivery		
Each Relay		
TOTALS: No relay		
TOTALS: 1 relay		
TOTALS: 2 relays		

SENDER INDICATE BY CHECK

Plain Lang. / Sig. Book, Con. Code, or Air Code / Code (Spd.) (See Table E-4-II) / Code
Urgent or Abb. Proced. / Priority / Routine / Deferred / Acknowledge
Radio (Key) R G F I / Radio (Voice) / Bridge Circuit / Visual Flag Other / Sound / Wire / Carrier
Exec. at / Dist. / Circuit

Additional delay of .... min. in delivery to:
Reason:
Actual Time of release by Umpire:
UMPIRES FILE No.

HEADING 260410 (DATE AND TIME OF ORIGIN GROUP)

TEXT:
BLUE IS EXPECTED TO BEGIN RAIDING OF ORANGE SUPPLY LINES SOON X PROCEED WITH YOUR DIVISION TO SINGAPORE X TOPOFF X STEAM TO LAT FOUR NORTH LONGITUDE ONE HUNDRED SEVEN EAST X CONCENTRATE YOUR COMMAND THERE X TASK UNIT TWO POINT ONE POINT TWO X CRUDIV FIVE C V NINE CURDIV SEVENTEEN DESDIVS SIXTEEN SEVENTEEN A O FOUR X COMPLETE CONCENTRATION BY TWELVE HUNDRED HOURS HOW ONE JULY X AVOID ACTION WITH MAJOR ENEMY FORCES PRIOR TO CONCENTRATION X DETAILED PLANS NOW EN ROUTE TO YOU VIA OFFICER MESSENGER IN AIRPLANE X

Fig. 91
Message, COMBATDIV 1 to COMCRUDIV 5

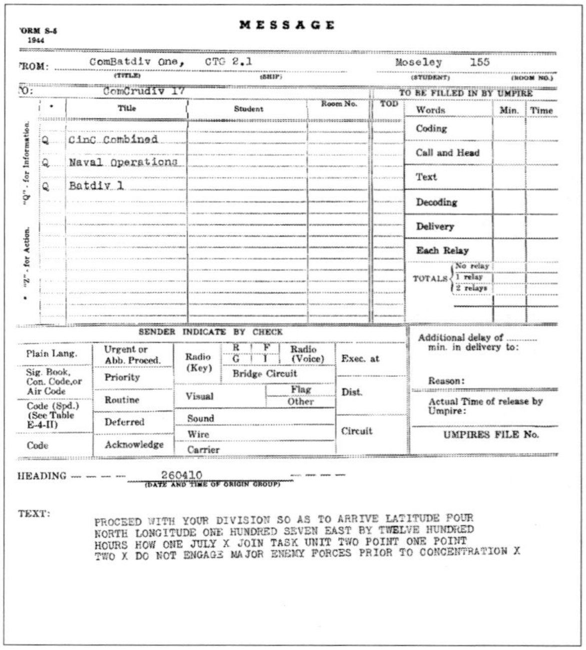

FORM S-5
1944

MESSAGE

FROM: ComBatdiv One, CTG 2.1 (TITLE) (SHIP) Moseley (STUDENT) 155 (ROOM NO.)

TO: ComCrudiv 17

•	Title	Student	Room No.	TOD
Q	CinC Combined			
Q	Naval Operations			
Q	Batdiv 1			

"Q" - for Information. "Z" - for Action.

TO BE FILLED IN BY UMPIRE

	Min.	Time
Words		
Coding		
Call and Head		
Text		
Decoding		
Delivery		
Each Relay		
TOTALS: No relay		
TOTALS: 1 relay		
TOTALS: 2 relays		

SENDER INDICATE BY CHECK

Plain Lang. / Sig. Book, Con. Code, or Air Code / Code (Spd.) (See Table E-4-II) / Code
Urgent or Abb. Proced. / Priority / Routine / Deferred / Acknowledge
Radio (Key) R G F I / Radio (Voice) / Bridge Circuit / Visual Flag Other / Sound / Wire / Carrier
Exec. at / Dist. / Circuit

Additional delay of .... min. in delivery to:
Reason:
Actual Time of release by Umpire:
UMPIRES FILE No.

HEADING 260410 (DATE AND TIME OF ORIGIN GROUP)

TEXT:
PROCEED WITH YOUR DIVISION SO AS TO ARRIVE LATITUDE FOUR NORTH LONGITUDE ONE HUNDRED SEVEN EAST BY TWELVE HUNDRED HOURS HOW ONE JULY X JOIN TASK UNIT TWO POINT ONE POINT TWO X DO NOT ENGAGE MAJOR ENEMY FORCES PRIOR TO CONCENTRATION X

Fig. 92
Message, COMBATDIV 1 to COMCRUDIV 17

FORM S-5
1944

MESSAGE

FROM: Combatdiv One (TITLE) CTG 2.1 (SHIP) Moseley (STUDENT) 155 (ROOM NO.)

TO: ComCrudiv 4

•	Title	Student	Room No.	TOD
Q	CinC Combined			
Q	Naval Operations			
Q	Batdiv 1			
Q	ComCardiv 2			

HEADING — — — — 260410 (DATE AND TIME OF ORIGIN GROUP) — — — —

TEXT:
BLUE IS EXPECTED TO BEGIN RAIDING ORANGE SUPPLY LINES SOON X
YOUR DIVISION PART OF MY TASK UNIT TWO POINT ONE POINT ONE X
C V EIGHT ASSIGNED TASK UNIT TWO POINT ONE POINT THREE X
C V NINE ASSIGNED TASK UNIT TWO POINT ONE POINT TWO X
ALL VESSELS PREPARE FOR EXTENDED TOUR AT SEA X DETAILS OF
CONCENTRATION BY OFFICER MESSENGER X

Fig. 93
Message, COMBATDIV 1 to COMCRUDIV 4

FORM S-5
1944

MESSAGE

FROM: Combatdiv One (TITLE) CTG 2.1 (SHIP) Moseley (STUDENT) 155 (ROOM NO.)

TO: ComDesron 2

•	Title	Student	Room No.	TOD
Q	CinC Combined			
Q	Naval Operations			
Q	Batdiv 1			

HEADING — — — — 260410 (DATE AND TIME OF ORIGIN GROUP) — — — —

TEXT:
BLUE WILL PROBABLY BEGIN TO RAID ORANGE SUPPLY LINES SOON X
PROCEED WITH YOUR DESRON LESS DESDIV FIFTEEN AND TWENTY NINE
TO MANILA X TOP OFF THERE X THEN PROCEED TO LATITUDE SEVEN
NORTH LONGITUDE ONE HUNDRED THIRTEEN EAST ARRIVING NOT LATER
THAN TWELVE HUNDRED HOW ON ONE JULY X YOUR DESRON LESS
FIFTEEN AND TWENTY NINE BECOMES PART OF TASK UNIT TWO POINT
ONE POINT THREE X

Fig. 94
Message, COMBATDIV 1 to COMDESRON 2

FORM S-5
1944

MESSAGE

FROM: Commandor Batdiv One (TITLE) CTG 2.1 (SHIP) Moseley (STUDENT) 155 (ROOM NO.)

TO: ComCarDiv Two

•	Title	Student	Room No.	TOD
Q	CinC Combined			
Q	Naval Operations			
Q	Batdiv 1			

HEADING — — — — 260410 (DATE AND TIME OF ORIGIN GROUP) — — — —

TEXT:
ORANGE SUPPLY LINES WILL PROBABLY SOON BECOME SUBJECTED TO BLUE
RAIDS X YOU COMMAND TASK UNIT TWO POINT ONE POINT THREE X C V
EIGHT CRUDIV TWENTY ONE DESRON TWO LESS DESDIV FIFTEEN AND
TWENTY NINE X CONCENTRATE VICINITY LATITUDE SEVEN NORTH LONGITUDE
ONE HUNDRED THIRTEEN EAST BY TWELVE HUNDRED FOX ONE JULY X AVOID
ENGAGING SUPERIOR ENEMY FORCES X DETAILED PLAN NOW EN ROUTE TO
YOU BY OFFICER MESSENGER IN PLANE X

Fig. 95
Message, COMBATDIV 1 to COMCARDIV 2

FORM S-5
1944

MESSAGE

FROM: Combatdiv One (TITLE) CTG 2.1 (SHIP) Moseley (STUDENT) 155 (ROOM NO.)

TO: ComCrudiv 21

•	Title	Student	Room No.	TOD
Q	CinC Combined			
Q	Naval Operations			
Q	Batdiv 1			
Q	AR-2, 3			
Q	AO-2, 4			
Q	Desdivs 16, 17			

HEADING — — — — 260410 (DATE AND TIME OF ORIGIN GROUP) — — — —

TEXT:
BLUE IS EXPECTED TO BEGIN RAIDING ORANGE SUPPLY LINES SOON X
YOU ASSUME TEMPORARY COMMAND ADDITIONAL UNITS AR TWO AND THREE
AO TWO AND FOUR DESDIVS SIXTEEN AND SEVENTEEN X TOP OFF DD X
SORTIE MANILA AND STEAM BEST PRACTICABLE SPEED TO LATITUDE
TWELVE NORTH LONGITUDE ONE HUNDRED THIRTEEN EAST X SEND AR TWO
AND FOUR AO TWO TO KAMRANH BAY UNDER ESCORT NOT LESS THAN TWO
DD X SEND AO FOUR AND DESDIVS SIXTEEN AND SEVENTEEN TO JOIN
TASK UNIT TWO POINT ONE POINT TWO AT LATITUDE FOUR NORTH
LONGITUDE ONE HUNDRED SEVEN EAST X THEY MUST ARRIVE THAT
POINT BY TWELVE HUNDRED HOW ON ONE JULY X HAVING COMPLIED
WITH FOREGOING YOU WILL JOIN COMCARDIV TWO WHO COMMANDS TASK
UNIT TWO POINT ONE POINT THREE X COM CV EIGHT NOW AT KAMRANH
BAY X CONCENTRATION MUST BE COMPLETED BY TWELVE HUNDRED HOURS
HOW ON ONE JULY X CONCENTRATION POINT TASK UNIT TWO POINT ONE
POINT THREE IS AT LATITUDE SEVEN NORTH LONGITUDE ONE HUNDRED
THIRTEEN EAST X

Fig. 96
Message, COMBATDIV 1 to COMCRUDIV 21

and San Bernardino Straits. Search Unit 3 was ordered to conduct continuous single-plane patrols over the Karimata, Gaspar, and Banka Straits and to the south toward the Sunda Strait. Task Group 2.2 was also to be prepared to operate with the 1st Mobile Fleet in Philippine waters on short notice and also to arm all patrol aircraft with bombs and torpedoes in case strikes were ordered. O-22's Logistics Plan, not surprisingly, echoed Vice Admiral OD's in terms of base facilities for repair and supply—especially the fuel situation at his various air bases—as well as for anchorages, landing strips, and defense forces. He instructed his task unit commanders to maintain gasoline levels at their bases at no less than a three-day supply. If difficulty arose in procuring fuel, the base of operations could be shifted to a nearby base and its unused fuel facilities. Admiral O-22 thought that this was practical in both the eastern and western approaches of the area covered by the operation, and he did not anticipate long-range search operations continuing for more than a week. In addition, Task Unit Commanders were authorized to communicate directly with Commander, Task Group 2.3 in Manila as to their fuel needs.[29]

Commander O-56, Commander, Task Unit 2.2.3, Search Unit 3 (Lt. Col. John Romlein, U.S. Army, Command and Staff Class of December 1945)

Upon receipt of Operation Plan No. 7-45, task units and naval patrol plane squadrons—plus all base personnel—were to proceed to their various bases. AV-1, the seaplane tender, was to proceed to Davao Bay on the night of 27 June. Naval Patrol Plane Squadron 2, with its six planes, was to proceed to Puerto Princessa and set up temporary bases for continuous single-plane patrols over Balabac Strait, while sending its six remaining planes to Davao for temporary duty under Commander, Naval Patrol Plane Squadron 6. Naval Patrol Plane Squadron 3 was to proceed to Tutu Bay and prepare for immediate long-range patrols; Naval Patrol Plane Squadron 4 was to remain at Tutu Bay and commence patrols with Naval Patrol Plane Squadron 1 at 0600 on 27 June. Naval Patrol Plane Squadron 5 was to proceed to Banka. Its commander, with five planes, was to operate from Banka and maintain continuous single-plane patrols in the Banka, Gaspar, and Karimata Straits, as well as southward to the Sunda Strait. The remainder of Naval Patrol Plane Squadron 5 was to prepare for long-range searches. Commander, Naval Patrol Plane Squadron 6 was to remain at Banka and act as the coordinator for long-range searches from that base as his squadron proceeded to Davao to prepare for long-range searches. Rear Admiral O-22 would remain at Davao and coordinate the searches from that base. Fleet Air Wing 1's Search Plan laid out the specific latitudes and longitudes through which extended air searches were to pass. Air searches were to commence

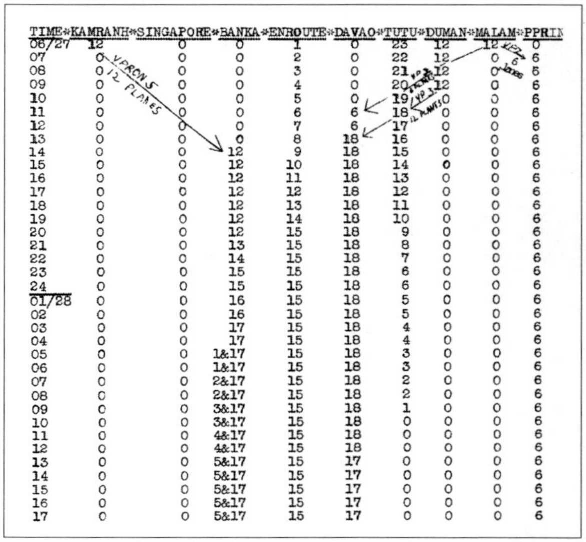

TIME	KAMRANH	SINGAPORE	BANKA	ENROUTE	DAVAO	TUTU	DUMAN	MALAM	PPRIN
06/27	12	0	0	1	0	23	12	12	0
07	0	0	0	2	0	22	12	0	6
08	0	0	0	3	0	21	12	0	5
09	0	0	0	4	0	20	12	0	6
10	0	0	0	5	0	19	0	0	6
11	0	0	0	6	6	18	0	0	6
12	0	0	0	7	6	17	0	0	6
13	0	0	0	8	18	16	0	0	6
14	0	0	12	9	18	15	0	0	6
15	0	0	12	10	18	14	0	0	6
16	0	0	12	11	18	13	0	0	6
17	0	0	12	12	18	12	0	0	6
18	0	0	12	13	18	11	0	0	6
19	0	0	12	14	18	10	0	0	6
20	0	0	12	15	18	9	0	0	6
21	0	0	13	15	18	8	0	0	6
22	0	0	14	15	18	7	0	0	6
23	0	0	15	15	18	6	0	0	6
24	0	0	15	15	18	6	0	0	6
01/28	0	0	16	15	18	5	0	0	6
02	0	0	16	15	18	5	0	0	6
03	0	0	17	15	18	4	0	0	6
04	0	0	17	15	18	4	0	0	6
05	0	0	1&17	15	18	3	0	0	6
06	0	0	1&17	15	18	3	0	0	6
07	0	0	2&17	15	18	2	0	0	6
08	0	0	2&17	15	18	2	0	0	6
09	0	0	3&17	15	18	1	0	0	6
10	0	0	3&17	15	18	0	0	0	6
11	0	0	4&17	15	18	0	0	0	6
12	0	0	4&17	15	18	0	0	0	6
13	0	0	5&17	15	17	0	0	0	6
14	0	0	5&17	15	17	0	0	0	6
15	0	0	5&17	15	17	0	0	0	6
16	0	0	5&17	15	17	0	0	0	6
17	0	0	5&17	15	17	0	0	0	6

Fig. 97 COMFAIRWING 1 search schedule, annex B, Operation Plan No. 7-45

at 0600 on 27 June; the General Plan was to maintain continuous single-plane searches over all of these areas.[30]

The remaining forty-nine naval patrol planes were to be deployed at the east and west ends of the operational area. Because a search from Davao and Banka and back again required approximately fifteen hours' flying time at a cruising speed of 150 knots airspeed, this would be a barrier patrol, starting from Tutu Bay. All patrols were to be westbound until eastbound searches could start, at such time as five naval patrol planes were searching the Banka, Gaspar, Karimata, and Sunda Strait areas; seventeen naval patrol planes would be in the Banka area for long-range purposes; fifteen naval patrol planes would be in the air at any one time, half eastbound and the other half westbound; seventeen planes would be in the Davao area for long-range patrols; and six would be in the Balabac Strait area. While at their bases, patrol planes were to be well dispersed and hidden in order to render Blue air attacks as "innocuous" as possible. If Blue air attacks became too frequent, planes would be further dispersed among additional air bases, coming into contact with each other only at the last base and when topping off fuel. Regardless of the starting point of a patrol, however, the various geographic points had to be passed over at the "proper" times and at "proper" intervals. Also, to avoid collisions while planes were flying in opposite directions in the Java Sea area, eastbound planes would fly at altitudes ending in five hundred feet and westbound planes would fly at altitude in even thousands. Kaoe Bay would be searched by westbound planes; commands were reminded that this search route was within range of Blue patrol planes from Darwin, Wyndham, Koepang, Surabaya, and Batavia. When passing these bases, advantage was to be taken of low altitude and low-visibility conditions as protection from Blue aircraft radar.[31]

From	To	Distance	Time	
DAVAO	Pt. A	350	1-40	
Pt. A	Pt. B	150	1-00	
Pt. B	Pt. C	570	3-50	
Pt. C	Pt. D	530	3-30	Westbound
Pt. D	Pt. E	160	1-05	
Pt. E	Pt. F	300	2-00	
Pt. F	Pt. G	160	1-05	
Pt. G	BANKA	100	0-40	
		2200 total	14-50 total	
Pt. G	SINGAPORE	210	1-30	
Pt. G	MEDAN	350	2-30	
BANKA	Pt. G	100	0-40	Eastbound
Pt. G	Pt. E	470	3-10	
Pt. E	Pt. N	300	2-00	
Pt. N	Pt. M	290	1-50	
Pt. M	Pt. L	290	1-50	
Pt. L	Pt. K	330	2-20	
Pt. K	Pt. J	170	1-10	
Pt. J	Pt. H	120	0-50	
Pt. H	DAVAO	280	1-50	
		2350 total	15-40 total	

Fig. 98
Table of distances and times at average speed of 150 knots

Accurate coverage of the Balabac Strait, Gilolo Passage, Ceram Sea, Flores Sea, Java Sea, and Karimata Strait for westbound planes and of the Karimata Strait, Java Sea, Makassar Strait, Celebes Sea, Molucca Sea, and Molucca Passage for eastbound planes was considered vital, though the patrol planes were to keep their presence unknown to the enemy as long as possible and avoid "unnecessary" risks. No offensive actions, other than the strafing of Blue submarines, were to be undertaken unless specifically directed by "competent" authority. According to Admiral OD, "100 percent coverage by single planes over Balabac, Karimata, Gaspar,

	BANKA	SINGAPORE	MEDAN	KAMRANH BAY	RANGOON	BANKOK	MANILA	DUMAN-QUILAS	MALAMPAYA	PEURTO PRINCESSA	TUTU BAY
BANKA		210	540	880	1240	940	1270	1000	1080	1000	1040
SINGAPORE	210		350	760	1050	720	1300	1200	1080	960	1080
MEDAN	540	350		850	310	600	1480	1480	1320	1300	1380
KAMRANH	880	760	850		850	520	680	900	600	580	840
RANGOON	1240	1050	310	850		310	1440	1720	1400	1380	1620
BANGKOK	940	720	600	520	310		1240	1450	1120	1080	1280
MANILA	1270	1300	1480	680	1440	1240		440	260	320	510
DUMANQUIL	1000	1200	1480	900	1720	1450	440		300	310	180
MALAMBAYA	1080	1080	1320	600	1400	1120	260	300		95	300
PEURTO P.	1000	960	1300	580	1380	1080	320	310	95		260
TUTU BAY	1040	1080	1380	840	1620	1280	510	180	300	260	
DAVAO	1270	1360	1620	1080	1840	1510	530	120	440	460	270

Fig. 99
Table of distances between main and subsidiary air bases used by Task Group 2.2 planes

and Banka Straits is absolutely necessary." The first plane would not leave Banka eastbound until 2400 on 27 June, and from that time on a plane would then leave every other hour. Westbound planes would leave Tutu Bay or Davao every hour commencing 0600 on 27 June until 2400 on that same day, when the other end would commence, and westbound planes would begin leaving every other hour starting at 0100 on 28 June. This, Admiral O-22 thought, would provide for a balanced schedule for the various areas. He wanted all commands to understand, however, that such a constant schedule, if under Blue observation, would soon "in all probability develop into a trap by Blue fighters, therefore, all planes will endeavor to remain unseen, but the entire time schedule may be shifted forward or back, and the search routes may be changed at any time to avoid dangerous areas." He also noted that if a plane was about to be forced down or destroyed, or enemy fighters approached it, that plane was to radio its position in plain language to its base immediately, along with the situation, if there was time.[32]

Vice Admiral OF, Commander of Task Unit 2.3.1 and of the Southwest Area Fleet, divided his command into six components. Rear Admiral O-9 commanded Escort Group No. 1, which consisted of two light cruisers, six destroyers of Destroyer Division 15 (plus two attached destroyers), and eight patrol vessels organized into Patrol Division 31. Captain O-30 commanded Escort Group No. 2, also consisting of two light cruisers, six destroyers from DESDIV 12 plus two attached destroyers, and eight more patrol vessels, these from Patrol Division 30. Commander O-31 commanded Escort Group No. 3, which was made up of six destroyers from Destroyer Division 13 plus two attached destroyers and eight patrol vessels from PATDIV 32. Commander O-75 led the Fuel Oil Supply Group of two fuel oilers, and Commander O-90 led the Aviation Gasoline Supply Group, two aviation gasoline tankers and two destroyers. In addition, there were Local Defense Forces of an unspecified nature.

Commander O-75, Commander, Task Unit 2.3.4, Fuel Oil Supply Group; Commander O-90, Commander, Task Unit 2.3.5, Aviation Gasoline Supply Group; and Commander, Task Unit 2.3.6, Local Defense Forces (Lt. Cdr. Harold Sommer, U.S. Navy, Command and Staff Class of December 1945)

In describing the General Situation to his forces, Admiral OF invoked the Emperor's determination to maintain Orange lines of supply and transportation. Admiral OF declared that Task Group 2.3's mission was to escort Orange shipping in the South China Sea between latitudes 15 degree north and 5 degrees south and longitudes 90 and 130 degrees east; to control movements of Orange and neutral ships; to maintain fuel oil, diesel oil, and aviation gasoline reserves at the Subsidiary Bases; and to defend Orange ports and bases.[33]

Escort Group Nos. 1 and 2 were to escort northbound and southbound convoys moving at ten-knot speeds between Singapore and Point J (latitude 15 degrees north and longitude 113 degrees east) as well as escort convoys between Singapore and Palembang. Escort Group No. 3 was to escort northbound and southbound convoys moving at speeds of sixteen knots between Singapore and Point J, while the Fuel Oil Supply Group was to maintain fuel and diesel oil reserves at the Subsidiary

Fleet Base in Singapore. The Aviation Gasoline Supply Group was to maintain the aviation gasoline reserve at the Subsidiary Air Bases at Rangoon, Banka, Bangkok, Puerto Princessa, Singapore, Medan, Davao, Malampaya Sound, and Dumanquilas. The Local Defense Forces were to defend the ports and bases in accordance with present plans, and the entire force—less the Local Defense Forces—was to be prepared on short notice from Commander, Task Group 2.3 to concentrate in Philippine waters. Vice Admiral OF also noted that this plan was to go into effect at 0600 on 27 June and that attached Logistics and Communications Plans would also be employed.[34]

Admiral OF's Movement Plan summarized the weekly number and frequency of convoys moving between the various Orange ports in the South China Sea area. The convoy designated to leave Singapore at 0500 on 1 July and head north was designated Convoy No. 1; the convoy heading south and arriving at Singapore at 0500 on 2 July was Convoy No. 2; and the one headed south and arriving at Singapore at 1700 on 2 July, Convoy No. 3. OF's plan to assemble his Task Units entailed Cruiser Division 16 departing Manila at 0600 on 27 June at a speed of twenty-five knots with a full mine load. This unit was to proceed to the Balabac Strait, mine those waters, and then detach two of its light cruisers to join Convoy No. 1; the division's other two light cruisers would join Convoy No. 2. His two fleet oilers would depart Manila with the first convoy departing that port after 0600 on 27 June to join the southbound escorted convoy at Point J. Patrol Division 30 would depart Manila at 0600 on 28 June at a speed of sixteen knots and join southbound Convoy No. 2, while DESDIVs 15 and 29 would depart Singapore at 0500 on 1 July with Convoy No. 1. DESDIV 15, with two additional destroyers, would continue with that convoy to the north, while two other destroyers from DESDIV 29 would join Convoy No. 2 to the south. Patrol Division 31 would depart Singapore with Convoy No. 1 at 0500 on 1 July, while Destroyer Division 13, with two attached destroyers, would leave Camrahn Bay at 0600 on 29 June and, at a speed of thirty knots, join and proceed with southbound Convoy No. 3. Destroyer Division 12 would depart Camrahn Bay on the same day and at the same time and speed, join southbound Convoy No. 2. Finally, PATDIV 32 would depart Camrahn Bay at 0400 on 29 June and proceed to the southbound Convoy No. 3. These routes were specified in terms of latitude, longitude, and lettered points; convoys were ordered to keep five miles to the right of their prescribed tracks.[35]

The Convoy General Plans ordered that unescorted vessels from Burma depart so as to arrive at Singapore twenty-four hours prior to the scheduled departure of the convoys. In addition, vessels from Banka, Billiton, and Sumatra were to arrive unescorted at Palembang twenty-four hours prior to the scheduled departure of their convoys. Vessels from Bangkok and Saigon were to follow an inshore route unescorted and then join the convoys at Point K. At Point J, convoys from the Orange homeland were to meet and exchange escorts. From Point K, vessels bound for Saigon and Bangkok were to proceed independently to their destinations. At Singapore, vessels bound for Burma would do the same, and those bound for Banka, Billiton, and Sumatra would be escorted to Palembang. CTG 2.3 was also to notify all concerned by dispatch of any schedule changes as far in advance of sailing times as practicable. In addition, the commanders of Escort Group Nos. 1, 2, and 3 were authorized to divert convoys from the specified routes in order to safeguard them from enemy attack, but CTG 2.3 had to be notified as soon as possible about newly estimated times of arrival.[36]

		NORTHBOUND		SOUTHBOUND	
Convoy No.	Escort Group.	Leave SINGAPORE	Arrive Point J	Arrive Point J	Arrive SINGAPORE
1.	2.3.1	0500 1 July	0500 5 July		
2.	2.3.2			0500 29 June	0500 2 July
3.	2.3.3			0500 30 June	1700 2 July
4.	2.3.1			0900 5 July	0900 9 July
5.	2.3.3	1700 3 July	0500 6 July		
6.	2.3.2	0500 8 July	0500 12 July		
7.	2.3.3			0500 7 July	1700 9 July
8.	2.3.3	1700 10 July	0500 13 July		
9.	2.3.1	0500 19 July	0500 19 July		
10.	2.3.2			0900 16 July	0900 16 July

Fig. 100
Convoy schedule

Admiral OF's Logistics Plan ordered escort vessels operating between Singapore and Point J—as well as between Singapore and Palembang—to take on maximum fuel loads prior to their initial sorties. OF also ordered that all vessels refuel to capacity each time they returned to Singapore and that at each fueling they advise the fueling station commander as to the time and amount of the next fueling. Additionally, the Fuel Oil Supply Group was to maintain fuel and diesel oil reserves at Singapore; the fuel oilers would base at Camrahn Bay and join the convoys to Singapore when delivering to that point. The Aviation Gasoline Supply Group would operate a tanker and a destroyer out of Manila and maintain aviation gasoline reserves at Malampaya Sound, Puerto Princessa, Dumanquilas, and Davao. Another tanker and another destroyer out of Camrahn Bay would maintain reserves at Rangoon, Medan, Banka, Bangkok, and Singapore. Movements here, as far as practical, would be made only at night. Also, in supplying aviation gasoline to Medan and Rangoon, maximum use was to be made of returning empty vessels by loading them with oil drums. Ships of Task Group 2.3 needing repair would be sent to Manila, Singapore, or Camrahn Bay, whichever was most convenient to that ship. Estimated times of ship repairs were also to be sent to Commander, Task Group 2.3. Included here as well was a Communication Plan.[37]

(b) Convoys between PALEMBANG and SINGAPORE.

Convoy No.	Escort Group	Leave SINGAPORE	Arrive PALEMBANG	Leave PALEMBANG	Arrive SINGAPORE
1.	2.3.2	1900 2 July	2100 3 July		
2.	2.3.2			2100 3 July	2300 4 July
3.	2.3.1	1200 9 July	1400 10 July		
4.	2.3.1			1600 10 July	1800 11 July

Fig. 101
Convoys between Palembang and Singapore

Conclusion

The Orange Statement of the Problem represented something different from the other Operations Problems. In this case, Orange actually outnumbered Blue in fleet carriers, an unusual scenario for a war game planned late in the Pacific War and taking place in the western Pacific. Consistent with wartime practices by the United States, however, the Blue carrier strikes, which were meant to destroy Orange merchant ships and shore facilities, were only a prelude to the drawing out of Orange air and surface forces to engage Blue air and surface forces. The only element left out by Blue and Orange was the role of submarines, and even they were to be employed as scouts by both sides. It was not envisioned, however, that they would be deployed to damage the enemy fleet before the major surface engagement.[38]

Still, both strategically and tactically speaking, this Operations Problems was quite consistent with late Pacific War strategies and fleet doctrines of both sides. The U.S. capture of the Philippines during the Pacific War was carried out specifically to cut Japan off from its supply of raw materials in the southwest Pacific and Southeast Asia, while the idea of seizing Formosa was considered as a way to provide a staging area for the invasion of the Japanese home islands or as a naval and air base from which to continue strangling Japan's economy and war-making potential. Though the strategic and tactical scenarios of Operation Problem 5 did not need to for strictly pedagogical purposes, both reflected the wartime U.S. Pacific Fleet's operational combination of interwar doctrine and Pacific War practice, as well as characteristic responses of the Imperial Japanese Navy.[39]

NOTES

1 Command and Staff Class of December 1945, "Operations Problem 5: Orange Statement and Staff Solution," 10 September 1945, pp. 1–3, NHC, folder 2536, box 125, RG 4, NWC.

2 Ibid., pp. 3–4.

3 Ibid., pp. 4–5.

4 Ibid., p. 6.

5 Ibid., p. 7.

6 Ibid., pp. 8–17.

7 Ibid., pp. 18–22.

8 Ibid., pp. 23–25.

9 Ibid., pp. 25–27.

10 Ibid., pp. 27–29.

11 Ibid., p. 29.

12 Ibid., pp. 30–32.

13 Ibid., pp. 32–34.

14 Ibid., pp. 36–37.

15 Ibid., pp. 36–38.

16 Ibid., pp. 38–40.

17 Ibid., pp. 40–41.

18 Ibid., pp. 41–43.

19 Ibid., pp. 43–46.

20 Ibid., pp. 46–51.

21 Ibid., pp. 51–53.

22 Ibid., pp. 53–57.

23 Ibid., pp. 57–63; see also pp. 65–70.

24 Command and Staff Class of December 1945, "Operations Problem 5: Orange—Section B," 3 October 1945, p. 1, NHC, folder 2536, box 125, RG 4, NWC.

25 Ibid., p. 2.

26 Ibid., pp. 3–4.

27 Ibid., pp. 5–6.

28 Command and Staff Class of December 1945, "Operations Problem 5: Orange—Section B," 2 October 1945, pp. 1–2, NHC, folder 2536, box 125, RG 4, NWC.

29 Ibid., pp. 2, 4.

30 Ibid., p. 5.

31 Ibid., p. 6.

32 Ibid., pp. 6–7, 9.

33 Command and Staff Class of December 1945, "Operations Problem 5: Orange—Section B," 3 October 1945, pp. 1–2.

34 Ibid., pp. 2–3.

35 Ibid., p. 4.

36 Ibid., pp. 5–6.

37 Ibid., pp. 7–8.

38 For the idea that naval aviation and submarine warfare up to the Pacific War were subordinated to a primarily battleship navy because of the power of surface line officers in the U.S. Navy, see William McBride, *Technological Change and the United States Navy, 1865–1945* (Baltimore: Johns Hopkins Univ. Press, 2000), pp. 111–212, especially 184, 195–96, and 209. What McBride fails to understand, however, is that elements of pre-1941 tactical doctrine were combined with new, wartime elements reflective of lessons learned in 1941–42; see Trent Hone, "U.S. Navy Surface Battle Doctrine and Victory in the Pacific," *NWCR* 62, no. 1 (Winter 2009), pp. 67–105; and Hone, e-mails to author, 28 February, 2 March, 3 March, and 14 March 2011. It was this newly combined tactical doctrine that the immediate postwar Naval War College was practicing.

39 Secondary sources on the battles around the Philippines and the South China Sea are numerous, but for one traditional account, see Clark Reynolds, *The Fast Carriers: The Forging of an Air Navy* (Annapolis, Md.: Naval Institute Press, 1968), pp. 253–300.

THE BATTLE
OF
SAVO ISLAND

IX *Operations Problem 5*

The History of the Maneuver, October 1945

The History of the Maneuver for Operations Problem 5 demonstrates this Pacific War doctrine and practice, especially in terms of air operations. The document begins with Advance Data for Move 1, with which the Chart Maneuver was scheduled to commence at 1400 on 4 October (actual). Move 1 was to be a twenty-four-hour move commencing at 0600 on the Maneuver Date of 30 June. On 28, 29, and 30 June, weather conditions were to be generally poor for flying in the primary areas of operation. Students were urged to complete all preparations for the commencement of Move 1 prior to 1130 on 4 October, positioning all forces and plotting information accurately on their boards as of 0600 on 30 June. Students were warned to avoid "unauthorized information"—that is, corridor gossip. Students were to give enough information on the Record of Move to portray what actually happened and why, not burdening the record with details of "no consequence." Records for each Move were to be submitted when flimsies for the next Moves were sent to the Master Plot. Students were to keep their Fuel Accounts up to date (but not to submit them until called for), make notes for the Critique, and use the periods between Moves to plan future dispositions and read in preparation for their theses.[1]

In the Preliminary Move, Blue Task Force 75, the Air Force, was divided into four Task Groups. These consisted of the Batavia Group, forty-eight Liberator naval patrol planes; the Brunei Bay Group, which comprised four destroyers, five seaplane tenders, and twenty-four Mariner naval patrol bombers; the Surabaya Group of twenty-four Mariners and three seaplane tenders; and the Saleh Bay Replacement Command Group. The Batavia Group carried out photographic reconnaissance of Orange airfields, oil refineries, mines, smelters, and harbors at Singapore, Banka, and Palembang on D–6, D–3, and D–2. They noted a few fighter planes in the air and on the airfields at Singapore and Banka and, commencing on D–1, this group made daily searches of the South China Sea between longitude 103 degrees east and longitude 110-40 minutes east (that is, 110° 40' E) toward Indochina. The Brunei Group operated from Balikpapan and commenced daily reconnaissance on D–1 of the Sulu Sea and South China Sea between longitude 110-40 and 115 east, up to latitude 11 north. Blue surface ships proceeded en route via the Sulu Sea toward Brunei Bay in order to establish a base there. The Surabaya Group was not alerted prior to D-Day, as it had no mission until that time, but the Saleh Bay Group dispatched twelve fighter planes to Commander, Batavia Group and an additional nine to Balikpapan for air cover at those bases. Commander, 7th Fleet also ordered CTF 75 to extend his search mission to include the entire South China Sea instead of just the southern sector.[2]

Task Force 78 was at anchor at Kaoe Bay during the Preliminary Movement period. Strike Group 1 furnished photographic reconnaissance from flights made on D–6 of the oil storage facilities, fuel depots, gasoline storage facilities, air facilities, mines, and harbors at Tutu Bay, Puerto Princessa, and Davao. Results from this photographic reconnaissance were reported to Commander, 7th Fleet and Commander, Air Force. No shipping had been seen at these harbors, but fighter aircraft had been observed on the ground and in the air at Puerto Princessa and Davao. "Considerable" activity had also noted in and around Manila. One Orange naval patrol plane was able to get a report off on Task Force 78 before it was shot down by Blue carrier fighters. Other Orange search planes approaching Kaoe Bay and the northwest over periods of two hours were shot down before they could make observations of TF 78. These Orange searches commenced on 27 June. Meanwhile, Task Force 79, in compliance with a request from CTF 78, fueled Mine Division 6, plus oilers and escorts. These ships were ordered to join Task Group 78.2 and accompany that unit through the Karimata Strait. The first refueling rendezvous was scheduled for D–5, but TF 79 stayed in port during the Preliminary Movement period.[3]

Orange forces began their Preliminary Moves as well. Task Group 2.1's components began concentrating, with Task Unit 2.1.1 concentrating at latitude 12-00 north, and longitude 113-00 east, TU 2.1.2 at latitude 4-00 north, longitude 107-00 east. At the same time, TU 2.1.3 was to concentrate at latitude 7-00 north and longitude 113-00 east, while TU 2.1.4 was to concentrate at Camrahn Bay. Task Group 2.2 saw its seaplane tender shifted to Davao Bay on the night of 27 June, and Naval Patrol Plane Squadron 2 sent six naval patrol planes to Puerto Princessa. The remaining six planes of the squadron proceeded to Davao, while Naval Patrol Plane Squadron 3 flew to Tutu Bay, Naval Patrol Plane Squadron 5 went to Banka, and Naval Patrol Plane Squadron 6 to Davao. Task Group 2.3 saw Cruiser Division 16 depart Manila at 0600 on 27 June for the Balabac Strait. The two oilers also left Manila at that time for the southbound convoy, and Patrol Division 30 left Manila at 0600 on 28 June, also to join the southbound convoy. Destroyer Division 13 and its two additional destroyers left Camrahn Bay at 0600 on 29 June to join the same convoy, and DESDIV 12 left Camrahn at the same time to join southbound Convoy No. 2. PATDIV 32 departed Camrahn at 0400 on 29 June to join the southbound Convoy No. 3.[4]

As noted above, Move 1 was a twenty-four-hour move that commenced at 0600 on 30 June. At that time, Blue submarines on patrol observed a large Orange Task Force leaving Camrahn Bay on an easterly course; they were not able to determine its composition or disposition. At the same time, three Orange light cruisers were seen proceeding from Camrahn on a course of 195 degrees true and a fleet oiler and a destroyer leaving the same location on a course of 070 degrees true. Meanwhile, Orange Task Group 2.1 was concentrating in the South China Sea, with Task Unit 2.1.1 fueling at its concentration point. Vessels of Task Unit 2.1.2 were headed for their concentration point but would not actually join for several hours, and the vessels of TU 2.1.3 were about four hours from meeting at their concentration point, missing a destroyer transiting from Singapore. Orange air forces were carrying out their assigned search missions and the convoy movements were being executed as planned. Blue's air searches were also being carried out as scheduled; Blue Striking Groups 1 and 2 left Kaoe Bay at 1800 on 30 June and proceeded to the Celebes Sea. An air search had been launched by one carrier operating independently at 1100 on 30 June, covering the northwest quadrant. At 1300, one of the Blue carrier search planes was shot down off Davao by Orange land-based fighters before it could observe the anchorage.[5]

At 0300, Blue naval patrol plane 63-P-2 detected Orange Cruiser Division 5 as it proceeded north from Singapore. The contact was developed and the search continued, but the patrol plane did not close to gunfire range. At 0430, Blue naval patrol plane 63-P-5 made contact on a slow-moving convoy of one large and two small ships headed southwesterly. This plane too developed the contact and continued the search but without closing to gunfire range. At 0300, a Blue submarine observed an Orange gasoline tanker and a destroyer leaving Manila on a southerly course, while at 0445 Blue tender-based naval patrol plane 63-P-4 detected Orange CRUDIV 17; as before, the contact was developed and the search continued but the plane did not come under fire. At 0515, Blue naval patrol plane 63-P-4 made contact with Orange fleet carrier CV-9 and six destroyers. The carrier was observed turning into the wind to launch. The Blue plane retired to evade Orange fighters and continue the search, and it was not fired on by the ships. At 0555, the same plane sighted a large convoy with escorts on course 215 degrees true at a speed of eight knots. At the same time, Blue naval patrol plane 203-P-2 made visual contact with a large surface group and observed a carrier turning into the wind to launch. This was actually Orange Task Unit 2.1.3. The Blue plane retired at maximum speed and continued its search. At this point, Move 1 ended after four hours of Actual Time and twenty-four hours of Problem Time.[6]

In Move 2, at 0600 on 1 July, Orange Task Unit 2.1.1 finished concentrating and dispatched destroyers DD-119 and DD-120 to escort tanker AO-2, which was cruising in the vicinity of the concentration point, to Camrahn Bay. By 1200, TU 2.1.2

had also completed concentration, and TU 2.1.3 did the same by 1400. Orange Task Group 2.2 continued scheduled air patrols between Davao and Banka at air speeds of 150 knots, and the Task Unit conducted three single-plane shore-based searches east of Surigao as well as a patrol south of Halmahera. TG 2.3 convoy movements continued in accordance with the convoy schedule, and a fuel supply group proceeded toward Singapore in company with Convoy No. 2. Blue TG 75.1 conducted its air search of Malacca Straits, and TG 75.2 received orders to move surface forces to Brunei Bay. Gas and oil delivery was also requested of CTF 79 for delivery at Balikpapan and Brunei Bay. TF 78 was en route to Kaoe Bay, specifically to the area off Sulu Archipelago in order to launch aircraft strikes on Davao and Tutu Bay. TF 79 remained at anchor at Saleh Bay and Surabaya, but at 1400 gasoline tankers AOG-1 and AOG-2 sailed, with destroyer leader DL-357 and destroyer DD-380 as escort. At 0600, Orange Convoy No. 3 picked up one bogey (i.e., unknown air contact) on radar, at a distance of forty miles and decreasing. The convoy went to General Quarters and stood by to repel an air attack. The radar contact was Blue search plane 63-P-6, which reported to CTF 75 about a convoy of ten merchant ships, six destroyers, and eight submarine chasers on course 215 at a speed of twelve knots.[7]

Orange Task Unit 2.1.3 also had a bogey on its radar screen, bearing 320 at a decreasing distance of forty miles, and a combat air patrol (CAP) was vectored to destroy it. The bogey was actually Blue search plane 201-P-1. At 0651, this plane reported that it was being attacked by Orange fighters. The Orange CAP destroyed the Blue plane. At 0750 Orange Task Unit 2.1.2 picked up Blue search plane 63-P-5 on radar, bearing 000 true and at forty miles, decreasing. Fleet carrier CV-9 vectored out its combat air patrol and declared an antiaircraft alert. This Blue search plane was also destroyed, but not before it got off a report of the contact and of being attacked by Orange fighters. Orange Cruiser Division 5 at 0800 picked up Blue search plane 63-P-3 on radar bearing 340 true at a decreasing distance of forty miles and trained its antiaircraft batteries. Five minutes later, CRUDIV 5 was sighted by the Blue search plane, which reported four heavy cruisers on course 050 at a speed of thirty knots. The plane was not destroyed and continued its search. By this time, the Blue commander of Task Group 78.2 planned to send three strikes—at 0900, 1000, and 1400—against Davao, with the targets being shipping, airfields, and parked aircraft. Accordingly, at 0900, TG 78.2 launched forty fighters, twenty-four scout bombers, and eighteen torpedo bombers and, an hour later, twenty fighters, thirteen scout bombers, and nine torpedo bombers. At 1400, the planned strike of thirty-four fighters, twenty-one scout bombers, and six light transport planes (VGs) was canceled and sent to Dumanquilas instead. Blue TG 78.1 furnished sixteen fighters as a combat air patrol for TF 78 and also a strike of thirty-two fighters and sixteen torpedo bombers against Tutu Bay. At 0920, TF 78 picked up Orange search plane 4-P-11 bearing 000 true, at a decreasing distance of fifty miles. Two fighters were vectored out to intercept and destroy the Orange search plane, which managed to get off a contact report before it was shot down. The report stated that two fleet carriers along with several cruisers and destroyers were on a course of 310 at "high" speed.[8]

At 0930, Orange TU 2.3.1 picked up Blue search plane 63-P-1 on radar, also bearing 000 true and at forty miles, closing. The Orange ships went to General Quarters and prepared for an air attack. The Blue plane reported ten to fifteen merchant vessels, two light cruisers, and several escorts on a course of 020 at a speed of ten knots and then continued its search. At 1010, Orange TU 2.2.4 also picked up a bogey, this one bearing 160 degrees at a decreasing distance of eighty miles, a speed of 160 knots, and an altitude of eight thousand feet. At the same time, the same unit received a report of "many" aircraft but took no action. In fact, this Task Unit failed to report the incoming strike to CTG 2.2 or to send land-based fighters or seaplanes to intercept or scout. Fifteen minutes later, the first Blue strike on Davao was intercepted by twelve Orange land-based fighter planes. In the ensuing aerial combat, Orange lost eight fighters but shot down four Blue fighters, two scout bombers, and two torpedo bombers. Five minutes after that, Blue Task Force 78 picked up a bogey on radar bearing 280 degrees true at a distance of thirty-eight miles and closing at an altitude of two thousand feet and a speed of 150 knots. Two fighters were vectored out to destroy Orange naval patrol plane 3-P-6, which was shot down after reporting that it was being attacked by Blue fighters. At the same time, Orange naval patrol plane 3-P-6 picked up a bogey bearing 100 at a decreasing distance of forty miles and succeeded in making a contact report to CTF 2 before being shot down.[9]

At 1040, the first Blue strike hit Davao. In the attack, Orange seaplane tender AV-1 was sunk, seven Emily seaplanes were destroyed on the water, fuel and ammunition dumps at the airfields were approximately 90 percent destroyed, antiaircraft defenses at the airfields were "about" 50 percent destroyed, and the airfields were knocked out of action for twelve hours. Blue lost another two fighters, one scout bomber, and one torpedo bomber to antiaircraft fire. At 1140, the second Blue strike hit Davao. Antiaircraft fire was meager, and there were few targets, since the airfield was out of commission. Blue lost two torpedo bombers in the strike. Search efforts continued, with Orange TG 2.1 observing a bogey on its radar screen at 1325 but losing the contact at 1335 without making an interception. Also at 1335, Blue search plane 13-S-10 from TG 78.1 observed eight to ten Orange Emily patrol planes on the water and two fighters on the airstrip at Dumanquilas. Ten minutes later, Orange Commander, Task Group 2.1 evacuated his seaplanes from Dumanquilas to Malampaya Sound. In the subsequent Blue airstrike, the Orange airfield was knocked out for twelve hours; Orange lost nine fighters and saw 75 percent damage to its fuel dumps and installations. Blue lost four fighters, two scout bombers, and three torpedo bombers to Orange fighters and antiaircraft fire.

Meanwhile, the Blue strike on Tutu Bay at 1550 found no worthwhile targets; at 1430 Blue submarine SS-198 observed a few Orange surface vessels on course 180 degrees at an increasing distance of eight miles.[10]

Move 3 began at 1800 on 1 July. At 2100, Orange search plane 6-P-10 made contact on a small convoy about 60 miles north of Asoembawa Island, headed north at low speed. At 2125, Orange search plane 2-P-11 made contact with a small convoy of one seaplane tender and two destroyers headed north from the Makassar Strait and trailed it until after daylight, at which point the convoy was observed heading into Tarakan. At 0100 on 2 July, Orange search plane 2-P-9 made contact again with this force of three ships but continued its search upon discovery that Orange search plane 2-P-11 was already trailing the Blue force. At 0240 on 2 July, Blue search plane 72-P-4 made radar contact with a large force in the South China Sea, headed south at high speed. Developing the contact revealed seventeen blips, some quite large. From the shape of the formation, the search plane crew determined that it was a carrier task force. An hour later, Blue search plane 72-P-4 made contact with a convoy headed northeasterly and developed it as twelve large and fourteen small blips. This was thought by Blue to be the slow convoy with its escorts. At 0445, Blue search plane 203-P-4 made radar contact on a large force headed in an easterly direction at high speed, believed by Blue to be a combatant group that included a carrier. At 0500, a Blue air group of forty naval patrol bombers and twelve fighters took off from Batavia, while an hour later a Blue strike of twenty fighters hit Banka. A second strike of twenty Blue naval patrol planes hit Palembang. This strike was unopposed and resulted in 50 percent of Orange oil installations there being destroyed. The strike on Banka was opposed by twelve Orange fighters, which shot down six Blue naval patrol bombers and four Blue fighters. Six Orange fighters were lost to the Blue fighters, as were two Blue naval patrol bombers to Orange antiaircraft fire. A few minutes later, Orange fighter planes protecting Camrahn Bay shot down two Blue patrol planes that had been attempting to reconnoiter and photograph the bay.[11]

At 0655, Blue search plane 13-T-5 made contact with Orange TU 2.1.3 and was able to develop the contact before being shot down. The Blue pilot reported a carrier, cruisers, and destroyers present. Unfortunately for Blue, two additional search planes were also shot down by Orange. A few minutes earlier, Orange carrier search planes had sighted Blue TG 78.1 and were able to report the presence of two carriers, several cruisers, and many destroyers. Two of these planes, 8-T-6 and 8-T-7, were shot down by the Blue combat air patrol, while two others escaped to northern Borneo. At 0700, Orange search plane 2-P-1 reported the presence of TG 75.2 with its seaplane tenders turning toward Tarakan in northern Borneo. At the same time, Orange search plane 3-P-5 made radar contact on TG 78.1 but was shot down before it could develop the contact. One minute later, Orange search plane 2-P-1 made a distant radar contact on TG 78.2 and was able to determine that it consisted of two carriers and "many other" ships in the formation before it too was shot down by Blue fighters. At 0530, a Blue strike of twenty-four fighters and twenty-four scout bombers took off from fleet carrier CV-13. Another Blue air strike, of twenty-four dive-bombers, each carrying two thousand-pound bombs; twelve scout bombers carrying the same load; and twelve escorting fighters on Puerto Princessa, put the airfield out of commission for twelve hours as well as destroyed 50 percent of the available fuel. This strike was intercepted by twelve Orange fighters and Blue lost two dive-bombers and six fighters while shooting down eight Orange fighters.[12]

At this time, Blue search plane 72-P-4 made radar contact on Orange TU 2.1.3, and at 0745 a Blue carrier strike from Task Group 78.1, which had launched at 0530, hit Orange 2.1.3, which had a CAP of eight fighters aloft. Six of the Orange fighters were shot down. The Blue strike consisted of eight fighters, fifteen torpedo bombers, and sixteen fighter-bombers (VFBs), each of the latter carrying two thousand-pound bombs. Blue losses were four fighters, two torpedo planes, and two fighter-bombers to the Orange combat air patrol and seven fighter-bombers to antiaircraft fire. However, they achieved three bomb hits and two torpedo hits on Orange fleet carrier CV-8, which was left dead in the water and sinking. At 0815, an Orange carrier strike launched at 0700 from fleet carrier CV-8, the strike consisting of twenty-four dive-bombers, three torpedo planes, and sixteen fighter-bombers. This force struck Blue TG 78.1, losing six dive-bombers, two torpedo planes, and eight fighter-bombers to the Blue CAP as well as twelve dive-bombers, one torpedo plane, and seven fighter-bombers to Blue antiaircraft fire. Blue fleet carrier CV-13 was struck by a thousand-pound bomb, which put the carrier's flight deck out of commission for seven hours. Because of heavy air losses, however, Orange succeeded only in sinking Blue light cruiser CL-25 and inflicting 50 percent damage on light cruiser CL-26.[13]

At 1400, a weather front that had developed between Johol (probably modern Jolo) and Palawan moved southwestward toward Borneo, just as a large strike from Orange Task Unit 2.1.1 proceeded to strike the Blue forces to the east of the Balabac Strait. This strike, however, was turned back by the weather front. At 1030, Blue search plane 72-P-2 made contact with Orange TU 2.1.2. Ordered to develop the contact, the Blue pilot reported one fleet carrier, seven cruisers, and eight destroyers before being shot down by the Orange combat air patrol. At 1130, the weather in Area B and the west half of Area C was announced as characterized by a very low ceiling, five miles visibility, frequent rain squalls, and flying conditions that were impossible for small planes except in the vicinity of ships and bases. At 1800, an Orange submarine stationed in the Makassar Strait reported seeing a large carrier task force moving south through the strait at about 1200 at an estimated speed of twenty-five knots. Move 3 concluded with the surviving ships of Orange TU 2.1.3 joining TU 2.1.1 and coming under the control of CTG 2.1 at 1800.[14]

At the beginning of Move 4, at 1800 on 2 July, Blue TG 75.1 was still based at Batavia, while TG 75.2 had seaplane tenders AV-6 and AV-12 at Balikpapan. Task Group 75.3 had not conducted any operations, and TG 75.4, the Carrier Replacement Group, had no planes in the air. In addition, Blue TG 78.1, Carrier Group 1, was steaming on course 348 in the Sulu Sea at a speed of twenty knots. TG 78.2 was at the same speed on a course of 185 en route to a fueling rendezvous in Java City, while TF 79 was at anchor in Saleh Bay, with the exceptions of TGs 79.2 and 79.6, which were at anchor at Surabaya. TG 79.3, less destroyers DD-382, DD-400, and DD-401, was en route on course 028 to Balikpapan. At the same time, Orange TU 2.1.1 was joining TU 2.1.3 to the westward of Palawan, these units steaming generally southwest in the Palawan Passage. Orange Task Unit 2.1.2, the Carrier Group, was on a course of 348 at a speed of eighteen knots off the east coast of the Malay Peninsula. Tanker AO-2 and its two escorting destroyers were steaming at nine knots on a course of 090, while Orange repair ships AR-2 and AR-3 were at anchor in Camrahn Bay. Orange TU 2.2.1 was conducting shore-based air searches in the Sulu Sea, the Celebes Sea, and east of the Philippines. It was also noted by both sides that barrier searches did not completely cover the Sulu Sea, making it possible for both to operate there undetected.[15]

About this time, Orange TU 2.2.2 conducted air patrols of Balikpapan and Surigao Strait, while TU 2.2.3 (Search Unit 3) maintained a barrier patrol of the Karimata and Gaspar Straits. Orange TU 2.3.1 (Escort Group 1) escorted Convoy No. 1 on a course of 033 at a speed of ten knots, while TU 2.3.2 was escorting Convoy No. 2 on a course of 213 at the same speed. Task Unit 2.3.3 escorted Convoy No. 3 to Singapore, where it was due to arrive by 0240 on 3 July. In addition, Orange TUs 2.3.4, 2.3.5, and 2.3.6 (the Fuel Oil Supply Group) dispatched gasoline tanker AOG-1 and destroyer DD-47 from Basquet Bay at 1800 on 2 July for Manila. The weather also improved significantly in the area of operations as Blue Task Group 75.2 left Balikpapan at 0100 en route to Linkas and Naval Patrol Plane Squadron 203 left the same anchorage at 0100 on 3 July, to arrive at Linkas by 0300 on that same morning. Blue TG 78.2 detected one bogey on radar, bearing 150 degrees true at forty miles, opening. TG 78.2 went to antiaircraft alert; the contact was Orange naval patrol plane 3-P-9. The Orange plane reported TG 78.2 at a distance of thirty-nine miles bearing 330 degrees true, held the contact and developed it, informing Orange that this was a Blue carrier group at high speed, now on course 220. At 0200, Orange TUs 2.1.1 and 2.1.3 slowed to twelve knots and refueled their destroyers. Forty minutes later, Orange Convoy No. 3 arrived at Singapore, where its escorts fueled before departing to escort Orange Convoy No. 5.[16]

At 0350, Orange Task Unit 2.1.2 picked up one bogey on radar, bearing 200 degrees true at a distance of fifteen miles, closing and flying low. TU 2.1.2 slowed to five knots in the hope of not being picked up, since it was in close proximity to some islands, but Blue naval patrol plane 73-P-4 reported it, on course 350 with medium spacing, before it continued on its search. At 0515, Orange TU 2.1.1 had a bogey on its radar screen bearing 215 at a distance of forty miles and on a parallel course. At the same time, the Orange Local Defense Commander reported to CTF 2 that he had many bogeys on radar bearing 175 degrees true at a distance of 80 miles and closing at high speed. All planes at Manila were alerted; they took off and cleared the Manila area, while twenty-seven fighters were ordered to intercept the incoming Blue strike. Two minutes later, Blue search plane 201-P-4 located "many" surface ships bearing 035 degrees true at a distance of thirty-five miles, and at 0530, this same plane reported two Orange battleships, two carriers, six or seven cruisers, and seven to ten destroyers on course 260 at a speed of ten to fifteen knots. The Blue plane was shot down by Orange Task Unit 2.1.1's CAP at 0532. Two minutes prior to this, Blue TG 78.2 had picked up one radar bogey bearing 000 degrees true, closing at a distance of forty miles. Simultaneously, Orange search plane 3-P-9 picked up TG 78.2 bearing 180 degrees at distance of thirty miles. This plane reported two Blue carriers, seven cruisers, and many destroyers in circular formation. After a report that it was being attacked by Blue fighters, it was not heard from again.[17]

At 0545, Blue struck Manila, with eight fighter-bombers, sixteen scout bombers, six torpedo bombers carrying torpedoes, six torpedo bombers each carrying a thousand-pound bomb, and eight fighter escorts. The air strike was intercepted by the twelve Orange Local Defense fighter planes as well as twenty-seven replacement fighters. Blue losses amounted to six fighters, six fighter-bombers, twelve scout bombers, and six torpedo bombers to aerial combat and antiaircraft fire. Orange losses amounted to six fighters from the Local Defense Force and ten from the Replacement Group, and Blue managed to sink gasoline tanker AOG-1 as well as damage two merchant ships in the harbor. Ten minutes later, at 0555, Orange Task Group 2.3.1 picked up one bogey on radar bearing 120 degrees true and closing, at a distance of eighteen miles. Antiaircraft batteries were alerted against Blue naval patrol plane 73-P-6, which reported the Orange body as several merchant ships escorted by two light cruisers, a large number of destroyers, and small escorts at low speed. The Blue plane then continued its search.[18]

At 0600 on 3 July, the beginning of Move 5, Blue TG 75.1 was conducting air searches from Batavia, while TG 75.2 was doing the same. Seaplane tenders AV-6 and AVP-12 were en route from Tarakan to Linkas, conducting searches with Naval Patrol Plane Squadrons 201 and 203, each of which had eleven planes. TG 75.3 did not conduct any air operations during Move 5. TG 75.4, still based at Saleh Bay, was ordered to supply ten fighters, five scout bombers, and eight torpedo bombers as replacement aircraft by 0824 to CTG 78.2, 170 miles northwest of Saleh Bay. Meanwhile, TG 78.1 continued operations in the Sulu Sea; TG 78.2, maintaining a combat air patrol, steamed on course 218 at a speed of twenty knots for a refueling rendezvous in the Java Sea; and TF 79, the Service and Escort Force, was at anchor at Saleh Bay. Certain elements of TF 79, however, were elsewhere. TG 79.6 (tankers AO-1 and AO-2) was at anchor at Surabaya, and TGs 79.2 and 79.5 (less provision ship AF-11)—were en route to rendezvous at 0600 on 4 July at 04-48 degrees south and 117-05 east. In addition, Task Groups 79.1, 79.4, and 79.8 all had orders to rendezvous with TG 78.2 at 0900 on 4 July.[19]

As to Orange forces, Task Units 2.1.1 and 2.1.3 were steaming in company at twelve knots on course 265, fueling destroyers and operating generally to the southwestward of Palawan Island. TU 2.1.2 was on course 200 at nine knots, fueling from AO-4, westward of Singapore. TU 2.1.4, minus tanker AO-2, had repair ships AR-2 and AR-3 in port at Camrahn Bay. TU 2.2.1 was conducting air searches from Manila and Singapore, including barrier patrols at the southwest tip of Celebes, from Cape Mandahka to Borneo, and at Point Maslembo. There was also a six-plane sector search in the Sulu Sea, another six-plane sector search covering the approaches to Surigao Strait, and a four-plane sector search covering the approaches to San Bernardino Strait. Additionally, this Orange unit was keeping two planes over the Sulu Archipelago between Cape Kante-Haven, Borneo, and Zamboanga on a continuous basis, as well as a one-plane search at the Makassar Strait. In addition to these searches, Task Unit 2.2.2 was conducting patrols at Balabac and Surigao, and TU 2.2.3 was maintaining a barrier patrol over the Gaspar and Karimata Straits. TU 2.3.1 (Escort Group 1) was escorting a northbound convoy on course 033 at a speed of ten knots, while TU 2.3.2, escorting Convoy No. 2, was approaching Singapore from the northeast. Task Unit 2.3.3 was at Singapore fueling escorts; TU 2.3.4, the Fuel Oil Supply Group, remained at Singapore throughout this move. In TU 2.3.5, the Aviation Supply Group, AOG-1 was sunk by air attack at 0715 on 3 July near the entrance to Manila Bay. Destroyer DD-47 was on the way to Cavite; gasoline tanker AOG-2 and destroyer DD-48 were at Singapore. Finally, TU 2.3.6, the Local Base Defenses, had received replacement aircraft from Camrahn Bay for the defense of Manila.[20]

At the start of the action in this Move, Orange naval patrol plane 3-P-4 picked up a radar contact of many surface ships bearing 100 degrees true at a distance of forty-five miles and closing. The pilot reported that he was sure one carrier was present in the formation, but his plane was shot down by the TG 78.1 CAP at 0640. This Task Group, in fact, had many bogeys on its radar screen, bearing 320 degrees true, closing at a distance of 70 miles and at a speed of 180 knots. Four Blue fighters were vectored, but all were lost to the east of Coron Bay after having shot down three Orange fighters. At 0705, TG 78.1 again had many bogeys on radar, this time at 350 degrees true, closing at a distance of forty miles. This air group was 78.1's own strike returning from an attack, but the strike force was being trailed by Orange bogeys. Orange fighters reported that three Blue fleet carriers, along with many cruisers and destroyers, were on course 260 at a speed of twenty knots at latitude 11-36 degrees north and longitude 121-10 degrees east. Two Blue fighters and six Orange fighters were lost in the ensuing engagement. Fifteen minutes later, an incoming Orange strike on TG 78.1 was intercepted by Blue fighters; by 0730, Blue had lost twelve fighters and Orange had lost ten fighters, seventeen scout bombers, and fifteen torpedo planes to a combination of Blue fighters and antiaircraft fire. Orange succeeded, however, in sinking Blue light carrier CVL-23 and inflicting what was determined to be 22 percent damage on battle cruiser CB-2 by a torpedo hit. That ship could still make twenty-five knots, but heavy cruiser CA-38 had also sustained a torpedo hit, by which its speed was reduced to seventeen knots. Additionally, Blue destroyer DD-365 was sunk by a kamikaze that struck at the waterline. Subsequent to this action, at 0800, TG 78.1 shot down four Orange scout planes.[21]

At the same time, Blue naval patrol plane 201-P-2 reported three merchant ships under escort at slow speed before the plane continued on its patrol. The group sighted was Orange 2.1.4, on course 090, which reported the plane at a speed of 130 knots bearing 370 degree true at forty miles. The plane, however, did not come within the range of antiaircraft guns. A few minutes later, Blue naval patrol plane 73-P-6 spotted Orange Convoy No. 1, which it had previously seen and reported. At 0825, Orange naval patrol plane 2-P-1 sighted Blue Task Group 78.1 bearing 045 degrees and closing at a distance of forty miles. After reporting the contact and that it was being attacked by Blue fighters, the Orange plane was shot down by two Blue CAP fighters. Five minutes later, Orange Task Unit 2.3.6 at Puerto Princessa picked up many bogeys on radar bearing 352 degrees true at a speed of 150 knots and headed south. Fighters from Malampaya, however, were not scrambled in time to intercept this Blue strike group. At 0905, Blue carrier-based aircraft were intercepted by four Orange fighters from Puerto Princessa. The Blue force lost one fighter, two scout bombers, and two torpedo bombers to aerial combat and antiaircraft fire, while Orange lost all four of its fighters. Blue also succeeded in putting one of the airfields out of commission for twelve hours and inflicted some additional damage on other installations.[22]

At 0930, TU 2.1.2 had a bogey, Blue naval patrol plane 72-P-3, on radar, bearing 340 degrees at a distance of forty miles. This plane was shot down by combat air patrol fighters after it reported the Orange unit bearing 160 degrees at a distance of forty miles and that it was itself under attack by enemy fighters. At 1025, an Orange strike from TU 2.1.1 was launched against Blue TG 78.1. The

Blue CAP of twenty-four fighters intercepted this strike, inflicting losses of fourteen fighters, twenty-one dive-bombers, and thirteen torpedo planes on Orange. Blue lost fourteen of its fighter planes, fleet carrier CV-13 to one torpedo and four bomb hits, and heavy cruiser CA-38 to two torpedo hits. Orange survivors of the strike, returning to Malampaya Sound, consisted of ten fighters, twenty-seven dive-bombers, and eleven torpedo planes, many badly damaged by antiaircraft fire. At 1250, another Orange strike of eight fighters, eighteen scout bombers, and eight torpedo planes took off from Malampaya Sound; it was intercepted at 1330 by ten Blue fighters. Blue lost six fighter planes in the ensuing aerial engagement and light carrier CVL-26 to one torpedo and one bomb hit. Orange lost five fighters, eleven scout bombers, and five torpedo planes to fighters and antiaircraft fire. This move concluded with Orange naval patrol plane 6-P-12 sighting many Blue surface ships bearing north at a distance of forty miles. However, the plane, which had sighted TG 78.2, was subsequently shot down by TG 78.2's fighters.[23]

Move 6 began at 1800 on 3 July, with the weather forecast as low clouds, rain squalls, and poor visibility. Aerial searches, therefore, were predicted to be ineffective. There was, however, a possibility of clearing by forenoon. At 1800, Orange fleet carrier CV-6 received a complete replacement carrier air group from Singapore. The twenty-four fighters remaining in TU 2.1.1's CAP were concentrated on fleet carrier CV-7; the intention was to put the remaining planes from Malampaya on this carrier as well, as early as possible on 4 July. At 1820, Orange naval patrol plane 3-P-10 obtained radar contact of a large Blue force on a northerly course in the Java Sea. From the disposition and speed of the force, the plane deduced it was a combatant group, and it was in fact Blue Task Group 78.2. The Blue unit had picked up three "snoopers" on its radar, which its combat air patrol chased off before the Orange planes were able to develop their contact. At 0100, an Orange naval patrol plane on barrier patrol made contact on a large force going south at high speed toward Sibutu Passage. The force sighted was TG 78.1, which had picked up the Orange plane, which in turn stayed out of gun range. At 0320, Blue naval patrol plane 91-P-5 made contact on a large force on a northeast course at a "moderate" speed that the pilot took to be a probable carrier task force. The force sighted was actually Orange TU 2.1.2, which had picked up the snooper but could not fire, as the plane did not close to gun range. At 0600, a Blue strike of thirteen heavy bombers from Batavia hit Banka and left the airfield "well cratered." Two bombers were lost to antiaircraft fire and Orange fighters, while Orange lost three fighter planes that were shot down. The Blue strike force did not observe any Orange planes on the ground, found antiaircraft fire light, and succeeded in putting the Orange airfield out of commission for twenty-four hours, with 90 percent of the base's fuel supplies destroyed and no operational planes left on the field.[24]

Move 7, which began at 0600 on 4 July, saw steadily improving weather; flight conditions were now considered "moderately good." At 0600, Orange naval patrol plane 4-P-5 sighted a convoy with three auxiliaries, six destroyers, and three light cruisers headed north for the Makassar Strait. This force was Blue TG 79.2, which made visual contact with the plane but did not get it into gun range. Ten minutes later, Orange naval patrol plane 5-P-6 sighted two destroyers headed east at high speed in the Java Sea. These were actually destroyer-minesweepers, which also sighted the Orange plane but did not have it in range. At the same time, Blue naval patrol plane 203-P-1 sighted a large convoy and escorts northbound off Camrahn Bay. This was Orange Task Unit 2.3.1, which likewise sighted the scout but could not open fire on it. At 0740, another contact was made, this time by Blue naval patrol plane 91-P-1 sighting a tanker and a destroyer in the Malacca Strait headed for the South China Sea. This Orange force sighted the Blue snooper, and ten minutes later Blue naval patrol plane 91-P-7 made radar contact on a large force in the South China Sea. The Blue plane closed sufficiently to see that it was a large Orange carrier task force heading southwest at high speed, but the Blue plane was then shot down by the Orange combat air patrol. At 0800, Blue naval patrol plane 91-P-4 sighted a northbound convoy of ten merchant ships with numerous small escorts; this was Orange TU 2.3.3. The snooper was also sighted by the Orange unit but did not close to gun range. Fifteen minutes later, Blue naval patrol plane 91-P-5 made radar contact on a large force but was shot down by carrier planes before being able to amplify its report. The force sighted was Orange TU 2.1.2.[25]

At this time, Blue Commander, 7th Fleet informed the Commander-in-Chief of the Pacific Fleet about Modification 2 to the 7th Fleet's Basic Operation Plan, changing the entire Blue strategic plan of operations. In a nutshell, Modification 2 called for TG 78.1 to join TG 78.2 in the Java Sea and the combined force to restrict its operations to harassing raids around the focal point of Orange convoys at Singapore. In this context, at 1400 an Orange naval patrol plane on barrier patrol at Makassar Strait sighted two small auxiliaries southbound. These were seaplane tenders, which also sighted the snooper but were not able to open fire. Orange had already had Task Group 2.1 launch a search at 1200 that reached the limit of its radius at about 1500 without making any contact, at which point the search returned to its carrier. At 1715, Orange naval patrol plane 3-P-5 made radar contact on a large force in the Java Sea, which the pilot was able to close on and develop as a large carrier task force on a southerly course, refueling at about ten knots. After getting off this report, however, the plane was shot down by the Blue CAP. The force in question was TG 78.2, which had had two snoopers on radar and was able to drive off the second plane after shooting down the first. Move 7 ended with Orange fleet carrier CV-9 taking aboard twenty-four dive-bombers and twenty-four torpedo planes—the remnants of the various air groups from Malampaya—throughout the day.[26]

In Move 8, which began at 1800 on 4 July, Blue's disposition included Striking Group 1 steaming on a course of 219 degrees at a speed of eighteen knots to join Striking Group 2. Blue dispositions also included TG 78.2 in the Java Sea en route to the Karimata Strait at a speed of twenty-five knots, while TGs 79.6 and 79.7 were at anchor at Saleh Bay, except for tankers AO-1 and AO-2 from TG 79.6, which were at Surabaya. Task Group 79.2 was operating with TG 75.2 (except for gasoline tanker AOG-1, at anchor in Balikpapan), while TGs 79.1, 79.4, and 79.8 were operating with TG 78.2 after the first two Task Groups finished refueling at Surabaya. Task Group 79.3, minus AOG-1 (apparently shifted at some point from 79.2), arrived at Tarakan at 1400, while TG 75.1 conducted a search in the South China Sea. TG 75.3, meanwhile, was being armed for an attack on the Orange carrier force, and TG 75.2 had moved Naval Patrol Plane Squadron 203 from Linkas to Balikpapan and sent eight planes of Naval Patrol Plane Squadron 201 to search to the west of northern Borneo. At the same time, TG 75.2's Destroyer Division 8 was en route to Balikpapan. Last, TG 75.4 was arming its reserve of replacement aircraft at Balikpapan in readiness for a coordinated air strike on the Orange carrier force.[27]

As concerned Orange forces, the entirety of Task Group 2.1 was steaming in company, except for TU 2.1.4. Fuel oil tanker AO-4 was at anchor in Singapore, and repair ships AR-2 and AR-3 were still at Camrahn Bay. Task Unit 2.2.1 conducted air searches in the South China Sea and eastward, while TU 2.2.2 conducted routine patrols over Balboa and Surigao Straits. At the same time, TU 2.2.3 maintained a barrier patrol over the Karimata and Gaspar Straits, and TU 2.2.4, the Mobile Base Unit, was dissolved at Manila on 3 July. TG 2.3 had TU 2.3.1 steaming on course 033 at a speed of ten knots escorting a northbound convoy, while TU 2.3.2 escorted Convoy No. 1 toward Banka at ten knots. TU 2.3.3 (Convoy No. 5) continued northward at sixteen knots; TU 2.3.4, the Aviation Gasoline Supply Group, proceeded to Banka with Escort Group 2; and TU 2.3.5, the Fuel Oil Supply Group, was moored in a river near Singapore.[28]

At 1800, Orange naval patrol plane 6-P-2 picked up many surface ships bearing 090 at a speed of twenty knots, a distance of thirty miles, and a course of 225. This force consisted of eight cruisers, two carriers, two battle cruisers, and "many" destroyers. This was TG 78.1, which had picked up the Orange plane bearing 270 degrees at the same thirty miles. The plane continued on its patrol as Blue naval patrol planes 72-P-1 and 72-P-3 picked up surface ships on a course of 015 degrees, bearing 200, distant thirty-five miles, and at a speed of twenty knots. Reported as three carriers, two battleships, eight to ten cruisers, and many destroyers, this force was reported as changing course at 1833 to 260 degrees with the same speed. It was Orange TG 2.1, which had these bogeys on radar bearing 195, distance forty miles, course 010, and at a speed of 170 knots. The Blue planes did not close to gun range, and the CAP could not locate them in the growing darkness. At 2200,

Orange search plane 6-P-2 reported four or five surface contacts at course 120, speed ten knots at latitude 50 degrees north and longitude 119-15 east. The Orange plane continued on its patrol. The force it sighted was Blue Task Group 75.2, which had picked up the bogey at the same latitude and longitude on a course of 090 and a speed of 120 knots. The plane did not come within antiaircraft-battery range.[29]

At 2320, the Orange Carrier Task Group, TG 2.1, picked up many bogeys on radar to the south at forty miles. The planes were flying low at 110 knots. TG 2.3 went to antiaircraft alert and began taking evasive action but withheld fire until the Blue force, twenty-four Catalina naval patrol bombers from Batavia, armed with a total of forty-eight torpedoes, came within range. The Blue strike force hit TG 2.1 at 2340 and obtained two hits on fleet carrier CV-6, which could subsequently make only 13.5 knots and had a list of fifteen degrees. The carrier's catapults were damaged, and its stability decreased to the point that the list could not be corrected. Heavy cruiser CA-9 also took a hit that caused 52 percent damage and slowed the ship to sixteen knots, while light cruiser CL-10 was 80 percent damaged and slowed to seven knots. Orange battleship BB-11 also suffered 10 percent damage but could still maintain twenty-five knots. The Blue force lost six of its naval patrol bombers. At 0200, Blue trailing planes reported the position of Orange TG 2.1 as well as the position of the Orange cripples, which were headed to Singapore.[30]

At the same time, Orange trailing planes made constant reports to their headquarters as they located the Blue Fast Carrier Task Force. At 0300, Orange Task Group 2.3 commenced launching its strike. Twenty minutes later, Blue launched nineteen torpedo bombers, twenty-eight scout bombers, and thirty-six fighter planes from Batavia to attack Orange TG 2.3. This strike was coordinated with an air attack from Blue TG 78.2, which launched its aircraft in time for the group to form up and start its flight at 0400. Fifteen minutes before, the Orange air strike had departed for its target, Blue TG 78.2. At 0535, TG 78.2 picked up the Orange strike on its radar. At 0600, a Blue strike of nineteen naval patrol planes—each carrying twelve five-hundred-pound bombs—attacked the airfields and installations at Banka, losing two planes to antiaircraft fire but badly damaging the airfield's installations and putting it out of commission for forty-eight hours. The end of Move 8 found the CAPs of both Task Groups defending against the attacking strikes.[31]

Move 9 began at 0602 and saw the Blue carrier strike consisting of twenty-four escorting fighters; thirty-four fighter-bombers, each carrying two five-hundred-pound bombs; twenty-five torpedo bombers carrying torpedoes; and thirty-nine scout bombers, each carrying two thousand-pound bombs. This force was intercepted by twenty Orange fighters and took losses of eight fighters, eleven fighter-bombers, nine torpedo bombers, and fourteen scout bombers. Orange lost twelve fighters, and fleet carriers CV-7 and CV-9 were both left sinking. The Blue strike from Batavia consisted of thirty-six fighters, twenty-eight scout bombers, and nineteen torpedo bombers. The strike claimed 50 percent damage on the previously hit Orange battleship BB-11, which now slowed to twelve knots. Orange battleship BB-12 was 25 percent damaged and slowed to twenty knots, while light cruiser CL-11 was 30 percent damaged and reduced to twenty-two knots. Orange lost sixteen fighters defending against this strike, totaling a loss of twenty-eight planes from the two Blue strikes. The Orange strike consisted of twenty escorting fighters; forty-eight scout bombers, each carrying two five-hundred-pound bombs; and twenty-four torpedo bombers carrying torpedoes. This force succeeded in sinking light carrier CVL-22 and in slowing fleet carrier CV-12 to twenty-five knots with a near miss. Blue light cruiser CL-53 was also sunk, and Blue lost sixteen fighters defending against the strike. Orange losses to aerial combat and antiaircraft fire were ten fighters, thirty scout bombers, and fourteen torpedo planes. Move 9 concluded the Maneuver; flimsies were submitted for maneuvers contemplated until 1800 on 5 July.[32]

The Critique

On 12 October, Commodore Bowdey issued instructions for the Critique that was to commence on 16 October. Staff and students studied copies of the Blue and Orange Operation Plans between 0830 and 1030, at which time all hands assembled in the auditorium in Pringle Hall, where the Staff led a brief discussion of the Staff and Student Solutions. Students involved outlined their plans, after which the conduct of the Problem was discussed. There were then Critiques of Alternate Moves as well as a separate discussion of the aerial phases of each move. The Critique continued until 1345 on 17 October; students were encouraged to take an active part and express their ideas in ways that promoted discussion and thereby benefited all concerned.[33]

Conclusion

As noted in the previous chapters, this Maneuver was the most aviation-oriented of the Pacific Basin Operations Problems analyzed here, and one in which both Blue and Orange lost carriers to each other's strikes. The Maneuver ended before the major Blue objective was attained, which was to destroy Orange merchant shipping and shore facilities to such an extent that Orange was forced to commit its heavy surface forces to battle. As stated above, concluding a Maneuver early was not unusual; it depended on when the Director thought that the applicable operational lessons had been taught. It is significant that while the History of the Maneuver played out a carrier duel, the Solution of the Problem was to actually destroy Orange heavy surface forces, especially battleships. As subsequent chapters will illustrate, most of the remaining Operations Problems were even more surface warfare oriented, but they were also highly reflective of late–Pacific War American naval doctrine and fleet practice.[34]

NOTES

1 Command and Staff Class of December 1945, "Advance Data for Move 1—Operations Problem 5, Blue and Orange," 18 April 1945, pp. 1–2, NHC, folder 2536, box 125, RG 4, NWC. The date indicates that this document was originally produced for the June 1945 Command and Staff Class but then was probably recycled, since the heading on the document cites the December 1945 class.

2 Command and Staff Class of December 1945, "Operations Problem 5: History of Maneuver," 5–15 October 1945, p. 1, NHC, folder 2540, box 126, RG 4, NWC.

3 Ibid., pp. 1–2.

4 Ibid., pp. 2–3.

5 Ibid., p. 4.

6 Ibid., pp. 4–5.

7 Ibid., pp. 6–7.

8 Ibid., pp. 7–8.

9 Ibid., pp. 8–9.

10 Ibid., p. 9.

11 Ibid., p. 10.

12 Ibid., p. 11.

13 Ibid., pp. 11–12.

14 Ibid., pp. 12–12a.

15 Ibid., p. 13.

16 Ibid., pp. 13–15.

17 Ibid., p. 15.

18 Ibid., p. 16.

19 Ibid., p. 17.

20 Ibid., pp. 17–18.

21 Ibid., pp. 18–19.

22 Ibid., pp. 19–20.

23 Ibid., p. 20.

24 Ibid., pp. 21–22.

25 Ibid., p. 23.

26 Ibid., p. 24.

27 Ibid., p. 25.

28 Ibid., pp. 25–26.

29 Ibid., p. 26.

30 Ibid., p. 27.

31 Ibid., pp. 27–28.

32 Ibid., pp. 28–29.

33 Command and Staff Class of December 1945, "Procedure for Critique of Operations Problem 5," 12 October 1945, p. 1, NHC, folder 2536, box 125, RG 4, NWC. Unfortunately, critiques at this time were not recorded, so all we have to analyze are occasional "critique instructions"; Dr. Evelyn Cherpak, Curator, Naval Historical Collection, U.S. Naval War College, Newport, R.I., e-mail to author, 19 July 2010.

34 Trent Hone, "U.S. Navy Surface Battle Doctrine and Victory in the Pacific," *NWCR* 62, no. 1 (Winter 2009), pp. 67–105; and Hone, e-mails to author, 28 February, 2 March, 3 March, and 14 March 2011.

X Operations Problem 6

An Amphibious Assault, October 1945

In early November 1945, Commodore Bowdey issued Instructions for Operations Problem 6. The Problem, of which the Operations Plan of Admiral BB, Commander, 13th Fleet, was part, was an Exercise in operational planning for an amphibious assault on an island in the Volcano Islands in the western Pacific. This Operations Problem entailed what NWC called the Staff Estimate of the Situation, actual preparation of the various plans in the assault, and then critiques by students and NWC Staff at the end of the Exercise.

Instructions

The Command and Staff classes had been divided into four staffs—Copper, Brass, Nickel, and Zinc—each of which was to act independently as the staff of Admiral BB, as well as that of Vice Admiral BZ, the Amphibious Force Commander. One student officer in each staff would be designated Chief of Staff, and it was his duty to organize and direct his group so as to "allot the work as evenly as possible in order to give each member of his staff full opportunity to develop his planning ability and best prepare the requirements. Thus students are advised to give thought and study to the duties assigned them." Bowdey listed the various publications available to each staff for the Exercise, including Commander-in-Chief, Pacific Fleet (CINCPACFLT)–CINCPOA bulletins; doctrine manuals on joint Army-Navy action, ship-to-shore movement, and landing operations; intelligence data on the Bonins and Volcanoes; communications; employment of various naval platforms; reconnaissance; and military symbols, abbreviations, and terminology. Each staff was to devise an Operations Plan for Vice Admiral BZ, as well as plans for Movement, Bombardment and Fire Support, Air Support, Logistics, Communications, Medical Affairs, Landing, and both Preliminary and Supporting Operations. NWC Staff were available for assistance and advice in the areas of Air Operations, Communications, Naval Gunfire, Movement and Landing, Logistics, Task Organization, Preliminary Operations, and General Matters.[1]

For the Critique that was to take place toward the end of the month, each staff was assigned various topics for the discussion of Student Staff Estimates, followed by the discussion of the Operations Plan and its Annexes. For the former, the Copper Staff was to study the Theater of Operations, Enemy Courses of Action, Recommendations for Preliminary and Support Operations, and the Air Support Plan. The Brass Staff was to study and prepare an outline on Relative Ground Fighting Strengths, Blue Courses of Action, the Movement Plan, and the Medical Plan, while the Nickel Staff was to study Relative Air Fighting Strengths, the Determination of Blue's Best Courses of Action and Decision, the Landing Plan, and the Logistics Plan. Finally, the Zinc Staff was to study Relative Surface Naval Fighting Strengths, Vice Admiral BZ's Operations Plan, the Bombardment and Fire Support Plan, and the Communication Plan. The four staffs were "reminded that written suggestions and constructive criticism relative to the scope, method and conduct of Operations Problem 6 are required."[2]

The General Situation

What followed was a description of the General Situation at the time the assault was supposed to take place. In this context, the member countries of the United Nations and the Axis were at war, with circumstances existing in the various theaters of operations consistent with what had actually occured on 1 February 1945 in World War II. Previous Blue operations in the Pacific Ocean Areas had been initiated to seize Orange bases, build other Blue bases, and develop a base system in support of Blue land, sea, and air forces for "subsequent offensive operations in the Western Pacific." The pattern of advance in the Pacific involved a succession of amphibious operations through the "Pacific Mandated Islands" and the Mariana Islands, along the New Guinea coast, and through the Philippines. The current line delimiting control by United Nations forces had only recently been attained and was still being consolidated. The Orange fleet, though reduced and "severely battered" in the battle for Leyte Gulf in late October 1944, was still considered a "fleet in being." Moreover, large Blue naval forces were deployed and engaged in covering and consolidating recently won positions and developing satisfactory bases for subsequent actions. In the Southwest Pacific Area (SWPA), the attack on Orange forces in the Philippines was being pressed to the maximum extent possible by United Nations forces, though most of the major combatant units of the Blue Pacific Fleet that had been used in the Philippines had recently been released to the Commander-in-Chief of the Pacific Fleet. In the South Pacific Area, things had been "stable" for quite some time. Accordingly, forces there had been reduced to the minimum required to ensure operation and maintenance of logistics facilities and preservation of lines of transportation and supply. However, Blue commitments in other theaters required that operations in the Pacific Ocean Areas be limited to the extent of forces now available.[3]

Assignment of the Problem

The Letter of Instruction itself for the Operations Problem was signed out by Fleet Admiral BA, Commander-in-Chief of the Pacific Fleet, to Admiral BB, Commander, 13th Fleet, on 1 February. Admiral BA ordered his subordinate to organize personnel assigned to him into a "joint planning staff" and prepare an Estimate of the Situation for the seizure of an island in either the Bonins or the Volcanoes. The purpose was to establish an air base suitable for medium-bomber and fighter operations against the Orange homeland as well as against lines of supply and transportation. For security reasons, the operation was to be known as Operation CINDER; a target date was to be set as early as "practicable" during the most favorable climatic period. The naval, air, and ground forces needed for the operation were listed. Admiral BA reiterated, however, that forces allocated for CINDER had to be held to the minimum consistent with accomplishing the task. Blue submarines and land-based air forces would support the operation as directed by the Commander-in-Chief of the Pacific Fleet. Current information on Orange naval, air, and ground forces was available from which BB and his staff could prepare an Estimate of the Situation: a decision for the accomplishment of the mission; a brief description of BB's concept, including an operations map that showed the maneuvering ashore; and a list of all forces required to carry out the proposed course of action, including the necessary assault shipping.[4]

Orange had been on the "strategical" defensive since December 1942, "defending its strong points in the Western Pacific to the utmost." Orange had defended every position with all available strength and resources, usually concentrating its defenses in and around air bases and leaving the remainder of Orange-held land areas generally unoccupied. The latest information from "all sources" indicated that Orange naval strength available for the defense of the Bonins and Volcanoes included six battleships, three fleet and two light carriers, two escort carriers, four heavy and five light cruisers, and twenty-nine destroyers. This was considered the current maximum surface strength of the Orange 1st Mobile Fleet; there was no definitive information on submarines. Midget submarines had attacked Blue shipping in advanced area harbors and were active in the Philippines. The estimate was that there was a total of "about" seventy boats available to Orange and that they would be active in the Bonins-Volcanoes.[5]

It was estimated that Orange had just over two thousand fighters, bombers, and reconnaissance aircraft in the Orange homeland, the Kuriles, Formosa, China, Celebes, Malaya, and Burma. There were also an estimated additional 1,250 ship-based aircraft on carriers, tenders, and other ships. Carrier air strength comprised sixty fighters, thirty-six torpedo planes, and seventy-two scout bombers. Blue information on Orange supply and repair for operational units was "meager," but it was known that, on one hand, Orange had cannibalized parts from crippled aircraft to a great extent but that, on the other, it had major supply and repair bases in Singapore, Formosa, and the Orange homeland. Orange also had an estimated two to five hundred transport aircraft, but little was known of their state of operational readiness. Orange ground forces were stationed in the Orange homeland, the Kuriles, Korea, Manchuria, China, French Indochina, Thailand, Burma, Malaya, the Dutch East Indies, the Philippines, Formosa, the Ryukyus, the Bonins and Volcanoes, Micronesia, New Guinea, the Bismarck Archipelago, and the Solomons. Many of these units had already been bypassed by Blue. Altogether, these units numbered about 5.4 million troops, but owing to Orange shortages in shipping, the Orange rate of reinforcement for them was considered "low."[6]

Operations Plan No. 13-45

In Operations Problem 6, Admiral BB, as Commander of both the 13th Fleet and the Pacific Fleet, noted that Blue forces had established air and naval bases in the Marshalls, the Marianas, the Admiralties, and the western Carolines and on Leyte and that they were continuing the occupation of the Philippines generally. Moreover, Blue submarine, surface, and air actions had continued to "aggravate" the "already critical" Orange shipping situation. Orange lines of communication to remaining bases in the Marshalls, the Carolines, and the Marianas had been cut, and those to the Philippines, the Dutch East Indies, and Malaya were being rendered "hazardous." The Orange homeland was being subjected to heavy bomber raids. Blue now needed island bases from which fighters could cover the bomber operations, however. Orange had numerous operational airfields in bombing range of Iwo Jima—the island selected for the assault—in an arc that extended from Okinawa to Tokyo, especially on the nearby island of Chichi Jima. Marcus Island was also within range but had not been recently operational. Many of Orange's ships that had survived the battle of Leyte Gulf with damage were under repair, repairs that Admiral BB assumed would take four or five months.[7]

Admiral BB took it for granted that CINDER would have cooperation from Commanding General (CG), 14th Air Force in the form of searches from bases in the China theater. Air forces of the Pacific Ocean Areas and Southwest Pacific Area would also institute and maintain long-range reconnaissance over the western Pacific, extending their searches as far as possible so as to relieve the 14th Air Force of these missions. At the same time, the Commanding General of the 20th Air Force would provide general support by attacking targets in the Orange homeland. These attacks would be intensified just prior to, and be coordinated with, Blue carrier strikes on the Orange homeland; details of this coordination were being arranged between the Commander-in-Chief of the Pacific Ocean Areas and the Commanding General of the 20th Air Force. Strategic Air Forces, Pacific Ocean Areas would also attack Iwo Jima and Chichi Jima with "maximum intensity," commencing "about" D–20. Commander, Submarine Force, Pacific Fleet (COMSUBPAC) would provide intelligence on Orange naval sorties from the Orange homeland, the East China Sea, and the South China Sea. Blue submarines would also maintain reconnaissance off Orange bases and along their approaches, provide lifeguard services as ordered by Admiral BB, furnish weather reports, and be prepared to concentrate for the support—either "strategical" or tactical—of the 13th Fleet, also as directed by Admiral BB. At the same time, Commander, Northern Pacific Force (COMNORPACFOR) would contain Orange naval and air forces in the North Pacific by conducting reconnaissance and harassing Orange positions in the Kuriles. Commander, Marshalls-Gilberts Area would also support the operation by protecting essential sea and air communications, providing services from facilities under his control for forces staging through the Marshalls-Gilberts Area, and routing shipping through the Marshalls-Gilberts.[8]

Admiral BB assumed that Orange would attempt to "interfere" with the assault on the selected island using aircraft, submarines in the vicinity of the Blue objective or along Blue lines of communication, and available units of the Orange fleet.

BB's orders to his forces were to capture, occupy, and defend Iwo Jima, in order to extend Blue control over the western Pacific and establish bases for future operations. His forces would be constituted into what he called "special groups": his fleet flagship, a Search and Rescue Group, an Anti-Submarine Warfare Group, a Logistic Support Group, and Service Squadron (SERVRON) 10. There would also be a Joint Expeditionary Force, which was to capture, occupy, and defend the island, as well as destroy or drive off any Orange forces attempting to interfere with the movement or landing. In addition, the Joint Expeditionary Force was to direct the operation of aircraft attached to it and those designated by other forces, and to route and provide escorts for all shipping moving to and from the objective.[9]

Also under BB's command was a Fast Carrier Task Force that was to prevent interference by Orange air action. The Fast Carrier Task Force was to protect the Joint Expeditionary Force and Iwo Jima from D-Day through D+2 from attack by Orange surface forces. Further, it was to furnish "direct day air support" and night fighter protection for the landing operations in the same period. From D+3 on, the Fast Carrier Task Force would operate in the China Sea Area in support of Operation CINDER as ordered by Admiral BB. Another of BB's subordinate commands, Forward Area, Central Pacific, was to protect shipping with its air and naval forces. Forward Area, Central Pacific was also to control anchorages and provide logistic support at Saipan, Guam, Eniwetok, and Ulithi, including the entry, berthing, maintenance, sortie, and movement of shipping from and between all forward areas and to Iwo Jima. BB set D-Day as the day of the initial landing. He would fly his flag on Guam until he shifted it to heavy cruiser CA-35.[10]

Annex A to the Operations Problem, the Air Plan, saw Orange air resistance at Iwo Jima being reduced initially to a minimum by repeated Blue shore-based air strikes from the Marianas, but Admiral BB thought that Orange "air reaction" to the landing would be "intense" notwithstanding, since Orange could launch strikes from bases in the homeland and elsewhere in range of the Volcanoes. Accordingly, BB requested air support from shore-based aircraft in the Central Pacific, Southwest Pacific, and China theaters, especially for searches. In addition, ten "support" carriers would accompany the Joint Expeditionary Force and furnish the majority of the direct air support at Iwo Jima. The Anti-Submarine Warfare Group—composed of escort carriers and destroyer escorts—would be temporarily assigned to the Commander, Joint Expeditionary Force during the movement and initial landings. Two other support carriers would accompany the Logistic Support Group. "Transport" carriers would be loaded with replenishment flight crews, aircraft, and limited supplies of spare parts for both fast and support carriers. These transport carriers would operate with the Logistic Support Group and be controlled by BB. In addition, the Fast Carrier Task Force would conduct strikes on the Tokyo area on D–3 and D–2 at the same time as Blue surface vessels were conducting their initial bombardment of Iwo Jima. On D-Day and D+1, the Fast Carrier Task Force would furnish air cover and direct day and night support. On D+3, as noted above, it would move to the China Sea Area.[11]

There were to be numerous shore-based air forces in support of CINDER. First, the Search and Reconnaissance Group was to furnish offensive screens of PB4Y Liberator naval patrol bombers against Orange search planes. In addition, it was to furnish offensive screens in advance of and on the flanks of the Fast Carrier Task Force and the amphibious forces from D–6 to D+3. These offensive screens would also be deployed against Orange picket boats and provide photo reconnaissance as needed. The Strategic Air Force, Pacific Ocean Areas was to conduct air strikes with "maximum intensity" against military installations in the Bonins and Volcanoes beginning on D–20 and continuing until the arrival of the Amphibious Support Force on D–3. Strategic Air Forces, Pacific Ocean Areas was also to destroy enemy shipping and aircraft, complete the photo reconnaissance of the area, and furnish "special" strike missions from D–3 until the Assault Phase was complete. Forward Area, Central Pacific would furnish antisubmarine coverage both day and night in the vicinity of friendly ports and along shipping routes. It was also to neutralize enemy bases in the Forward Area and provide air-sea-rescue and air-evacuation services. The Southwest Pacific Area would search from Darwin, Morotai, and Leyte as well as strike Orange air installations on Formosa so as to cover the Fast Carrier Task Force's strikes on the Bonins and Volcanoes. The 20th Air Force would provide air strikes by the 20th Bomber Command on Kyushu and the Ryukyus beginning

D–15, carry out strikes on Tokyo by B-29 Superfortresses of the 21st Bomber Command beginning D–14, and furnish search and reconnaissance to locate enemy picket boats. BB ordered that these attacks be intensified and coordinated with the Fast Carrier Task Force's strikes on the Orange homeland.[12]

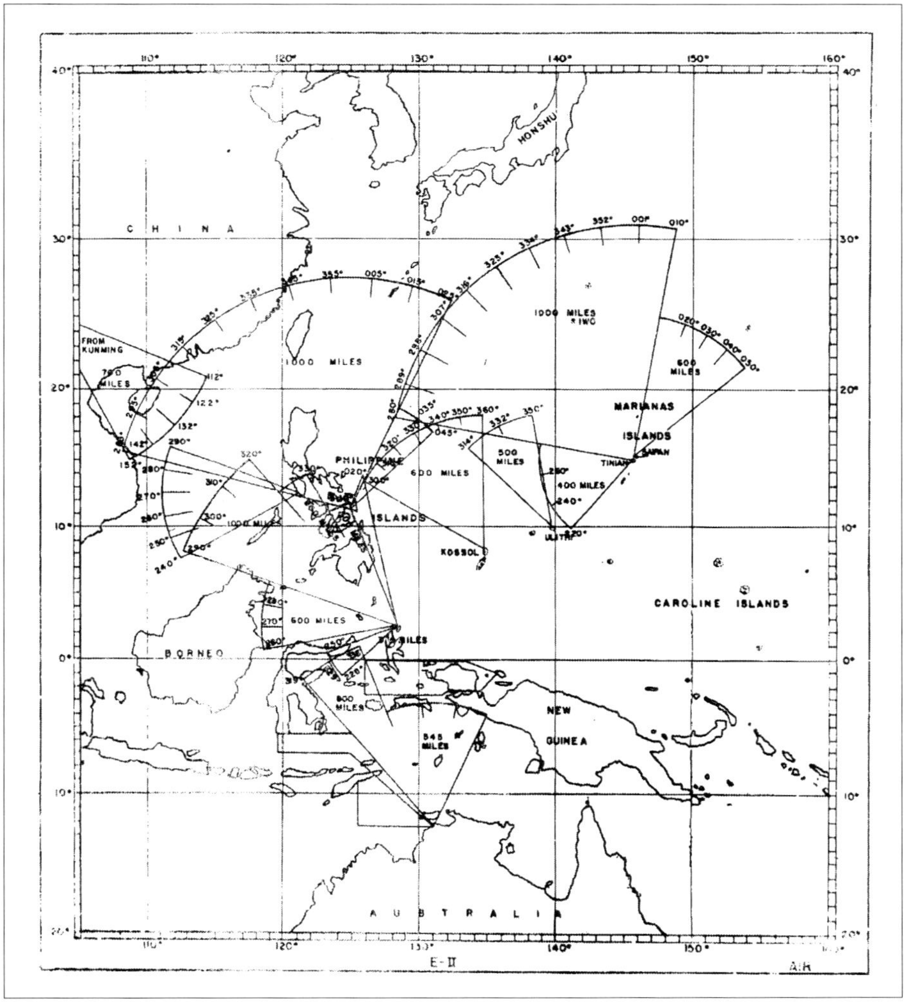

Fig. 102
Air Search Plan, Appendix I to Annex A, Operations Plan No. 13-45

The forces available to Admiral BB were immense. In addition to his heavy cruiser flagship, 13th Fleet's Search and Reconnaissance Group was composed of fifty-four PB4Y Liberator naval patrol bombers, twelve PB2Y3 Coronado naval patrol bombers, sixty-three PBM Mariner naval patrol bombers, and twenty-four Ventura naval patrol bombers, all serviced by eleven seaplane tenders and organized as Fleet Air Wing 1. BB's Anti-Submarine Warfare Group consisted of an escort carrier and nine destroyer escorts organized into Anti-Submarine Units Able and Baker, while the Logistic Support Group comprised six escort carriers, a light cruiser, an ammunition ship, a tanker, an attack cargo ship (AKA), a general stores issue ship (AKS), a fleet ocean tug (ATF), eleven destroyers, and six destroyer escorts. Service Squadron 10 included six destroyer tenders, seven mobile floating dry docks, three miscellaneous auxiliary ships (AGs), two fuel oil tankers, four gasoline tankers, three barracks ships (APLs), five repair ships of various kinds, seventeen ocean tugs, six Liberty ships, twelve oil storage ships, thirteen stores-issuing barges, 155 various yard patrol craft (YPs), and one destroyer-minesweeper.[13]

The Fast Carrier Task Force was broken down into five Fast Carrier Task Groups, totaling twelve fleet carriers (two of them able to operate aircraft at night), five light carriers, eight fast battleships, one battle cruiser, three heavy cruisers, nine light cruisers, three antiaircraft cruisers, and seventy-seven destroyers. BB's Forward Area, Central Pacific organization was immense in itself, comprising three heavy cruisers, fifteen destroyers, twelve destroyer escorts, sixteen minesweepers (AMs), and one hospital ship (AH). Forward Area, Central Pacific included Air Defense Command–Marianas, which had available 119 day and night fighter planes, thirty-six torpedo bombers, seventy-two transport aircraft, and twenty utility aircraft. In addition, there was the Air Sea Rescue Unit, twenty-five naval patrol bombers serviced by a seaplane tender, and Service Squadron 12, made up of one transport (AP), two survey ships (AGSs), three net-laying ships (ANs), two salvage vessels (ARSs), one barracks ship, one floating dry dock, three Coast Guard cutters (CGCs), and ten YPs. An Air Evacuation Unit had twelve transport aircraft and six naval patrol bombers. Also in the Forward Area were the Western Carolines Sub-Area, with another eighty-four fighter planes and eighteen torpedo bombers organized as Air Defense Command–Palaus; Air Defense Command–Ulithi, with twelve fighters and eighteen torpedo bombers; the Utility Squadrons, six torpedo bombers and nine utility planes (JMs); and naval forces as assigned by the Commander, Forward Area, Central Pacific (COMFWDAREA). The Blue Joint Expeditionary Force was yet another huge armada. This force included eleven escort carriers, seven prewar modernized battleships, seven heavy cruisers, fifty-six destroyers, twenty-six destroyer escorts, three amphibious force flagships (AGCs), forty-five attack transports, eighteen attack cargo ships, six high-speed transports (APDs), nine destroyer-minesweepers, sixteen minesweepers, three salvage vessels, two fleet ocean tugs, one rescue ocean tug (ATR), one landing craft repair ship (ARL), one battle-damage repair ship (ARB), two gasoline tankers, twelve tank landing craft (LCTs), sixty-three infantry landing craft (LCIs) of various types, forty-eight tank landing ships (LSTs), twenty-four medium landing ships (LSMs), twelve support landing craft (LCSs), twelve district motor minesweepers (YMSs), fourteen submarine chasers, and eighteen patrol craft of various types. The embarked troops numbered over seventy-two thousand Marines of the V Amphibious Corps (VAC), stationed prior to embarkation at Guam, Saipan, and Tinian.[14]

Not surprisingly, the Logistic Plan was also extensive. First, it was determined that 13th Fleet would be supported logistically by Commander-in-Chief of the Pacific Ocean Areas through Commander, Service Force, Pacific (COMSERPAC); Commander, Air Forces, Pacific (COMAIRPAC); Commander, South Pacific Force (COMSOPAC); and to some limited extent the Commander-in-Chief of the Southwest Pacific Area (CINCSWPA). Services for 13th Fleet units in the Marianas and western Carolines would be provided by COMFWDAREA, while services in the

Marshalls-Gilberts Area would be provided by Commander, Marshall-Gilberts Area. Support shipping that was under control of COMSERPAC, or Commander, South Pacific Force, or another agency would transfer to the operational control of Commander, 13th Fleet upon arrival in designated ports or rendezvous areas. Ammunition ships were to load to capacity, and provision, general stores, and cargo ships were to load to "Fleet Issue Loading." Fleet tankers, however, were to lift only half-capacity cargoes of diesel oil and aviation gasoline, but with maximum loads of fuel oil. Standard stocks of lubrication oils in drums and compressed gases were to be carried as prescribed by "existing circumstances." It was also directed that Commander, Service Squadron (COMSERRON) 10, or his representatives, would administer services afloat at anchorages in the Marianas, the western Carolines, and Eniwetok. Moreover, the Logistic Support Group would furnish logistical support to both the Joint Expeditionary Force and the Fast Carrier Task Force through the facilities provided by Commander, Service Force, Pacific and Commander, Air Forces, Pacific.[15]

Logistical items such as ammunition, fuel, lubricants, clothing, general and medical stores, and dry and fresh provisions to last between sixty and 120 days—depending on the item—were being brought from Manus in the Admiralties, Ulithi in the Carolines, Eniwetok in the Marshalls, and Guam in the Marianas. All ships returning to port for resupply were first to transfer all remaining stocks to ships remaining in the operational area; a cargo ship assigned to the Logistic Support Group was to be available to transfer supplies at sea; tankers were to carry deck loads of provisions for issue as required when refueling ships. There were also to be available three AKSs from Ulithi and the Marianas, each holding about nine thousand items representing support for thirty thousand men for thirty days. Naval Supply Depot (NSD) Manus was to supply two additional shiploads of supplies on this order, NSD Guam another two; some supplies would be obtained from NSDs Saipan and Eniwetok to top off quantities. Aviation spare parts were to come from the various aviation supply ships, barges, depots, and annexes of Commander, Air Forces, Pacific, including two supply ships that would rotate between the fleet's anchorage at Ulithi and the operational area. There would also be an aviation supply barge that could replenish an entire Fast Carrier Task Group and a division of escort carriers at a time, and Aviation Supply Depots served as emergency sources of shore-based supply at Oahu, Manus, and Roi, in the Marshalls. Replacement aircraft for the carriers would also be available from aircraft pools, with the primary one at Guam and secondary ones at Ulithi, Manus, Pearl Harbor, and the Marshalls. Four escort carriers provided by Commander, Air Forces, Pacific would carry out aircraft replenishment to the fast carriers in the operational area after obtaining aircraft from the designated pools; escort carriers assigned to the Joint Expeditionary Force would obtain replenishment aircraft from transport escort carriers or the

Guam pool, depending on the Fast Carrier Task Force's demands on the transport escort carriers. Additionally, there would also be Commander, Air Forces, Pacific representatives in the Forward Area who would assign replacement pilots and air-crew through their type command.[16]

All fleet oilers were under the Commander-in-Chief, Pacific Fleet, except those under overhaul or assigned to the North Pacific Force. Six oilers were under "urgent overhaul," which left twenty-seven from the 13th Fleet, two from the South Pacific Force, and six from local sources in the Marshalls, the Marianas, and Palau. Fuel—nearly seven million barrels from Pearl Harbor, Eniwetok, Ulithi, and the Marianas—would be supplied by COMSERPAC, and COMSERRON 10's representative at Eniwetok would allocate fueling facilities for forces staging through the Marshalls. Two gasoline tankers under the control of Commander, Joint Expeditionary Force would only carry diesel fuel for refueling the small ships of the force; Commander, Logistic Support Group was responsible for the replenishment of these two ships. Emergency reserve supplies of fuel oil, diesel oil, and aviation gasoline were in SERVRON 10's floating storage at the Ulithi, Marshalls, and Marianas bases. Lubricating fuels would also be available from the fleet oilers, and limited quantities would be available from SERVRON 10's floating storage as well.[17]

Ammunition was dealt with initially by the Commander-in-Chief of the Pacific Fleet, who was responsible for the special loading of ammunition for all classes of ships. There would be ammunition ships at Ulithi, as well as one operating with the Logistic Support Group, that could transfer ammunition under way. Ammunition magazines were maintained at Saipan, which also had an ammunition ship available to Commander, Joint Expeditionary Force. There was a Naval Ammunition Depot at Guam, and Commander, Service Force, Pacific was to ensure that special gear necessary for transferring ammunition at sea was available to the fleet, light, and escort carriers as well as to the heavy and light cruisers in the Forward Area. Commander, Destroyers, Pacific (COMDESPAC) was to ensure the same type of gear was available for the destroyers in the Forward Area. Commanding officers of these ships would be responsible for its installation; instructions were the responsibility of the type commanders. In addition, COMSERPAC was to ensure that logistics ships that were assigned were properly outfitted for at-sea transfer.[18]

Commander, Service Force, Pacific would have two fleet oilers at Eniwetok to provide potable water to the amphibious forces as the Commander, Joint Expeditionary Force specified. Combatant ships and auxiliaries that could do so would need to be prepared to supply potable water to smaller vessels when required. For the repair of surface ships, Admiral BB had several facilities available and was prepared to change locations to meet operational requirements if his subordinate commanders so requested. Specifically, the Forward Area was to have a fleet ocean tug, two rescue ocean tugs, a landing craft repair ship, an internal combustion engine repair ship (ARG), an ocean tug (ATO), and six Army Transport Service (ATS) ships. One battle-damage repair ship and one landing craft repair ship would accompany the Joint Expeditionary Force, and repair and salvage ships then in transit would report to Commander, Joint Expeditionary Force for assignment upon arrival in the Forward Area. Four hospital ships would be available for the operation, two at Saipan by D–2 and two at Ulithi by D-Day. Commander, Joint Expeditionary Force would arrange with COMFWDAREA to sail hospital ships to the objective as required; Commander, Joint Expeditionary Force would be responsible for discharging the hospital ships back to the Marianas. These two commanders would additionally oversee air evacuation of casualties from the objective back to the Marianas. Commander, Forward Area, Central Pacific would be responsible for evacuation of casualties from the Marianas to the rear areas.[19]

Medical stores would be available from the ships in the Logistic Support Group, and COMSERRON 10 would maintain at Ulithi supplies of blood plasma, penicillin, drugs to fight malaria, surgical instruments, and expendable supplies in one provision store lighter (YF) and three barges with supplies for up to 100,000 and 150,000 men, respectively, for thirty days. Another twelve cargo ships would each have medical supplies for a hundred thousand men for thirty days. In addition, each fleet oiler would carry "standard" loads of medical supplies, enough for sixty days and equivalent to a load for a fleet carrier or battleship; or for a heavy cruiser, light cruiser, light carrier, or escort carrier; or four destroyers or destroyer escorts. COMFWDAREA was also responsible for the reception and hospitalization of casualties at staging and convalescing areas in the Marianas, with 1,500 beds available at Saipan and another 3,500 at Guam. Medical care for the civilian population at the objective was the responsibility of Commander, Joint Expeditionary Force during the Assault Phase of the operation, through Military Government units. After the Assault Phase, the Island Commander (ISCOM) would be responsible for the medical care and hospitalization of the civilian population. Unit commanders would direct the burial of personnel at sea or in cemeteries ashore. Ships returning casualties to the Marianas or the rear areas would report the number of stretcher case prisoners, and Army and Navy ambulatory wounded, and Army and Navy stretcher wounded; the names of units embarked; and the number of officer and enlisted troop casualties. These reports were to be made within forty-eight hours—but only if security permitted—to COMFWDAREA, the Commandant of the 14th Naval District (COM 14) at Oahu, or the commander of the South Pacific Force.[20]

Prisoners of war were to be evacuated to the Hawaiian Area by any available shipping, while wounded prisoners of war were to be evacuated to the South Pacific Area or the nearest hospital, as the physical condition of the prisoner required. Ships carrying prisoners of war for evacuation to the Hawaiian Area would furnish COM 14 prior to arrival with information—again, security permitting—on the number of Japanese officer and enlisted prisoners, the number of Korean prisoners, and the number of stretcher cases. Full use of captured material would be made on the spot when it was essential to current operations. Captured documents were to be forwarded to the Joint Intelligence Center, Pacific Ocean Areas (JICPOA) at Pearl Harbor as soon as possible; every effort was to be made to preserve captured documents and materials intact so that they could be "profitably" examined by intelligence personnel.[21]

The Logistic Support Group was to be commanded by a rear admiral designated by Commander, Service Force, Pacific and would consist of combatant and auxiliary vessels made available by COMSERPAC and COM13FLEET. Commander, Logistic Support Group was responsible for furnishing logistical support to units of the 13th Fleet as directed by COM13FLEET. Further, this officer was to arrange with Area Commanders, Service Force, or type commanders for the replenishment of his group in order to meet the logistical requirements of 13th Fleet units. Commander, Logistic Support Group was additionally to supervise replenishment operations at sea and, as concerned fleet oilers, employ Eniwetok as the reloading point until directed otherwise by COM13FLEET. Commander, Logistic Support Group was to recommend other reloading points if they became more practical, advise about reloading at sea from chartered tankers, and maintain a minimum of six fleet oilers in the assigned operating area after D+4. He was to make provisions for the transfer of ammunition at sea and maintain "moderate" stocks of replacement ammunition in "appropriate" vessels in order to augment the supplies available to fleet units.[22]

Commander, Logistic Support Group was to effect the delivery of spares and "fly away" change packs for carrier air groups through replacement carriers and other available ships. He was to consolidate relief pilots and other aircrew from departing transport carriers with those remaining in the operating area and have the departing carriers report to COMFWDAREA for reloading. His duties further included ensuring that dry and refrigerated provision ships returned to the appropriate ports for discharge of remnant cargoes and reloading and that the maximum possible amount of provisions were transferred from oilers to ships being fueled. He was additionally to ensure that standard stocks of medical supplies were maintained on the necessary vessels and that vessels maintained "normal" stocks of supplies when in port. This officer was further to ensure the distribution of replacement personnel from pools aboard his ships to units and ships requiring replacements, and he was to keep those units informed about the availability of the replacements. Commander, Logistic Support Group was also to oversee the escort vessels for units of his group that were returning to reloading points. Moreover, he could delegate any of these duties to a subordinate commander not in company with him when necessary.[23]

Commander, Service Squadron 10 was to provide direct support to units of the 13th Fleet in all ports where he was a representative; maintain sources of replenishment for units of his group at Forward Area ports; and issue capacity loads of provisions to fleet oilers when reloading. He was also to maintain liaison with related shore-based supply agencies of the Army and Navy. In addition, he was to keep the commanders of the 7th Fleet, Logistic Support Group, and Forward Area, Central Pacific apprised of the logistical situations at the reloading points. He was also to provide upkeep and battle-damage repair facilities for all operating forces to the fullest extent possible. Further, COMFWDAREA was ordered to maintain established supply levels at all of his bases and exercise general supervision over the logistic support agencies within his area—including the loading and routing of the replacement transport escort carriers. He was also to inform the commanders of the 7th Fleet, Logistic Support Group, and major task forces about the movement of ships engaged in logistical support for the combat area. COMFWDAREA additionally routed and sailed garrison and maintenance shipping echelons to the objective as requested by the Commander, Joint Expeditionary Force. He was additionally to be prepared to make emergency air deliveries to the objective area as might be necessary.[24]

Finally, Admiral BB gave directions to the Commander, Marshalls-Gilberts Area, who was to route and sail garrison and maintenance shipping echelons to the Forward Area or the objective as required by the commander of Forward Area, Central Pacific or the Joint Expeditionary Force. This officer was also to furnish "maximum" logistical support to 13th Fleet units staging through his area. He additionally had oversight of established "Logistic Rendezvous Areas"—especially the area of operations for the Logistic Support Group, designated HEARTBURN.[25]

The Fast Carrier Task Force was to replenish fully prior to the departure from facilities under the control of COMSERRON 10 at Ulithi. In addition, Commander, Fast Carrier Task Force was to arrange a detailed schedule of replenishment for CINDER with Commander, Logistic Support Group and distribute final schedules to the interested commands. All ships were also to be given opportunities to practice replenishing at sea. Replenishment facilities for the Joint Expeditionary Force were to be provided at Eniwetok for units mounting from or staging through that port. These facilities would be administered by a representative of COMSERRON 10. A representative in the Marianas would—in conjunction with COMFWDAREA—provide additional replenishment facilities at Saipan and Guam. To arrange replenishment not covered in the Operations Plan, the commander of the Joint Expeditionary Force or of the Fast Carrier Task Force would advise Commander, Logistic Support Group of his requirements. Commander, Logistic Support Group was authorized to allocate fleet oilers for temporary duty with assault forces in the vicinity of the objective and to shift operating areas when the military situation required.[26]

Operations Plan No. 13-45 attached reference material, especially on the enemy: military installations, an anchorage chart, a beach diagram, photos of landing beaches, outlines of Orange defenses, and "special air and gunnery target" maps.[27] In addition, there was a very detailed Communication Plan. The Communication Plan would become effective as soon as Operations Plan No. 13-45 went into effect, except for task organizations under Commander, Forward Area, Central Pacific, until COMFWDAREA reported to COM13FLEET for duty. The Communication Plan would become effective for task organizations of the Joint Expeditionary Force as soon as that force departed the ports where it was initially organized. Additionally, the Communication Plan would be effective for training or rehearsals upon the order of Task Force or Task Group Commanders.[28]

Except as authorized by the Pacific Command, Zebra Time was to be employed as the time of origin of messages as well as for contact and amplifying reports. A number of communication security measures were to be observed by all ships in port. First, units were to preserve radio silence at staging and assembly points, except for warnings or defensive measures against enemy forces. Fixed circuits were to be used for "essential" dispatches, and traffic volume was to be controlled and held to a minimum by Task Force Commanders or the Senior Officer Present Afloat (SOPA). Task Force Commanders and SOPA were to supervise radio traffic even more actively at advanced bases. Any message originated by forces afloat had to be released by the TF Commander or SOPA, sent ashore by boat or visual means, and transmitted by the local shore radio station, which would provide copies. Advanced bases had even more explicit instructions. At Eniwetok, for example, minimum use of harbor circuits was prescribed, as was the use of low transmitter power both day and night. Ship-to-shore circuits were not to be used; radio traffic was to be sent ashore by hand or visual means. The senior 13th Fleet commander or SOPA was to arrange for VHF communications in the anchorage. Saipan was instructed to coordinate rehearsal communications with Marine Corps units then rehabilitating on Saipan, in order to confuse Orange radio surveillance. VHF, VHF voice radio (MN), and TBS communications among ships of the Joint Expeditionary Force were to be kept to a "normal" level. Maximum use was to be made of visual communications (semaphore, flashing light, signal flags), and local shore stations and afloat units were not to use fleet channels.[29]

Use of TBS was prescribed for units at Ulithi, as were visual means. In addition, certain kinds of equipment, such as VHF and MN, for use between various kinds of ships, such as carriers and destroyers, were ordered. Traffic between flag officers, battleships, and cruisers was to be on secondary TBS or selected circuits of what was called "Fleet Common" VHF. Personal visual messages could be sent only by flag and commanding officers. Units at Ulithi were not to use high-frequency ship-to-shore channels. Outgoing traffic was to be delivered to Island Commanders by boat, visual, or approved circuits for forwarding by fixed circuits or air transportation to Guam. There was a particular stress on security, especially a prohibition against revealing "under any circumstances . . . ship names, organizations, or other classified information in the clear on voice or other radio circuits." Along these lines, flag officers were responsible for guarding harbor circuits for warnings and "red alerts," which would be rebroadcast on TBS and Fleet Common. The Atoll Commander (ATCOM), Ulithi was to maintain radar guard unless otherwise directed and broadcast warnings and alerts on certain channels. Communications control at Guam was to be promulgated by the Commander-in-Chief of the Pacific Fleet. At the objective, SOPA would, after the Assault Phase was complete, ensure "complete uniformity" with measures taken at Saipan and Ulithi, unless modified by him or another "competent" authority.[30]

Movement reports were to be transmitted via fixed or point-to-point circuits whenever possible, and call signs were issued to ensure that messages concerning ship movements were automatically delivered to those who needed to know. Task Force Commanders and Officers in Tactical Command were to prescribe conditions of radio silence. This was especially true of commanders afloat and those at advanced bases "which are particularly vulnerable to enemy interception, such as those in the Marianas, the Western Carolines and the Marshalls." These units were to maintain conditions of radio silence to "frustrate interception by enemy listeners and must adhere to the spirit of such conditions under all circumstances." Radio silence could be lifted only in actual or imminent contact with the enemy to report the contact or issue tactical orders. Radio silence could also be broken to transmit information that was "vital" to operations that were in progress or immediately pending. In general, breaking radio silence was permitted when the necessity for immediate transmission outweighed the importance of denying knowledge of the presence and position of Blue forces and no other means of communications sufficed. Officers in command and controlling communication were to take full advantage of alternatives to radio traffic. Until communications at the objective had been turned over to the Landing Force Commander or Island Commander, fixed circuits would be controlled by Commander, Joint Expeditionary Force or the Attack Force Commander.[31]

All ships were also instructed to ensure they could shift TBS frequencies promptly when ordered and modify TBS for use with Continuous Wave (Morse code). In addition, force commanders were to establish Continuous Wave VHF circuits as they considered necessary. Aircraft frequencies were also laid out in detail.[32]

Since Continuous Wave Combat Calls were not enciphered, their continuous use allowed the enemy to listen at a distance and identify with "comparative" ease all the Blue units involved in the operation; care had to be taken. The same radio discipline had to be observed with voice radio. As the theater of operations approached the Orange homeland, increased interference and deception were to be expected. "All commands must intensify training of radio operators in working through interference and in identifying enemy deception." Orange was making a "determined" effort to interfere with ship-to-shore and command communication channels by means of deception from shore stations in East Asia. Thus far, Orange had been unable to "cope" with Blue authentication systems, making proper authentication on all Blue radio circuits of "primary" importance. Similarly, friendly aircraft were to conform to prescribed approach procedures in terms of Identification Friend or Foe and other forms of recognition.[33]

The Officer in Tactical Command was to control the use of radar and IFF. In the assignment of radar guard ships, Officers in Tactical Command were to give "due consideration" to opportunities for maintenance and ensure that all radar equipment was operative when in contact with the enemy. OTCs were also to conduct frequent drills to ensure that surface ship IFF was in "efficient" operating condition. There were procedures for radar countermeasures and silence, and OTCs were to prescribe the chain of visual responsibility when in Cruising Dispositions so as to provide for the most "expeditious" handling of communications. OTCs were to be informed if night cruising orders could not be completely distributed by sunset. Moreover, friendly ships in waters patrolled by Blue submarines were to carry out sonar recognition procedures as prescribed by CINCPOA. More exactly with radar countermeasures, the overall control of these countermeasures was vested in COM13FLEET. The commanders of the Joint Expeditionary Force and Fast Carrier Task Force were authorized to employ radar countermeasures for tactical situations as long as they did not affect operations in other areas. The Communication Plan also recognized that it was frequently impractical for one officer, the OTC, to exercise control of radar countermeasures over a fleet disposition of widely separated groups operating with a large degree of tactical independence. Accordingly, in some cases each Task Group Commander would control radar countermeasures for his command, and each CTG was encouraged in doing so to employ Pacific Command plans and directives.[34]

Unless another organization was prescribed by commanding or senior officers, control of radar countermeasures was to be vested in the Combat Information Center, where a Radar Countermeasures Control Officer, qualified to execute countermeasures for the ship or unit concerned, was to be stationed. In the absence of other instructions, all ships so equipped were to maintain continuous intercept watches when within range of enemy aircraft searches. Radar intercept operators were to immediately report all intercepts that could not be definitively categorized as friendly to the Combat Information Center for evaluation. If the Radar Countermeasures Officer could not determine the nature of an intercept, it was to be reported to the OTC for dissemination to the entire force.[35]

OTCs of forces under attack were to control radar jamming, assigning frequencies to jammer-equipped units. Units without monitoring equipment for their jammers were to be assigned definite frequencies for barrage jamming by Task Force or Task Group Commanders controlling radar countermeasures operations, and the force or group Radar Countermeasures Officer was to be kept informed of the frequency and station used by each unit. Units directed to participate in barrage jamming were to be assigned frequencies evenly spaced throughout the desired band, with at least one spot jammer held in reserve to fill any gaps. It was not desirable for the Radar Countermeasures Control Unit to participate in jamming, since this would prevent maximum use of his intercept receiver for viewing the overall effort.[36]

No radiation of jamming energy was to occur until the signal to commence was received. Otherwise, more information might be given to the enemy about the presence of the Blue Force than was possible to deny him by jamming. It was also necessary to consider the direction of maximum radiation of energy in assigning jamming missions to the various units since the choice of antennae on any ship was limited and some were directional. Window and radar deception would also be controlled by the OTC, in accordance with the instructions of the Task Force Commander. Ships equipped were to make maximum use of HF or VHF direction finders to obtain bearings of enemy tactical or radar transmissions.[37]

In addition, communications in the Forward Area, Central Pacific in connection with the routing of convoys and support shipping were the responsibility of COMFWDAREA, except at the objective during the Attack Phase. Convoy and routing communications at the objective were to remain the responsibility of SOPA until the objective was turned over to COMFWDAREA. A Convoy and Routing Officer, Forward Area at Guam was to inform those who needed to know about the routing of support shipping. Movement reports were to conform to Pacific Command procedures. Arrival reports were to be made in the form of daily summaries forwarded by SOPA to CINCPACFLT and COM13FLEET. All movement report radiograms were to carry special operating signals in headings to indicate their operational significance.[38]

Weather communications were taken into account, by ordering weather stations and designated weather-reporting units to report by radio indications of gale-force winds, heavy swells, or storms that might endanger ships and aircraft or otherwise impact operations. For these reports, radio silence could be broken, taking operational priority into account, unless tactical or other considerations prohibited such breaches. The aerological officer of each shore air base was also to prepare brief summaries of search plane weather observations each evening and forward these summaries to the commanders of the 13th Fleet and the Forward Area, Central Pacific by operational priority via the appropriate broadcast circuit.[39]

Operations Plan No. 13-45 addressed public relations. A program for radio broadcasts was planned to provide the listening public of major U.S. networks news of the campaign, consistent with security and noninterference with operations. To this end, as many ships as practicable were to be equipped with speech equipment for transmission of news broadcasts. Commander, Joint Expeditionary Force was to provide for transmission a minimum of five thousand words of press copy on D-Day and ten thousand words on each subsequent day by direct Navy radio if practical, if the tactical situation permitted, and if this did not interfere with operational traffic.

Finally, cryptographic aids were outlined. Commander, 13th Fleet assigned and would distribute a Combined Assault Code for use in the operation and a "Shackle Cipher" that was not to be used in emergency turns or lifeguard communications with submarines but was changed each day to increase security.[40]

Conclusion

For fairly obvious reasons, the Naval War College Staff had the student officers recreate, to a great extent, the setting of the Iwo Jima operation. Since no History of the Maneuver survives for this Exercise—as was common in many of the Operations Plans—we do not know how closely the game-floor execution adhered to the actual operation. It was fully characteristic, however, for the students to replay amphibious assaults of late in the war. An even more productive scenario would have been Okinawa, because of that operation's even larger scale and the added problem of kamikazes. Still, any amphibious assault maneuver served the pedagogical purposes of the Exercises and Operations Problems by requiring the students to plan the massive logistical effort this phase of naval warfare required. Logistics was one of the key areas emphasized by Admiral Pye and his officers at this time, as it would be by Admiral Spruance in the next two academic years.[41]

NOTES

1 George Bowdey, "Instructions for Operations Problem 6," 6 November 1945, pp. 1–5, NHC, folder 2540, box 126, RG 4, NWC.

2 Bowdey, "Instructions for Critique—Operations Problem 6," pp. 1–2, ibid.

3 Bowdey, "The General Situation for Operations Problem 6," 4 October 1945, pp. 1–5, NHC, folder 2540, box 126, RG 4, NWC.

4 BA to BB, "Letter of Instructions for Operations Problem 6," 4 October 1945, pp. 1–2, NHC, folder 2540, box 126, RG 4, NWC. See also "Enclosure A to Letter of Instructions: Forces Assigned to Blue Pacific Task Forces," pp. 1–3, NHC, folder 2540, box 126, RG 4, NWC.

5 "Enclosure B to Letter of Instructions," p. 1, NHC, folder 2540, box 126, RG 4, NWC.

6 Ibid., pp. 2–3. See also "Standard Nomenclature of U.S. Naval Vessels," pp. 1–6; "Amphibious Warfare Terms and Definitions," pp. 1–4; and "Abbreviations for Senior Commands in the Pacific Theater"; all in "Letter of Instructions for Operations Problem 6."

7 "The General Situation, Operations Problem 6," 4 October 1945, pp. 1–2, NHC, folder 2540, box 126, RG 4, NWC.

8 Ibid., p. 2.

9 Ibid., p. 3.

10 Ibid., pp. 3–5.

11 "Annex A, Air Plan, Operations Problem 6: Operations Plan No. 13-45," 4 October 1945, p. A-1, NHC, folder 2540, box 126, RG 4, NWC.

12 Ibid., pp. A-2 to A-3.

13 "Annex B, Task Organization, 13th Fleet (less Joint Expeditionary Force), Operations Problem 6: Operations Plan No. 13-45," 4 October 1945, pp. B-1 to B-3, NHC, folder 2540, box 126, RG 4, NWC.

14 Ibid., pp. B-3 to B-6; and "Annex C, Forces Assigned Joint Expeditionary Force, Operations Problem 6: Operations Plan No. 13-45," 4 October 1945, pp. C-1 to C-3, NHC, folder 2540, box 126, RG 4, NWC.

15 "Annex D, Logistic Plan, Operations Problem 6: Operations Plan No. 13-45," 4 October 1945, p. D-1, NHC, folder 2540, box 126, RG 4, NWC.

16 Ibid., pp. D-1 to D-3.

17 Ibid., pp. D-3 to D-4.

18 Ibid., p. D-4. Type commanders and their staffs, usually in the United States, were responsible for the manning, training, and repair of all units of a "type" (e.g., destroyers, patrol aircraft) in a given area (e.g., the entire Navy, or the Pacific Fleet), preparing them after construction, during workup, and during major repairs and overhaul in the United States for assignment or return to combat commanders.

19 Ibid., pp. D-4 to D-5.

20 Ibid., pp. D-5 to D-6.

21 Ibid., p. D-7.

22 "Annex D, Appendix I, Operating Instructions, Operations Problem 6: Operations Plan No. 13-45," 4 October 1945, p. D-1-1, NHC, folder 2540, box 126, RG 4, NWC.

23 Ibid., pp. D-1-1 to D-1-2.

24 Ibid., pp. D-1-2 to D-1-3.

25 Ibid., p. D-1-3.

26 "Annex D, Appendix II, Replenishment Instructions, Operations Problem 6: Operations Plan No. 13-45," 4 October 1945, pp. D-2-1 to D-2-2, NHC, folder 2540, box 126, RG 4, NWC.

27 "Annex E, Enemy Information, List of Source Material, Operations Problem 6: Operations Plan No. 13-45," 4 October 1945, p. E-1, NHC, folder 2540, box 126, RG 4, NWC.

28 "Annex F, Communication Plan, Operations Problem 6: Operations Plan No. 13-45," 4 October 1945, p. F-1, NHC, folder 2540, box 126, RG 4, NWC.

29 Ibid., pp. F-1 to F-2.

30 Ibid., pp. F-2 to F-3.

31 Ibid., p. F-4.

32 Ibid., pp. F-5 to F-6.

33 Ibid., pp. F-6 to F-7.

34 Ibid., pp. F-7 to F-8.

35 Ibid., pp. F-8 to F-9.

36 Ibid., p. F-9.

37 Ibid., p. F-10.

38 Ibid., pp. F-10 to F-11.

39 Ibid., pp. F-11 to F-12.

40 Ibid., p. F-12.

41 The historiography on American amphibious operations in the Pacific War is substantial, but a good rendition that illustrates the complexity of amphibious assaults is George Dyer, *The Amphibians Came to Conquer: The Story of Admiral Richmond Kelly Turner,* 2 vols. (Washington, D.C.: U.S. Navy Dept., 1972). For the assertions by Admirals Spruance and Turner that the Pacific War was fought around amphibious assaults, see William McBride, *Technological Change and the United States Navy, 1865–1945* (Baltimore: Johns Hopkins Univ. Press, 2000), pp. 184, 195. For the new emphasis on logistics at the postwar Naval War College, see John B. Hattendorf, B. Mitchell Simpson III, and John Wadleigh, *Sailors and Scholars: The Centennial History of the U.S. Naval War College* (Newport, R.I.: NWCP, 1984), pp. 178–89.

XI New Maneuver Rules and More Exercises for a New Class, January 1946

The Naval War College's Command and Preparatory Staff Courses both graduated classes in December 1945. In January 1946, a combined class of junior and midgrade officers began the last six-month course before the College transitioned back to a normal, peacetime, full-year course. The members of this combined group, known simply as the Command and Staff Class, were issued Maneuver Rules for their Exercises and Operations Problems. The new Maneuver Rules were primarily carried over from the previous ones, but there were changes. There were changes in titles of Maneuver personnel, for instance. The Recorder was now known as the Historian of the Maneuver, and the Gunfire Scorers were known as Damage Computers.[1]

Maneuver Rule Changes

There were also more elaborate tables. For example, in Section D—Visibility, Audibility, Detection, and Smoke Screens—a new table broke out aircraft of various types seen in different numbers and at different altitudes. In the same section, the table on maximum radar-search ranges now included an entire task force as a potential object of search and in Section E, Communications, there was much more detailed information on the time it would take to draft and decode messages; these activities were now divided into Administrative Losses (of time) and Operational Losses. It was still to take one minute to draft every six words of an administrative message in plain language and one minute for the sender to encrypt and the recipient to decrypt every two words in the same type of message. Operational Losses of time meant that where messages were nominally sent or received at ten words per minute, time of transmission could be doubled or tripled by various types of delays.[2]

In Section G, Torpedo Fire, the January 1946 Rules approached torpedo types and characteristics in much more detailed ways. These weapons were now divided into Aircraft, Submarine, and Surface Craft Torpedoes—as they were in the fleet itself—as well as "Fictitious" torpedoes for the three platforms, no doubt an attempt to capture new and experimental weapons. In the same section, torpedo hits were categorized in a much more detailed fashion as well, with different percentages

Class Seen	Number in Group	Col. 1 Same Altitude or higher than observer			Col. 2 Lower Than observer		
		High Vis.	Norm Vis.	Low Vis.	High Vis.	Norm Vis.	Low Vis.
VP, VPB	1	6	5	1.7	4	3	1
" "	2-6	8	7	2.3	5	4	1.3
" "	7 plus	9	8	2.6	6	5	1.7
VF, VSU, VBS, VTB	1	5	4	1.3	3	2	.7
" " " "	2-6	7	6	2	4	3	1.0
" " " "	7 plus	8	7	2.3	5	4	1.3

Fig. 103
Sighting aircraft

Object of Search	Searching Ship			
	Large	Intermediate	Small	SS
TASK FORCE	65,000	60,000	42,000	24,000
LARGE SHIP	50,000	45,000	38,000	20,000
INTERMEDIATE SHIP	43,000	38,000	35,000	18,000
SMALL SHIP	35,000	30,000	24,000	13,000
SS	9,000	8,000	7,000	6,000

(g) The maximum ranges of main battery fire control radars are equal to those of surface search radars.

(h) The maximum ranges of double purpose fire control radars are

Large Target (or over 7 planes)	Intermediate Target (or 3-7 planes)	Small Target (or 1-2 planes)
38,000 yds.	35,000 yds.	24,000 yds.

(i) The maximum ranges that may be obtained from airborne search radar are as follows:

Object of Search	Range in Miles	
Aircraft	7	
Schnorchel (Sea state 0 to 1	9	
Surfaced submarine	22	
Ship (Small)	35	
Ship (Large & intermediate)	40	
Task Group or Force (4 or more ships)	50	

Fig. 104
Theoretical maximum radar ranges

awarded for damage depending not only on ship type but on whether the ship was at anchor, under way, or maneuvering. In Section I, Aircraft, times to transport aircraft on elevators of an aircraft carrier were given in greater detail as to the type of carrier, type of aircraft, and specific elevators. In addition, times required for changing from one condition of readiness to another were given in a different manner, as seen in figure 108, and aerial combat losses were figured a bit differently in terms of the formula for losses among both attackers and defenders. Finally, as the figures 110–12 illustrate, damage to ships from aerial attack, including by rockets and in terms of bomb penetration of armored decks, was recorded in a much more detailed way.[3]

Task Force Disposition Exercise

In early January 1946, the new Command and Staff Class of June 1946 went through an Exercise entitled "Exercise in Developing Task Force Dispositions." The Schedule of Employment and the Statement of Problem evolutions, which took place between 5 and 11 January, combined the battles of Saipan, Leyte Gulf, and Iwo Jima into a scenario that demonstrated how task force dispositions were to operate in the future. The scenario again assumed a Blue war against Orange. Commander, 5th Fleet was the Blue Officer in Tactical Command of the operation, and Blue Task Force 58, the Fast Carrier Task Force, was supporting an amphibious assault against an Orange-held island in the Volcano group. The General Plan was for 5th Fleet to destroy enemy naval and air forces in the area of operations so as to prevent

Type	Speed	Range	14" Penetrative Hit Value	
		AIRCRAFT TORPEDOES		
E-1	33.5	6000	2.3	
E-2	40(x) 30(y)	2500 5000	1.6	
		SUBMARINE TORPEDOES		
F-1	30	4000	2.3	Wakeless
F-2	47(x) 32.5(y)	4500 9000	2.4	
F-3	46	11000	3.7	Wakeless
		SURFACE CRAFT TORPEDOES		
G-1	47(x) 34(y) 27(z)	6000 9200 14000	3.0	
G-2	46	18000	3.4	Wakeless
G-3	50(x) 40(y)	26000 40600	4.5	Wakeless
		FICTITIOUS TORPEDOES (AIRCRAFT)		
V-1	55	5000	7.5	Wakeless
		FICTITIOUS TORPEDOES (SUBMARINE AND SURFACE CRAFT)		
S-1	60(x) 50(y)	30000 40000	7.5	Wakeless

Fig. 105
Torpedo types and characteristics

Ships	At Anchor	Underway	Maneuvering
BB	75%	50	40%
CVB, CV	75%	50	45%
CVL	65%	45	35%
CA	65%	45	35%
CL	50%	40	30%
DD	50%	40	20%
CVE	60%	45	25%
Aux or M/V	60%	45	25%

Fig. 106
Torpedo damage by ship type and position

	CVB			CV			CVL		CVE	
Type Air Plane	Fwd Elev	Deck Edge Elev	Aft Elev	Fwd Elev	Deck Edge Elev	Aft Elev	Fwd Elev	Aft Elev	Fwd Elev	Aft Elev
VF VFB	2	2	2	2	1	2	2	2	1	1
VSB	2	1	2	1	1	1	1	1	1	1
VTB	2	1	2	1	1	1	1	1	1	1

Fig. 107
Aircraft carrier elevator transport times

FROM CONDITION		TO CONDITION	
	X	XI	XII
XIII	30 minutes	20 minutes	10 minutes
XII	20 minutes	10 minutes	**********
XI	10 minutes	**********	**********

Fig. 108
Flight deck readiness conditions

Attackers	Defenders	Attackers Loss	Defenders Loss (LD)
(n) VF	(m) VF	m X .1	n X .15
(n) VF	(m) VSB,VTB VOS	LD X.2	n X .5
(n) VF	m VP/VPB	n X .3	n X .3

Fig. 109
Aerial combat losses

Target Size	Rocket Attack	Dive or Glide Bombing	Low Level Bombing	Sight Bombing 11,000	7000	3000
Large	35	16	32	8	15	30
Intermediate	30	9	20	5	9	20
Small	25	6	15	3	5	14
DD and SS	15	3	10	1	3	9

Fig. 110
Hit percentages and ship damage by bombs and rockets

Orange interference with Blue amphibious operations and to "reduce" Orange sea power. TF 58 was specifically to prevent major surface attacks on the Joint Expeditionary Force, and Commander, Task Force 58 was to exploit any opportunity to bring about a fleet action. If such an action occurred, he was to engage "decisively," furnishing his own airpower, antisubmarine patrol, and air spotting, as well as delivering air strikes in coordination with the Fast Carrier Task Force. Commander, 5th Fleet (COM5FLEET), however, also had to provide for the contingency of an Orange surface attack against the Joint Expeditionary Force, Task Force 51. The contingency plan entailed the formation of a Striking Force, Task Force 59, under the command of Vice Admiral BL, Commander, Battleship Squadron 2, with ships from TF 58 and the Joint Expeditionary Force.[4]

In the Exercise, the amphibious operation proceeded as planned, and TF 58 was operating in four Task Groups northwest of the newly seized island. Two Fast Carrier Task Groups were engaged in air support operations of the Joint Expeditionary Force, one group had just struck the Orange homeland, and the fourth was conducting a sweep against the Ryukyus. Blue submarines were also deployed to observe any Orange surface main-force sortie, and Blue Fleet Air Wing 1, based on tenders, had moved up to the objective and was also scouting for approaching Orange surface forces. On the third day of the operation, COM5FLEET received reports from the commander of Task Group 58.4—which was retiring from its strike on the Orange homeland—and from a submarine of Submarine Division (SUBDIV) 121 off the Bungo Channel that an Orange force of battleships, cruisers, destroyers, and carriers was deploying from the Inland Sea. COM5FLEET therefore put his plan into effect. Since it was closest to the Orange force, TG 58.4, Carrier Group 4 (CARGRU 4), became the basis of Task Force 59 and was reinforced with ships from the two Fast Carrier Task Groups carrying out air support operations. These two Task Groups continued to fly air support for operations ashore, but with lighter screens as their escort ships went to TF 59. With these reinforcements TF 59 was to consist of three fleet and one light carriers, five battleships, two battle cruisers, nine light cruisers, two antiaircraft cruisers, forty-three destroyers, and one prewar modernized destroyer. CTF 59 also had Fleet Air Wing 1 under his operational control, with one seaplane tender, twelve Catalina naval patrol bombers, and twenty-four Mariner naval patrol bombers, as well as twelve submarines of SUBDIVs 121 and 122. His air strength included 135 Hellcat and 108 Corsair fighters, forty-five Helldiver scout bombers, and fifty-four Avenger torpedo bombers, as well as six observation scout planes embarked in his battleships and thirty-six scout observation planes operated by his cruisers.[5]

	2000# GP		1600# AP		1000# AP		1000# SAP		1000# GP		500# SAP		500# GP		250# GP		100# GP	
	Dive or Glide	Hori-zontal	Dive or Glide	Hori-zontal	Dive or Glide	Hori-zontal	Dive or Glide	Hori-zontal	Dive or Glide	Hori-zontal	Dive or Glide	Hori-Zontal	Dive or Glide	Hori-zontal	Dive or Glide	Hori-zontal	Dive or Glide	Hori-zontal
TARGET TYPE A	NO	NO	NO	YES 11,000	NO	YES 14,000	NO	NO	NO	NO	NO	NO	NO	NO	NO	NO	NO	NO
B	NO	NO	NO	YES 11,000	NO	YES 14,000	NO	NO	NO	NO	NO	NO	NO	NO	NO	NO	NO	NO
C	NO	NO	NO	YES 13,000	NO	YES 17,000	NO	NO	NO	NO	NO	NO	NO	NO	NO	NO	NO	NO
D	NO	NO	YES 350K 4,000	YES 8,000	NO	YES 12,000	NO	NO	NO	NO	NO	NO	NO	NO	NO	NO	NO	NO
E	NO	NO	YES 300 3,500	YES 6,000	YES 350K 4,000	YES 8,000	NO	NO	NO	NO	NO	NO	NO	NO	NO	NO	NO	NO
F	NO	NO	YES	YES 2,500	YES	YES 3,000	YES 300 2,000	YES 4,200	NO	NO	YES 350 3,000	YES 7,000	NO	NO	NO	NO	NO	NO
G	NO	NO	YES	YES 2,400	YES	YES 3,000	YES 300 2,000	YES 4,000	NO	NO	YES 350K 3,000	YES 6,800	NO	NO	NO	NO	NO	NO
H	YES	YES 2,500	YES	YES 1,500	YES	YES 2,200	YES	YES 3,600	NO	NO	YES 300 2,000	YES 4,500	NO	NO	NO	NO	NO	NO
J	YES	YES	YES	YES	YES	YES	YES	YES	YES	YES	YES	YES	YES	YES	YES	YES	YES	YES
K	YES	YES	YES	YES	YES	YES	YES	YES	YES	YES	YES	YES	YES	YES	NO	NO	NO	NO
L	YES	YES	YES	YES	YES	YES	YES	YES	YES	YES	YES	YES	YES	YES	YES	YES	YES	YES

*Figures in blocks gives minimum altitude from which a bomb must be dropped to penetrate. In case of dive and glide bombing figures are given for minimum speed and altitude.

Fig. 111
Bomb penetration of armored decks

Note: P - Penetrative hit.
N - Non-Penetrative hit.

	2000# GP		1600# AP		1000# AP		1000# SAP		1000# GP		500# SAP		500# GP		250# GP		100# GP	
	P	N	P	N	P	N	P	N	P	N	P	N	P	N	P	N	P	N
TARGET TYPE A	-	10	15	4	12	3	-	4	-	5	-	2	-	4	-	2	-	1
B	-	10	15	4	12	3	-	4	-	5	-	2	-	4	-	2	-	1
C	-	10	15	4	12	3	-	4	-	5	-	2	-	4	-	2	-	1
D	-	10	16	4	14	3	-	4	-	5	-	2	-	4	-	2	-	1
E	-	15	18	4	16	3	-	4	-	8	-	2	-	5	-	2	-	1
F	-	20	20	5	18	4	20	5	-	10	15	3	-	6	-	3	-	1.5
G	-	20	20	5	18	4	20	5	-	10	15	3	-	6	-	3	-	1.5
H	30	20	20	5	18	4	20	5	-	10	15	3	-	6	-	3	-	1.5
J	50	30	40	10	35	8	40	10	45	20	35	6	40	12	25	6	20	3
K	50	30	40	10	35	8	40	10	45	20	35	6	40	12	-	6	-	3
L	90	60	70	20	60	16	65	20	70	40	40	12	50	24	30	12	20	6

Fig. 112
Bomb damage percentage of loss to target ship

So reinforced, TF 59 was thought "more than" adequate to destroy the Orange force. Vice Admiral BL envisioned a daylight engagement against enemy forces that very day and sortied to rendezvous with CARGRU 4. His battle plan and instructions were sent to his subordinate units as a naval message that estimated the Orange surface force as three battleships, two carriers, four heavy cruisers, five light cruisers, and twenty destroyers. Blue submarines were ordered to attack these Orange forces as they retired into the Inland Sea; naval patrol aircraft were to locate the Orange force and maintain contact. BL was assuming that the enemy objective was the Blue Joint Expeditionary Force and that the Orange force could be decisively engaged immediately, but he also surmised that the engagement might take place in range of Orange land-based air forces and that Orange submarines might be deployed.[6]

He expected TF 59 to destroy the Orange force in several stages. There would be initial air strikes on the enemy carriers. Subsequent action entailed air strikes and then assaults by light forces—i.e., light cruisers, destroyers, and destroyer escorts, as opposed to heavy forces (battleships, battle cruisers, and heavy cruisers) —on the Orange battle line. The latter attack would be supported by Blue battle-line gunfire in "normal action" at long ranges. His forces would then utilize "all weapons" at their various ranges in pursuit, in order to prevent an Orange surface attack on the Joint Expeditionary Force. BL ordered his battle line commander, Rear Admiral BA, to reduce enemy battle-line forces at long range and then destroy the remnants at close range. Vice Admiral BC would command the Center of the light forces, while Rear Admirals BD and BF would command the Right and Left Flanks, respectively. These three components would make simultaneous attacks on the Orange force and then operate at discretion. The initial attacks on the Orange surface forces were to be coordinated by Admiral BC, whose Flank Group was to be in the Van. Rear Admiral BG, Commander, Carrier Group (COMCARGRU) 4, after sinking the Orange carriers, would defend all the Blue Task Groups against enemy air attack while Blue surface forces were engaged with the Orange battle line. Retreating Orange forces were then to be attacked to the "limits of logistics" or until enemy land-based air forces made pursuit "unprofitable."[7]

Of note in these orders, in addition to the obvious references to late Pacific War battles, is that detailed assignments of Task Groups were not made by the instructors; each student could allocate forces as he saw fit. In addition, students were to draw diagrams of the various task force dispositions, diagrams that were to illustrate the position of each surface ship or of each small-ship unit in close formation; the position and schedule of air patrols and air task units; fighter-direction ships; and "radar guards," including radar pickets and radar-intercept ships. Students were to assume that ships were equipped with radar and that battleships, cruisers, carriers, and fighter-direction destroyers were fitted with radar-intercept equipment. Students were to illustrate the assignment of surface forces and carriers into a Task Force 59 Task Organization and then draw up Battle Disposition, Approach Disposition, and Day Cruising Disposition Diagrams for Task Force 59.[8]

The next steps in the students' process of preparation for game play were organized administratively, and apparently arbitrarily, under the heads of "Requirements A, B, and C." The first two are best considered together.

The Staff Solution: Requirements A and B

The Staff Solution to the Task Force Disposition Exercise also reflected American naval doctrine of the late Pacific War. On 11 January, the Staff Solution was put forward by Vice Admiral BL, Commander, Battleship Squadron 2 and CTF 59. Admiral BL compared specific strengths and weaknesses of his force platforms to those of the Orange forces. BL had had these data compiled already, was quite familiar with them, and had specific comparisons readily at hand. Since he had to consider these comparative capabilities in his Battle Plan, his Battle and Approach Dispositions developed hand in hand with the Battle Plan itself.[9]

Admiral BL listed the strengths of the battleships to be their heavy gun batteries, their great ability to resist damage, and the moderate to high speeds of the newer ships. The weaknesses were the large targets the battleships offered and the fact that the older battleships were "slow and sluggish." Battle cruisers also had the advantage of heavy gun batteries, good antiaircraft batteries, and high speed, but they also made for large targets and had only "moderate" damage-resisting qualities. Heavy cruisers had strong "intermediate" gun batteries, effective antiaircraft batteries, and high speed, and presented only "medium" targets, but they had low resistance to damage. Light cruiser advantages and disadvantages were equivalent, with the added advantage of "rapid firing intermediate" gun batteries. BL noted, however, that Orange cruisers were armed with torpedoes and that Orange light cruisers could carry and lay mines.[10]

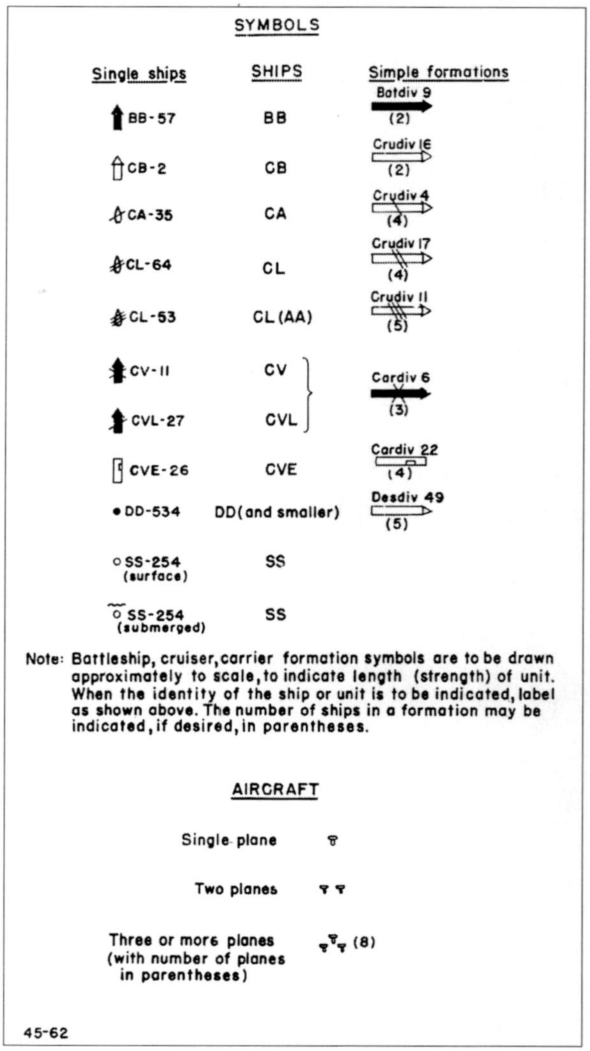

Fig. 113
5th Fleet ship and aircraft symbols

Admiral BL's aircraft carriers represented the Blue striking force with the highest speed, but the high-speed advantage did not apply to the escort carriers. Moreover, he found carriers to have a long list of disadvantages: they were large targets, had weak gun batteries, and had only "moderate to low" damage-resistance qualities. Aviation operations could be stopped by even moderate damage to the flight deck or by heavy weather; the escort carriers were, like the older Orange battleships, "slow and sluggish." Destroyers he found "highly vulnerable" even to "light" guns and bombs, and their effectiveness as weapon platforms was "markedly" reduced by heavy weather. On the other hand, he found the Blue destroyers to have strong torpedo batteries, good sonar gear and depth charges, and effective "light intermediate" guns and antiaircraft batteries; they were small targets and highly maneuverable as well. Submarines, for their part, enjoyed strong torpedo batteries and "invincibility" when submerged, but on the surface they were vulnerable to all forms of attack and beset by "poor mobility" when submerged. Admiral BL thought that the strengths and weaknesses of all of the types in his force—older and newer battleships; battle, fleet, light, and escort carriers; light and antiaircraft cruisers; older and newer destroyers; destroyer escorts; and smaller types, such as patrol vessels, minelayers, and motor torpedo boats (MTBs)—had to be taken into account, because of the need for the "most effective mutual support" possible in the forthcoming operations.[11]

The TF 59 Battle Disposition was next given. Task Group 59.1, the battle line, consisted of his older destroyer as the armed guard ship and five battleships, organized into Battleship Divisions 6 and 9. Task Group 59.2, the Right Flank, consisted of the four light cruisers of Cruiser Division 17 and the nine destroyers of Destroyer Squadron 60 while TG 59.3, the Center, consisted of the two battle cruisers of CRUDIV 16 and the seventeen destroyers of DESRONs 24 and 25. TG 59.4, the Left Flank, consisted of the four light cruisers of CRUDIV 14 and the nine destroyers of DESRON 54. Task Group 59.5, the Carrier Group (Carrier Division 8), consisted of three fleet and one light carriers; two antiaircraft cruisers and one light cruiser were attached, as were nine destroyers from DESRON 47. Admiral BL's Battle Disposition showed a deployment to the right but the organization and disposition would do equally well for a deployment to the left. In either case, BL saw the light forces in the Center reinforcing either the Left or Right Flank; he argued that maneuvers of station units or Task Groups need not affect the disposition axis or the deployment course of the task force.[12]

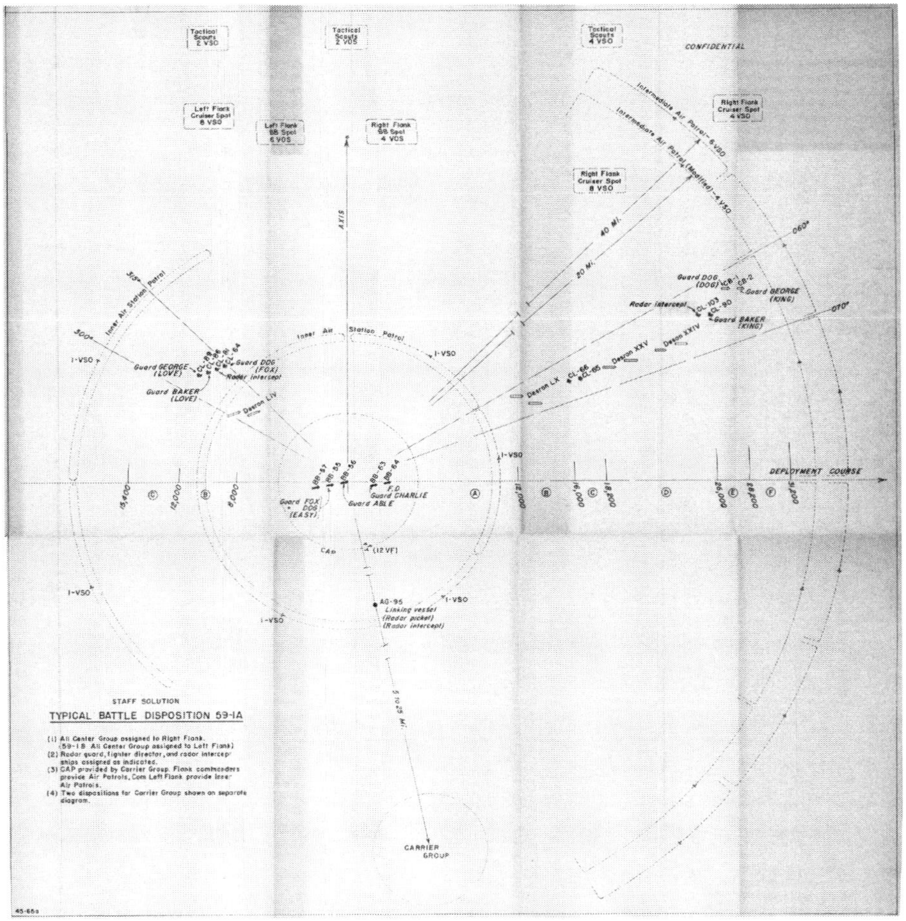

Fig. 114
Typical Battle Disposition

Admiral BL speculated on the possibility of a "Detached Wing" of ships that could deal with various enemy formations. He had seven heavy ships assigned to TF 59—three battleships that could make twenty-seven knots, two that could make 32.5 knots, and two "comparatively" lightly armored battle cruisers. He surmised, however, that it would be more advantageous to have all of his battleships concentrated into the battle line and that his battle cruisers would be better reserved

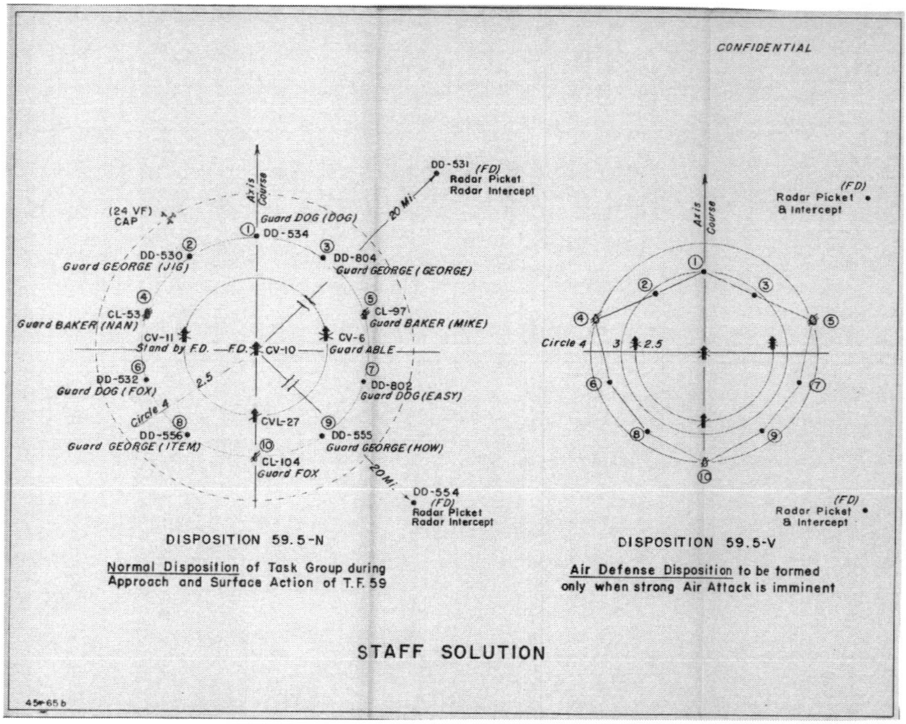

Fig. 115
Normal and Air Defense Dispositions

to act as an "effective" means of "exerting the desired heavy pressure" on Orange center light forces at contact or on the enemy flank during the battle itself. The OTC could detach Battleship Division 9, with the two fast battleships, for separate operations if the situation demanded. That would leave the battle cruisers as the Detached Wing, but he planned to use them to strengthen the Van and force an early deployment of Orange light forces. If he found a need for a Detached Wing, it would be Battleship Division 9 itself.[13]

Admiral BL evenly apportioned his flank groups, each of what he thought was the "minimum" strength acceptable, with light forces in the Rear and the latter group of ships being essentially defensive. He saw the Center ships—the two battle cruisers and the two destroyer squadrons—reinforcing the flank group in the Van. He was going to put his light forces in the Van and force the Orange commander to commit his own light forces by having the Blue battle cruisers attack the Orange cruisers. Should the Orange commander reverse his deployment and withdraw his light forces, Blue would be able to position itself between the Orange force and its home bases. In addition, BL argued that any Orange attempt to avoid superior Blue light forces could be countered by Battleship Division 9 in the Rear.[14] Vice Admiral BL was conscious of various Orange threats to Blue. First was the possibility of strong shore-based air cover. What BL was really concerned about, however, was the presence of long-range torpedoes on Orange destroyers, light cruisers, and some of the heavy cruisers. He planned to mitigate this threat by so disrupting Orange light forces early that they could not initiate a torpedo attack. Given the reported range of Orange torpedoes, though, the stationing of the Blue defensive light forces required "special consideration." The placement of these elements had to be a compromise between the best theoretical position for the effective interception of attacking light forces, on the one hand, and, on the other, one that could be supported by the battle line and would not interfere with Blue light forces. Given the range of the Orange torpedoes, he decided to move the limit of the Van's defense area out to sixteen thousand yards, while the Rear continued to employ a standard defense area of twelve thousand yards.[15]

Last, BL intended to employ his Carrier Group to maintain what he termed a tight "Task Force Air Defense Disposition" until the last possible moment before shifting to his Approach Disposition. He hoped that by this time Orange carrier air strength would have been "largely neutralized" by Blue air strikes. If that was the situation, he would be able to reduce the screen of the Carrier Group in the Battle and Approach Dispositions to "minimum effective" strength, that of three cruisers and nine destroyers. With two destroyers stationed as radar pickets on the disengaged side and ten ships deployed as a screen, the Carrier Group would have to depend on the other battle groups for early radar warning from the engaged side. Unfortunately, the carriers would not have much maneuvering room, but he saw the arrangement as a compromise between flight operations and defense of the carriers against air and submarine attack.[16]

As mentioned earlier, this Exercise resulted in a fairly typical disposition of ships and a characteristic set of assumptions about battle given what was known in early 1946 about both American and Japanese naval capabilities in the latter stages of the Pacific War. Clearly, considering what actually happened during the battle of Leyte Gulf, it was a good idea to address such worst-case scenarios as the need to defend amphibious shipping from surface forces. Certainly, "Admiral BL's" planning reflects what the Navy knew by then to have been the characteristics of Japanese Long Lance torpedoes. The whole scenario reflected interwar

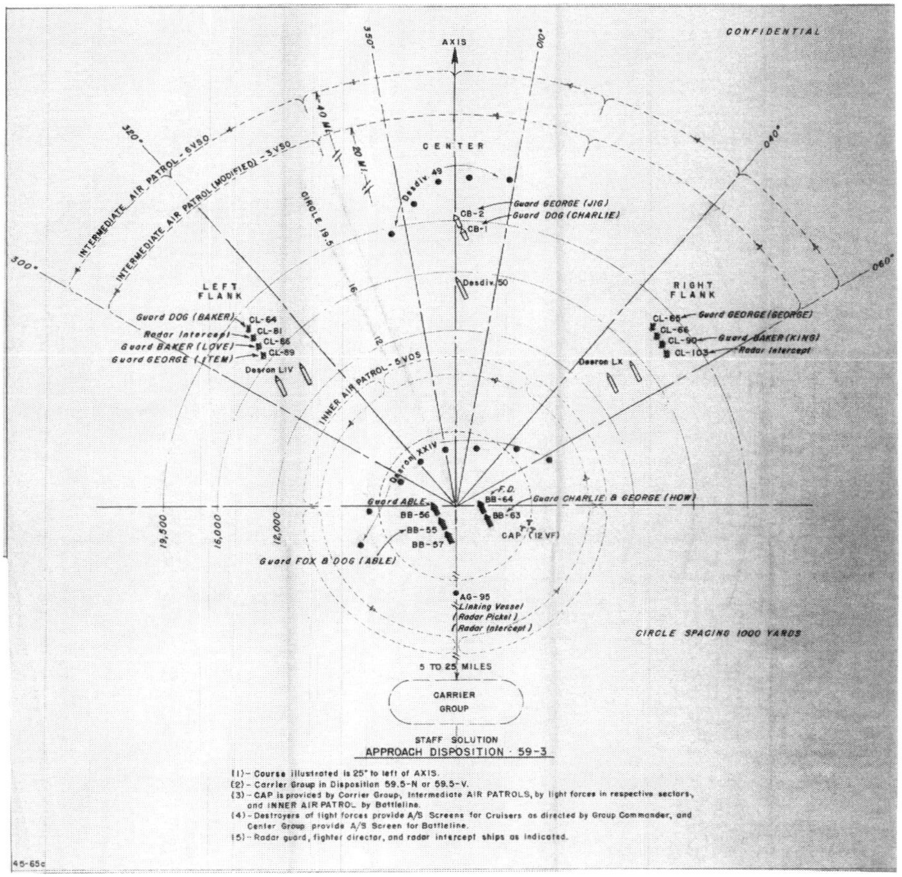

Fig. 116
Approach Disposition

American naval doctrine supplemented by lessons from the Pacific War. The scenario might, however, have been based on the possibility of fighting a future enemy in the Pacific at more northern latitudes, perhaps opposing a Soviet battle fleet near the Arctic Circle. The exercise scenario of an attack in the Volcano group did not involve weather so heavy as to preclude the use of carrier airpower, as might be expected in the far north. Still, by this time, knowledge of what had happened at Leyte Gulf would have been widespread among American naval officers, who might well have concluded that while carrier air forces could be expected to damage Orange forces, including carrier- and shore-based air assets, significantly, surface forces had to be configured so as to meet a surge of enemy battle groups. Naval aviators would have argued that carrier forces could destroy all of Orange air and surface power, but it is highly unlikely that all American naval officers would have subscribed to that idea, given the events of October 1944.[17]

The Staff Solution: Requirement C

On the same day, another Staff Solution, concerning a Day Cruising Disposition, presented a different problem to the students, and one even more akin to recent Pacific War experience, as opposed to interwar thinking. In this scenario, Carrier Group 4 was returning from a strike on the Orange homeland. The Staff posed "complementary military problems" for this air operation and had developed cruising dispositions for a variety of operating conditions. Now, the student playing Commander, Task Group 58.4 had to develop a Task Organization and Cruising Disposition suitable for these situations. Task Group 58.4 contained one light and three fleet carriers of CARDIV 8, two battleships of BATDIV 9, two battle cruisers of CRUDIV 16, two detached antiaircraft cruisers, and eighteen destroyers of DESRONs 47 and 54. An Orange air attack was the most likely threat, with Orange submarines as a possible but less serious concern. Contact with Orange surface forces, however, was not considered likely, because of TG 58.4's air patrols, radar, and speed. The Cruising Disposition, therefore, had to provide for defense against Orange air attack while also allowing the group to conduct flight operations and providing against any submarine or surface threat. The problem was that positioning the ships for effective air defense was not necessarily compatible with the efficiency of flight operations. A disposition suitable for air defense was also acceptable against submarine attack, but the requirements of defense against surface attack conflicted with those of air defense. As a result, CTG 58.4 decided not to give any special consideration to surface attack, given its slight possibility. In addition, he considered that the purpose of his surface ships was to "exploit" the principal weapon of the group—the carrier aircraft—and that the sole purposes of the carriers themselves were to transport, support, and direct their air groups. Therefore, the role of the surface escorts was entirely defensive.[18]

The student playing the role of CTG 58.4 noted that because of the effectiveness of enemy suicide-plane tactics, it was current policy to station an entire destroyer squadron as a radar picket unit during carrier strikes. If the danger of an enemy air attack was great, doctrine called for at least a pair of destroyers as a daytime radar picket; nonetheless, the student determined that no more than four destroyers could be spared from the carrier screen. However, each radar picket destroyer would have a Radar Picket Combat Air Patrol. The duties of "Radar Guard" ships involved "considerable" duplication, so he assigned many of them to the same ships while issuing Special Task Group Instructions to distribute others to specific ships. The Task Group Guide was generally a carrier at the center of the disposition, because of the primary importance of flight operations. In light of the danger from suicide planes, however, the student in this case placed a battleship in the center of the disposition the Guide. Defensive support for the escort carriers, with their "weak" antiaircraft batteries, was addressed. An antiaircraft light cruiser would often be near them in the formation and, since they had even greater antiaircraft firepower than the cruisers, one of the fast battleships was stationed astern. Finally, since a "tight" disposition was best for antiaircraft defense but carriers required maneuvering room for flight operations, the outer screen was to be beyond the range of the antiaircraft batteries of the heavy ships, so as to permit the latter to "develop" their most effective antiaircraft fire. Antiaircraft fire and sonar detection decreased with separation of the ships; a compromise had to be found to accommodate ship defense with flight operations.[19]

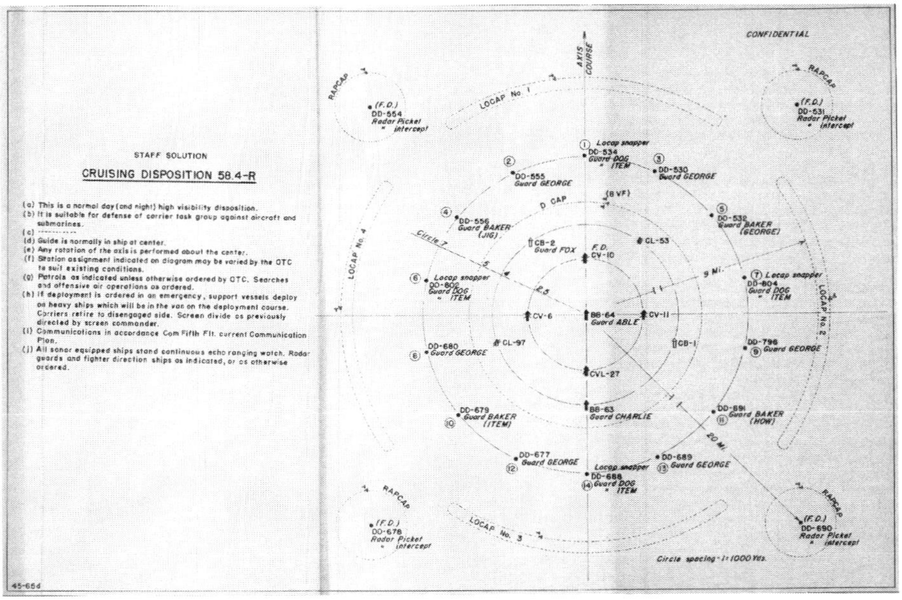

Fig. 117
Cruising Disposition

Aircraft Search Exercise

The next week, an Aircraft Search Method Exercise was run. The purpose was to acquaint the student with forms of search operations by aircraft that had arisen during the war but were not yet found in standard publications. The authors of the exercise welcomed suggestions from student officers who had knowledge of the subject, noting, however, that a "thorough" knowledge of the standard Maneuvering Board (a special plotting sheet, with associated nomograms, routinely used at sea for quick, graphical solutions of relative-motion problems) was a "valuable adjunct" to any form of search. The Exercise additionally looked closely at the concept of visibility—long used by seafaring people without an adequate definition but now considered, for the purposes of meteorological entries in a ship's log, to be the distance at which "large prominent" objects were just visible. The problem was the use of terms such as "large" and "prominent." While a destroyer was large and prominent compared to a whaleboat it was small and inconspicuous in comparison to the liner *Queen Mary.* Visibility was a "very important" factor in planning searches by either surface ships or aircraft, but the peacetime use of the term was "misleading" and "false." So many factors went into it that no "set amount" could be determined and accepted: "Simply because a mountain can be seen a hundred miles away does not at all imply that a ship on the surface can be sighted by an aircraft at the same distance, and the acceptance of such an assumption may lead to disastrous results."[20]

One concept of visibility was Maximum Range, or Sighting Distance, at which the object of the search could be seen. This distance was a function of the size of the object, the density of the haze in the atmosphere, the height of the observing aircraft, the azimuth of the sun, the alertness of the lookouts, and even the cleanliness of the aircraft's windows. Allowances could be made for these various factors and their sum subtracted from the visibility to give the Sighting Distance, expressed in terms of probability of sighting the object. At the Sighting Distance, the probability of sighting was almost "nil," whereas if the plane was immediately over the object, the probability was, of course, practically 100 percent. From the "extreme" Sighting Distance, the probability of sighting increased as the distance to the object decreased, and the point where the probability was 50 percent was known as the "Effective Visibility." This was the distance at which the chance of sighting an object in any search situation was "average," or 50 percent. At the same time, if the value of the range of Effective Visibility was arbitrarily reduced below that required for a 50 percent probability, the probability of sighting was increased. Students were warned, however, that there was no assurance that everything within the Effective Visibility Range could be sighted. The use of radar in aircraft search operations did not change these definitions, but certain factors concerned with visual search could be neglected: the density of haze, the bearing of the sun, and the alertness of the lookouts. (For purposes of the Exercise, the authors asserted, it was "always assumed that equipment was functioning properly.")[21]

Since it would frequently be necessary or desirable to establish an aircraft patrol along a certain line in order to detect the passage of a ship, the Station Patrol described in the fleet's *Scouting Manual* for surface ships could be used with aircraft. However, both methods were inefficient, since they required large numbers of platforms; moreover, if a single ship or plane failed to maintain its position or a search pattern, the value of the patrol was "seriously reduced." A conventional form of

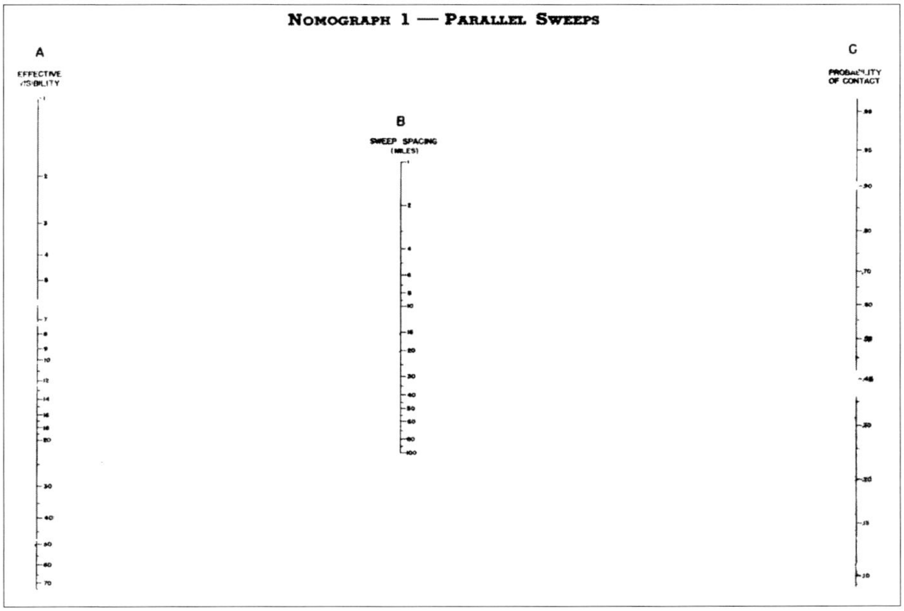

Fig. 118
Nomograph 1—Parallel Sweeps

parallel search was ineffective when the object was moving, as was "readily apparent." (See figure 119, which indicates areas in which an enemy ship could be seen on the assumed course and speed but remain undetected during the period of the search. Reversing the direction of the flight [figure 120] increased the effectiveness of the search but still left open areas.) A "barrier patrol" could close the open areas. Figures 121 and 122 demonstrate the methods for laying out such a patrol. Figure 121 illustrates what the instructors were discussing above, while figure 122 required a slightly longer flight to cover the same area. This longer-flight patrol method, however, had the advantage that even if the speed of the enemy had been "grossly" overestimated, the barrier would still be effective, and if underestimated, the zone where the enemy probably would not be found could be "consolidated" into one block rather than being scattered in small sectors throughout the entire area being searched.[22]

The patrol was illustrated for the Exercise (figure 123) with the actions of a single plane, though barrier patrols usually required more than one. The length of the barrier line, perpendicular to the most probable course of the enemy unit sought, was determined by the total length of the area across which the transit might be attempted. This might be, for instance, the distance between two islands, between the mainland and an island, or the farthest out to sea that an aircraft could fly. When the length of a barrier was too long for a single aircraft, it might be possible to set up barrier lines from each side toward the middle. When there was no danger that transit would be attempted close to shore, the barrier line could be drawn only where the transit was expected. "Shortening the barrier line always saves flying time." For instance, if a plane with airborne search radar had a "reliable" detection range of thirty-six miles for ships and eighteen for submarines, a flight could be repeated every twelve hours for the continued interception of twelve-knot ships or six-knot submarines. But by flying more than one plane simultaneously, the width of a barrier would be increased and "many disadvantages" of night operations, such as the problem of identifying ships, could be avoided.[23]

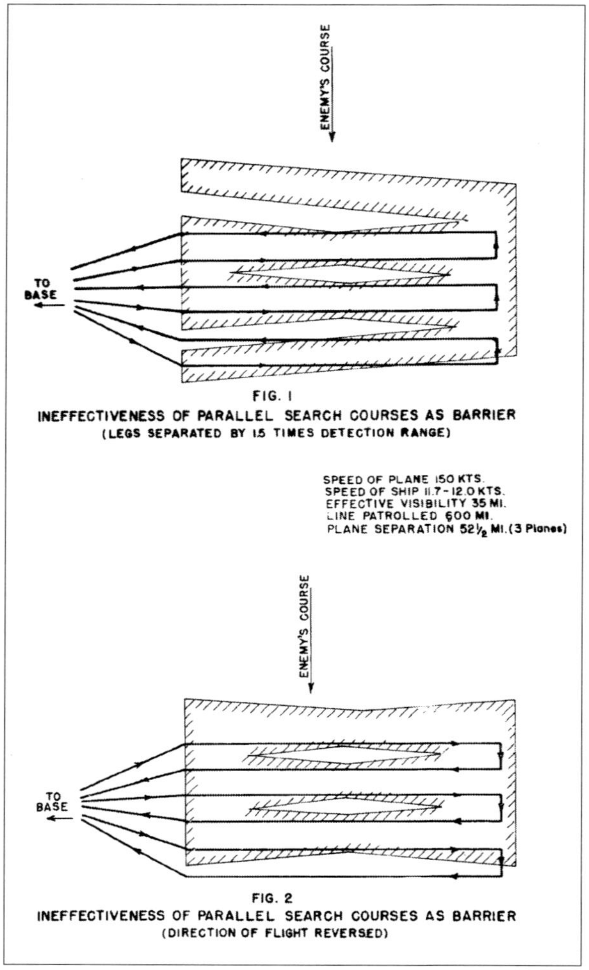

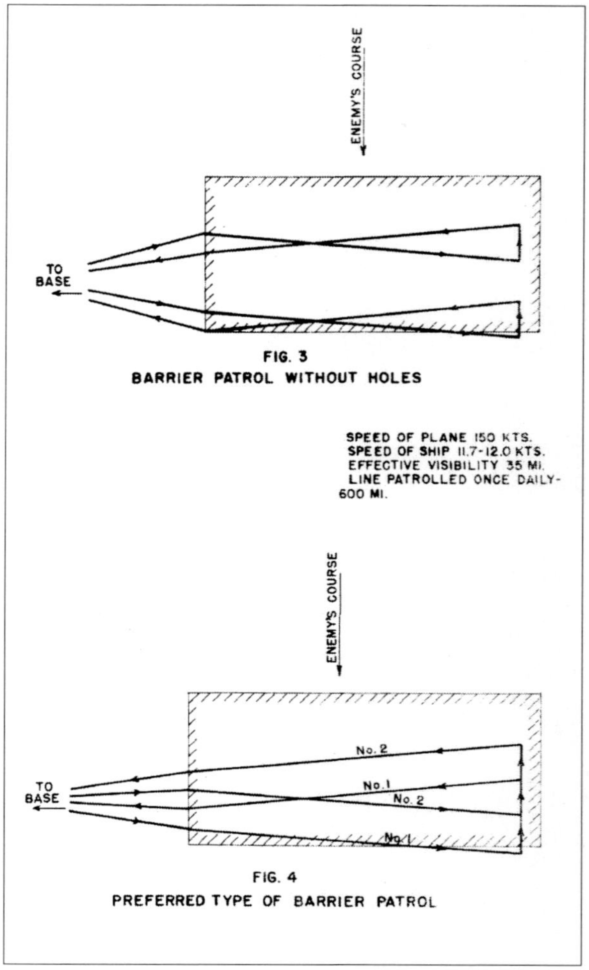

Fig. 119 (Top)
Ineffectiveness of parallel search courses as barrier

Fig. 120 (Bottom)
Ineffectiveness of parallel search courses as barrier

Fig. 121 (Top)
Barrier patrol without holes

Fig. 122 (Bottom)
Preferred type of barrier patrol

Figure 124 illustrated the most "profitable" method of combining simultaneous flights of planes in crossing tracks with certain distances between them. If surface ships were the objects of the search, enough planes were needed to make the width of a barrier equal to or slightly greater than an estimated day's run of the ships being sought. This type of plan could also be used to lay out a barrier against a submarine, though there was less probability that a submarine might pass through a barrier submerged if half of the flying was done at night; in this case, the instructions were to fly the barrier patrol every twelve hours and cover a width equal to or greater than half a day's run. If submarines were the object of the search, there was some advantage in making the width of the barrier somewhat wider than the minimum distance indicated for surface ships of the same speed. (Students were

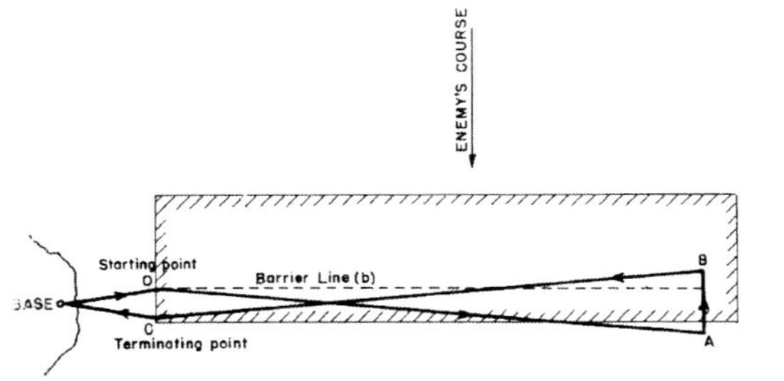

FIG. 5
BARRIER PATROL BY A SINGLE PLANE

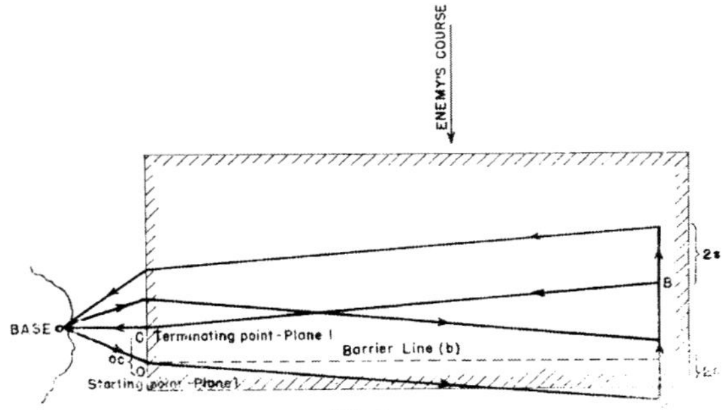

FIG. 6
CONSTRUCTION OF BARRIER PATROL BY TWO PLANES

Fig. 123 (Top) Barrier patrol by a single plane

Fig. 124 (Bottom) Construction of barrier patrol by two planes

referred to the February 1944 and August 1944 copies of the *United States Fleet Anti-Submarine Bulletin,* available in the Naval War College Archives.) There were also general instructions for parallel searches by aircraft, which could be used whenever the position of the object was uncertain and when its course and speed were unknown. If the position was known, no search was necessary; if the course and speed were both known, a different type of search, based on barrier patrol methods or a Maneuvering Board Solution, was considered more effective. To determine the spacing of the planes, Nomograph 1, seen in figure 118, was used. To employ this method, two of the three factors of course, speed, and position had to be known. Students were to obtain Effective Visibility criteria from the Maneuver Rules and use various scales of Probability of Contact to determine the spacing of the aircraft. (Students were also referred to Aerial Escort Plans contained in such references materials as *Anti-Submarine and Escort Convoy Instructions* for the latest types of antisubmarine screens that employed aircraft as screening units.)[24]

The Chart Maneuver

Another Exercise that same week acquainted students with the principles, forms, and methods of naval searches as well the mechanics of a Naval War College Chart Maneuver. The Exercise was introduced by assertion that although aircraft and radar had changed the methods of naval search operations, the principles had not been changed. Instead, new technologies had been adapted to "fundamental" principles and forms, and, therefore, an officer who understood the latter would be able to use the former more proficiently. Also, weather, the failure of equipment, or the non-availability of aircraft might make it necessary to return to earlier forms of search.[25] Student officers, especially if they were to be engaged in strategic planning, needed to become acquainted with the principles, forms, and methods of naval search and on the use of ships, planes, and radar in them.[26]

The strategic and operational situations again assumed that Blue and Orange were at war and that the Blue Fleet had been assembled in Hawaiian waters prior to engaging the Orange fleet. The Orange fleet, in turn, had been assembled in the Caroline and Marshall Islands to resist Blue movement westward from Hawaii. The Orange Operations Plan called for forming a Striking Force under Rear Admiral

OJ, the Striking Force to be composed of two battleships in Battleship Division 3; one fleet carrier, with thirty-two Zeke fighters, sixteen Val dive-bombers, and sixteen Kate torpedo planes; two heavy cruisers of CRUDIV 8; eight destroyers of Destroyer Divisions 9 and 22; and three fleet submarines of SUBDIV 19. OJ's mission was to gain information on the movement of the Blue Main Body west of Hawaii and destroy any "inferior" enemy formations encountered.

Admiral OJ's ships were disposed with the battleships and screening destroyers in column, the carrier and its supporting cruisers and destroyers on a different course, and the submarines on yet a different course to scout ahead. It was assumed that all Orange ships had two-thirds of their maximum fuel supply. Instructions included directing how each student was to move his group of ships on the Master Plot and create Chart Maneuvers with flimsies. Students were also required to submit drawings of each disposition of each group of ships under their charge. Students were to plot all tracks of vessels and planes as well as consult with liaison officers and instructors about any matters in doubt. In addition, there was some basic plotting instruction: indicating errors, keeping Records of Move Forms for

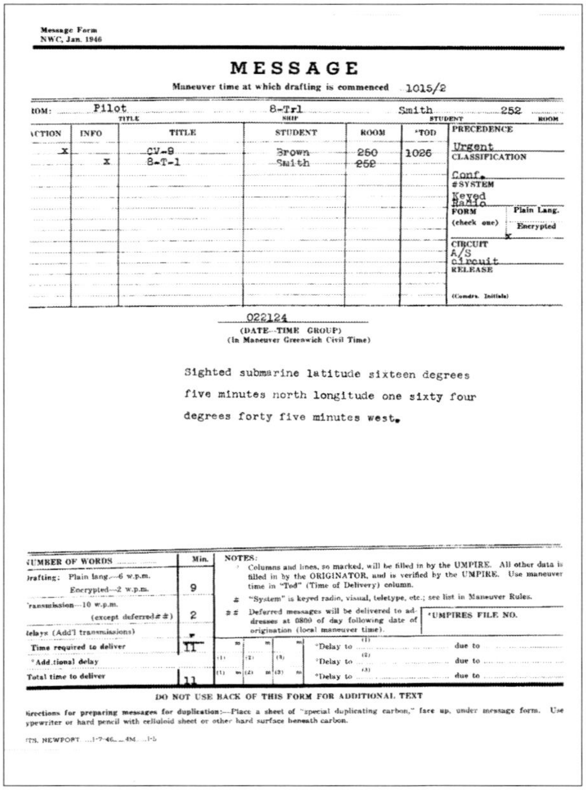

Message Form
NWC, Jan. 1946

MESSAGE

Maneuver time at which drafting is commenced 1015/2

FROM: Pilot (TITLE) 8-Trl (SHIP) Smith (STUDENT) 252 (ROOM)

ACTION	INFO	TITLE	STUDENT	ROOM	*TOD
x		CV-9	Brown	250	1026
	x	8-T-1	Smith	252	

PRECEDENCE: Urgent
CLASSIFICATION: Conf.
#SYSTEM: Keyed Radio
FORM (check one): Plain Lang. / Encrypted x
CIRCUIT: A/S circuit
RELEASE: (Comdrs. Initials)

022124
(DATE—TIME GROUP)
(In Maneuver Greenwich Civil Time)

Sighted submarine latitude sixteen degrees five minutes north longitude one sixty four degrees forty five minutes west.

NUMBER OF WORDS	Min.
Drafting: Plain lang.—6 w.p.m. Encrypted—2 w.p.m.	9
Transmission—10 w.p.m. (except deferred##)	2
Delays (Add'l transmissions)	
Time required to deliver	11
*Additional delay	
Total time to deliver	11

NOTES:
* Columns and lines, so marked, will be filled in by the UMPIRE. All other data is filled in by the ORIGINATOR, and is verified by the UMPIRE. Use maneuver time in "Tod" (Time of Delivery) column.
\# "System" is keyed radio, visual, teletype, etc.; see list in Maneuver Rules.
\#\# Deferred messages will be delivered to addresses at 0800 of day following date of origination (local maneuver time).

*UMPIRES FILE NO.

*Delay to (1) .... due to ....
*Delay to (2) .... due to ....
*Delay to (3) .... due to ....

DO NOT USE BACK OF THIS FORM FOR ADDITIONAL TEXT

Directions for preparing messages for duplication:—Place a sheet of "special duplicating carbon," face up, under message form. Use typewriter or hard pencil with celluloid sheet or other hard surface beneath carbon.

NTS. NEWPORT ...1-7-46...4M...1-5

Fig. 125
Message 022124

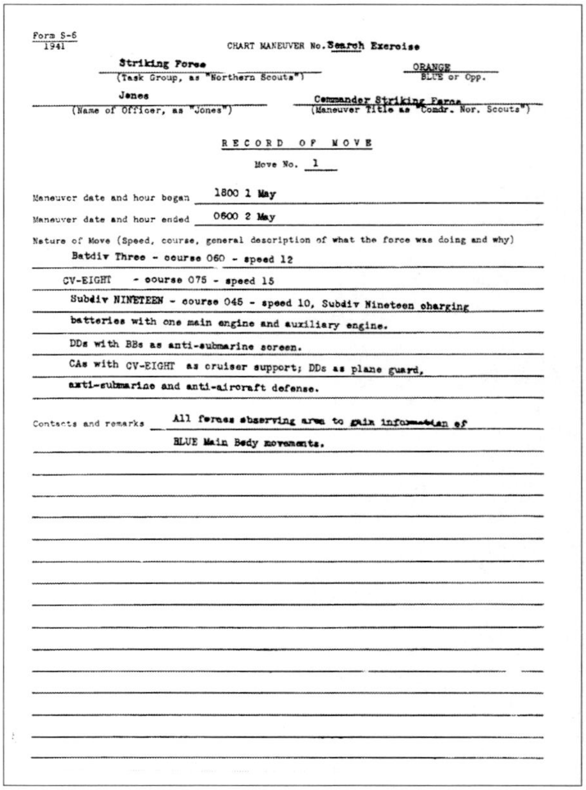

Form S-6
1941

CHART MANEUVER No. Search Exercise

Striking Force (Task Group, as "Northern Scouts") — ORANGE (BLUE or Opp.)

Jones (Name of Officer, as "Jones") — Commander Striking Force (Maneuver Title as "Comdr. Nor. Scouts")

RECORD OF MOVE

Move No. 1

Maneuver date and hour began 1800 1 May

Maneuver date and hour ended 0600 2 May

Nature of Move (Speed, course, general description of what the force was doing and why)

Batdiv Three - course 060 - speed 12
CV-EIGHT - course 075 - speed 15
Subdiv NINETEEN - course 045 - speed 10, Subdiv Nineteen charging batteries with one main engine and auxiliary engine.
DDs with BBs as anti-submarine screen.
CAs with CV-EIGHT as cruiser support; DDs as plane guard, anti-submarine and anti-aircraft defense.

Contacts and remarks All forces observing area to gain information of BLUE Main Body movements.

Fig. 126
Record of Move 1

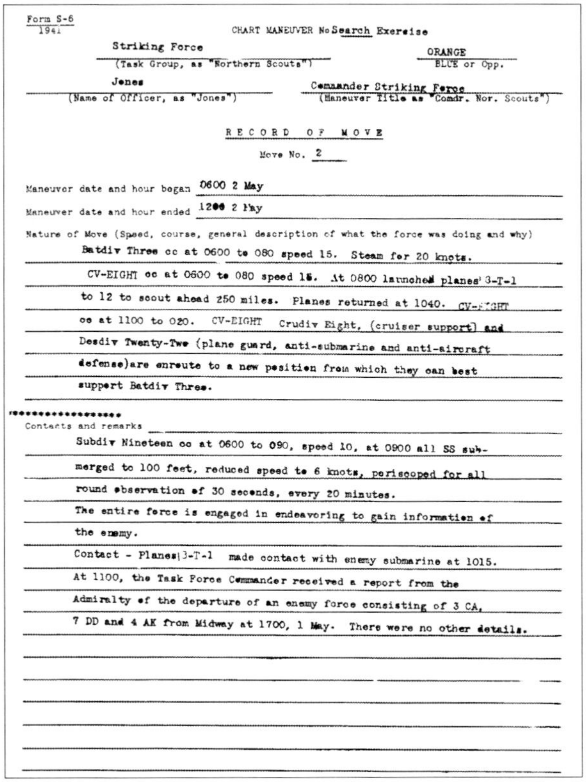

Form S-6
1941

CHART MANEUVER No Search Exercise

Striking Force
(Task Group, as "Northern Scouts")

ORANGE
BLUE or Opp.

Jones
(Name of Officer, as "Jones")

Commander Striking Force
(Maneuver Title as "Comdr. Nor. Scouts")

RECORD OF MOVE

Move No. 2

Maneuver date and hour began 0600 2 May

Maneuver date and hour ended 1200 2 May

Nature of Move (Speed, course, general description of what the force was doing and why)

Batdiv Three cc at 0600 to 080 speed 15. Steam for 20 knots.
CV-EIGHT cc at 0600 to 080 speed 15. At 0800 launched planes 3-T-1
to 12 to scout ahead 250 miles. Planes returned at 1040. CV-EIGHT
cc at 1100 to 020. CV-EIGHT Crudiv Eight, (cruiser support) and
Desdiv Twenty-Two (plane guard, anti-submarine and anti-aircraft
defense)are enroute to a new position from which they can best
support Batdiv Three.

Contacts and remarks

Subdiv Nineteen cc at 0600 to 090, speed 10, at 0900 all SS sub-
merged to 100 feet, reduced speed to 6 knots, periscoped for all
round observation of 30 seconds, every 20 minutes.
The entire force is engaged in endeavoring to gain information of
the enemy.
Contact - Planes 3-T-1 made contact with enemy submarine at 1015.
At 1100, the Task Force Commander received a report from the
Admiralty of the departure of an enemy force consisting of 3 CA,
7 DD and 4 AK from Midway at 1700, 1 May. There were no other details.

Fig. 127
Record of Move 2

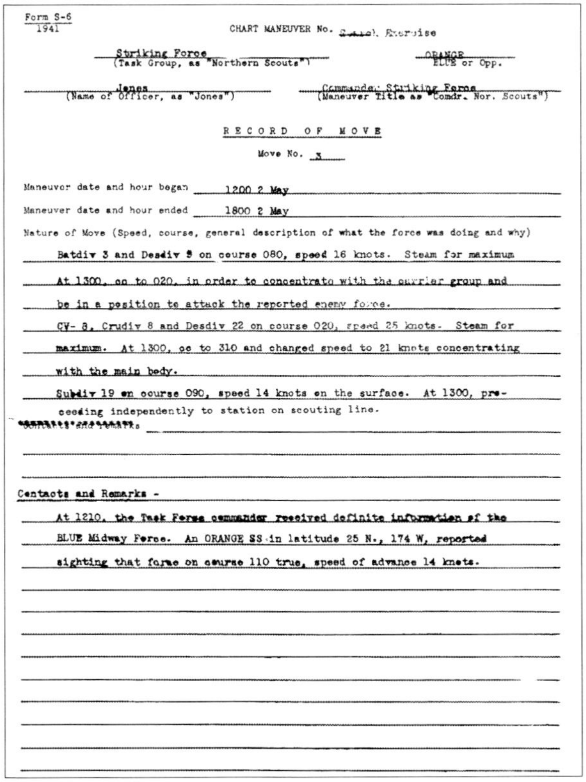

Form S-6
1941

CHART MANEUVER No. Search Exercise

Striking Force
(Task Group, as "Northern Scouts")

ORANGE
BLUE or Opp.

Jones
(Name of Officer, as "Jones")

Commander Striking Force
(Maneuver Title as "Comdr. Nor. Scouts")

RECORD OF MOVE

Move No. 3

Maneuver date and hour began 1200 2 May

Maneuver date and hour ended 1800 2 May

Nature of Move (Speed, course, general description of what the force was doing and why)

Batdiv 3 and Desdiv 9 on course 080, speed 16 knots. Steam for maximum
At 1300, cc to 020, in order to concentrate with the carrier group and
be in a position to attack the reported enemy force.
CV- 8, Crudiv 8 and Desdiv 22 on course 020, speed 25 knots. Steam for
maximum. At 1300, cc to 310 and changed speed to 21 knots concentrating
with the main body.
Subdiv 19 on course 090, speed 14 knots on the surface. At 1300, pro-
ceeding independently to station on scouting line.

Contacts and Remarks -

At 1210, the Task Force commander received definite information of the
BLUE Midway Force. An ORANGE SS in latitude 25 N., 174 W, reported
sighting that force on course 110 true, speed of advance 14 knots.

Fig. 128
Record of Move 3

the surface ships and submarines separately from those for aircraft, tracking battleship fuel consumption on Fuel Work Sheets, and handling dispatches (messages).[27]

In the Demonstration of the Chart Maneuver, Move 1 established the Orange force's formations, courses, and speeds. The scenario then moved to the Blue commander in Hawaii, ordering Naval Patrol Plane Squadron 12, consisting of twelve Catalina naval patrol bombers at Kauai, to intercept the Orange force, to form and maintain a barrier patrol, and to confirm the presence of the Orange Striking Force. Naval Patrol Plane Squadron 12's barrier patrol would consist of one plane at a time; students were to chart the plane's movements. Move 2 consisted of the Orange fleet carrier CV-8 launching twelve torpedo planes in pairs to conduct searches out to a radius of 250 miles. One of the torpedo planes sighted a Blue submarine, which caused the Orange commander to change the course of the carrier force. The same day, the Orange Task Force Commander received a message from the Orange Admiralty that an enemy surface force of three heavy cruisers, seven destroyers, and four cargo ships had been sighted by a submarine departing Midway Island; the submarine had not been able to maintain contact. In Move 3, Orange

Form S-6
1941

CHART MANEUVER No. Search Exercise

Striking Force
(Task Group, as "Northern Scouts")

ORANGE
BLUE or Opp.

Jones
(Name of Officer, as "Jones")

Commander Striking Force
(Maneuver Title as "Comdr. Nor. Scouts")

RECORD OF MOVE

Move No. 4

Maneuver date and hour began 1800, 3 May

Maneuver date and hour ended 0600, 4 May

Nature of Move (Speed, course, general description of what the force was doing and why)

Batdiv 3 and Desdiv 9 on course 020, speed 16 knots throughout the move. At 0345, CV-8, Crudiv 8 and Desdiv 22, which had been on course 310 at 21 knots, rendezvoused with the battleship group and joined that group, taking up their course and speed.

Subdiv 19 continued proceeding independently to their initial scouting stations. At 0100, the 3 SS of this division were on station and commenced a search from the flank, patrol method.

Contacts and remarks

Fig. 129
Record of Move 4

Form S-7
1930

CHART MANEUVER No. Search Exercise

ORANGE Striking Force
(Task Group, as BLUE Main Body)

Smith
(Name of Officer, as Jones)

FUEL ACCOUNT FOR MOVE

ORANGE Batdiv Three
(Organization unit, as BLUE ~~Batron~~ Batdiv One)

VESSEL BB-6

VESSEL	BB-6
On hand, beginning of move 1	3000.0
Received: Amount	
Place	
From (Ship)	
Time (Hours)	
Expended	52.8
On hand, End of Move 1	2947.2

Total during Maneuver up to End of this Move

Expended	52.8
Received	

VESSEL	
On hand, beginning of move 2	2947.2
Received: Amount	
Place	
From (Ship)	
Time (Hours)	
Expended	34.2
On hand, End of Move 2	2913.0

Total during Maneuver up to End of this Move

Expended	87.0
Received	

VESSEL	
On hand, beginning of Move 3	2913.0
Received: Amount	
Place	
From (Ship)	
Time (Hours)	
Expended	42.6
On hand, End of Move 3	2870.4

Total during Maneuver up to end of this Move

Expended	129.6
Received	

Fig.130
Fuel account for Moves 1–3, Orange BATDIV 3

submarines I-153, I-155, and I-156 were entering an area near Necker Island with low visibility, rain, and squalls because of a weather front. One of these submarines made another contact report on the Blue surface force but was again not able to maintain contact. Nevertheless, the submarine reports caused the Orange Task Force Commander to concentrate his surface ships in an attempt to use his submarines to find the Blue force and have the surface units attack and destroy it.[28]

In Modified Move 3, Orange fleet carrier CV-8 reduced its speed and fell back behind BATDIV 3 and Destroyer Division 9 as the latter two units received search orders from the Orange Task Force Commander. Submarine Division 19's boats were ordered to remain on the surface and report all contacts until forced to submerge. CV-8 and BATDIV 3 later rendezvoused, and the submarines began flank patrol searches that eventually resulted in a contact, though the submarine making the report was not heard from again. Accordingly, in Move 4, Destroyer Division 9 was ordered to take up a scouting position and maintain contact with the Blue force in preparation for a Main Body attack the following day.[29]

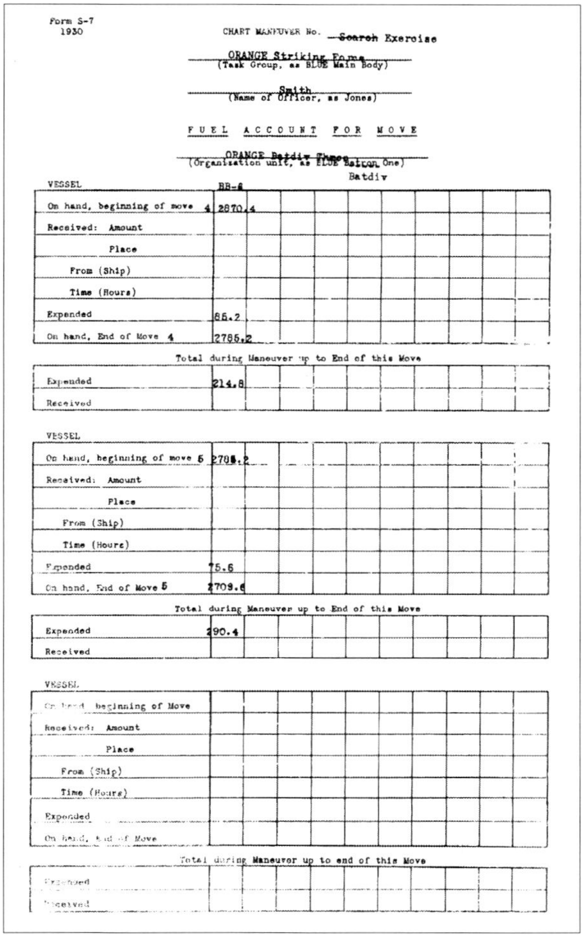

Form S-7
1930

CHART MANEUVER No. ~~Search~~ Exercise

ORANGE Striking Force
(Task Group, as BLUE Main Body)

Smith
(Name of Officer, as Jones)

FUEL ACCOUNT FOR MOVE

ORANGE Batdiv ~~Three~~
(Organization unit, as BLUE ~~Batron~~ One)
Batdiv

VESSEL	BB-6
On hand, beginning of move 4	2870.4
Received: Amount	
Place	
From (Ship)	
Time (Hours)	
Expended	85.2
On hand, End of Move 4	2785.2

Total during Maneuver up to End of this Move

Expended	214.8
Received	

VESSEL	
On hand, beginning of move 5	2785.2
Received: Amount	
Place	
From (Ship)	
Time (Hours)	
Expended	75.6
On hand, End of Move 5	2709.6

Total during Maneuver up to End of this Move

Expended	290.4
Received	

VESSEL	
On hand, beginning of Move	
Received: Amount	
Place	
From (Ship)	
Time (Hours)	
Expended	
On hand, End of Move	

Total during Maneuver up to end of this Move

[illegible]	
[illegible]	

Fig. 131
Fuel account for Moves 4–5, Orange BATDIV 3

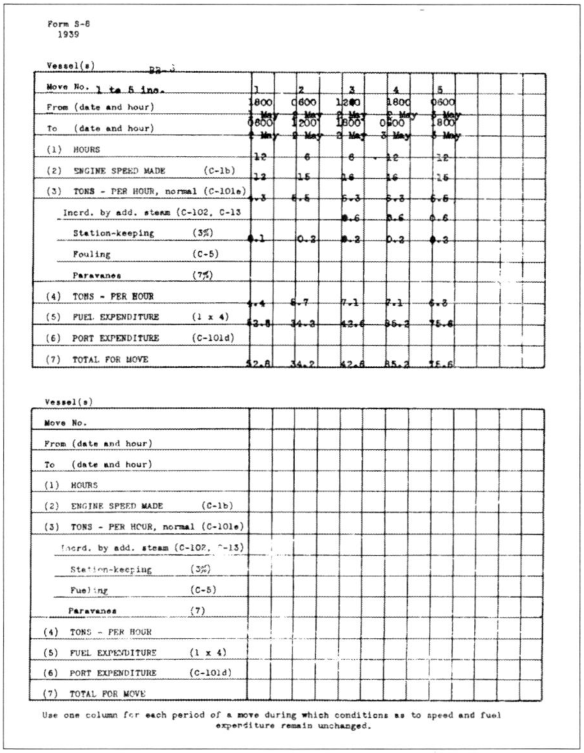

Form S-8
1939

Vessel(s) BB-3

Move No. 1 to 5 inc.	1	2	3	4	5
From (date and hour)	1800 1 May	0600 2 May	1200 2 May	1800 2 May	0600 3 May
To (date and hour)	0600 2 May	1200 2 May	1800 2 May	0600 3 May	1800 3 May
(1) HOURS	12	6	6	12	12
(2) ENGINE SPEED MADE (C-1b)	12	15	16	16	15
(3) TONS - PER HOUR, normal (C-101e)	4.3	5.5	6.3	6.3	5.5
Incrd. by add. steam (C-102, C-13			0.6	0.6	0.6
Station-keeping (3%)	0.1	0.2	0.2	0.2	0.3
Fouling (C-5)					
Paravanes (7%)					
(4) TONS - PER HOUR	4.4	5.7	7.1	7.1	6.3
(5) FUEL EXPENDITURE (1 x 4)	52.8	34.2	42.6	85.2	75.6
(6) PORT EXPENDITURE (C-101d)					
(7) TOTAL FOR MOVE	52.8	34.2	42.6	85.2	75.6

Vessel(s)	
Move No.	
From (date and hour)	
To (date and hour)	
(1) HOURS	
(2) ENGINE SPEED MADE (C-1b)	
(3) TONS - PER HOUR, normal (C-101e)	
Incrd. by add. steam (C-102, C-13)	
Station-keeping (3%)	
Fueling (C-5)	
Paravanes (7)	
(4) TONS - PER HOUR	
(5) FUEL EXPENDITURE (1 x 4)	
(6) PORT EXPENDITURE (C-101d)	
(7) TOTAL FOR MOVE	

Use one column for each period of a move during which conditions as to speed and fuel expenditure remain unchanged.

Fig. 132
Fuel Work Sheet, Search Exercise

In Move 5 Commander, Destroyer Division 9 reported contact with an enemy force consisting of three heavy cruisers, four cargo ships, and seven to eight destroyers on a course of 110 true and a speed of fourteen knots. He reported that he would continue to scout, but as a result of this report, Rear Admiral OJ decided to change course to 110 true and match the speed of fourteen knots. He later changed course to 250 to avoid getting too close to Kauai by daylight and so he could launch a daytime strike of twelve dive-bombers and twelve torpedo planes on Blue. On recovering these planes, the Orange force was ordered to retire westward at high speed; SUBDIV 19 was directed to form a patrol line to cover the retirement of the Striking Force. Subsequently, when the Blue force was reported to have changed course to 090 true, Admiral OJ determined that its position by 0700 would be accurately enough known to locate it by means of the Scouting Line.[30]

This Exercise also entailed demonstrating proficiency in establishing aerial barrier patrols and patrol methods for searching from the flanks. For instance, in this case, formulae were used to establish Points A and O of an aerial barrier patrol; if a

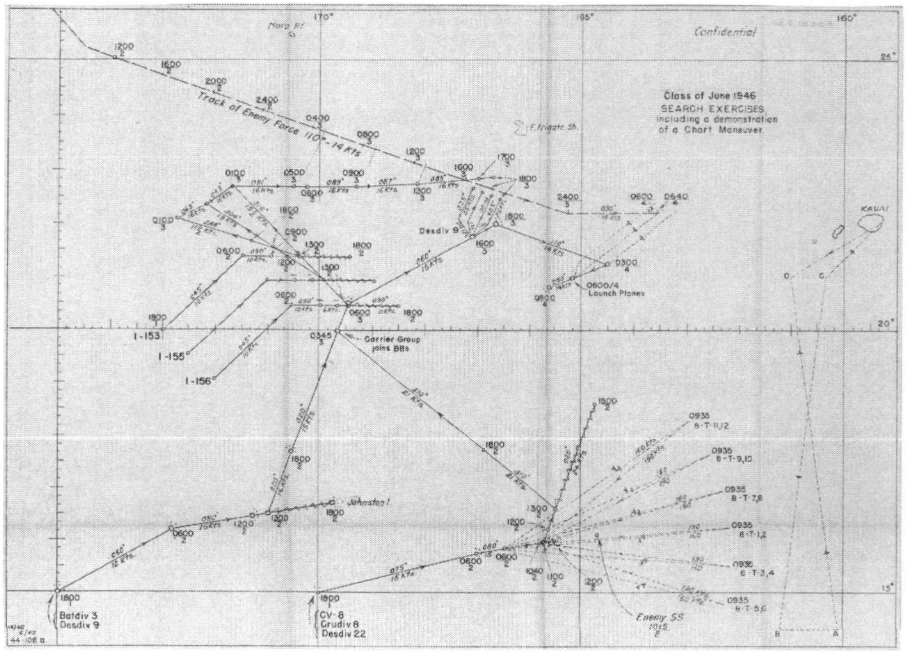

Figure 133
Search Exercises

naval patrol plane flew courses of 173, 270, and 007 degrees for a total distance of 973 miles, the student would take into account the force and direction of the wind to calculate that the plane could complete the patrol in 7.39 hours; he would also demonstrate his ability to determine that given a barrier width of 144 miles, because an enemy ship steaming at fifteen knots could cross the barrier in 9.6 hours, the barrier patrol would have to be repeated at least that often. In the Exercise, the Task Force Commander specified a rendezvous for the scout ships and the Blue force; the scout ships were to head immediately for their assigned initial positions on the Scouting Circle. The scout's initial position was therefore advanced 7.25 miles, the maximum "practical" scouting distance, beyond the Blue position circle drawn through the initial scouting position. The search then continued in the "usual manner," with each scout following in the track ahead. Destroyers also took their station on the Scouting Line in "normal order"—in other words, the second ship in the column being the Right Flank Ship and the third and fourth ships in column being, respectively, next from the left and the Left Flank Ship. Unless otherwise stated, the bearing of the Scouting Line was always normal to the scouting course. It also was noted that had the Exercise been continued, the second

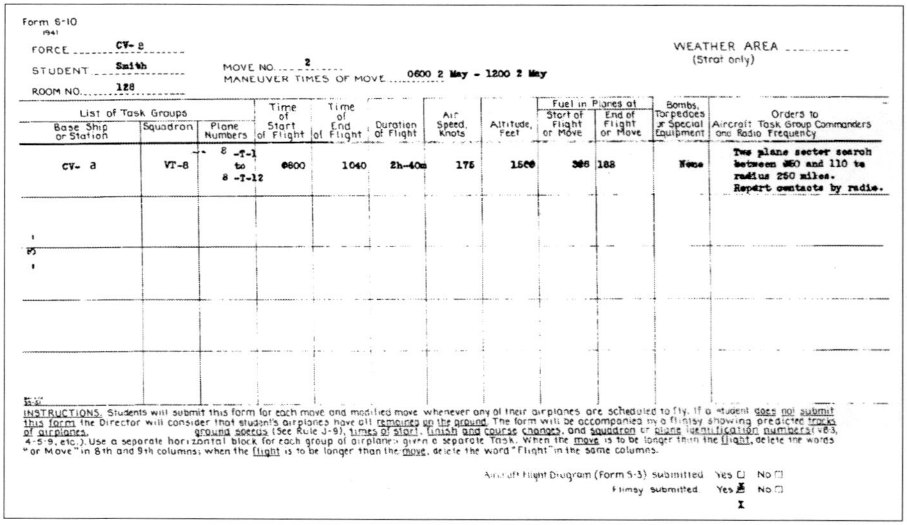
Form S-10
1941

FORCE CV- 8
STUDENT Smith
ROOM NO. 128
MOVE NO. 2
MANEUVER TIMES OF MOVE 0600 2 May - 1200 2 May
WEATHER AREA
(Strat only)

List of Task Groups: Base Ship or Station	Squadron	Plane Numbers	Time of Start of Flight	Time of End of Flight	Duration of Flight	Air Speed, Knots	Altitude, Feet	Fuel in Planes at Start of Flight or Move	Fuel in Planes at End of Flight or Move	Bombs, Torpedoes or Special Equipment	Orders to Aircraft Task Group Commanders and Radio Frequency
CV- 8	VT-8	8 -T-1 to 8 -T-12	0800	1040	2h-40m	175	1500	386	188	None	Two plane sector search between 050 and 110 to radius 250 miles. Report contacts by radio.

INSTRUCTIONS. Students will submit this form for each move and modified move whenever any of their airplanes are scheduled to fly. If a student does not submit this form the Director will consider that student's airplanes have all remained on the ground. The form will be accompanied by a flimsy showing predicted tracks of airplanes, ground speeds (See Rule J-9), times of start, finish and course changes, and squadron or plane identification numbers (VB-3, 4-S-9, etc.). Use a separate horizontal block for each group of airplanes given a separate Task. When the move is to be longer than the flight, delete the words "or Move" in 8th and 9th columns; when the flight is to be longer than the move, delete the word "Flight" in the same columns.

Aircraft Flight Diagram (Form S-3) submitted Yes ☐ No ☐
Flimsy submitted Yes X No ☐

Fig. 134
Aircraft Flight Form, Move No. 2

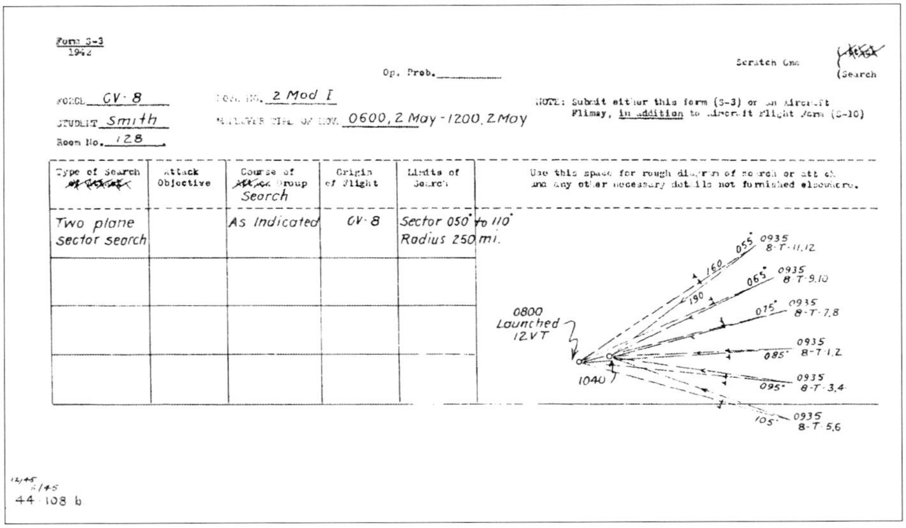

Fig. 135
Aircraft flight plan, Move No. 2, Mod. I

scout from the right would have made contact with the Blue force shortly after 1800.[31]

Conclusion

The theoretical pedagogy of war gaming aside, the very basis of these Disposition and Search Exercises made perfect sense in the early postwar period, especially since search methods and communications about contacts had been so weak at the beginning of the war and there were still areas where senior officers such as Admirals Spruance and Pye always saw need for improvement.[32]

What is most interesting about this last Exercise is that the Blue and Orange forces were situated so close to Hawaii. Not since 1942 had the United States been threatened in or near Hawaiian waters, but in this scenario Orange had a fully operational fleet and control of Micronesia, such that the Combined Fleet could deploy to the area around Midway and even toward Hawaii. Yet this was essentially a surface operation in the Central Pacific. Moreover, severe weather in the Central Pacific probably would not have hampered carrier operations as much as it would in the North Pacific. Thus, this operation still reflected late–Pacific War American naval doctrine that had been developed in the interwar period but refashioned after the 1941–42 battles so as to take carriers into account.[33]

It could, however, be argued that this Exercise was held more to demonstrate Japanese naval doctrine than American practices. Blue had no carriers, while Orange had one, but it was kept in subordinate roles, such as scouting, as were the submarines. Later in the Exercise the Orange carrier launched dive- and torpedo bombers against the Blue surface fleet before the Orange task force retired westward, but this was done primarily to cover the retreat of the heavy surface forces.[34] The exercise was also true to past form in demonstrating the propensity of the Imperial Japanese Navy to disperse forces, which Orange did at the beginning of the Exercise.[35] Thus, another purpose of the Exercise could have been to teach the officers the potential folly of dividing carrier and surface forces rather than amalgamating them into mutually supporting, combined-arms naval task forces, as the U.S. Navy had so often done between 1942 and 1945.[36]

NOTES

1 "Maneuver Rules," January 1946, pp. i to c-9, NHC, folder 2561-I, box 130, RG 4, NWC.

2 Ibid., pp. d-4, d-11, e-3.

3 Ibid., pp. g-1 to g-4, i-2 to i-3, i-9, i-13 to i-15.

4 Command and Staff Class of June 1946, "Exercise in Developing Task Force Dispositions: Schedule of Employment and Statement of Problem," 5 January 1946, pp. 1–3, NHC, folder 2563, box 131, RG 4, NWC. For fleet operations during the battles of Saipan, Leyte Gulf, and Iwo Jima, see Thomas Buell, *The Quiet Warrior: A Biography of Admiral Raymond A. Spruance* (Boston: Little, Brown, 1974), pp. 257–304; H. P. Willmott, *The Battle of Leyte Gulf: The Last Fleet Action* (Bloomington: Indiana Univ. Press, 2005); and George Dyer, *The Amphibians Came to Conquer: The Story of Admiral Richmond Kelly Turner,* 2 vols. (Washington, D.C.: U.S. Navy Dept., 1972), pp. 969–1051. For U.S. Pacific Fleet wartime operations plans and battle orders that matched this scenario almost exactly, see Trent Hone, "U.S. Navy Surface Battle Doctrine and Victory in the Pacific," *NWCR* 62, no. 1 (Winter 2009), pp. 67–105; and Hone, e-mails to author, 28 February, 2 March, 3 March, and 14 March 2011.

5 Command and Staff Class of June 1946, "Exercise in Developing Task Force Dispositions: Schedule of Employment and Statement of Problem," pp. 3–4 and 6–7.

6 Ibid., p. 4.

7 Ibid., pp. 4–5. Admiral BL's orders clearly reflect the combined interwar and wartime doctrine encountered elsewhere in the study. They are highly similar, for instance, to orders by both Adm. Raymond A. Spruance and Adm. William F. Halsey at the battles of the Philippine Sea and Cape Engaño, respectively. At the battle of the Philippine Sea, Admiral Spruance directed that the enemy carriers be knocked out first, as a prelude to air strikes on Japanese heavy surface forces and, perhaps, in preparation for a surface engagement. At Cape Engaño, Admiral Halsey intended to use his carriers to sink the Japanese carrier force, weaken the Japanese surface force with his airpower, and then have Task Force 34, the Pacific Fleet battle line, either engage the remaining Japanese force at night or mop up cripples the next day; see Ivan Musicant, *Battleship at War: The Epic Story of the USS* Washington (New York: Harcourt Brace Jovanovich, 1986), pp. 257–59, 295–303; Trent

Hone, "The Evolution of Fleet Tactical Doctrine in the U.S. Navy, 1922–1941," *Journal of Military History* 67 (October 2003), pp. 1107–48; Hone, "U.S. Navy Surface Battle Doctrine and Victory in the Pacific," pp. 67–105; and Hone, e-mails to author, 28 February, 2 March, 3 March, and 14 March 2011. They are even highly similar to Adm. Husband Kimmel's plan to ambush the Japanese Fleet in the Central Pacific in 1941; see Hone, "Evolution of Fleet Tactical Doctrine in the U.S. Navy," pp. 1138–40; and Edward Miller, *War Plan Orange: The U.S. Strategy to Defeat Japan, 1897–1945* (Annapolis, Md.: Naval Institute Press, 1991), pp. 286–312.

8 Command and Staff Class of June 1946, "Exercise in Developing Task Force Dispositions: Schedule of Employment and Statement of Problem," pp. 5, 8–9.

9 Command and Staff Class of June 1946, "Exercise in Developing Task Force Dispositions: Staff Solution, Requirement A (Battle Disposition) and Requirement B (Approach Disposition)," 11 January 1946, p. 1, NHC, folder 2563, box 131, RG 4, NWC.

10 Ibid., pp. 1–2.

11 Ibid., pp. 2–3.

12 Ibid., p. 3.

13 Ibid., p. 4.

14 Ibid., pp. 4–5.

15 Ibid., p. 5.

16 Ibid., pp. 5–6.

17 For an analysis of what naval aviators thought carrier airpower was really capable of in what they called "three-dimensional" warfare by the latter part of the Pacific War, see Clark Reynolds, *The Fast Carriers: The Forging of an Air Navy* (Annapolis, Md.: Naval Institute Press, 1968), pp. 156–300, and William McBride, *Technological Change and the United States Navy, 1865–1945* (Baltimore: Johns Hopkins Univ. Press, 2000), pp. 182–210. For limitations placed at times during the interwar period on the carriers and submarines by the Gun Club, see Thomas Hone and Trent Hone, *Battle Line: The United States Navy, 1919–1939* (Annapolis, Md.: Naval Institute Press, 2006), pp. 90–125, and Craig Felker, *Testing American Sea Power: U.S. Navy Strategic Exercises, 1923–1940* (College Station: Texas A&M Univ. Press, 2007), pp. 33–75. For the combination of interwar and wartime doctrine that took greater carrier power into account by 1943–45 but still noted limitations in that carrier power, see Hone, "U.S. Navy Surface Battle Doctrine and Victory in the Pacific," pp. 67–105; and Hone, e-mails to author, 28 February, 2 March, 3 March, and 14 March 2011.

18 Command and Staff Class of June 1946, "Exercise in Developing Task Force Dispositions: Staff Solution, Requirement C (Day Cruising Disposition)," 11 January 1946, pp. 1–2, NHC, folder 2563, box 131, RG 4, NWC.

19 Ibid., pp. 2–3.

20 Command and Staff Class of June 1946, "Other Aircraft Search Methods," 12 January 1946, preface and p. 1, NHC, folder 2564a, box 131, RG 4, NWC.

21 Ibid., pp. 1–2.

22 Ibid., pp. 2–4.

23 Ibid., pp. 4–5.

24 Ibid., pp. 6–7.

25 This idea—that new technologies developed in the last war might not always work in the next one—was a major theme of Admiral Spruance's when he became President of the Naval War College. He repeatedly noted that fighting at night or in northern latitudes, in particular, might deprive the U.S. Navy of its carrier forces and make heavy surface forces such as battleships and heavy cruisers a necessity; see Friedman, *Digesting History: The U.S. Naval War College, the Lessons of World War Two, and Future Naval Warfare, 1945–1947* (Newport, R.I.: NWCP, 2010), pp. 143–49, 156–61, 183–96, 199–207, 306–307.

26 Command and Staff Class of June 1946, "Search Exercises, including a Demonstration of a Chart Maneuver," 12 January 1946, p. 1, NHC, folder 2564, box 131, RG 4, NWC.

27 Ibid., pp. 2–5.

28 Ibid., pp. 6–9.

29 Ibid., pp. 6–11.

30 Ibid., p. 12.

31 Command and Staff Class of June 1946, "Staff Solution of Search Exercises, including a Demonstration of a Chart Maneuver," 17 January 1946, pp. 1–2, NHC, folder 2564-B, box 131, RG 4, NWC.

32 For the need for improvement in American naval search methods and contact reports in early battles, such as Midway, see Gordon Prange, *Miracle at Midway* (New York: McGraw-Hill, 1982), pp. 132–397. For Spruance's continued concern about expertise in naval search and communications methods, see Friedman, *Digesting History*, pp. 156–61.

33 Hone, "U.S. Navy Surface Battle Doctrine and Victory in the Pacific," pp. 67–105.

34 Command and Staff Class of June 1946, "Search Exercises, including a Demonstration of a Chart Maneuver," 12 January 1946, p. 1. This dimension of the Exercise, whereby carrier aircraft and light forces were used to cover the retreat of heavy surface forces, however, was an example of interwar U.S. Navy tactical doctrine; it had been developed by 1940 for use in a surface engagement that began to go badly; see Hone, "Evolution of Fleet Tactical Doctrine in the U.S. Navy," pp. 1123–24.

35 Hone, "U.S. Navy Surface Battle Doctrine and Victory in the Pacific," pp. 67–105; and Hone, e-mails to author, 28 February, 2 March, 3 March, and 14 March 2011. For the Imperial Japanese Navy, see Jonathan Parshall and Anthony Tully, *Shattered Sword: The Untold Story of the Battle of Midway* (Washington, D.C.: Potomac Books, 2005), pp. 3–59.

36 Douglas Smith, *Carrier Battles: Command Decision in Harm's Way* (Annapolis, Md.: Naval Institute Press, 2006), pp. 209–55.

XII Convoy Protection and Naval Communications
January 1946–February 1946

On 18 January 1946, Commodore Bowdey provided some instructions for all officers concerning the preparation of Solutions to the Problem in general. If typewritten, the documents had to be double-spaced, on only one side of a sheet, and bound with sketches or plotting sheets.[1]

Solutions were reviewed by one or more members of the Staff and pertinent remarks made. The Solutions were returned to the student officers as soon as the Solutions of the class as a whole had been reviewed. When a student officer stated that a procedure was to be in "accordance with doctrine," a résumé of the applicable doctrine was to be included, as well as definitive evidence that the subordinate who was to carry out the doctrine also knew what the doctrine was. Similarly, if reference was made to an annex, its substance was to be included in the Solution unless the Statement of the Problem permitted it to exist in name only.[2]

Demonstrative Problem: Protecting Convoys

On the same day, a "Demonstrative" (that is, demonstrational) Problem began. This Exercise had three functions. First, it was to familiarize students with a publication entitled *Operational Functions of Command* and demonstrate its use in the Solution of a Problem. In addition, the Exercise was to "stimulate" student interest in the South China Sea theater and, finally, help students acquire knowledge about the capabilities and limitations of certain ships and aircraft, via their proxies in the Blue and Orange navies. The General Situation was set in December 1945, in a scenario of continued warfare in the Pacific. Orange and Purple (the USSR) were still in a state of neutrality with respect to each other. The Yellow (Chinese) army had retreated to its western mountains, and Orange was still in complete control of the eastern portion of China, as well as all of French Indochina, Thailand, Malaya, and eastern and southern Burma. Red (British) troops were penetrating Burma in a campaign that, if successful, would open transportation between India and "Chunking" (that is, Chungking, modern Chongqing) and thus improve the logistical support of the Yellow army. Orange still controlled Formosa and the Pescadores, but Blue air forces based in the Philippines—as well as the Blue 3rd and 7th Fleets—had made Formosa of little use to Orange. The same Blue forces had attempted heavy air attacks on the Pescadores, but Orange ground and fighter protection was so strong there that little damage had been done, at high cost to the Blue air forces.[3]

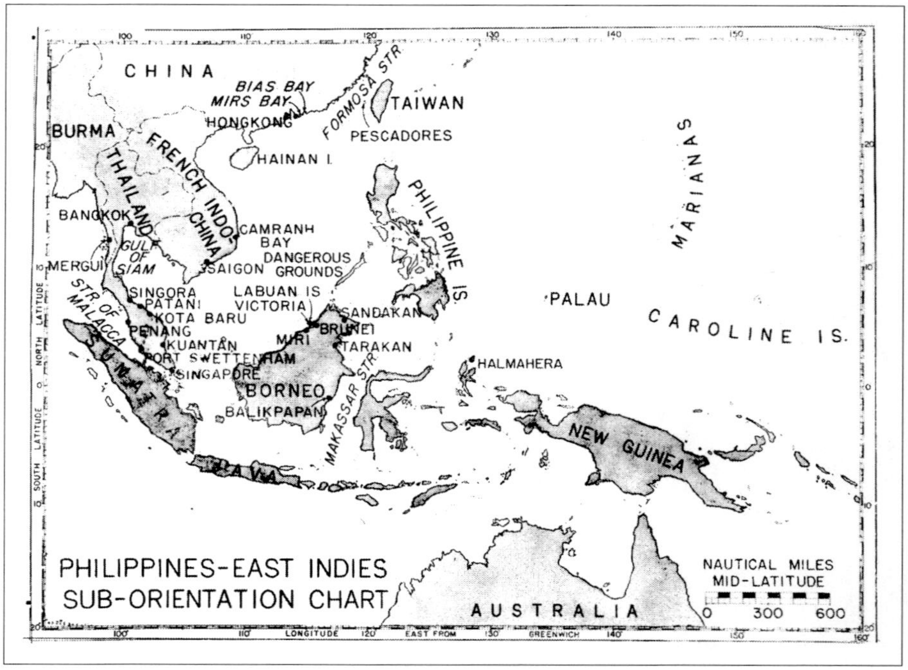

Map 8
Philippines–East Indies Sub-Orientation Chart

Blue, however, now had complete control of the sea area south of 18 degrees north and east of 120 degrees east in the western Pacific. Orange still held the northern part of Luzon, but with poor conditions of supply. Blue army forces on Luzon were larger than what was required to clear Orange from the area, and Blue now controlled the remainder of the Philippines as well. For the purposes of the Exercise, all of this information was considered well known by senior Blue officers in the South China Sea area. Students were also informed that Blue was using Manila as both an army and navy base, as the site of the headquarters of the commander of Blue Task Force 73, and as an administrative subdivision of the 7th Fleet called the "Control Force," under Vice Admiral BE. He in turn commanded ships and aircraft known as "Organization of Task Force 73, Control Force." Admiral BE had forty-eight Liberator naval patrol planes at Fleet Air Base, Manila under his command but no submarines. He also had occasional use of a seaplane tender as a floating headquarters for land-based squadrons in the Calamian Islands, especially around Coron Bay. BE's command was "amply" supplied with fuel, ammunition, and other necessities. It was in this General Situation that Admiral BE received Operation Plan No. 15-45 from Admiral B, Commander, 7th Fleet (COM7FLEET), and from which he had to make his Estimate of the Situation for the tasks assigned to him.[4]

In the General Situation, Commander, Southwest Pacific was planning an invasion of continental Asia, starting with a seizure of Hainan Island so that the Blue 6th Army could be based there for operations on the continent. The Orange Combined Fleet was stationed at bases at Amami Oshima and had refused to be drawn out into an action in the western Pacific. In addition, there was a striking force in the

Pescadores, composed of two *Ise*-class hybrid battleship/carriers, four *Atago*-class heavy cruisers, and eighteen *Takanami*-class destroyers. There was also a division of cruisers, a few destroyers, and some minelayers at Singapore; an unknown number of Orange submarines, based in the Pescadores and operating in the northern areas of the South China Sea; Orange fighters and patrol planes operating from the same island group; and a division of Orange ground troops on Hainan, backed by aircraft at numerous airfields on the island and the continent. The Blue 3rd Fleet was operating east of the Philippines, seeking an opportunity to destroy the Orange Combined Fleet; it also had the mission of covering lines of transportation, communication, and supply between the Blue homeland and Manila. Submarines of the Blue Pacific Fleet were also operating in the South China Sea but would be withdrawn from certain areas on the commencement of the operations as detailed below. It was additionally planned that after the seizure of Hainan, Blue army air forces would protect the shipping between the Philippines and China from attack by Orange land-based air forces. It was assumed that Orange strength in the Pescadores and at Singapore would remain as it was and that the Blue Pacific Fleet could prevent the Orange Combined Fleet from entering the South China Sea, though it was not considered possible to maintain surveillance of Orange bases and forces in the Pescadores for some time.[5]

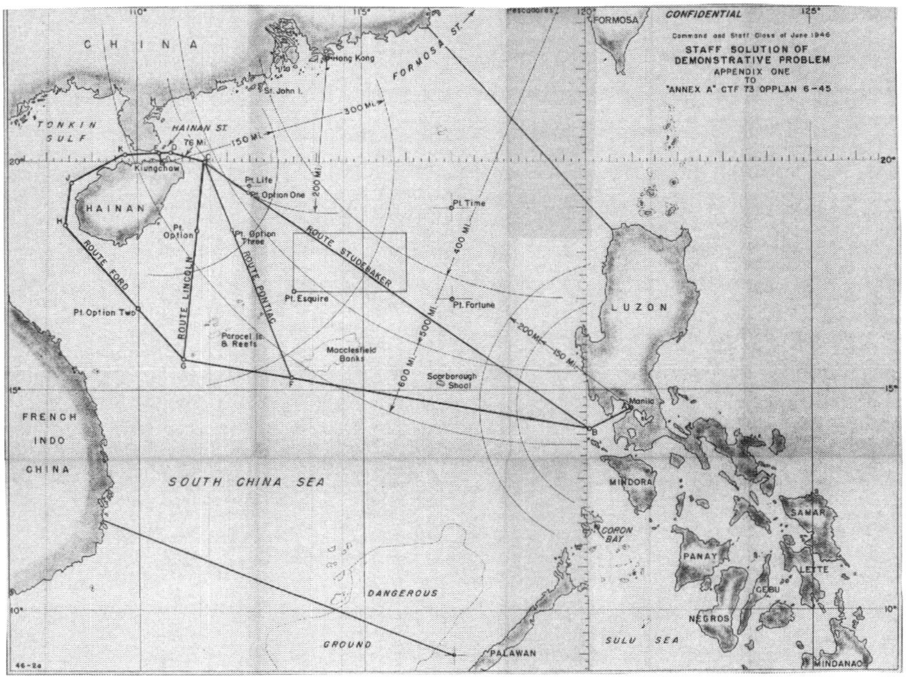

Fig. 136
Demonstrative Problem, Blue Statement, Staff Solution of Demonstrative Problem

In seizing Hainan, the Blue Main Body, under Vice Admiral BA, would cover Vice Admiral BF, commander of the 7th Amphibious Force (COM7PHIBFOR), in the latter's transit from Manila to Hainan; conduct bombardment of Hainan; and furnish fire support to the troops ashore. Rear Admiral BC, commander of Blue's Fast Carrier Task Force, would furnish air cover for both the Main Body and the 7th Amphibious Force; bomb designated targets on Hainan and the continent; and coordinate these operations with Rear Admiral BD, the commander of Fleet Air Wings. This latter would conduct air searches and patrols as well as bomb in coordination with the Fast Carrier Task Force. Vice Admiral BE's Control Force would escort the 7th Amphibious Force to Manila. In addition, the Control Force would cover the units of Captain BL, Commander, Mine Force, whose outfit would conduct minesweeping and demolition off the Hainan beaches. The Control Force would also furnish fire support during the invasion, screen amphibious ships in the assault areas, and protect lines of transportation and supply between the Philippines and China. It was assumed that Hainan would be completely under Blue control by 31 January 1946 and that the facilities at the port of Kiungchow (modern Qiongzhou) would be able to unload twenty ships in sixty hours. It was also thought, however, that 175 ships per month would be needed to sustain the base on Hainan.[6]

Planning for the operation was to be complete by 14 January, and no other 7th Fleet Task Forces would be available for the operation except for that of Captain BJ, commander of the Patrol Force, which was to conduct antisubmarine and counter-mine patrols off the entrance of Manila Bay before and after the departure of the invasion force to Hainan. This patrol command would later become the Hainan Local Defense Group, under Admiral BE. The only other 7th Fleet force would be under Rear Admiral BH, commander of the Service Force, which was to furnish all logistical support. It was also planned that 6th Army would relieve the 7th Amphibious Force once Hainan was taken and secured, at which point the 7th Amphibious Force would be returned to Manila; the timing would be jointly decided by COM7FLEET and the Commanding General, 6th Army. The Submarine Force, under Captain BS, was to patrol between Luzon and Naochow Island (today's Naozhou Dao), as well as between Golo Island, the Philippines, and French Indochina; to report, trail, and attack Orange forces attempting to interfere with the operation; and then to withdraw from the invasion area and maintain observation of the Pescadores when directed by Admiral BE. Rear Admiral BG, commander of the Philippine Sea Frontier, was to protect the harbors, bases, and other naval establishments in the Philippines and, by 1 January, prepare and issue a Sortie Plan for the invasion force.

Students were to create the Estimate of the Situation and the Operational Plan for the Control Force, with annexes detailing movement, search, communications, and logistics. Staff Solutions were to be issued; students had a choice of adopting those or creating their own.[7]

The Blue Estimate

Three days later, a student playing Vice Admiral BE summarized the General Situation and the various tactical dispositions of the Blue and Orange forces before the invasion of Hainan. He outlined Task Force 73's forces as two battle cruisers, four heavy and four light cruisers, eight escort carriers, thirty-six destroyers, twelve destroyer escorts, eight destroyer minesweepers, forty-eight Catalina naval patrol planes, and what would become the Hainan Local Defense Group of minesweepers and patrol craft. BE's specific mission from COM7FLEET was to protect the lines of transportation and supply between Manila and Hoihow Bay once Hainan had been seized and secured. Admiral BE noted that he had been handed a very specific mission and that his superior had not given much latitude in carrying it out. It did not really start until after Hainan was securely under Blue control and its advance base facilities were operational. At that point, BE's mission would be strictly in the context of assisting COM7FLEET to prepare for the invasion of the Asian continent.[8]

The basis of Admiral BE's plan was "properly" escorted convoys moving in "successive" passages between Manila and Hoihow Bay. Orange hope for successful prosecution of the war lay in its ability to maintain its lines of transportation and supply in the South China Sea; Blue's seizure of the Philippines allowed Blue in a short period of time to sever the flow of raw materials from the Dutch East Indies, French Indochina, Malaya, Manchuria, and China, as well as disrupt Orange logistical support to its army in China. Presumed success by Red (again, British, not Soviet) armies in opening logistical routes between India and China would further assist Yellow against Orange. But Orange's possession of Formosa and the Pescadores allowed it to protect to a limited extent its lines of communication in the northern portion of the South China Sea, so Orange's mission would be to deny Blue additional bases in the South China Sea. Given that the Blue seizure of Hainan would even further degrade Orange's control of the South China Sea, Admiral BE assumed that Orange's immediate reaction to the loss of Hainan would be attempted neutralization of the Blue facilities there.[9]

BE then outlined the four "logical" routes between Manila and Hainan Island, with steaming distances ranging from 735 to 1,010 miles. The eastern and western termini of all four routes were within reconnaissance range of the Pescadores, the western termini of all of the routes were within range of Hong Kong, and three of the routes were within range of Camrahn Bay. The western portions of all four routes also passed through waters that could be mined, but these zones could also be covered by Blue army air units from Hainan. Moreover, the eastern approaches could be covered by the Philippine Sea Frontier and TF 73's Patrol Force. It was also assumed that Blue army air units from Hainan would neutralize Orange air units from Hong Kong and continental Asian bases. Admiral BE chose the third route—south of Scharborough (Scarborough) Shoal, Macclesfield Bank, and Parcel Island (the Paracel Islands) and Reefs via Tonkin Gulf and the Hainan Straits—as the best for his convoys, since its western terminus was over six hundred miles from the Pescadores and much of the route was through minable waters.[10]

As regards the power of his forces and those of the enemy, BE considered his personnel well trained, with "high" offensive spirit, and assumed the same of Orange. He thought Orange would resort to suicide tactics if the opportunity arose.

Comparing various weapons platforms, he thought the large-caliber guns on Orange's hybrid battleship/carriers were negated by the larger number of smaller-caliber guns on his battle cruisers. Orange also had superiority in large-caliber guns on its heavy cruisers but was outnumbered in light cruisers and destroyers. In addition, BE's battle cruisers held superiority in speed over Orange's hybrid battleship/carriers and equality of speed in heavy cruisers, light cruisers, and destroyers. BE's escort carriers, however, were slow. Each side also had adequate base facilities by which to support their respective forces. BE noted that atmospheric conditions favored the employment of smoke screens during the contemplated period of operations. He expected contact with Orange forces during the contemplated period, since Orange lines of transportation to Singapore, Malaya, the Dutch East Indies, and French Indochina all traversed his prospective convoy routes. He further expected support from Hainan, in terms not only of logistics—especially the servicing of aircraft—but of Blue army air units flying antisubmarine patrols and searches and fixed defenses at Hainan ensuring the security of the base.[11]

BE outlined other strengths and weaknesses of his and the Orange forces. For instance, his convoys could only move at slow speeds and might find themselves in bottlenecks at certain points of the route. His ship-based air was superior to Orange's. He did not know the strength of Orange land-based air, and he was also concerned about Orange torpedoes, submarines, and mines, as well as the fact that Orange could choose the time and place of attack. Still, he pointed out, Orange would have difficulty in concentrating its forces. Admiral BE also had the advantage that the Blue 3rd Fleet would be containing the Orange Combined Fleet, but BE had to handle any Orange forces from the Pescadores or Singapore. He was "handicapped" by the sizes and speeds of the convoys as well as the slow speed of the escort carriers, but confident that his naval patrol planes would provide effective air search, antisubmarine coverage, and countermining operations. BE felt that in a limited sense he had the strategic initiative, since the commencement of his convoy operations was dependent on his decision and it was the prevention of Blue convoy operations that would cause Orange to begin its countermeasures and play its hand. He nevertheless worried about how prescriptive his mission from COM7FLEET was and how dependent Hainan would be on his convoys getting through, given that base's limited port facilities. He was additionally worried that it was necessary to divide his forces: since Orange air, surface, and submarine forces were his "physical objective," he had to maintain forces in readiness even if Orange remained at its bases. Even Orange forces that had not been deployed were a potential threat to the Blue convoys.[12]

Charged with his mission of preventing the Orange Combined Fleet from entering the South China Sea, Vice Admiral BE delineated four general lines of action: to evade Orange, contain Orange, destroy Orange, or some combination of the three.

His specific targets were the heavy surface striking units in the Pescadores; the Singapore Force of cruisers, destroyers, and destroyer-minesweepers; land-based air units in the Pescadores; land-based air in continental Asia; and the submarine force in the Pescadores and South China Sea. In order to protect the convoys, he could have his forces either patrol the selected convoy route, escort the convoys themselves along the route, or try to cover the selected route. Since there were, however, four potential convoy routes between Manila and Hoihow Bay and any one of them would entail a wide dispersal of forces, there could be only "moderate" antisubmarine protection, "meager" antiaircraft protection, and little or no protection from surface attack, because of the difficulty of concentrating his forces.[13] Nine convoys per month would be needed to supply the base at Hainan, given the number and size of the ships and the speed of the convoys. Three Escort Groups and a Hunter Killer Group could provide antisubmarine and some antiaircraft protection but none from surface attack, especially if surface forces from the Pescadores and Singapore coordinated their attacks. Trying to cover the selected route, on the other hand, would provide good protection from the surface threat but not the submarine or air threats. The addition of the Hunter Killer Group would only partially alleviate the submarine threat. However, by combining the convoy escorts with the three Escort Groups, a Covering Group, and the Hunter Killer Group, he could optimize the employment of his forces. Essentially, there would be twenty-ship convoys with six escorts each, but protection from threats was never more than "adequate" or "moderate."[14]

Admiral BE also surmised that any future expansion westward by Blue would entail an even more likely response by Orange, because of the key lines of transportation and supply in the South China Sea. Among Orange assets, the Pescadores Force was the greatest threat, since it had the majority of Orange surface forces, as well as embarked aircraft. BE thought that this force could, by day or night, destroy any one of his convoys or inflict serious damage even if it did not deploy the hybrid battleship/carriers. Moreover, through the use of submarines, patrol planes, and surface ships, BE thought, the Orange Pescadores Force could evade the Blue Covering Group and fall on the convoys. The Singapore Force was not as strong, but a strike from it at night could be "costly," and it too could employ search assets to evade the Covering Group and hit the convoys. His worst-case scenario was a night attack by both forces in coordination, something he noted would be "very costly," but he did not know whether Orange could carry that out. Orange submarine attacks, especially at night, were highly likely. BE was not, however, greatly concerned about Orange land-based air, because of the strength of Blue army air and his own embarked air assets. Mines, however, were to be expected.[15]

Given all of these potential scenarios, Admiral BE thought that in an encounter between his Covering Group and the Orange Pescadores Force his "probable" losses would be "acceptable" and that meanwhile the convoys could take evasive action. If the Orange Singapore Force was met, the convoys would take evasive action while the Covering Group took things in hand. If, however, the Covering Group was not able to intervene, BE thought, a night attack on the convoys by the Orange Singapore Force would be "disastrous." A combined Orange force might easily destroy the convoys as well as the Covering Group, unless the latter was able to intercept and destroy each force separately. In that scenario, he argued, he could keep his losses acceptable. But all of these situations made him highly dependent on air searches by his naval patrol squadrons. Interestingly, he asserted that with his convoy organization, destroyers, and escort carriers organized into the Hunter Killer Group, losses to submarines could be kept to a low level. Also, because of all of his various army and navy air and mine countermeasure assets, he saw threats from Orange land-based air and mines as low, though they were likely to be laid.[16]

In a later session, Vice Admiral BE illustrated in more detail how he was going to formulate the tasks for his command and apportion fighting strength to them. Given the Orange missions of destroying Blue lines of communication between Manila and Hainan and of preventing further Blue expansion into the South China Sea, BE would deploy his forces along the various convoy routes from Manila once Hainan was secured. He was particularly concerned about Orange surface forces and land-based air assets in the Pescadores and about the surface force at Singapore. He also noted that Orange submarines were operating in the South China Sea from the Pescadores. To counter these forces, he was going to form the three Escort Groups, each of a light cruiser, an escort carrier, and eight destroyers. The Covering Group would consist of the two battle cruisers, four escort carriers, four heavy cruisers, one light cruiser, and twenty destroyers. He did not think that his Escort Groups could handle the Orange surface striking forces, but he was confident that the Covering Group could, because of its speed, embarked air strength, and the guns of the battle cruisers.[17]

If BE was confident that the Covering Group could take care of the Orange surface forces from either the Pescadores or Singapore, he foresaw that it would be in "great difficulties" if it faced both at once. However, they could be taken on if the Covering Group's air strength could damage the enemy surface forces and so allow the Covering Group's surface forces to destroy them "in detail." He additionally thought that the assignment of six or more escorts to each convoy, augmented by escort carrier aviation, would "radically" reduce the Orange submarine threat. His combat air patrol and antiaircraft defenses were "adequate" against Orange land-based air forces. Mines were a threat, but the patrols of the Local Defense Forces in Manila and Hoihow Bay could reduce it by sweeping operations as convoys entered and departed port. In any case, alertness "by lookouts for floating mines while [a] convoy is en route is my defense against that form of attack."[18]

The Control Force was divided among five groups. First was the Covering Group, which was to cover the convoys en route, intercepting and destroying Orange air, submarine, and surface forces. Next came Escort Groups 1, 2, and 3, which were to carry out the actual escort functions, defending the convoys from enemy submarine, air, and minor surface attacks. Finally, there was the Hunter Killer Group, which was to defend the convoys from submarine attack specifically between Manila and Hoihow Bay. These tasks were determined by what Admiral BE thought would be Orange attack priorities. He saw the Pescadores Force as intent on destroying the Blue convoys by air and surface attacks, with priority going to cargo vessels and then covering forces and escorts. The Pescadores Force was to destroy Blue convoys but evade covering forces if they were present. He assessed that the Pescadores Force might attack in coordination with the Singapore Force, if the latter were so ordered, but the primary mission of the Singapore Force seemed to be minelaying at the western terminus of the convoy routes. The Orange Submarine Force would search for the convoys, destroy Blue shipping en route (giving cargo ships priority over combatants), and lay mines along their routes. BE, however, did not know the strength of the Orange Submarine Force. Finally, he expected the Pescadores Air Force to conduct air searches to help locate the convoys, trail and report Blue movements, and attack shipping en route, again with cargo ships as the priority targets.[19]

Given the Blue missions, Vice Admiral BE was going to restrict the Covering Group geographically so as to facilitate its support of the convoys and Escort Groups. This limitation, he argued, would permit the coordination of air searches and provide the best covering positions for the convoys. The convoy movement schedule would require a detailed schedule to be incorporated into his Movement Plan. Accordingly, he designated Task Groups and Units for various assignments, with the Covering Group designated Task Group 73.1, under Rear Admiral BM, and Escort Groups 1, 2, and 3 as Task Groups 73.2, 73.3, and 73.4, respectively, under Rear Admirals BO and BP and Captain BF, respectively. The Hunter Killer Group was designated Task Group 73.5, under Captain BG. He then laid out, in accordance with COM7FLEET's Operation Plan No. 15-45, his Communications Plan. He next noted the importance of Manila to the logistical support of his forces. Given the importance of the Service Force for logistical support to the Covering and Hunter Killer Groups while at sea, procedures and arrangements were to be worked out with Commander, Service Force, prior to the commencement of operations.[20]

In his Intelligence Estimate, after rehearsing strengths, weaknesses, and current general positions and dispositions of the various Blue and Orange forces in the South China Sea region, BE made clear the intelligence needed to support his convoy-escort mission. He needed to know the strength, composition, and movement of the Orange Pescadores Force as well as of the land-based air at that base. He also needed to know the number, location, and "approximate" movements of the Orange Submarine Force, as well as the numbers, types, and movements of the Singapore Force. He thought that any attempt to obtain intelligence on the Pescadores Force by aerial reconnaissance would be too costly but that the Blue Submarine Force could collect it. Air searches and barrier patrols might be needed if the Orange Pescadores Force evaded Blue submarines, but these would have to be carried out by Fleet Air Wings, which had already established regular patrols of the area. He could periodically and for limited times augment these patrols with aircraft and radar picket lines of the Covering Group as well as the forty-eight Catalina naval patrol bombers of Fleet Air Wing 15 based at Manila and Coron Bay. These assets would give him adequate coverage of this theater of operations. BE held that Orange would not be able to interfere with southern barrier patrols and thought that naval patrol planes' own armament would provide sufficient defense against Orange fighters intercepting the northern patrols. Additionally, he did not think that Orange would be able to prevent all of the Blue submarine observation missions. Therefore, the barrier patrols, the daily searches by the Covering and Escort Groups, and the Submarine Force's missions should provide all of the intelligence he needed.[21]

Admiral BE had to take precautions against Orange counterintelligence, including air searches and radio interference. Against the first threat, he would deploy the combat air patrols of the Covering and Escort Groups, as well as radar picket combat air patrols. Further, the various groups would maintain radio silence at sea. Moreover, each Group Commander was instructed to practice the highest condition of "radar silence" possible. Still, he did not think that operating these convoys secretly for an extended period of time was possible, because of the continuing need to sweep mined waters and the constant sailing of convoys back and forth. Therefore, he was counting on "maximum efficiency" from the Fleet Air Wings and Blue army air to deny Orange forces any strike opportunities, and on the same level of effectiveness from the Philippine Sea Frontier and the Hainan Local Defense Forces in their countermining operations. Task Group 73.6, the Northern Patrol Group, would consist of thirty naval patrol planes from Fleet Air Wing 15, while the Southern Patrol Group would have another eighteen naval patrol planes. The Hainan Local Defense Group would consist of Mine Divisions 5 and 19, assigned by the Patrol Force.[22]

In a Synthesis of the Operation Plan, Admiral BE restated the strategic and operational situations, the status of both enemy and friendly forces, and the key objective of turning Hainan into a base for the invasion of continental Asia. He went into the details on the disposition of his forces, the route the convoys would take, the alternate routes, and the roles of the Patrol Force and Hainan Local Defense Force. Units and unit missions were detailed down through the Task Group, with additional plans for Movement, Intelligence, Logistics, and Communications. There were also emergency provisions for towing ships if necessary. D-Day was set for 1 February 1946.[23]

Naval Communications Exercise: The Truk Operation

Just a few days later, a Naval Communications Exercise set in September 1944 at Rabaul was held, involving elements of Blue 3rd Fleet and an Orange force. The Blue elements were led by Vice Admiral BF, Commander of TF 31—the Raiding Force—as well as of Air Forces, 3rd Fleet. BF's forces included the Striking Group, Task Group 31.1, consisting of four battleships organized as Battleship Division 5; two fleet carriers organized into Carrier Division 2; four heavy cruisers as Cruiser Division 6; eight light cruisers as Cruiser Divisions 9 and 11; twenty-six destroyers organized as Destroyer Squadrons 5, 12, and 14; and nine destroyer leaders as Destroyer Squadron 21. In addition, there was the Air Support Group, TG 31.2, under Captain B-15; it consisted of thirty-six scout planes and forty-nine naval patrol planes organized into Fleet Air Wings 1 and 10. Finally, Admiral BF's forces contained the army's 3rd Bombardment Group of twenty-four dive-bombers and TG 31.3, the Submarine Group, twelve submarines organized into Submarine Squadron 6, under Captain B-14.[24]

In the General Situation, Blue was extending its line of control in the Pacific westward, through the line Kiska–Marcus–Truk–Rabaul. Blue's 3rd Fleet was to capture Truk and use it as an operational base. Orange forces included several battleships, fifteen to twenty cruisers, and about thirty-five destroyers at Truk. Orange cruisers and destroyers could lay mines, as could their embarked aircraft. Twenty to twenty-five Orange submarines were also known to be operating around Truk, where there were also an "indeterminate" number of seaplanes, land planes, auxiliaries, and defense craft. Orange shore defenses on Truk consisted of about fourteen six-inch guns with ranges of twenty-five thousand yards, twenty-four mobile three-inch antiaircraft guns, and about a hundred mobile antiaircraft machine guns. Orange aircraft strength was determined as eighteen Catalina naval patrol bombers, twenty Hellcat fighters, ten Avenger torpedo planes, and twelve Helldiver dive-bombers. Ponape had five Mariner naval patrol bombers but no fighters or bombers, while Eniwetok had thirteen Flying Fortress land-based bombers and four Mariners. There were an estimated further thirty Flying Fortresses and eight Mariners elsewhere in the Marshalls and another fourteen naval patrol planes in the Marianas.[25]

Operational Plan No. 3-44 called for Task Force 52 to cover another force, TF 51, as it captured Wotje and Jaluit. Task Groups 52.1 and 52.2 were additionally to raid Marcus Island and inflict "maximum destruction" on air and shore installations in the Marshalls while the Blue Amphibious Force seized Truk itself. For the purposes of the Exercise, a number of assumptions were given—for instance, that all available

Orange combatant units would defend Truk and that their strength was as reported above. It was also assumed that a major day action would take place, in excellent weather and with Orange submarines present. Orange attrition attacks would take place at night by air, surface, and submarine forces, and Blue surface forces would be discovered by Orange. The initial step, therefore, was for the Striking Group to destroy or drive off Orange sea forces and reduce Orange air strength in the Truk area. The Striking Group was to be assisted in these operations by the Air Support Group, specifically by coordinating aerial torpedo attacks by Blue carrier forces on Orange heavy ships, and with beacon assistance from the Submarine Group. There was also to be reconnaissance of Truk by land-based planes from Kavieng and seaplanes stationed at various points. The Submarine Group was to provide reconnaissance, trail Orange forces, and, as much as possible, attrite Orange surface forces as they sortied from Truk. The Submarine Group was also to attack damaged Orange units returning to Truk, taking care to remain clear of Blue surface forces operating around Truk. Orange battleships were the primary targets.[26]

In another Exercise, the same force, again commanded by Vice Admiral BF, was organized into Rear Admiral B-1's battle line, consisting of four battleships of Battleship Division 5; Rear Admiral B-3's Center Force, four heavy cruisers and nine destroyer leaders of Cruiser Division 6 and Destroyer Squadron 21, respectively; Captain B-20's Right Flank Force of two light cruisers from Cruiser Division 9 and nine destroyers from Destroyer Squadron 12; and Rear Admiral B-2's Carrier Group, two fleet carriers of Carrier Division 2, four antiaircraft cruisers of Cruiser Division 11, and eight destroyers from Destroyer Squadron 14. The General Situation was the same as that of the previous exercise. This time, a large Orange surface force consisting of five battleships, seven or eight heavy cruisers, eight to ten light cruisers, and thirty to thirty-five destroyers was reported to be in Truk Harbor. In addition, about twenty to twenty-five Orange submarines were operating around Truk, which also had eighteen naval patrol planes, twenty fighters, ten torpedo planes, and twelve dive-bombers in aerial strength. Thirty-one additional naval patrol planes and forty-three land-based bombers were in the Marshall, Mariana, and Caroline Islands.[27]

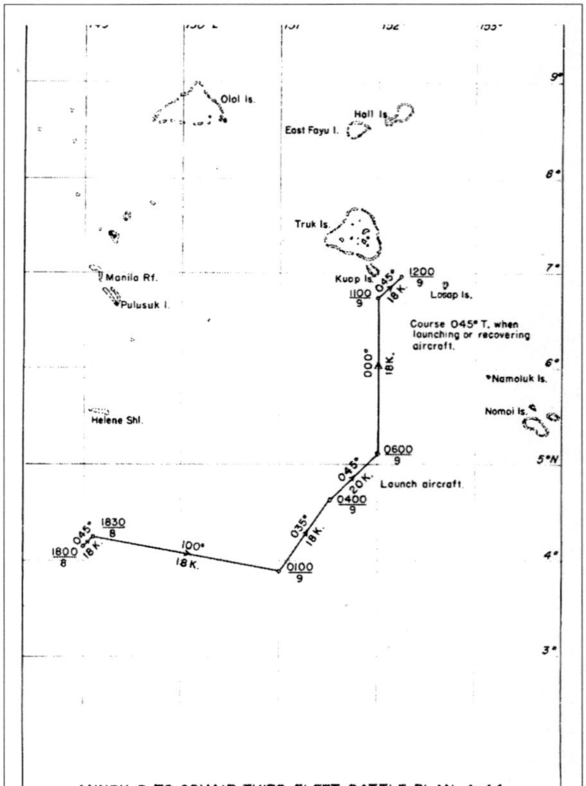

ANNEX B TO COMAIR THIRD FLEET BATTLE PLAN 4-44
Contemplated route for T.F.-31 during night 8-9 Sept.

Map 9
Annex B to COMAIR 3rd Fleet Battle Plan 4-44

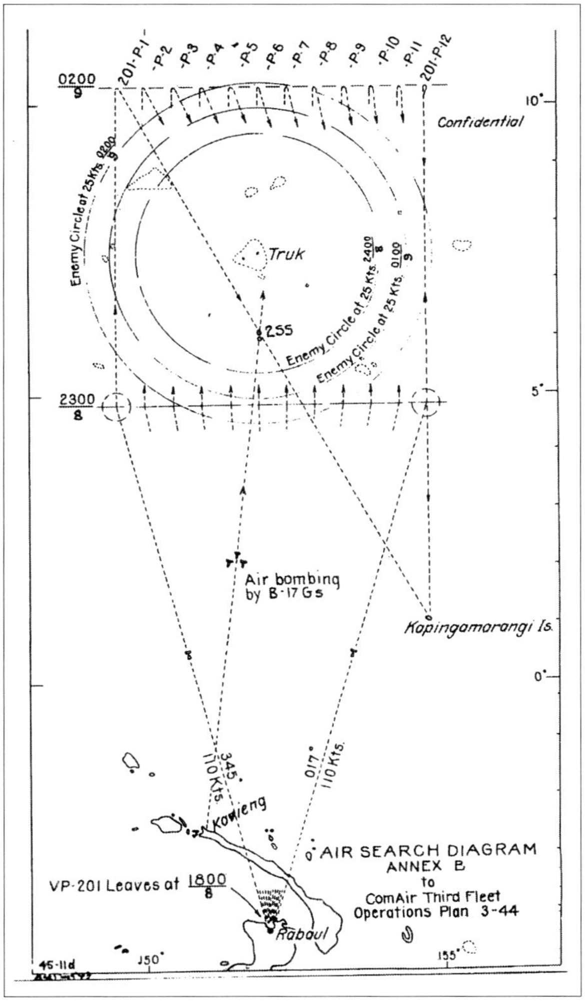

Fig. 137
Air Search Diagram

Blue submarines were going to operate in the area, while Blue land-based aircraft would bomb Truk and scout for Orange surface forces prior to the "main engagement." It was now assumed that Orange surface forces would sortie at night, that these forces would seek engagement, and that Orange forces would be in the strength reported. It was also assumed that all Orange units would engage in the defense of Truk and that the encounter itself would take place during the day. Blue submarines were expected to make night attrition attacks on Orange surface forces, as would Blue land-based aircraft. It was further assumed that Orange air, surface, and submarine forces would attack at night, that these attacks would include submarine strikes, and that Orange light forces and submarines would mine the waters ahead of the Blue surface force. Blue air and surface forces were again to combine so as to destroy or drive off Orange sea and air forces through repeated air strikes and daylight surface engagements. There was more attention here to battle-line action, specifically to engage the enemy battle line first at "extreme" ranges, then close as rapidly as possible, and continue firing until the enemy battleships were reduced. The Blue battle line was to furnish its own air spotting and inner antisubmarine patrol; the Left Flank Force had particular orders to guard the battle line against enemy light force attacks while its cruiser planes provided air spotting and battle-line intermediate antisubmarine patrol. Destroyers of the Left Flank Force would also make torpedo attacks against the enemy battle line and would be supported by the cruisers when directed by the OTC.[28]

The Carrier Group was to make dawn air strikes against Truk's planes and air installations with 25 percent of its fighters and 50 percent of its dive-bombers. Remaining fighters and dive-bombers would be used for combat air and antisubmarine patrols, respectively. The Carrier Group was also to make repeated strikes against Orange air and surface forces throughout the day, furnish combat air patrols for itself as well for as the battle line, and furnish its own antisubmarine patrol. When ordered, it was additionally to operate on the disengaged side of the battle line, outside Orange's main-battery gun range. All planes were ordered to report sightings of submarines and drifting mines, and all aircraft armed with depth charges, as well as destroyers, were to attack submarines "vigorously." The plan ended by detailing the Communication Plan; the Cruising, Approach, and Battle Dispositions; and the assignment of Rear Admiral B-1, Commander, Battleline, as the second in command of the force.[29]

Communications: Message Drafting Exercise

In early February 1946, the Command and Staff Class of June 1946 undertook another Naval Communications Exercise. This one included having the students draft messages about various scenarios, with attention to classification, precedence, method of transmission, addressees, and authorized abbreviations. Most of the scenarios themselves were quite obviously dictated by recent Pacific War experiences, in this case the operations of TF 58, but with a different geographic focus. Here, carrier air strikes were to take place against Japanese naval forces in the North Pacific. The students were to draft messages reporting a sighting of "many enemy planes" ninety miles from TF 58, as well amplifying contact reports about an enemy surface force of three carriers, four heavy cruisers, two light cruisers, and eighteen destroyers three hundred miles from the task force. Next, students were to draft messages indicating the movement of TF 33—consisting of the fleet carriers *Enterprise* and *Saratoga;* the fast battleships *Washington* and *Alabama;* the heavy cruisers *Chicago, St. Paul,* and *Pittsburgh;* the light cruiser *San Diego;* and Destroyer Squadrons 24 and 25—from Pearl Harbor to Adak, in the Aleutians. The final Naval Communications Exercise concerning the Pacific was an order to Task Group 58.4, under the command of Vice Admiral BZ, Commander, Carrier Division 3, to sortie from Adak with the fleet carriers *Yorktown* and *Lexington,* the light carriers *Belleau Wood* and *Cowpens,* the heavy cruiser *Vincennes,* the light cruisers *Biloxi* and *Duluth,* and Destroyer Squadrons 21 and 25. TG 58.4 was to proceed at twenty-eight knots to within air-strike distance of Kashiwabara Bay and destroy an enemy force consisting of two fleet carriers, three heavy cruisers, and twelve destroyers anchored there. Task Group 58.4 was then to retire back to Adak at the conclusion of the operation.[30]

Conclusion

Like many of the Operations Problems, these Exercises gave attention to naval forces comprising air, surface, and submarine elements. As had been consistently the case in the Pacific by 1943–45, Orange forces focused on surface units, while Blue employed combined-arms task forces. It certainly made sense to war-game such strategic scenarios as convoy protection. It could also be argued that there was valuable instruction in having the students practice actions in which friendly surface forces might have to engage enemy surface forces at night, in daytime absent friendly carrier forces, and in northern latitudes. Given the concerns about Japanese surface forces by Admirals Halsey and Spruance even late in the war, it is clear why the organization and employment of a battle line was so greatly emphasized. Like the U.S. Pacific Fleet late in the war, the early postwar Naval War College curriculum saw carriers as the new focal point of American naval power but not as the sole source of that power in all scenarios.[31]

NOTES

1 George Bowdey, "General Instructions Applicable to the Preparation of All Solutions," 18 January 1946, p. 1, NHC, folder 2566-G, box 131, RG 4, NWC.

2 Ibid.

3 Command and Staff Class of June 1946, "Demonstrative Problem: Blue Statement," 18 January 1946, p. 1, NHC, folder 2566, box 131, RG 4, NWC.

4 Ibid., pp. 1–2.

5 Ibid., pp. 3–4. This scenario obviously never happened, but protecting convoys and ensuring the logistical support of amphibious assault operations would become a primary concern of Admiral Spruance's future study when he became President of the Naval War College in March 1946; see Friedman, *Digesting History: The U.S. Naval War College, the Lessons of World War Two, and Future Naval Warfare, 1945–1947* (Newport, R.I.: NWCP, 2010), pp. 143–49, 156–61, 183–96, 199–207.

6 Command and Staff Class of June 1946, "Demonstrative Problem: Blue Statement," p. 4.

7 Ibid., pp. 5–6, 10.

8 Command and Staff Class of June 1946, "Demonstrative Problem: Staff Solution," 21 January 1946, pp. 1–5, NHC, folder 2566-A, box 131, RG 4, NWC.

9 Ibid., pp. 6–7.

10 Ibid., pp. 7–9.

11 Ibid., pp. 12–14.

12 Ibid., pp. 15–17.

13 Ibid., pp. 17–20.

14 Command and Staff Class of June 1946, "Demonstrative Problem: Staff Solution," 22 January 1946, pp. 20–22, NHC, folder 2566-B, box 131, RG 4, NWC.

15 Ibid., pp. 23–26.

16 Ibid., pp. 27–29.

17 Ibid., p. 30.

18 Command and Staff Class of June 1946, "Demonstrative Problem: Staff Solution," 23 January 1946, p. 31, NHC, folder 2566-C, box 131, RG 4, NWC.

19 Ibid., pp. 32–33.

20 Ibid., pp. 33–36.

21 Command and Staff Class of June 1946, "Demonstrative Problem: Staff Solution," 24 January 1946, pp. 37–39, NHC, folder 2566-D, box 131, RG 4, NWC.

22 Ibid., pp. 39–40.

23 Command and Staff Class of June 1946, "Demonstrative Problem: Staff Solution," 25 January 1946, pp. 41–47, NHC, folder 2566-E, box 131, RG 4, NWC. See also Command and Staff Class of June 1946, "Demonstrative Problem: Staff Solution," 26 January 1946, pp. 48–63, NHC, folder 2566-F, box 131, RG 4, NWC, for the final form of the Execution Phase of this Operational Plan.

24 Command and Staff Class of June 1946, "Naval Communications," 2 February 1946, p. 1, NHC, folder 2569-G, box 131, RG 4, NWC.

25 Ibid.

26 Ibid., pp. 2–3.

27 Ibid., p. 4.

28 Ibid., pp. 4–5.

29 Ibid., pp. 5–6.

30 Command and Staff Class of June 1946, "Naval Communications Message Drafting Exercise," 2 February 1946, p. 1, NHC, folder 2569-H, box 131, RG 4, NWC.

31 For the preponderance of American carrier airpower by 1943–44, see Clark Reynolds, *The Fast Carriers: The Forging of an Air Navy* (Annapolis, Md.: Naval Institute Press, 1968), pp. 51–155. For the concerns about scenarios in which carrier airpower could not handle all aspects of remaining Japanese naval forces, even in 1944, see Trent Hone, "U.S. Navy Surface Battle Doctrine and Victory in the Pacific," *NWCR* 62, no. 1 (Winter 2009), pp. 67–105; and Hone, e-mails to author, 28 February, 2 March, 3 March, and 14 March 2011.

NORTH

XIII *Operations Problem 1 and Board Maneuvers*

February 1946

Later in February, another Exercise, Operations Problem 1, began. The General Situation for the Orange Statement established that the war between the United Nations and Black (Germany) had culminated in the unconditional surrender of Black and that United Nations forces had succeeded in driving Orange forces almost entirely from Burma. Contingents of the Red (British) Fleet were now operating in the Pacific, the Bay of Bengal, and the Strait of Malacca with Blue naval forces. Orange and Purple (the USSR) were not at war, but Purple had repudiated its treaty of friendship and nonaggression with Orange and the treaty would expire in April 1946.

Operations Problem 1: The Orange Statement

Orange had ordered its commanders to avoid any "embarrassing" incidents with Purple, but this did not apply if Purple acted in a way "inimical" to Orange security. Blue naval strength in the Central Pacific had greatly increased, especially in aircraft carriers and other combatant ships, and Blue submarines were active in all waters adjacent to the Kurile Islands and the Orange homeland. Since Orange had been forced from the Marshall Islands, Blue naval forces had increased the frequency of their air and surface force raids on Formosa and the Orange homeland. Blue naval forces had also occupied Saipan, Guam, and Tinian in the Marianas; had taken Iwo Jima in the Volcanoes; and were in the process of capturing Okinawa in the Ryukyus. Blue had additionally occupied most of the Philippines and was about to capture the remainder of Luzon and Mindanao. Further, Orange agents reported significant activity in Blue's northwestern Pacific ports, as well as the assembling and training of an expeditionary force of about forty thousand troops in southwestern Alaska and on the Aleutians, specifically at Dutch Harbor. All of the Aleutian Islands were held by Blue, and the number of troops there was in excess of what was needed for the defense of the region. Moreover, there were indications from such ports as Vladivostok that Purple ground facilities were being prepared for planes and that these ports had seen a large number of Purple ground forces arrive, as well as Blue air and supply officers. This information had been widely distributed among the Orange Fleet's officer personnel.[1]

Most of the Orange Combined Fleet, under Admiral OA, was operating in the western Pacific, north of 20 degrees north latitude. A new force, the North East Area Fleet, had been constituted under Vice Admiral OB, Commander of the Orange 5th Fleet. The 1st Mobile Fleet, under Vice Admiral OZ, was responsible for the area directly south of 40 degrees north. OB's mission was to maintain control of the sea areas west of longitude 165 degrees east and north of latitude 40 degrees north "in order to assist in the protection of the ORANGE Homeland and the KURIL ISLANDS." The mission had been approved by both Admiral OA and the Orange Naval General Staff, though to date the mission had entailed merely the patrolling of adjacent waters against submarines, sporadic surface raiders, and intermittent land-based air attacks from the Aleutians. However, new orders had been received from the Naval General Staff to Vice Admiral OB, citing "fairly reliable" reports that a large Blue expeditionary force was being organized in the Puget Sound area, consisting of about twenty attack transports (APAs) and twenty AKAs, to be escorted by two prewar modernized *New Mexico*–class battleships, four *Oakland*-class antiaircraft cruisers, one *Independence*-class light carrier, eighteen destroyers, and eight destroyer escorts. Fuel oil tankers or other service ships were also believed present, and it was assessed that the force's destination was the Aleutians. There would be an attempt to find out when this force was to leave, but meanwhile it was known that the Blue Commander-in-Chief was also organizing a new task force that would include one battleship, three heavy and three light cruisers, two fleet carriers, and eighteen destroyers, all of the latest type. The new task force had not yet been organized and the fleet units were all still operating with their current formations. Orange thought that it could concentrate very quickly, take on fuel and other provisions at Pearl Harbor, and be ready for distant operations in a short time.[2]

This Maneuver's Operation Plan, designated No. 45-45, assessed that the Blue Expeditionary Force was meant to establish an advanced base in the western Pacific and ordered Vice Admiral OB's North East Area Fleet and Vice Admiral OZ's 1st Mobile Fleet to prevent that and maintain control over their assigned sea areas.[3] Not surprisingly, Orange forces looked much like Japanese forces had by 1944. The North East Area Fleet, designated Task Force 51, with OB in command, comprised two battleships organized into Battleship Division 3, under Rear Admiral OC; three heavy cruisers organized into Cruiser Division 5, under Rear Admiral OD; four light cruisers organized into Cruiser Division 10, under Rear Admiral OE; eleven destroyers organized into Destroyer Squadron 2, under Rear Admiral OF; and twelve destroyers organized into Destroyer Squadron 10, under Rear Admiral OG. There was a Service Force of two fuel oil tankers, and Vice Admiral OB also had under his operational command Rear Admiral OR's 12th Air Fleet, consisting of one light carrier with twenty-seven fighter planes and nine torpedo planes, as well as another ninety-six Zeke fighter planes and seventy-eight land-based Betty medium bombers. Rear Admiral OS's Patrol Wing 5 also had eighty-four Emily seaplanes, which were at OB's disposal. All sea- and land-based planes had the latest search radar and radio direction finders, and two of their bases had high-powered radio stations. All naval patrol planes and naval patrol bombers also had search radar, and OB was told by the College staff to assume that Blue planes were similarly equipped. He was additionally told that lost planes could be replaced from the Orange homeland within twenty-four hours but that the total number of serviceable planes could not be increased. He did have, however, a 50 percent excess in pilots and plane crews at all bases, something not consistent with conditions in Japanese forces after 1942.[4]

Additional forces under Admiral OB's operational control included a submarine detachment from the 6th Fleet—Submarine Squadron 1, under Rear Admiral OJ, consisting of twelve boats operating in the North Pacific–Aleutians area. Also organized was a command known as the Ominato Guard District Force, which was to cover the sea frontiers north of 40 degrees north and west of 165 degrees east. The Commandant of the District Forces, Vice Admiral OH, commanded all naval and air forces based within the district except those under the operational control of OB. Vice Admiral OH was directly subordinate to the Chief of the Naval General Staff and was to provide logistical support to the North East Area Fleet. All bases in the area were sufficiently stocked at the moment with all types of fuel and consumable supplies that would be needed by the afloat forces. All bases were also well defended with coastal and antiaircraft batteries, patrol and fighter aircraft, local naval defense craft, and "comprehensive" communication facilities. All of these fleet units—less the submarines on patrol—were either at their bases for upkeep or operating in adjacent waters. Ships at sea—again minus the submarines and local defense forces—were to be prepared to return to their bases, fuel, and make themselves ready for "distant service" on forty-eight hours' notice. Ships and submarines actually at their bases, minus the Local Defense Forces, were also to be ready for sea within forty-eight hours; OB had taken this forty-eight-hour time lag into account in his plans. For the Exercise, students were to complete Vice Admiral OB's Estimate of the Situation, Directive, and—if necessary, as annexes to the Operation Plan—a Movement Plan, Intelligence Plan, Search Plan (including air and surface phases), Communications Plan, and Logistics Plan. Also possibly needed were an Assault Plan, Battle Plan, Sortie Plan, Cruising Dispositions, Approach Plan, Retirement Plan, and Mining Plan.[5]

Operations Problem 1: Duties and Instructions

Prior to the Statement of the Orange Staff Solution for Operations Problem 1, duties and instructions for various personnel were enunciated in detail that affords insight into how these maneuvers were conducted. For instance, before the start of the Board Maneuver, Student Commanders were to study the Maneuver Rules, with particular emphasis on specific points in Sections A, B, C, D, E, F, G, and I. During the Maneuver itself, they were to make out two copies of the Move and Gunfire Sheet before gunfire started; the students were to keep the carbon copy and hand the original over to the Move Umpire. When gunfire was indicated, the students were to make four copies of the sheet, keeping the third copy for themselves and sending the original and the other two copies to the Move Umpire. The Move Umpire was also to provide blank forms stapled together in sets of two and four. Students were instructed to use hard pencils with "sharp points," so as to make the third copy legible, and to use separate forms to record secondary-battery gunfire. The Student Commanders were instructed not to delay "unnecessarily" in filling out and turning in the Move and Gunfire Sheets, since doing so held up the Maneuver. If they needed help, they were to ask Staff members to assist. Student Commanders were also to use care in filling in the coordinates for Columns 13 and 15; the Move Umpire was to check these and send forms back for correction if necessary. Target courses were to be filled in by the umpires, and coordinates were to be entered to the nearest reference *(X* and *Y)* lines.[6]

Student Commanders were told that when they received notice of damage from the Damage Computer (a Maneuver Staff member), they were to note the restrictions on their speed as well as note having received gun or torpedo fire. If the student had already handed in a Move and Gunfire Sheet, it was not to be corrected. Instead, corrections were to be included on the sheet for the next move. Each correction, however, was to be recorded on the board. Torpedo-firing data, in particular, were to be obtained from the plots of enemy vessels. The Student Commanders were told that they could obtain "fairly" accurate "set-ups" by using their Range and Target Angle Indicators but that they were to stay clear of the opposing side's ships and of the board when actually moving their ships. They were further instructed to be careful when walking around the board not to step on chalk lines, since those would be needed for the Critique! Finally, when commencing fire on an enemy ship, the Student Commander was to place an arrow alongside the target in time to give the opposing player the warning one would ordinarily get from a shell splashing nearby. An arrow five inches long was to be used to indicate main-battery fire, an arrow two inches long was to indicate secondary-battery fire, and the aggregated fire of a destroyer division was to be indicated by one arrow. When the Student Commander ceased fire, he was to draw another arrow alongside the target, this one with a line through the shaft. All of these lines were to be drawn in white chalk only.[7]

Prior to the start of the Maneuver, the Move Umpire was also to study the Maneuver Rules, particularly those in Sections A and F that referred to his role. He was also to draw sample copies of all forms used during the Board Maneuver and become familiar with the methods of filling them out properly. Further, he was to study the Staff Solutions of Board Maneuver Exercises 1 and 2 for details in making out the Move and Gunfire Sheets and the Torpedo Fire Forms. He was then to post lists of Student Commanders and Umpires for ready reference, as well as learn all those names, since he would be dealing with those persons directly during the Maneuver. The Move Umpire was additionally to obtain Force Damage Summary forms from a rack near the Maneuver Room and fill in the data in its two left-hand columns. These forms were for the Move Umpire to record damage to ships and divisions. Data for these columns could be found in the Statement of the Problem or, if not, in *Fleets.* Move Umpires were to be particularly interested in maximum speeds allowed for formations and for various forms of damage.[8]

Still prior to the Maneuver, the umpire was to obtain a Check Off Sheet and fill out the two left-hand columns from details the Student Commanders provided for the Maneuver. The umpire was to list controlling ships from the Maneuver and then assign a number, starting with 1, to each Student Commander who controlled ships. The Student Commander was to be notified that he was to write this number in the upper right-hand corner of his Move and Gunfire Sheet, along with his name and unit designation. The number itself was to be used for checking purposes during the Maneuver. The Move Umpire was also to supervise the Marine Corps orderlies in stapling Move and Gunfire Sheets in sets of two and four for the Student Commanders. Sets of four were to be used when the gunfire started; about a hundred of them would be needed, placed on tables in properly labeled piles.[9]

When the Maneuver started, students were to deliver the Move and Gunfire Sheets and all other forms they were required to make out to the Move Umpire. He was to check the forms carefully to see that all of the data were correctly filled out and that students were complying with the Maneuver Rules effective for the conditions of the Problem. He was especially to check that speeds were correct for conditions of formation, the sea, and damage. He was also to check that computations had been correctly made and, if gunfire was indicated, to see whether targets were clearly designated. The Move Umpire was further to check that the data on the numbers and calibers of guns being employed were properly filled out, as well as the coordinates of the Student Commander's own ships and the target ships. In general, the Move Umpire was to ensure that Student Commanders' actions were accurately recorded and their intentions were made clear to the Damage Computing Group. Forms that were incorrectly made out or indicated that a Student Commander was not complying with the Maneuver Rules were to be returned for correction; in that case, the student was advised to correct his movements on the board. In case of a disagreement between the Student Commander and the umpire, the Director of the Maneuver was to be notified. As soon as the forms were checked and found to be correct, they were to be checked off. At that point, first copies (that is, the first, and thus clearest, carbon copies) were to be turned over to the Reduction Recorder, while the originals went to the Range and Target Angle Computer, in numerical order.[10]

The duties of the Fire Effect Computer before the start of the Maneuver were to thoroughly study Section F of the Maneuver Rules, especially the paragraphs dealing with the computation of Gunfire Damage. He was also to focus on copies of the Staff Solution for Board Maneuver Exercise 2. In addition, the Fire Effect Computer was to obtain specially prepared copies of Fire Effect Tables for 1946 from the Naval War College Archives. These applicable tables were tabbed for the problem so as to speed up the process. The Fire Effect Computer was then to fill out Ship's Data Forms for ships and divisions of the forces, with data found in the Statement of the Problem or in *Fleets*. Following this, this officer was to fill out the left-hand columns of the Force Damage Summary, listing ships in order of tonnage. Data for this form were to be obtained from the Move Umpire's copy of the Ship Damage Table during the Maneuver. Finally, before the Maneuver, the Fire Effect Computer was to prepare Ammunition Expenditure Forms for the ships and divisions, obtain sample Ship Damage Tables, study these, and note how damage impacted the types of Spots and Control that were available.[11]

During the Maneuver and when the gunfire started, the Fire Effect Computer was to obtain copies of the Move and Gunfire Sheets with the Flyleaf attached from the Reduction Recorder. The Flyleaf was to be removed and stapled to the appropriate original and then received by the Range and Target Angle Computer. Column 23 of this form was to be filled out, and the Range and Target Angle was to be used in Column 16. The Fire Effect Computer was to use the Plane or Radar Spot unless damage indicated that the Top Spot had to be used or the range was beyond the Radar Spotting Range in the case of destroyers. This officer was then to complete computation for Column 24, the Fire Effect Inflicted, by multiplying Columns 20, 21, and 23. If Enfilade Fire was indicated, the Fire Effect Computer was to compute Column 22 using the appropriate multiplier for the percentage of Fire Effect obtained on the enfiladed ship. The computation was then to be completed for the Fire Effect of the firing ship on the enfiladed ships by multiplying Columns 22 and 23. The original of the form was then to be passed to the Damage Computer without delay. The Fire Effect Computer was to use originals of the Move and Gunfire Sheet after the Damage Computer had finished recording damage and ammunition expenditure for one gun per ship in Column 25. He was then to complete filling in the columns on the Ammunition Expenditure Form. Next, the Fire Effect Computer notified the Chief Damage Computer when the expenditure was 75 percent and 100 percent allowance per gun, the allowances to be found in *Fleets*. The Fire Effect Computer sent the original Move and Gunfire Sheet to the Historian upon completion. First copies were to be retained for checking purposes. The

Fire Effect Computer was to remember to initially compute Percentage of Damage, then record the ammunition expended. He was also to assist the Reduction Recorder when he had completed his other tasks.[12]

Before the start of the Maneuver, the Damage Recorder was also to study Section F of the Maneuver Rules and the other sections of the Rules that were applicable to the computation and effect of damage. In particular, he was to study copies of the Staff Solution for Board Maneuver Exercises 2 for examples. He was also to study the Ship Damage Table, noting the losses due to damage and filling out the left-hand columns of the Force Damage Summary using data from the Statement of the Problem. He was to then list ships and damage by tonnage; obtain green, blue, and red pencils from the Drafting Room; and have lists of Student Commanders and Umpires posted for "handy" reference. Additionally, the Damage Recorder was to prepare a supply of Ship Damage Tables by stapling them together in pairs with carbons between, about fifty pairs per Maneuver. Finally, the Blue Damage Recorder was to obtain data from the Orange Fire Effect Computer, and vice versa for the Orange Damage Recorder.[13]

During the Maneuver, the Damage Recorder was to receive the original of the Move and Gunfire Sheet from the Fire Effect Computer and record the damage to ships for each Move, adding in previous damage if any, and then pass the original back to the Fire Effect Computer, who would need the sheet for Ammunition Expenditure. The Damage Recorder was then to mark off the amount of damage on the appropriate line and label it with the Move number. When all of the damage had been recorded and the Ship Damage Table was delivered to the Move Umpire and Student Commander concerned, the Fire Effect Computer was to fill in the amount of damage sustained with colored pencils, green for Above Water Damage, blue for Under Water Damage, and Red for Total Damage. Damage of 5 percent through 15 percent was to be recorded in the Ship Damage Table as 10 percent, while damage from 15 percent to 25 percent was to be recorded as 20 percent. However, the record of damage was to be kept to 1 percent on the Force Damage Summary. In addition to the Student Commanders, the Director, the Move Umpire, the Reduction Recorders, the Fire Effect Computers, and the Communication Umpire were all to monitor the amount of damage. The Director was to specifically be notified when damage to a ship or division reached 80 percent. In addition, the Move Umpire received a copy of the Ship Damage Table, which he passed to the Reduction Recorder and the Fire Effect Computer. These two officers, after noting the Total Damage on their Force Damage Summaries, passed the table on to the Historian. The Communication Umpire was to be notified when Above Water Damage to ships and divisions totaled 30 percent, 50 percent, and 70 percent and when Total Damage reached 80 percent. When the Maneuver ended, the Damage Recorder was to turn completed Force Damage Summaries over to Drafting Room personnel, who were then to prepare enlarged Summaries for the Critique. During the Critique, the Damage Recorder was to be prepared to answer questions relating to the damage received, using original Ship Damage Summaries for this purpose.[14]

Prior to the Maneuver, the Reduction Recorder was also to study thoroughly Section F of the Maneuver Rules as well as the Table of Reductions and Additions to Multiplier "M." He was to cross out rules that did not apply for the Maneuver, in this case those on sea condition and sun glare. He was to then fill out the Ship's Data Form for the ships and divisions of force to which he was assigned, using data found in the Statement of the Problem, or *Fleets* if necessary. He was additionally to staple together about fifteen pairs of Fire Distribution Sheets, after filling in the ships and the destroyer divisions. Listing the ships in order of tonnage, he was then to fill out the left-hand columns of the Force Damage Summary and list the ships in order of tonnage on that document as well. He was additionally to obtain a supply of Gunfire Data Flyleafs and paper clips and then study the Staff Solution of Exercise 2 for practice in assigning reductions. Once the Maneuver started, he was to obtain first copies of the Move and Gunfire Sheets from the Move Umpire as soon as the latter checked them; it did not matter in what order the Reduction Recorder received these.[15]

The Reduction Recorder was subsequently to make out the Fire Distribution Sheet for his force and give the carbon copy to the Reduction Recorder for the opposing force. Lines were to be drawn between the firing ship and the target ship, labeled with the number and calibers of guns firing. The Reduction Recorder was to record changes of course greater than forty-five degrees and speed changes of five knots or greater. He was also to record his own ship's course, if under fire. Filling out Columns 17, 18, 19, and 20 of the Gunfire Data Flyleaf that was clipped to the appropriate Move and Gunfire Sheet, he was to pass the forms immediately to the Fire Effect Computer. The Reduction Recorder was to then keep the Force Damage Summary up to date with the latest Ship Damage Tables received from the Move Umpire. Reduction Recorders, ensuring that they were not holding up the completion of these forms for the Damage reports, were to check the Maneuver Board for Masked and Enfilade Fire before each move was made. When in doubt, they were to consult with the Chief Damage Computer and make the appropriate marks in Column 17. They were to work on forms in numerical order according to the number in the upper right-hand corner. These forms were to be passed to the appropriate Range and Target Angle Computer in the same order so that computing work could be maintained "efficiently."[16]

Prior to the start of the Maneuver, this officer was to also thoroughly study Section F, draw a Range and Target Angle Indicator form from the Drafting Room, and study the instructions about the form's use. He was also to complete the Ship's Data Form for his forces and bring his copy of the 1946 Fire Effect Tables for use into the Maneuver Room. Once the Maneuver started, he was to receive the original of the Move and Gunfire Form from the Move Umpire and work on those forms in numerical order. Using the Range and Target Angle Indicator, he was to determine the Range and Target Angle and enter that information in Column 16 of the form and note in Column 17 whether the target vessel was out of range. Noting in Column 17 if Bow or Stern Fire was effective, he was to fill in Column 21 with the number of guns firing on each target. The originals of the forms were then to be passed to the Fire Effect Computer without delay. Afterward, the Range and Target Angle Computer was to assist the Fire Effect Computer in maintaining the Expenditure of Ammunition Tally Sheet when the Damage Computer was furnished with the original.[17]

The final set of Instructions had to do with the Range and Target Angle Indicator itself. This device was to indicate to the student the Bow, Stern, and Broadside Fire as well as the Target Bearing, Range, and Fire Effect Target Angle for gunfire. It also indicated the target angle for torpedo-fire problems. To use the Indicator, the Firing Ship was always to be in the center of the chart. On the board, the student was to determine the quadrant of the Firing Ship by noting the approximate compass bearing of the target from the Firing Ship. If the target was north and east, it was in Quadrant 1; if the target was north and west, it was in Quadrant 2; south and east indicated Quadrant 3, and south and west Quadrant 4. On the board, the student was to then count the east–west or horizontal spaces between the Firing Ship and the target as well as the north–south or vertical spaces between the two ships. He was then to count the spaces from the horizontal and vertical lines nearest to the Firing Ship and the target. On the Range and Target Angle Chart, the student would locate the Target Ship by counting the same number of horizontal and vertical spaces obtained from observing the Maneuver Board. A mark was to be placed at the intersection of the horizontal and vertical lines that bordered the outboard space. To measure range, the student was to swing the Range and Target Angle Computer to the target mark so that the center line of the range arm intersected the mark. The range division within which the target mark fell was the range to the target to the nearest thousand yards. Odd numbers were not marked, because of space restrictions; the range from 4,500 to 5,500 yards was expressed as five thousand yards, and the range arm was so graduated for student convenience.[18]

To determine whether the student was firing Bow, Stern, or Broadside guns on the target, the student was to turn the center disk until one's Own Ships' Course Marker *(B)* coincided with one's Own Ships' Course on the compass rose chart. If the center line of the range arm fell in the sixty-degree *B* sector, the student was firing Bow guns on the target. If the center line of the range arm fell in the *S* sector, Stern guns were being fired; when the center line was in the 120-degree space, Broadside guns were being fired. Target Bearing was read from the chart Compass Rose; the direction of the center line of the range arm was the true bearing on the target. To determine the Fire Effect Target Angle, the student was to orient the Compass Rose on the end of the range arm with the north–south (vertical) lines of the chart. He was then to turn the upper disk of the Target Ships' Course; the center line of the range arm would fall within the zero-degree, forty five–degree, or ninety-degree sector, which would be the Fire Effect Target Angle. Similarly, the target angle for torpedo fire was determined by counting the number of five-degree spaces from the Target Ships' Course line to the center line of the range arm. Angles could be read to within one degree by estimating the value of the fraction of five-degree space subtended. By the use of this Indicator, the student was, in effect, measuring range and bearing from the corner of the thousand-yard square that was the nearest to the Firing Ship and Target Ship. Actual range and bearing measurements made from the board with a range tape and protractor might differ by as much as a thousand yards or five degrees. For Fire Effect measurement purposes, this was not material, but for more accurate bearing and target angle measurements for torpedo fire, the student was advised to use a standard Maneuvering Board plot of relative target movement.[19]

Operations Problem 1: The Orange Staff Solution

On 15 February 1946, the students put forth an Orange Staff Solution in which they reiterated the General Situation of an impending Blue strike in the North or Central Pacific. This time, however, Orange admitted that the Blue Fleet was so much more powerful than its own that any comparison of the two forces was of "doubtful value." Laying out again the dispositions of the various Blue and Orange forces, the Orange Staff assessed that Blue was going to attempt to establish an advanced base in the western Pacific and assumed that Vice Admiral OB could not expect any reinforcements for his North East Area Fleet. Admiral OB also assumed that the Orange Naval General Staff would inform him when the Blue forces departed. In addition, OB was free to take action against Purple if Purple assisted Blue in any way against Orange. Admiral OB surmised that he could deploy all of his forces against Blue's move, as he was confident that the Ominato Guard District Force could defend Orange bases against Blue strikes. OB's mission of preventing Blue from establishing an advanced base in his area of operations was clear, but the fact that Blue now "dared" contemplate seizing an advanced base in Orange homeland waters made plain how large a naval force it had. Still, Blue was thought to be at a "tremendous" disadvantage because of the distance from its Primary Bases. Establishment of a base west of the Aleutians would greatly alleviate this problem, though not entirely. As such, an advanced base would itself have to be supplied from three thousand miles away. Because of this obstacle for Blue, Admiral OB made the assumption that the actual base would be on Purple territory, to be used by Blue to seize some island in the Kuriles. The relatively small size of the Blue Expeditionary Force ruled out an attack on the Orange homeland. Seizure of an island in the northern Kuriles would, however, allow Blue to bomb the Orange homeland, give air cover to and replenish surface forces, and maintain relatively short lines of supply and communication with its own rear area.[20]

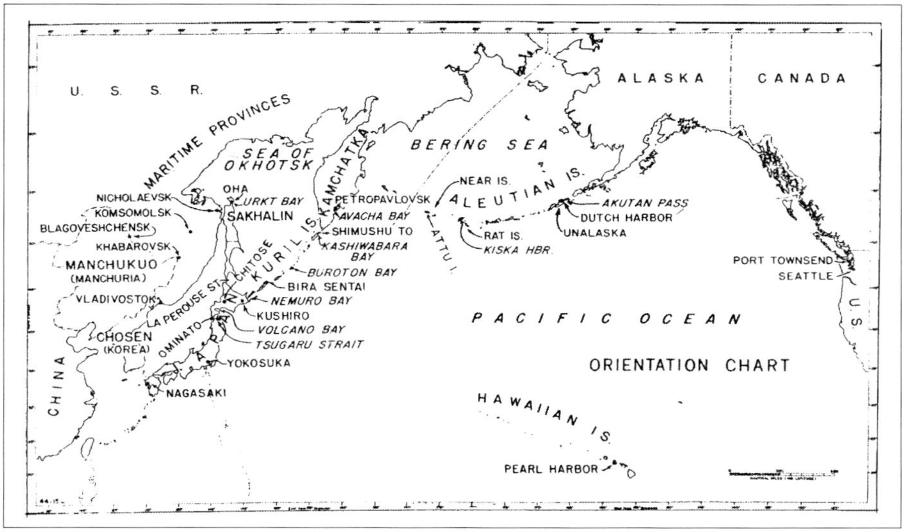

Map 10
North Pacific Orientation Chart

Admiral OB was not certain of the exact Blue objective, but thought that he had narrowed it down enough to interpose his forces between the Blue force and most of the likely objectives. OB also thought he had an advantage in interior lines of communication and supply, which were just under a thousand miles in length. He did not want to precipitate a conflict with Purple, but he also did not intend to allow Blue to establish a base in Purple territory. Vice Admiral OB was willing to violate Purple neutrality in order to prevent this, but he saw preventing Blue from getting to Purple territorial waters at all as a much more satisfactory solution. However, the imbalance in favor of Blue was clear: OB was outnumbered in everything except for heavy cruisers. The comparison in carriers was "disproportionately" in Blue's favor; Orange only had one training carrier, with a complement of thirty-six planes, while Blue had two fleet carriers and one light, with 212 fighters, scout bombers, and torpedo bombers. Admiral OB considered that he had an advantage in submarines, as Blue submarines had "not been very effective in the northern waters." Blue battleships and auxiliaries had very effective antiaircraft batteries, and its auxiliaries in particular could not be "recklessly" attacked by Orange aircraft. Virtually all of OB's 294 aircraft were land-based, scattered in the Kuriles, but he assumed that Rear Admiral OR would be able to concentrate the aircraft of his 12th Air Fleet quickly. OB did not think that Blue land-based aircraft would be much of a factor. Moreover, since his land-based bombers had twice the range of carrier-based ones, OB thought that if the Blue force were to be located, "it should be practicable to attack him as much as a thousand miles from the bases."[21]

The Blue Expeditionary Force of forty thousand would be superior in number and equipment to the land-based forces of the Ominato Guard District Force, but only after it landed. He considered that Blue and Orange land forces had generally equal combat efficiency, the "slightly" superior technical aspects of Blue forces being "offset by the fanatical zeal and willingness of Orange forces to die for the Emperor." He also thought that Orange naval forces were more acclimated to the northern area, especially given Blue's long transit time to the area of operations. Moreover, any Blue ships that were damaged would be in a precarious position so far from home.

Because of its harbor facilities, natural anchorages, land for airfields, and beaches suitable for landings, Avacha Bay was a most likely landing site for Blue. The Kurile island of Etorofu also had good landing beaches, as did the island of Paramushir, though the latter was hilly and rocky, thus advantaging the defender. Shumushu had an "excellent" landing site that could be the target of airborne troops. The islands of Onekotan, Shiashkotan, and Matua offered good landing beaches but no terrain for airfields. The same was true for Urup. So OB returned to Etorofu as the best possibility for a Blue landing, since it would be an advanced base with anchorages and airfields. Kunashir and Shikotan also offered good possibilities but not as many as Etorofu.[22]

The dominant aspect of the climate of the Aleutians-Kuriles-Kamchatka area generally was that its weather was unpredictable and bad. Rain, mist, fog, and snow could be found in all seasons, with snow predominating in the winter and fog in the summer, and conditions could change significantly in just a few miles. Since Blue would strike in the summer, a mean temperature of about fifty degrees Fahrenheit could be expected, with some rain and humidity but considerable fog or heavy overcast. Skies would be clear only for short periods of six to eight hours. Surface navigation would be difficult, and air operations might have to be curtailed at "critical" periods. In the first two weeks of July, sunrise would generally be at about 0300, while sunset would typically be just after 1900.[23]

Concerning lines of transportation and supply, Orange bases in the Kuriles were supplied from the large naval bases at Kure, with supply ships traveling through the Sea of Japan into the Pacific Ocean and then to their destinations. He considered these supply ships to be well protected by escort vessels and land-based aircraft, and he assumed that any Blue advanced base in the Aleutians would offer similar protection to Blue supply ships headed to Etorofu. He was concerned that while his own bases were well stocked with supplies, they lacked natural resources and had limited repair facilities. The base at Kashiwabara, on Paramushir Island, could, for instance, dry-dock only very small ships; ships of cruiser size or above would have to head to homeland bases for major repairs. All of his bases had "ample" airstrips and anchorages, and Kashiwabara and Ominato both had high-powered radio stations, but all of these bases also had limited ship-to-shore radio facilities.

Blue's bases west of Dutch Harbor were also limited in these categories, and none were known to be able to dry-dock ships at all. Blue would need airstrips if it were to raid the Kuriles with land-based aircraft. Orange had a more promising base complex, but Admiral OB thought base facilities were "very much limited" for both sides. Communications were likewise similarly limited. Radio communications were vital for both sides; cables were insufficient for seaborne naval operations, even where commanders were based ashore. Airmail services, however, in spite of the climate, were "excellent" for both sides.[24]

Orange battleships were superior to those in Blue's Escort Force and perhaps its Covering Force as well, though Blue had greater overall strength in this category. Blue was also superior in dive-, scout, and torpedo bombers, while Orange held the advantage in naval patrol, heavy bomber, and fighter planes. Orange could replace its planes much more easily, but since Blue had virtually all of its air strength on board carriers, it could deploy its air forces without regard to land bases. Further, Blue had better antiaircraft batteries in all of its ships and double Orange's destroyer strength, though OB's aircraft-equipped submarines would restore some balance. Orange ships also carried a greater number of torpedoes, including reloads, though Blue ships had more tubes for initial strikes. In addition, Admiral OB thought that fog and deployed submarines were in Blue's favor, but his own interior position and land bases would help him, as would his radar, his radio direction finders, and the region's nightlong twilight conditions. Vice Admiral OB's major weaknesses were the short range of his land-based fighters, fog, and "excessive" fuel consumption in his destroyers. Blue's major weaknesses would be an inability to replace plane losses easily, its distance from bases, the restrictions on its movements because of the need to protect convoys, and logistical problems with fuel and other consumables, again because of the distance from Blue's Primary Bases. Given these relative strengths and weaknesses, OB saw Blue significantly "handicapped" because of the need to operate so far from its bases and protect a slow convoy of ships while carrying out an amphibious operation against a hostile shore. Therefore, Admiral OB was confident that he could defeat the Blue force despite Blue's strategic initiative, especially since Blue might lose that initiative once it was wedded to protecting its fleet in the amphibious assault area.[25]

Vice Admiral OB thought that Blue had three options for approaching the area of operations. The first was for the Blue Expeditionary Force to evade Orange forces while the Blue Covering Force closed to contain Orange units. Another option was for Blue to arrive in Orange-controlled waters with its entire force concentrated. The final option was to destroy all appreciable Orange air and surface opposition before the landing took place. OB did not think, however, that Blue could completely evade Orange forces, nor did he think that the Blue Covering Force was strong enough to contain concentrated Orange units. OB admitted that Blue had previously conducted similar operations against Orange in the Central Pacific with "phenomenal" success, but he did not think that a Blue Fast Carrier Task Group that was limited to carrier airpower could predominate in the Orange theater of operations. Since Admiral OB's forces could do serious damage to a lightly escorted convoy and the Blue Expeditionary Force was the focus of the operation, he assumed that Blue would have to concentrate all of its forces and move in strength, especially to maximize its carrier airpower. In spite of what had happened in the Central Pacific, Admiral OB did not think that Blue carrier forces could knock out his air bases for long periods of time. Again banking on the idea that his aircraft losses could be easily replaced and that Blue would not be seeking a "showdown" until after the advanced base was secured, Vice Admiral OB decided that Blue could damage his airpower but not to such an extent that he could not continue to fight.[26]

Given the likelihood of a Blue approach in concentrated strength, OB listed the likely Blue objectives in order of probability as the Komandorski Islands, Avacha Bay on the Kamchatka Peninsula, Siberia, the northern Kuriles, the southern Kuriles, and the Orange homeland. The first three choices required the "connivance" of Purple, which was unlikely. The northern Kuriles were a more likely target, because of their anchorages and good airfields as well as their distance from the Orange homeland. He saw the southern Kuriles lacking in large, protected anchorages—even Etorofu lacked any anchorages—and as too close to the Orange homeland. He could not entirely rule out an attack on the southern Kuriles, but upon reconsideration of the factors listed above, he now thought that Kashiwabara Bay, Avacha Bay, or Buroton Bay in the southern Kuriles were the most likely, in that order. OB had the freedom to take action outside this assigned area of operations in order to carry out his orders, which specifically were to deny Blue control of the area by destroying, driving off, or diverting the Blue Expeditionary Force. He wanted to leave all possible courses open, but he obviously favored the first option, that of destroying the Blue force.[27]

Admiral OB argued that the possibility of doing so would be largely determined by the degree to which Blue had concentrated its forces and the locality in which Orange attacked. He would have to fight through the Escort Force to get at the convoy, but he thought he had "considerable" superiority over the Blue Escort Force. Further, destroying the Blue Expeditionary Force meant avoiding the Covering Force as much as possible, though he did not think this could be done entirely. If the Blue force could be attacked shortly after leaving Dutch Harbor, the Escort and Expeditionary Forces might be caught without the protection of the Covering Force, but the farther east he went, the weaker his forces would be. His submarines would be at the end of their supply lines, his land-based air forces would at some point be unable to participate in the strikes, and his search capabilities were dispersed. Since Blue would encounter all of these problems itself the closer it got to its objective, OB thought that his advantage lay in keeping his forces concentrated near his bases and homeland waters. This would mean fighting a concentrated Blue force, but with greater advantages and ability to exploit any mistakes Blue might make. He was counting heavily on submarine and air attacks to whittle down Blue strength (a strategy mirroring that of interwar Imperial Japanese Navy Decisive Battle Doctrine).[28]

Vice Admiral OB was open to the idea of driving off the Blue Expeditionary Force, but he wanted to destroy as much of the convoy as possible, since driving it off would "merely postpone the evil day." He was even more skeptical about diverting the Blue Expeditionary Force. Given the distances involved, the only way of doing so he envisioned was to attack a Blue base or the Blue Covering Force. Either would take him too far from his air bases; OB's bombers would not have fighter cover, his supply lines would be too long, the Covering Force could do an end run around the Orange forces to attack the objective, since it was faster, and in any case Blue bases would be heavily defended. Given all these conditions, he rejected the idea of trying simply to divert the Blue force. The choice between destroying and driving off the Blue Expeditionary Force would greatly depend on the weather and the intelligence information he received from his submarines and aircraft. If his submarines and aircraft could locate the Blue force early enough and damage it enough, he could interpose his Main Body between the Blue force and the objective areas and so accomplish his mission. He could disrupt a Blue landing, the Blue forces having evaded him early on, since his dispersed forces would eventually catch them. Destroying the Blue Expeditionary Force would be costly, but it could be done, after which he could retrieve and repair his crippled ships.[29]

As for how exactly to destroy the Blue Expeditionary Force, Admiral OB divided his task into two phases, Search and Attack. In the initial phase, as soon as he received information that the Blue units had departed their bases, he would initiate a Search Plan to locate these forces. The follow-on phase was, of course, to destroy the Blue Expeditionary Force as soon as it came into range of his planes and his other forces, and then to destroy as many other Blue units as possible. In the Search Phase, it was important for him to know about the strengths, dispositions, courses, speeds, and destinations of the various Blue forces. If the objective was in the northern Kuriles, he was superior to either the Escort or Covering Force but not both. Also, since Blue forces would probably be concentrated, the Blue Commander had the strategic initiative. It would take about five days for the Blue Expeditionary Force to transit from Seattle to Dutch Harbor, another six to ten days if the troops at Dutch Harbor were to be deployed, and another eleven days for all of the ships of the Blue Expeditionary Force to arrive in Alaska, if they did not proceed from Seattle as a convoy. OB noted the possibility of the convoy picking up troops in the Central Pacific and transiting directly from Seattle to the Kuriles.

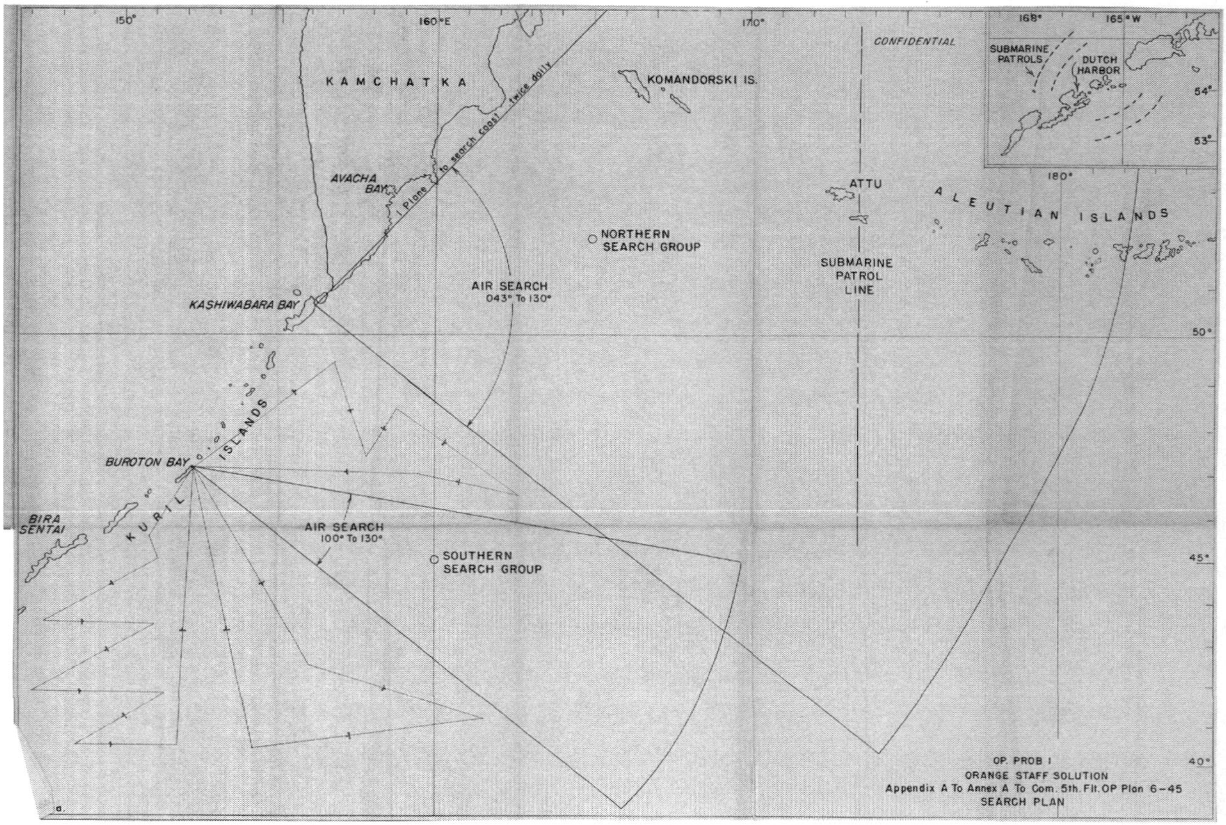

Fig. 138
Operations Problem 1, Orange Staff Solution, Search Plan

He did not think this was likely, but his Search Plan had to take the possibility into account. He further deduced that the Blue forces would take routes either north or south of the Aleutians or attempt far northern and southern routes as ways of evading Orange forces.[30]

Admiral OB thought that a northern route was the most likely, because of what he saw as Blue's "critical" logistical situation, the cover of "nasty" weather, and the fog of the Bering Sea. He surmised that Blue forces would have air cover from land bases all the way to the target and that the Blue Covering Force could easily join up with the Escort and Expeditionary Forces after refueling at Kiska. However, the Blue Commander might assume this route was the most heavily patrolled by Orange forces. The other route was slightly longer, had better weather for cover, and had the same advantage of land-based air coverage, but it would also entail the Blue force refueling in the western Aleutians in a protected anchorage rather than in the open sea. Dividing forces so as to try to shake the Orange forces did not seem a very good possibility, and OB largely discounted it.[31]

He counted heavily on his twelve submarines, but since they had "low mobility" it was necessary to keep a division of four submarines deployed at Dutch Harbor and rotate the others in as necessary. He also placed great reliance on his eighty-four naval patrol planes, because of their range, radar, and durability, but it might be necessary to supplement them with his sixty bombers, because of the losses he assumed the former would take from fighters. His single light carrier was effective only for extending the search range for his Surface Group, and it in turn had to be protected by that unit. He saw the light carrier potentially protecting some of the search aircraft with its fighters and land-based fighters with drop tanks perhaps doing the same. He also saw about two-thirds of his four heavy cruisers, six light cruisers, and twenty-three destroyers supplementing his submarines and aircraft in the Search Phase, but the remainder had to protect his battleships. He did not want any of his surface forces, however, searching so far out that they could not quickly concentrate with the Main Body, which would form as soon as the Blue forces were found by submarine, aircraft, or surface ships. The submarines' primary mission was to gain information on the deploying Blue forces, but they were to attack if they found opportunities that did not interfere with the primary mission. Search aircraft were also primarily to locate and track the enemy forces. However, these planes could also make low-visibility torpedo attacks if the situation permitted and bomb strikes if supported by fighter units. A Northern Search Group consisting of the light carrier, three light cruisers, and two destroyer divisions was to be formed, as was a Southern Surface Group of three light cruisers and two destroyer divisions.[32]

The final phase of Vice Admiral OB's plan was the actual destruction of the Blue Expeditionary Force. He had an inferior force; in fact, OB assumed that if the Blue and Orange forces were to meet in the mid-Pacific, equidistant from their respective bases, his force would be "annihilated." But with Blue far from its bases and perhaps even encountering ammunition or fuel shortages, OB thought that his submarines and aircraft could weaken Blue prior to a "decisive engagement." If Blue was going to attempt to destroy Orange air strength and surface ships with carrier airpower, OB's ace in the hole was the North Pacific weather. He assumed that the weather would greatly inhibit Blue carrier operations but that his own land-based air forces would not be similarly impacted. Orange forces could not destroy the Blue forces by land-based air attack alone, because of the strength of the Blue ships' antiaircraft batteries, but distance, weather, and "luck" would enable the Orange Main Body to close with a weakened Blue force and destroy it with guns and torpedoes. He envisioned the Attack Phase of his plan gradually proceeding from his Search Phase, each group of attacking forces striking on its own initiative as the situation permitted. He personally would command the Main Body, what he called the Attack Group, supported by the Submarine Group, the Air Group, and the Northern and Southern Search Groups. The Submarine Group would remain in its designated patrol lanes; the Air Group would dispatch kamikaze attacks if necessary and destroy enemy troops during their assault; and the Surface Search and Attack Groups would remain concentrated at Kashiwabara Bay until the Blue forces had been definitively located.[33]

Vice Admiral OB would command Task Force 51, the North East Area Fleet, and the 5th Fleet, as well as the Attack Group for this operation. He left his air and submarine forces in command of their senior type commanders as the most "natural" setup, and he planned on giving these officers the "widest latitude." He would perform all necessary coordination from his flagship at Kashiwabara Bay during the preliminary phase, then command all of the surface forces in the decisive phase once they were concentrated. He appointed Rear Admiral OR, commander of the 12th Air Fleet and of the Air Group, as his Deputy Commander, since OR was the next senior officer. Rear Admiral OJ, commander of Submarine Squadron 1, was to command the Submarine Group; Rear Admiral OF, commander of Destroyer Squadron 2, was to command the Northern Search Group; and Rear Admiral OG, commander of Destroyer Squadron 10, was designated as the commander of the Southern Search Group.[34]

OB's Communications Plan entailed radio silence at sea until lifted by his orders, with new information being broadcast to the deployed units at regular intervals. He did not see any particular problems with logistics, as Commander, Ominato District Guard Force was to oversee all such operations and all forces were operating from their regular bases and tenders. The surface forces also had at their disposal two tankers. The "crux" of the plan, however, was intelligence, since it was based on so many assumptions about the enemy. For this reason, though he wanted to give his subordinate commanders maximum latitude, he prescribed search operations minutely, since finding the Blue forces early on was the key. He did not think any Movement Plan was necessary, as his forces could be ready on forty-eight hours' notice and, in any case, he was ordered to carry on normal operations until further notified by the Naval General Staff. His Movement Plan, therefore, would involve only the movements of his air, submarine, and surface forces after the initial contact with the enemy.[35]

The Orange Statement and Staff Solution are intriguing from at least two perspectives. First, if the role of the students was to emulate wartime Japanese strategy and operations, Vice Admiral OB did so perfectly. The Orange plan entailed a dispersion and then concentration—it was hoped—of forces at the decisive moment to defeat the Blue forces. Orange also placed great reliance on heavy surface forces backed by submarines and land-based air, if only because that was all it had remaining at this point in the "war." Naval War College students at this point must have been aware in some detail of how the Japanese had operated. For instance, OB's staying ashore only until the Blue forces were located and then embarking at sea as commander of the Attack Group was an almost exact emulation of Admiral Isoroku Yamamoto, who in 1942 during the battle of Midway had embarked on the fast battleship *Yamato* at the head of his Main Body instead of staying ashore, where more extensive communication facilities had been available. Reliance on heavy surface forces for a "decisive battle" after Blue had been whittled down by submarine and air forces had been Japanese naval doctrine until Yamamoto insisted on changes for Pearl Harbor and subsequent operations, and it remained the dominant doctrine of the Imperial Japanese Navy for the rest of the Pacific War in spite of Yamamoto's influence. There was even the classic idea that "luck" or providence would somehow see Orange through, something definitely seen in Japanese plans for the battles of the Philippine Sea and Leyte Gulf, if not before. The only major change in this war game was the shift from the Central to the North Pacific, perhaps reflecting an initial concern by the Naval War College about conducting operations against the Soviet Union—a concern certainly shared by Admiral Spruance when he took over as President of the College in March 1946.[36]

The Schedule of Employment, the Board Maneuver, and Concepts of Postwar Surface Warfare

Later in February, the Command and Staff Class of June 1946 took part in another exercise, this one also having to do with Maneuver Rules that applied to the Board Maneuver and the "mechanics of simulating ship movements on the Maneuver Board." The Staff and students were divided into six groups; the Exercise would have three phases in which the groups would simulate specific movements of, in rotation, groups of ships designated as Battleship Division 1, Cruiser Division 5, and Destroyer Squadron 2. The Exercise was set up so that each student would have an opportunity to handle each collection of ships in each maneuver. Four Staff officers were assigned as the commanders of each group of ships, and one as the Director of the Maneuver.[37]

In Exercise 1, the students were to simulate on Move and Gunfire Forms a number of movements by Orange battleships, cruisers, and destroyers. Three groups of ships were posited, steaming at fifteen to sixteen knots in columns on course 000 degrees at various distances from each other. Given wind, sea, and visibility information, the students were first to maintain the Orange units on their current course and speed. Then they were to change the speeds and courses of the various units and divide the formations—such as splitting Destroyer Squadron 2 into Destroyer Divisions 15 and 31, on different headings and speeds. In subsequent moves, Orange Battleship Division 1 was also divided, with battleship BB-9 continuing on course 000 at a speed of twelve knots and BB-10 slowing to 12.5 knots and changing course to 045. All groups then rotated to give students maximum exposure to the multiple ship movements.[38]

Exercise 2 was meant to familiarize students with the gunfire aspects of the Maneuver Board. The heart of the Exercise was a simulated engagement between the

8/14/41 **Exercise 2**
Desdiv 15

Use separate blank for each salvo. Prepare in Duplicate.

Fired at 0701 (See Rule G-8) During Move 1 (Scratch one)

Firing Ships	Type torpedo and Speed	What Tubes fired NOTE A	Number torpedoes fired	Point of aim NOTE B	Sight angle. Base torpedo course. (Scratch one) NOTE C	Spread (degrees)	Depth (feet)	For Use of Umpire: Damage at beginning of move	Not under fire. Under Fire. Behind Smoke or Exceptional Conditions	Number of torpedoes that run	Firing Range	Track Angle	Results Obtained
Desdiv 15	G-3 50 Kts	All	9	----	BTC 257° (TA 55° SP 22 Kts SA 21° Brg 236 BTC 257°)	8°	22						

Signature ____________

NOTE A Enter tubes fired as "Stbd", "Port", "Ford", "Aft", "All", "Deck" or "Submerged" as may be appropriate after giving due consideration to the position and course of the firing vessel at the time of firing.
NOTE B Not required if base torpedo course plan of firing is used.
NOTE C Scratch either "Sight angle" or "Base torpedo course". Make entry in this column in degrees from 0 to 360.
The sight angle should be the angle you desire used between the line of sight and the track you desire your base torpedo to run on.

Fig. 139
Torpedo Fire Blank, Orange Forces, Exercise 2, Move 1

Orange force from Exercise 1 and a Blue force of the same number and types of ships: two battleships, four heavy cruisers, and eight destroyers for each side. Groups 1, 3, and 5 controlled the Blue forces and Groups 2, 4, and 6 the Orange. Blue forces were divided into Battleship Division 6, Cruiser Division 6, and Destroyer Squadron 21. Groups were again rotated to maximize exposure, and emphasis was placed on each student's becoming proficient at filling out the various parts of the Move and Gunfire Forms and employing the Table of Reductions and Additions to Multiplier "M." Forms would be inspected by a Staff officer, and the Director of the Maneuver would then have students make the move on the Maneuver Board. Again, a Staff officer was assigned as the Director, while three others were in charge of the Blue and Orange battleships, cruisers, and destroyers, respectively.[39]

In the General Situation of this Exercise, Orange is already engaged with Blue. Blue has one spotting plane from each battleship on station, as well as one from each of its heavy cruisers. Battleship Division 6 and Cruiser Division 6 are in columns, with battleship BB-55 and heavy cruiser CA-32 designated as Guides. Destroyer Squadron 21 is also in column, with DESDIV 41 in the lead. All ships have Radar Director Control and Radar Spotting available. In Move 1 for the Blue forces, Blue battleship BB-55 opens fire with its main battery on Orange battleship BB-9 and with its secondary battery on Orange DESDIV 15. BB-55 then fires its main battery on Orange battleship BB-10 and its secondary battery on Orange DESDIV 31. Orange Destroyer Divisions 15 and 31 subsequently open fire on Blue battleships BB-55 and BB-56 with five-inch guns. At the same time, Blue Cruiser Division

TABLE OF REDUCTIONS AND ADDITIONS
TO MULTIPLIER "M."

Symbol	CAUSE	Rule No.	Reductions or Additions	Remarks
a	ABOVE WATER DAMAGE	F-6		For each 10 percent life lost reduction to "M" is -.1
b	OPEN FIRE	F-7	-.5	Applies for move during which fire is opened.
c	SHIFT FIRE	F-8	-.3	Does not apply if new target is within 1500 yards of old target.
d	FIRE DISTRIBUTION	F-9(1)(2)		BB to CL firing on more than one target: -.2; Ships other than above firing on more than one target: -.3
e	OVER CONCENTRATION	F-9(4)		See table below
f	* SHIP UNDER NORMAL FIRE	F-10(1)	0	Not less then two thirds or more than four thirds of own guns.
g	* SHIP UNDER A CONCENTRATION	F-10(2)		Ship being fired upon by main batteries of two or more ships and more than four thirds of own guns. Ship with turrets: -(n-1); Ship with exposed guns: -(n)
h	* SHIP UNDER LESS than NORMAL FIRE	F-10(3)	+.1	Number of guns less than two thirds of own guns.
i	* SHIP NOT UNDER EFFECTIVE FIRE	F-10(4)	+.2	Number of guns, two or less.
j	SHIP UNDER MINOR FIRE	F-10(5)	-.1	Ship appreciably smaller than own type. See table below
k	AIR ATTACK	F-11		Reduction As decided by Director
l	SHOCK EFFECT	F-12	-.5	Reduction for move caused by BOMB, MINE or TORPEDO HIT.
m	TARGET COURSE CHANGE	F-13		See table below
n	TARGET SPEED CHANGE	F-14	-.1	For each 5 Kt. change of speed.

e — Over concentration:

Condition	Larger than 6"	6" or less
Daylight Top or Plane Spot	-(n-3)	-(n-2)
Radar Spot	-(n-2)	-(n-1)
Other Conditions	-(n-1)	-(n)

"n" equals number of ships concentrating

* Applies only if ships engaged are of the same type, and guns are of main battery.

j — Ship appreciably smaller than own type:

Caliber.	Number Guns.
6"–10"	8 or more
Less than 6"	16 or more

m — Target course change:

TARGET COURSE CHANGE	Reduction: Radar Dir. Control	Reduction: Optical Dir. Cont.	Reduction: Pointer Fire or Local Cont.
45°-59°	0	-.1	-.2
60°-74°	-.1	-.2	-.4
75° or more	-.2	-.3	-.6

46-18a

Symbol	CAUSE	Rule No.	Reductions or Additions	Remarks
o	SEA CONDITION	F-15		Moderate Sea: -.1; Rough Sea: -.2; Heavy Sea: -.4
p	* GUN or FUNNEL GASES	F-16(a)	-.1	Applies if gun or funnel gases drift across line of fire.
q	* TARGET HID by SMOKE SCREEN	F-16(b)		Reduction proportional to time except if Indirect Fire is used, then reduction is -.3
r	* SUNGLARE	F-16(c)	-.2	Two hours after sunrise and two hours before sunset. Target within 15 degrees of SUN
s	* TARGET SILHOUETTE	F-16(d)	+.2	At twilight, when target bears with 30 degrees of SUN. Other conditions, as ruled by DIRECTOR.
t	NIGHT FIRE EFFECT	F-17		See table below
u	MASKED FIRE	F-19		Ship within 1,000 yards of firing ship. Reduction proportional to time masked.
v	MELEE	F-20		Night or Day — Close range. Reduction decided by DIRECTOR.
w	ENFILADE	F-18		Ship within 1,000 yards of point of aim and within 100 yards of line of fire. See table below
x	RADAR SPOTTING † RANGES	F-2		See table below

* No Fire Control Radar available.

t — Night fire effect:

Condition	Reduction
(a) Radar Control and <u>with</u> Effective Illumination.	-.1
(b) Radar Control but <u>without</u> Effective Illumination.	-.2
(c) <u>Without</u> Radar Control but <u>with</u> Effective Illumination:	
Ranges less than 5,000 yards.	-.3
Ranges over 5,000 yards.	-.4

w — Enfilade:

Distance between Target and Enfiladed Vessel.	Percent of Final Multiplier of Firing Ship on her Target.
0–200 yds.	80
201–300 yds.	70
301–400 yds.	60
401–500 yds.	50
501–600 yds.	40
601–700 yds.	30
701–800 yds.	20
801–1,000 yds.	10

x — Radar spotting ranges:

Caliber of Gun	Limiting Range
16"-14"-12"	30,000 yds.
8"	23,000 yds.
6"	18,000 yds.
5"	15,000 yds.

† <u>Radar Director Control</u> is effective to maximum range of the gun.

46-18b.

Fig. 140
Tables of Reductions and Additions to Multiplier "M"

6's heavy cruisers, CA-32, CA-36, CA-37, and CA-38, open fire on Orange heavy cruisers CA-8, CA-7, CA-6, and CA-5, respectively.[40]

In Move 1 for the Orange forces, Orange BATDIV 1 has BB-9 as its Guide and a spotting plane from each ship on station. Orange CRUDIV 5 has heavy cruiser CA-5 as its Guide and a spotting plane from each ship on station, while Orange DESRON 2 is being led by DESDIV 15, followed by DESDIV 31. All ships are in columns at various speeds and intervals, like the Blue forces. Orange battleship BB-9 opens fire with its main battery on Blue battleship BB-55 and with its secondary battery on Blue DESDIV 41. Orange battleship BB-10 fires with its main battery on Blue battleship BB-56 and with its secondary battery on Blue DESDIV 42. Blue DESDIVs 41 and 42 then open fire with five-inch guns on Orange battleships BB-9 and BB-10, respectively. At the same time, Orange CRUDIV 5's heavy cruisers CA-5, CA-6, CA-7, and CA-8, open fire with their main batteries on Blue CRUDIV 6's heavy cruisers, CA-38, CA-37, CA-36, and CA-32, respectively. In the next phase of the Move, Blue DESDIV 41 opens fire on Orange battleship BB-9 with its main batteries and fires a salvo of 10-G-1 torpedoes at Orange BATDIV 1, with a torpedo speed of thirty-four knots and a spread of one degree. At the same time, Blue DESDIV 42 opens fire on Orange battleship BB-10, while BB-9 and BB-10 return fire on DESDIVs 41 and 42 with five-inch guns.[41]

In Move 2, Blue Battleship Division 6 continues its previous course and speed, with battleship BB-55 continuing to focus main-battery fire on Orange BB-9 and secondary fire on Orange DESDIV 15. BB-56 continues main-battery fire on BB-10 and secondary-battery fire on Orange DESDIV 31, but both BB-55 and BB-56 have sustained 1.7 percent Above Water Damage in Move 1. In addition, Orange DESDIVs 15 and 31 continue firing on BB-55 and BB-56 with twenty-four five-inch guns each, as in Move 1, while Blue CRUDIV 6's heavy cruisers, CA-32, CA-36, CA-37, and CA-38, continue main-battery fire on Orange CRUDIV 5's heavy cruisers, CA-8, CA-7, CA-6, and CA-5, respectively. However, all of Blue's heavy cruisers have 4 percent damage from Move 1. At this point, Orange DESDIV 15 opens fire on BB-55 and fires a spread of nine torpedoes at Blue BATDIV 6, with a spread of one degree and a speed of fifty knots. Orange DESDIV 31 then opens fire on BB-56, as BB-55 and BB-56 both continue to fire on Orange DESDIVs 15 and 31 with ten five-inch guns.[42]

As Move 2 continues, Orange BATDIV 1 continues on its previous course and speed, with BB-9 continuing main-battery fire on BB-55 and secondary-battery fire on Blue DESDIV 41. Orange battleship BB-10 also continues its main-battery fire on BB-56 and secondary fire on Blue DESDIV 42, but both of Orange's battleships by this time have sustained 3.1 percent Above Water Damage. Blue DESDIVs 41 and 42 continue firing on BB-9 and BB-10 with twenty five-inch guns each, while Orange CRUDIV 5's heavy cruisers, CA-5, CA-6, CA-7, and CA-8, continue main-battery fire on Blue CRUDIV 6's heavy cruisers, CA-38, CA-37, CA-36, and CA-32, respectively. All of Orange's heavy cruisers, however, have sustained 3.6 percent damage by this time. Blue DESDIVs 41 and 42 continue to fire on BB-9 and BB-10, respectively, while DESDIV 41 fires ten 10-G-1 torpedoes, with a spread of one degree and a speed of thirty-four knots, at BB-9. Also by this time, DESDIVs 41 and 42 have received 1 percent and 0.7 percent damage, respectively. Move 2 ends with BB-9 and BB-10 firing their secondary batteries on Blue DESDIVs 41 and 42, respectively, with two five-inch guns each.[43]

In Move 3, Blue battleship BB-55 continues main-battery firing on Orange battleship BB-9 and secondary battery firing on Orange DESDIV 15, while Blue battleship BB-56 continues main-battery firing on Orange battleship BB-10 and secondary battery firing on Orange DESDIV 31. Blue battleships BB-55 and BB-56 have by this time each received 5.4 percent Above Water Damage, and Orange DESDIVs

UNIT: BLUE Batdiv 6, Crudiv 6, Desron 21 — BATTERY: MAIN / SECONDARY (Delete One) — MOVE AND GUNFIRE SHEET — MOVE NO. 3 — Wind NE, Force 4, Sea Moderate

STUDENT: Exercise 2 — MANEUVER TIME: 0706-0709

MOVE DATA — GUNFIRE DATA

FIRING SHIP or DIVISION (1)	SPEED at START (2)	CHANGE of SPEED (3)	SPEED at END (4)	AVERAGE SPEED (5)	LOSS on account of SEA (6)	Final AVERAGE SPEED dur move (7)	COURSE at START of move (8)	CHANGE of COURSE (9)	COURSE at END of move (10)	GUNS FIRING Cal (11)	GUNS FIRING No. (12)	FIRING SHIP Coordinate Location x y (13)	FIRE DISTRIBUTION by arrow (14)	TARGET SHIP Coordinate Location x y course (15)	RANGE and TARGET ANGLE (16)	REMARKS Indicate-Bow, Stern Fire, Type of Spot, etc. (17)	REDUCTIONS to MULTIPLIER and SYMBOL (18)	Total REDUCTION (19)	MULTIPLIER (20)	NUMBER of GUNS (21)	FINAL MULTIPLIER (22)	NORMAL FIRE EFFECT (F.E. Tables) (23)	FIRE EFFECT INFLICTED (24)	AMMUNITION EXPENDED (25)
Batdiv 6																								
BB 55	23	0	23	23	-1	22	000	R60	060	16/45	9		→ BB9	315	26/45	BB9 turns 135° Des15 firing 24 5"	r1/o, r1/j, r2/m -1/a	÷5	.5	9	4.5	1.	4.5	3.1
BB 56										16/45	9		→ BB10	315	26/45	BB10 turns 135° Des15 firing 24 5"	r1/o, r1/j, r2/m -1/a	÷5	.5	9	4.5	1.	4.5	3.1
Crudiv 6																								
CA 32	25	-5	20	22.5	-2	20.5	045	L60	345	8"/55	9		→ CA8	225	12/45	CA8 slows 5 kts CA8 c/c 135°	r1/o, r2/m, r1/n, r2/a	-.6	.4	9	3.6	3.	10.8	6.8
CA 36										8"/55	9		→ CA7	225	12/45	CA7 slows 5 kts CA7 c/c 135°	r1/o, r2/m, r1/n, r2/d	-.6	.4	9	3.6	3.	10.8	6.8
CA 37										8"/55	9		→ CA6	225	12/45	CA6 slows 5 kts CA6 c/c 135°	r1/o, r2/m, r1/n, r2/a	-.6	.4	9	3.6	3.	10.8	6.8
CA 38										8"/55	9		→ CA5	225	12/45	CA5 slows 5 kts CA5 c/c 135°	r1/o, r2/m, r1/n, r2/a	-.6	.4	9	3.6	3.	10.8	6.8
Desron 21																								
Desdiv 41	25	0	25	25	-4	21	045	0	045	5"/38	20		→ BB9	315	8/0	BB9 c/c 135°	r1/o, r2/m	-.3	.7	20	14	.12	1.7	20.2
Desdiv 42											[illegible]		→ BB10	315	8/0	BB10 c/c 135°	r1/o, r2/m	-.3	.7	20	14	.12	1.7	20.2
Filled in by Student															RANGE T.A Computer	Reduction Recorder			RaT.A. Computer			F.E. Computer		

Unless otherwise noted under Remarks by student or Reduction Recorder, fire is assumed to be Radar Director Control, Broadside, Radar Spot.

*For Secondary Battery Fire fill in these spaces only. Use separate form.

46-1

FIRE CONTROL SYMBOLS

Method		Spot
Radar Director Control	Rad. Cont.	Radar — R.
Optical Director Control	Opt. Cont.	Plane — P.
Pointer Fire	Point	Top — T.
Local Control	Loc.	

Fig. 141
Move and Gunfire Sheet, Blue Forces, Exercise 2, Move 3

15 and 31 are firing on them with twenty-four five-inch guns each. Also by this time, Blue CRUDIV 6 has slowed to five knots, with heavy cruisers CA-32, CA-36, CA-37, and CA-38 continuing main-battery firing on Orange CRUDIV 5's heavy cruisers, CA-8, CA-7, CA-6, and CA-5, respectively. All of the Blue ships by this time have received 22 percent Above Water Damage. In addition to the five-inch gunfire, Orange DESDIVs 15 and 31 have fired a salvo of nine torpedoes at Blue BATDIV 6, again with a spread of one degree, a speed of 50 knots, and an aim of hitting the center of Blue battleship BB-55. Orange DESDIVs 15 and 31 have received 3 and 1.5 percent Above Water Damage during Moves 1 and 2, and Blue battleships BB-55 and BB-56 return fire at the Orange destroyers with ten five-inch guns each.[44]

As the move proceeds, Orange battleship BB-9 continues its main-battery fire on Blue battleship BB-55 and its secondary fire on Blue DESDIV 41, while Orange battleship BB-10 does the same against Blue battleship BB-56 and Blue DESDIV 42. At this point in the encounter, Orange battleships BB-9 and BB-10 have each received 12.1 percent of Above Water Damage, and Blue DESDIVs 41 and 42 continue their fire on the Orange battleships with twenty five-inch guns each. In addition, Orange CRUDIV 5's heavy cruisers, CA-5, CA-6, CA-7, and CA-8, have slowed to five knots and are continuing main battery fire against Blue CRUDIV 6's heavy cruisers, CA-38, CA-37, CA-36, and CA-32, respectively, but have each received 21.6 percent Above Water Damage. Blue DESDIVs 41 and 42 continue firing on the Orange battleships BB-9 and BB-10, fire a spread of ten torpedoes at BB-9, and have received 5 and 4.7 percent Above Water Damage, respectively. Orange battleships BB-9 and BB-10 are still firing their secondary batteries at Blue DESDIVs 41 and 42 with two five-inch guns each. As the Exercise ends, Orange DESDIVs 15 and 31 continue their fire on the Blue battleships, fire another spread of torpedoes at Blue battleship BB-55, and have received 11 and 9.5 percent Above Water Damage, respectively. Blue battleships BB-55 and BB-56 are also still firing their secondary batteries at Orange DESDIVs 15 and 31 with ten five-inch guns each.[45]

Conclusion

This Maneuver is another intriguing look at Naval War College simulations.[46] Actual wartime operations that came close to those depicted included the naval battles of Guadalcanal in November 1942 and Surigao Strait in October 1944.[47] The maneuvering of battleships, cruisers, and destroyers had been practiced countless times by both Blue and Orange at the Naval War College from 1911 to 1941, even when naval aviation and submarine warfare began to erode the dominance of the gunships.[48] Here again, in the immediate postwar period, there are no deployed aircraft carriers. In fact, the only aircraft involved are spotter planes from the battleships and cruisers, and the main firepower is that of gun batteries and torpedoes. Aircraft from carriers do not appear even in an air-defense mode, much less an offensive one, and submarines are absent. However, the purpose of the Maneuver was to familiarize students with the Maneuver Board and its associated forms; the ship platforms and weapons systems employed were considered less important than the board mechanics and the experience of decision making itself.[49]

NOTES

1 Command and Staff Class of June 1946, "Operations Problem 1: The Orange Statement," 7 February 1946, pp. 1–2, NHC, folder 2570, box 131, RG 4, NWC.

2 Ibid., pp. 2–3.

3 Ibid., p. 4.

4 Ibid., pp. 5–6. For the state of Japanese air forces after 1942, see Mark Peattie, *Sunburst: The Rise of Japanese Naval Air Power, 1909–1941* (Annapolis, Md.: Naval Institute Press, 2001), pp. 168–201.

5 Command and Staff Class of June 1946, "Operations Problem 1: The Orange Statement," pp. 7–8, 10.

6 Command and Staff Class of June 1946, "Operations Problem 1," app. 1, "Duties of the Student Commanders during Board Maneuver," 13

February 1946, p. 1, NHC, folder 2567-A, box 131, RG 4, NWC.

7 Ibid.

8 Command and Staff Class of June 1946, "Operations Problem 1," app. 2, "Duties of Move Umpire during Board Maneuver," 13 February 1946, p. 1, NHC, folder 2567-A, box 131, RG 4, NWC.

9 Ibid.

10 Ibid.

11 Command and Staff Class of June 1946, "Operations Problem 1," app. 5, "Duties of Fire Effect Computer," 13 February 1946, p. 1, NHC, folder 2567-A, box 131, RG 4, NWC.

12 Ibid.

13 Command and Staff Class of June 1946, "Operations Problem 1," app. 6, "Duties of Damage Recorder," 13 February 1946, p. 1, NHC, folder 2567-A, box 131, RG 4, NWC.

14 Ibid.

15 Command and Staff Class of June 1946, "Operations Problem 1," app. 6, "Duties of Reduction Recorder," 14 February 1946, p. 1, NHC, folder 2567-A, box 131, RG 4, NWC.

16 Ibid.

17 "Instructions for Use of Range and Target Indicator," 14 February 1946, p. 1, NHC, folder 2567-A, box 131, RG 4, NWC.

18 Command and Staff Class of June 1946, "Operations Problem 1," app. 4, "Duties of Range and Target Angle Computer," 14 February 1946, p. 1, NHC, folder 2567-A, box 131, RG 4, NWC.

19 Ibid.

20 Command and Staff Class of June 1946, "Operations Problem 1: Orange Staff Solution," 15 February 1946, pp. 1–5, NHC, folder 2570, box 131, RG 4, NWC.

21 Ibid., pp. 5–8.

22 Ibid., pp. 8–12.

23 Ibid., pp. 12–13.

24 Ibid., pp. 13–14.

25 Ibid., pp. 14–16.

26 Ibid., pp. 17–18.

27 Ibid., pp. 18–21.

28 Ibid., pp. 22–23. For the Decisive Battle Doctrine, whereby Japanese submarines and then land-based air forces would whittle down the American battle fleet as it advanced across the Pacific from Hawaii and then finish it off with heavy gunnery in the area of Micronesia, see David Evans and Mark Peattie, *Kaigun: Strategy, Tactics, and Technology in the Imperial Japanese Navy, 1887–1941* (Annapolis, Md.: Naval Institute Press, 1997), pp. 199–446, and Sadao Asada, *From Mahan to Pearl Harbor: The Imperial Japanese Navy and the United States* (Annapolis, Md.: Naval Institute Press, 2006), pp. 161–86.

29 Command and Staff Class of June 1946, "Operations Problem 1: Orange Staff Solution," pp. 23–24.

30 Ibid., pp. 24–28.

31 Ibid., pp. 28–29.

32 Ibid., pp. 29–32.

33 Ibid., pp. 32–36.

34 Ibid., p. 37.

35 Ibid., pp. 37–40. For the full Operational Plan, see ibid., pp. 41–55.

36 For Yamamoto's conduct at the battle of Midway, see Jonathan Parshall and Anthony Tully, *Shattered Sword: The Untold Story of the Battle of Midway* (Washington, D.C.: Potomac Books, 2005), pp. 19–38, 397–430. For an example of Japanese strategy and tactics from the summer of 1944 on, see H. P. Willmott, *The Battle of Leyte Gulf: The Last Fleet Action* (Bloomington: Indiana Univ. Press, 2005), pp. 35–78. For the idea of the Naval War College looking in early 1946 at possible operations in the northern Pacific against the Soviet Union, see Edward Marolda, e-mail to author, 13 October 2010. For Spruance's views on the shift to the north for war games and the idea that the U.S. Navy in general had to learn to fight in more northerly latitudes, see Hal M. Friedman, *Digesting History: The U.S. Naval War College, the Lessons of World War Two, and Future Naval Warfare, 1945–1947* (Newport, R.I.: NWCP, 2010), pp. 143–49, 156–61, 183–96, 199–207.

37 Command and Staff Class of June 1946, "Schedule of Employment and Board Maneuver Exercises: Exercises 1 and 2," 18 February 1946, pp. 1–3, NHC, folder 2567, box 131, RG 4, NWC.

38 Ibid., pp. 3–5.

39 Ibid., pp. 6–7.

40 Ibid., pp. 7–9.

41 Ibid., pp. 9–10.

42 Ibid., pp. 10–11.

43 Ibid., pp. 11–12.

44 Ibid., pp. 12–13.

45 Ibid., pp. 13–15.

46 Edward Miller, *War Plan Orange: The U.S. Strategy to Defeat Japan, 1897–1945* (Annapolis, Md.: Naval Institute Press, 1991), pp. 122–322. See also John Kuehn, *Agents of Innovation: The General Board and the Design of the Fleet That Defeated the Japanese Navy* (Annapolis, Md.: Naval Institute Press, 2008), pp. 162–79.

47 U.S. Navy Dept., *The Battle of Guadalcanal, 11–15 November 1942* (Washington, D.C.: Naval Historical Center, 1994), pp. 1–84; and Willmott, *Battle of Leyte Gulf*, pp. 136–216.

48 Michael Vlahos, *The Blue Sword: The Naval War College and the American Mission, 1919–1941* (Newport, R.I.: NWCP, 1980), pp. 55–96, 113–61.

49 Francis McHugh, *A Manual of War Games* (Newport, R.I.: Naval War College, 1960), pp. 1–3; McHugh, *Fundamentals of War Gaming* (Newport, R.I.: Naval War College, 1966); McHugh, "War Gaming and the Navy Electronic Warfare Simulator," *NWCR* 19, no. 7 (March 1967), pp. 22–32; and McHugh, "Eighty Years of War Gaming," *NWCR* 21, no. 7 (March 1969), pp. 88–90.

NORTH

XIV Operations Problem 2

The Blue and Orange Statements, February 1946

On 19 February 1946, Operations Problem 2 began with the Blue Statement. In this war game, Orange continued to be at war with the Allied nations, though not with Purple (the USSR). Purple, however, had communicated to Blue its growing apprehension about Orange preparations in Manchuria and the Orange homeland for offensive operations against Purple's Maritime Provinces. This activity was confirmed by certain Yellow (Chinese) and Purple "unspecified countermeasures." At the same time, the Blue Pacific Fleet, under the command of Admiral BN, was concentrated in the Eastern Marshalls, while the Orange Combined Fleet was concentrated in home waters, Orange having been expelled from New Guinea, the Carolines, and the Marianas. Blue surface and air forces had attempted raids into western Micronesia as a way to entice the Combined Fleet to fight, but so far unsuccessfully. Admiral BN had also been informed by the Commander-in-Chief of the Blue Fleet that negotiations were under way with Purple to allow Blue occupation and use of Purple Far Eastern ports and bases in anticipation of the outbreak of hostilities between Purple and Orange. Blue use of these facilities would be a preliminary to an attack on Orange forces in the Kuriles, including the occupation of Avacha Bay.[1]

Six weeks later, operations under Admiral BF, commander of the Blue 9th Fleet and Commander, Northern Pacific, began in the region. Both Blue and Orange had suffered losses in these operations. The Blue 9th Fleet, however, had replaced its losses and been slightly reinforced; it continued to be based at Attu and Kiska, with the Blue Expeditionary Force on Kiska as well. Admiral BA (who had replaced BF) then received an urgent report from Admiral BN of an Orange expeditionary force concentrating in the northern Kuriles, probably for a strike at Avacha Bay. In response, BA issued an Operation Plan that entailed organizing Task Force 91, the Raiding Force, under the command of Vice Admiral BC, Commander, Air Forces, 9th Fleet. The mission was to prevent any extension of Orange control north of the Kuriles by destroying Orange amphibious and naval strength in the northern Kuriles area. BC's forces included two battleships, organized into Battleship Division 6 and commanded by BC himself; two heavy cruisers, organized into Cruiser

Division 6 and commanded by Rear Admiral BE; four light cruisers, organized into Cruiser Division 12 and commanded by Rear Admiral BF; eight destroyers of Destroyer Squadron 21, commanded by Captain B-2; the eight destroyers of Destroyer Squadron 22, commanded by Captain B-3; the eight destroyers of Destroyer Squadron 24, commanded by Captain B-4; and one fleet carrier, referred to as Carrier Division 1 under Rear Admiral BG and embarking thirty-six Hellcat fighters, thirty-seven Helldiver scout bombers, and eighteen Avenger torpedo bombers. Blue land-based naval patrol planes in contact with the Orange force estimated its strength as two *Nagato*-class battleships; one *Shokaku*-class fleet carrier, with thirty-six Zeke fighters, twenty-seven Judy dive-bombers, and twenty-seven Jill torpedo planes embarked; four *Nachi*-class heavy cruisers; six prewar modernized *Natori*-class light cruisers; and twenty-three destroyers, fifteen of the *Fubuki* class and eight of the *Kamikaze* class.[2]

Vice Admiral BC's force sortied from Kiska and circled, conducting "appropriate" air searches and patrols while maintaining in readiness on the carrier's flight deck a strike of planes in anticipation of the coming engagement. Antisubmarine and combat air patrols were also launched, after which one of the scout planes located the Orange force. In light of the weather conditions and hours of daylight, Vice Admiral BC made sure that all of his ships were fueled before leaving Kiska. His ships were all equipped with radar, as were his scout and torpedo bombers. In addition, the four Kingfisher observation scout planes embarked on the battleships and the twenty-four Seahawk scout planes on his cruisers were also equipped with radar. He paid particular attention to Office of Naval Intelligence publications 222-J, *Supplement to Orange Fleet, 1945,* and *Airplane Characteristics, January 1945,* especially concerning the Orange *Natori* class's Type G-3 torpedoes and the ability of Orange aircraft to carry rockets. Mines were not considered a threat, but BC's carrier fighters were restricted to one thousand pounds of ordnance each. In the Exercise, a number of battle plans had been devised by Vice Admiral BC, and the developing situation was recognized as very similar to one of those plans.[3]

BC determined that he had the strategic initiative and that the Orange force had a merely "offensive-defensive" mission. He also determined that, as had been the case in many other Exercises of this nature, a surface action was "imminently probable," in spite of the presence of aircraft carriers. He was confident that there were no other Orange forces for him to contend with and that he had planned for all possible contingencies. As an example, his present Cruising Disposition offered "excellent" antiaircraft defense characteristics as a supplement to launching his aircraft to defend against air strikes. The numbers of carrier aircraft were roughly equal, though BC had ten more scout bombers than his adversary and Orange had nine more torpedo planes. His scout bombers could carry heavier bomb loads than Orange's. The primary mission for his carrier aircraft was, of course, to knock out

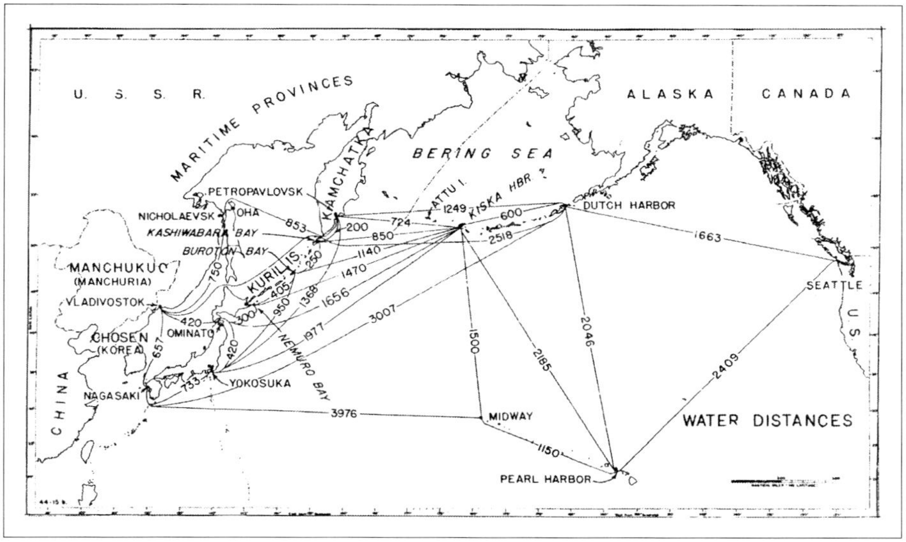

Map 11
North Pacific Ocean Water Distances

the Orange carrier, but they were also to "reduce the speed of the Orange battleline." Given aircraft characteristics, BC was going to have his fighters and scout bombers focus on damaging or destroying the enemy carrier while his "more effective" torpedo bombers focused on the Orange battleships. He was limited, however, in the number of fighters he had and therefore the number he could keep for combat air patrols as opposed to escort and strike duties. Accordingly, he determined—given the importance of knocking out the enemy carrier—to have only sixteen of his thirty-six fighters remain for a CAP. Surprisingly, given his concern about CAP assets, he assigned twelve of the remaining fighters to carry bombs and thus supplement his scout and torpedo bombers. His assumption, however, was that these fighters could jettison their ordnance loads if needed to engage fighters over the Orange carrier.[4]

Having only one carrier, Admiral BC was going to launch his strikes in two waves. If he ordered his first wave to take slow and indirect routes to the enemy force, the second wave could catch up for a combined, coordinated assault, and he could launch both before Orange planes got to him. Nevertheless, he decided to launch the two strikes separately but have them coordinate so that fewer of his fighters were needed to escort the bombers. His pilots were to make sure the Orange carrier was no longer operational before they focused on the battleships, cruisers, and destroyers. Ultimately, BC was most concerned about whether his torpedo bombers could slow down the Orange battleships enough to allow the topedo bombers to shift to the cruisers and employ rockets against the destroyers. Vice Admiral BC thought that if attacked he could survive an Orange air strike because of his force's more effective antiaircraft system. Conversely, he was confident that Orange weaknesses in antiaircraft defense meant that his aircraft would "effect considerable other damage or loss to his [Orange's] task group."[5]

Admiral BC saw a daylight engagement as "imminent" and a night assault by Orange light forces unlikely. He considered various factors about a daytime surface battle. First, his twenty-four destroyers of the 2,100-ton class had a high speed (thirty-six knots) and a heavy torpedo armament (ten tubes). However, he did not think that his destroyers could strike unless supported by his cruisers and until the Orange battleships were under the guns of his own battle line. In constructing his Approach and Battle Dispositions, therefore, he surmised that in case he ordered a destroyer torpedo attack, those ships should be in the Van so they could launch at a greater range and not be under enemy fire for an unnecessarily long time. This, however, would leave his Rear Flank unprotected, and at least four of his destroyers had to remain with the carrier as escorts and plane guards. Likewise, he thought that the Orange commander would put most of his own light forces in the Van, leaving heavier ships to perhaps attempt a reverse action and fall on the Blue Rear Flank. Admiral BC further thought that the Orange commander would assign ten to fifteen of his *Fubuki*-class destroyers to his Van, leaving four to his carrier and from four to nine for his Rear. Moreover, the fifteen *Fubuki*s, along with two light cruisers acting as flagships of destroyer squadrons, would dispose 104 five-inch guns to the sixty on twelve of his destroyers and 167 torpedoes to BC's 120. The Orange ships also had torpedo reloads and four heavy cruisers to Blue's two. In effect, the Orange ships had more speed and more guns, though Blue's heavy cruisers had better armor and his guns could penetrate more effectively in certain situations. The four Orange light cruisers not assigned as flagships were inferior to BC's. They were organized into a separate division possessing only twenty-eight guns above five-inch caliber, as well as sixty-four torpedoes. Vice Admiral BC therefore thought that his light cruisers could deal with them, though his destroyers could not.[6]

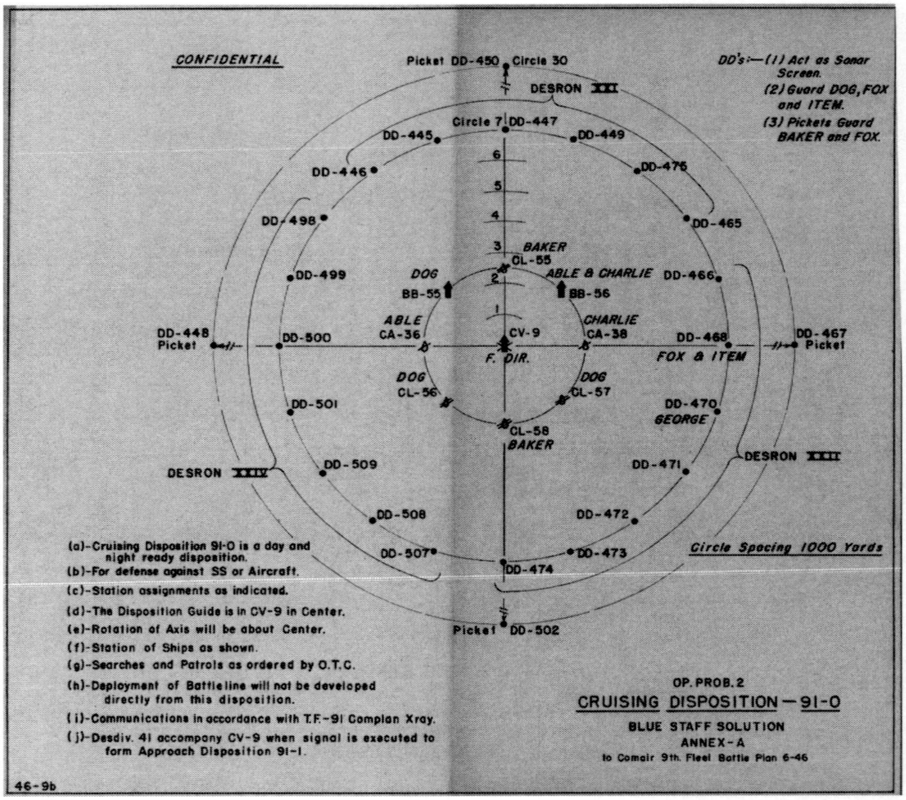

Fig. 142
Operations Problem 2 Cruising Disposition, Blue Staff Solution

In fact, Admiral BC considered his own light cruisers comparable to Orange's heavy cruisers. The latter had speed and gun-range advantages, along with the

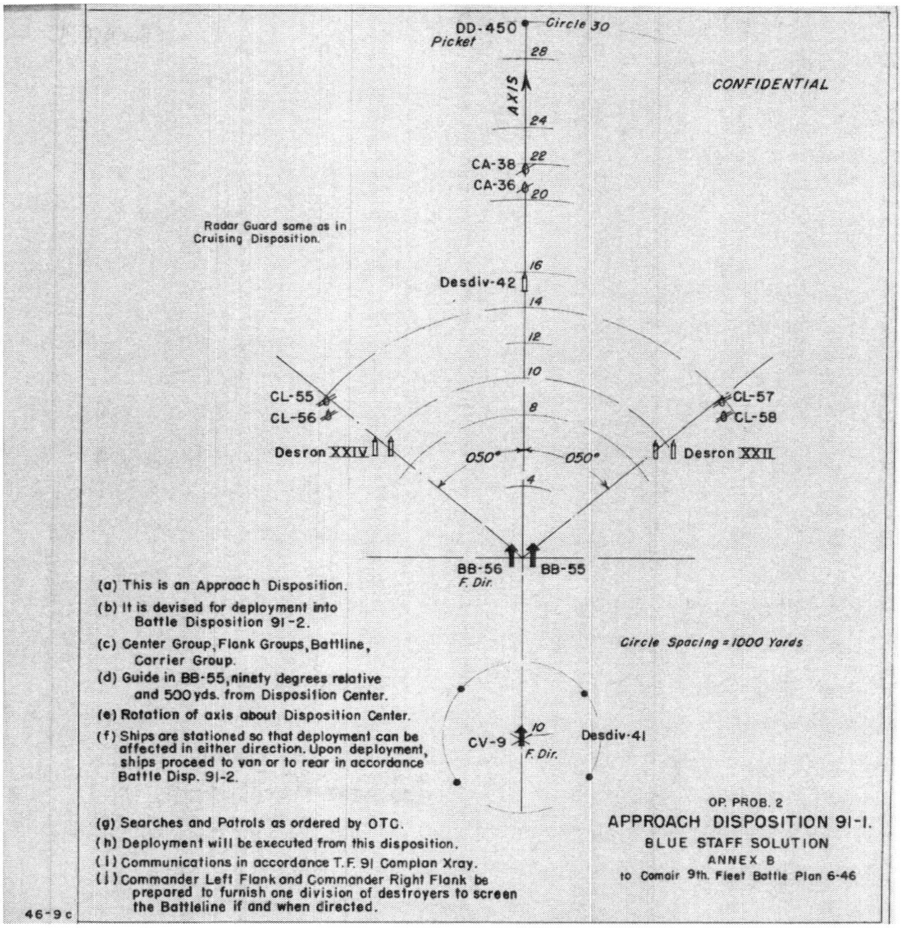

Fig. 143
Operations Problem 2 Approach Disposition, Blue Staff Solution

torpedo edge, but his light cruisers had two more main-battery guns per ship, a greater rate and volume of main-battery fire, more secondary guns, and better side armor. He expected to take "severe punishment" at ranges from thirty-one down to twenty-six thousand yards, "if" Orange could hit him; to receive "considerable" punishment between twenty-six and fifteen thousand yards; and then to "have it all my way" under fifteen thousand yards, if he had enough guns left. Given these circumstances, Vice Admiral BC decided to assign two light cruisers and eight destroyers to defend his Rear Flank against what he saw as a probable assault by Orange light forces. With four destroyers assigned to his carrier, the rest of his light forces—two heavy cruisers, two light cruisers, and twelve destroyers—would go to his Van and would probably be opposed by an Orange force of four heavy cruisers, two to four light cruisers, and ten to fifteen destroyers. However, even with this disposition he did not feel confident in carrying out a torpedo attack on the Orange battleships, unless those ships had been significantly slowed by the torpedo-plane attack. If that proved the case, additional air strikes could be employed to damage further the Orange light forces in the Van in preparation for his surface attack. If his airpower could not damage the Orange light forces, he felt his own light forces, if "wisely" employed, could "take the measure of Orange light forces." Attuned to the Orange Type G-3 torpedoes, he cautioned his flank commanders, "Always be mindful of long range torpedoes. Possibly Orange DDs are so equipped." Concerning the battlewagons, his *North Carolina*–class ships had a slight speed advantage but fewer secondary guns and a shorter secondary-battery horizontal range than their Orange counterparts. Still,

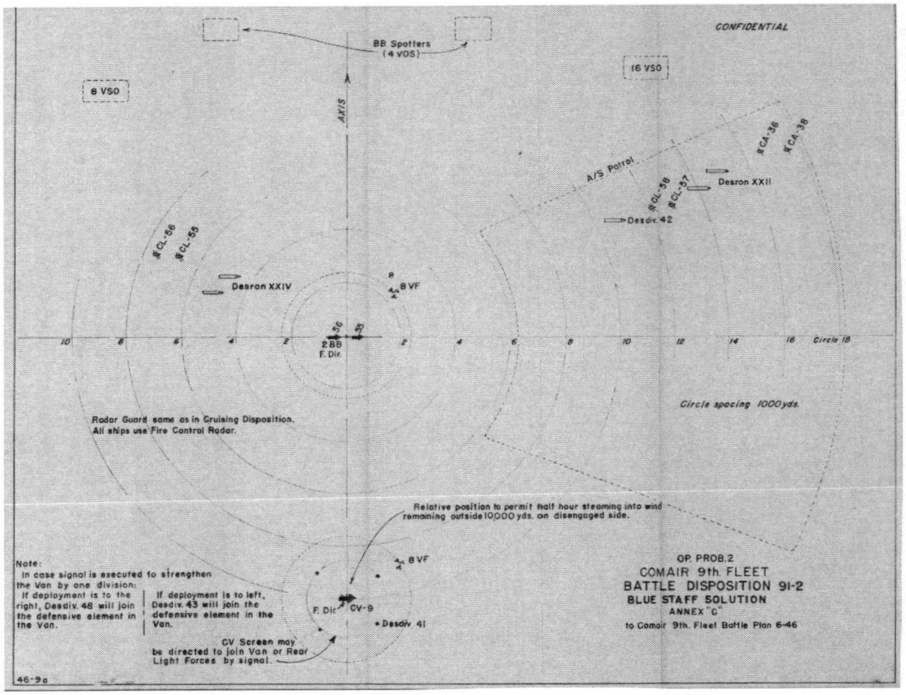

Fig. 144
Operations Problem 2 COMAIR 9th Fleet Battle Disposition, Blue Staff Solution

his battleships were newer, had more and larger guns with greater rates of fire, heavier shells, more effective armor, better compartmentation, and better fire control systems. His battleships additionally had more extensive and more effective antiaircraft weaponry. Since the Orange force had been sighted already, BC's scout and strike planes could keep him well advised about Orange's course, speed, and disposition as well as the estimated damage from Blue's initial air strikes.[7]

Essentially, Admiral BC was counting on one of his light cruisers to provide "effective" defense against four to six Orange destroyers while estimating that one of his heavy cruisers could do the same against three to four destroyers. Any one of his modern destroyers he saw as able to take on two Orange destroyers, but he was concerned about his torpedo planes damaging the Orange battleships after the Orange carrier was knocked out and then, ideally, whittling down the Orange Van light forces before his battleships engaged the Orange battle line. He would, if necessary, use his Van light forces to take out their Orange counterparts and then reverse course so as to employ his Rear Flank destroyers for a torpedo strike. In all this he was mindful, however, of the Orange long-range torpedoes with which he thought Orange destroyers were equipped.[8]

In the final organization of his Operation Plan, Vice Admiral BC designated Battleship Division 6, under Rear Admiral BD, as the battle line; Cruiser Division 6 and Destroyer Squadron 21 (less Destroyer Division 41), under Rear Admiral BE, as the Center Group; two light cruisers and Destroyer Squadron 22, under Captain B-10, as the Right Flank Group; the other two light cruisers and Destroyer Squadron 24, under Rear Admiral BF, as the Left Flank Group; and his fleet carrier and Destroyer Division 41 as the Carrier Group. Rear Admiral BE would also be the commander of the Van. There was continued emphasis on each OTC's taking initiatives as "favorable opportunities presented." In sum, Admiral BC envisioned an air strike knocking out the enemy carrier, and then whittling down the Orange heavy surface ships and light forces. At the same time, his own combat air patrol, battleships, and light forces would defend his carrier from air attack. Afterward, his bombers would defend the carrier from submarine attack and scout for additional enemy forces, and his battleship and cruiser scout planes would augment his antisubmarine patrols. His light forces would then launch a torpedo attack against the Orange battle line, as described above, after which Blue battleships would engage their Orange counterparts. He would be in tactical command, while Admiral BD would command the Blue battleships. The Van and Rear Flank forces would be commanded by the senior officers of those ships, respectively, and then later by Commander, Center Group. He thought his personnel were all "adequate," even "excellent." He also did not see any logistical problems, and he felt that he had adequate scouting, intelligence, and security for his force in terms of fighter protection from enemy scout planes, of radar manned by well-trained Combat Information Center teams, and of destroyers for antisubmarine protection.[9]

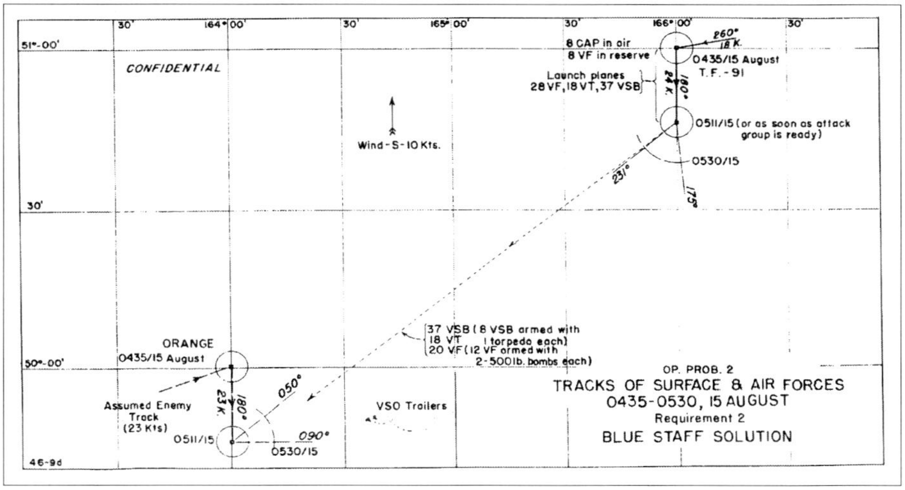

Fig. 145
Operations Problem 2, Blue Staff Solution, Requirement 2, Tracks of Surface & Air Forces

Again, this Operation Plan is a clear combination of both interwar and wartime doctrine that became prevalent in the U.S. Pacific Fleet by 1945. In focusing the Blue carriers on the Orange carrier before they became a support force for a surface engagement, Admiral BC did rely on his airpower's ability to destroy enemy airpower and significantly reduce Orange surface strength, but he nonetheless assumed the necessity of a surface battle. In addition to the pedagogical theory behind these exercises, the northerly latitude explains why surface ships and daylight engagements were emphasized rather than carrier operations, or even a night surface battle, which would have been so typical of the Imperial Japanese Navy. In the end, like the latter wartime Pacific Fleet commanders such as Spruance and Halsey, Admiral BC did not envision having enough carrier forces to destroy the Orange force entirely. That Orange was given just one carrier and a preponderance of surface forces was a fair representation of Japan, but it was assumed that the United States, even with a very large carrier force to carry out the majority of its strikes, might have to employ significant surface assets to destroy the enemy force.[10]

The Blue Staff Partial Solution

The next day, the "Blue Staff Partial Solution" was delivered. Vice Admiral BC saw nothing in the developing tactical situation that caused him to think he could not carry out his mission, which he divided into two phases—finding the Orange force and then destroying it. Since Orange had been located by this time, BC set about his second task—again assuming that Orange would also seek an engagement, that a daylight surface battle would be fought, that enemy submarines would be present, and that the weather would remain as at present. He thought the Orange force might be a raid or a convoy escort, as opposed to an intercept force, though why is unclear, since BC additionally assumed that Orange knew about the Blue Expeditionary Force.[11]

Next, Admiral BC estimated the ranges at which the guns on his heavy surface ships could best penetrate the armor of Orange's comparable vessels. Admiral BC also continued to assume that logistics would not be a problem—surprisingly, since Kiska was 420 miles away. Damaged ships, for instance, would have to travel that distance for temporary repairs, with escorts assigned only when the situation permitted, but this did not apparently weigh heavily on his mind. Conversely, Orange's bases, all well stocked, were only three hundred miles away from the area, but that fact also did not get a great deal of attention from the Blue Commander. Moreover, BC again neglected to outline his afloat train, a critical innovation developed by the U.S. Navy in the Pacific that had allowed it to conduct long-range operations in 1943–45.[12]

Because he knew the location of the Orange force, Vice Admiral BC assumed that it also knew his. He also assumed that each force would maneuver to close the range and that each commander would know the other's movements in a very timely way. For his part, he planned to use the floatplanes from his battleships and cruisers to keep constant tab on the enemy (in fact, a very typical Orange method of scouting). He calculated that the two forces—if they continued maneuvering as they had been—would be in maximum gun range of thirty-two thousand yards in less than two hours and that his aircraft would have whatever sun there was at their backs during their strikes. Because his carrier aircraft had generally done better than Orange's in previous engagements, because he had more battleship guns that could throw heavier shells, and because his ships had better antiaircraft protection and speed, he came to the conclusion that "[I have] definite overall superiority and feel sure that I can destroy the enemy."[13]

As no hostile air or surface forces were in a position to participate in the coming engagement other than those in company with the Orange force, and since his own Cruising Disposition afforded adequate defense, he was going to maintain this disposition for antiaircraft protection until he had launched all of his aircraft and countered Orange's air attack on his carrier. He would then move into the Approach Disposition and then the Battle Disposition when the time came.[14]

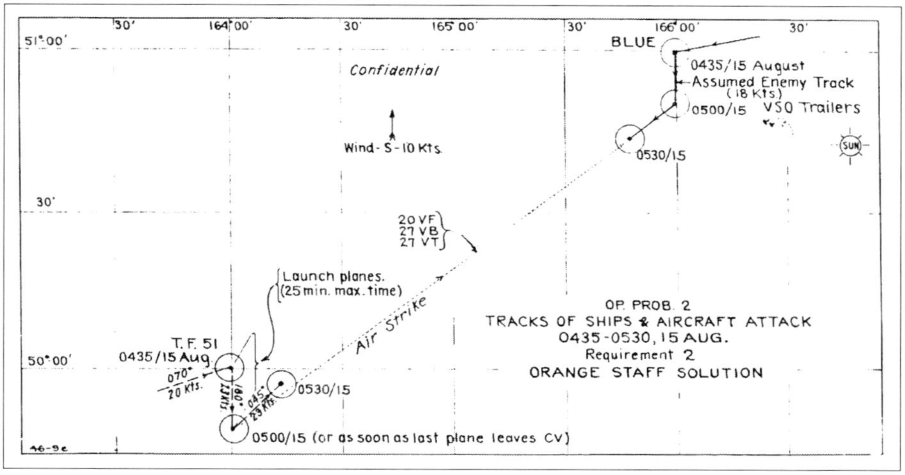

Fig. 146
Operations Problem 2, Orange Staff Solution, Requirement 2, Tracks of Ships & Aircraft Attack

The Orange Statement

The Orange Naval Ministry responded to the concentration of the Blue Fleet in the eastern Marshalls by ordering Admiral OA, commander of the 1st Mobile Fleet, to base his force in the Orange homeland and prepare to maintain control of the eastern approaches to the Orange homeland and the Kuriles. The Chief of the Naval General Staff informed Admiral OA that the Supreme War Council had decided to occupy Avacha Bay, to forestall Blue occupation. OA based his force—less Vice Admiral OB's 5th Fleet—in the Inland Sea. OB's force was based in Hokkaido and was tasked with ensuring Orange control of the northeastern approaches to the homeland and the Kuriles. In recent engagements by the 5th Fleet against Blue forces, Blue had succeeded in advancing forty to fifty transports to the Attu–Kiska region; thereafter, both sides had replaced losses sustained in six weeks of combat. OB now organized the greater part of his surface forces as the Attack Force, under the command of Rear Admiral OC, Commander, Battleship Division 3. Admiral OC had under his command his own two battleships; four heavy cruisers of Cruiser Division 5, under Rear Admiral OD; and four light cruisers of Cruiser Division 10, under Rear Admiral OE. Also under his command were the eleven destroyers of Destroyer Squadron 2, under Rear Admiral OF; twelve destroyers of Destroyer Squadron 10, under Rear Admiral OG, with twenty-six Pete floatplanes embarked; and one fleet carrier, under Captain O-1, with thirty-six fighters, twenty-seven dive-bombers, and twenty-seven torpedo planes. OC was ordered to destroy Blue surface forces west of Attu and north of latitude 40 degrees north. Orange land-based naval patrol planes reported Blue surface strength at Kiska as two *Washington*-class fast battleships, one *Essex*-class fleet carrier, two *San Francisco*–class heavy cruisers, four *Cleveland*-class light cruisers, and twenty-four *Fletcher*-class destroyers. OC subsequently sailed from Buroton Bay to take up a position along the Kiska–Paramushiro line.[15]

Given the fog and low visibility in the area, Rear Admiral OC had his force proceed in a Circular Disposition, conduct "appropriate" air searches and patrols, and keep carrier planes spotted on deck for potential strikes on the Blue force. When OC's scout planes located the Blue force, it too was in a Circular Disposition.

All of OC's ships had been topped off with fuel and his ships, carrier bombers, and floatplanes all had radar. Additionally, his carrier aircraft carried rockets, and his cruisers all had Type G-3 torpedoes. Given the open-ocean nature of the engagement, OC did not consider mines a problem. He needed—for purposes of the Exercise—to determine enemy strength, plan air operations accordingly, and then determine when and where to make contact with the opposing force. Communications were to entail standard TBS and VHF circuits, observing standard radio and radar-security measures.[16]

The Orange Partial Staff Solution

The "Orange Partial Staff Solution" placed the Blue force ninety-seven miles from Rear Admiral OC's task force. No other friendly surface forces were in the area, but neither were any other enemy forces. Admiral OC understood the Blue force to be comparable to his. He was not able to ask instructions of his superior, but he had Vice Admiral OB's Operation Plan. The enemy force was within striking distance of his carrier aircraft, as was he of the Blue carrier-air forces. He surmised that Blue could strike Paramushiro within the next twelve hours with carrier planes and by midnight of that day with surface craft. Judging by "its position and composition this force seems to be just what I was sent to sea to destroy." OC surmised that the Blue force he faced was a portion of the Blue Expeditionary Force assembling in the Attu–Kiska area to occupy Avacha Bay and meant to raid the Orange 5th Fleet in the northern Kuriles and perhaps stop the Orange Expeditionary Force there from seizing Avacha Bay. If the Blue force turned out to be stronger than initially reported, he would consider that fact at the time. OC assumed that his presence now would force the Blue Fleet either to retire or to attack his force rather than raid the 5th Fleet. He also assumed that, given the proximity of the two forces, air attacks were "inevitable," and—in a change from previous Exercises for this class—OC took the air battle to be the determining factor for the outcome of the engagement.[17]

Blue's two battleships, two heavy cruisers, and four light cruisers carried a total of twenty-eight floatplanes, and its fleet carrier had thirty-seven fighters, thirty-six scout bombers, and eighteen torpedo planes. Since he was three hundred miles from his bases and the Blue force was another ninety-seven miles away from him, OC determined that his land-based planes could not be employed for "several hours" and that the battle would have been decided by that time. Blue too would be unable to employ land-based air forces, being three hundred miles from its bases in the Aleutians. He next estimated that while Blue personnel were "in general" skillful and well trained, and had good morale, they were "definitely inferior to my Orange men in discipline, physical stamina, fighting spirit and happy fellowship in death. I, therefore, have a decided advantage in combat efficiency of personnel."[18] OC estimated Blue carrier-based air strength at ninety-one planes to his own ninety. He also estimated that Blue aircraft could carry twice the bomb load of Orange planes and that Blue fighters could carry bombs. Still, he had nine more torpedo planes than his adversary. As concerned surface forces, his battleships had lighter armor, slightly fewer guns in the main battery, and slightly slower speed than Blue's battleships, but these factors were outweighed by a superior secondary battery. In addition, Orange outnumbered Blue in heavy cruisers, and, of course, its heavy cruisers, light cruisers, and destroyers not only carried the outstanding Type G-3 but had reloads for their tubes. Blue surface forces had none of the latter advantages.[19]

Admiral OC went on to state that he was inferior in overall air strength, weight and flexibility of bomb load, and ruggedness of aircraft but that he had advantages in the speed and range of planes. His battleships were one knot slower in speed than their Blue counterparts and had less operational life remaining, one fewer major-caliber gun each in their main batteries, and weaker antiaircraft suites. Still, he thought that his guns could penetrate Blue battleships at ranges above twenty-eight thousand yards and, firing at ninety-degree target angles (that is, at the beams of the targets), from zero to nineteen thousand yards. Blue, on the other hand, could penetrate his battleships at all ranges and target angles, except from eleven to twenty-eight thousand yards at a target angle of forty-eight degrees. Much of this analysis was theoretical, and penetrating range might vary within several thousand yards, depending on list, roll, and yaw—that is, on where and at what angle individual shells struck a target's armor. Nevertheless, Admiral OC took the Blue battleships to have a comparative Fire Effect advantage. As to heavy cruisers, OC outnumbered the Blue commander four to two. Again, theoretically, Blue appeared to be able, salvo for salvo, to inflict more damage on Orange heavy cruisers, which he thought could be penetrated at all ranges presenting either ninety- or zero-degree (bow-on or stern-on) target angles. His heavy cruisers could penetrate Blue's heavy cruisers at the same target angles only at ranges under eighteen thousand yards. Given these realities of taking damage at longer ranges, OC assumed his superiority in this category of ships was more like three and a half to two rather than four to two. His heavy cruisers had a slight speed advantage, but Blue's had heavier armor and better antiaircraft defenses.[20]

His light cruisers he considered no match for Blue's. Although he had a six-to-four numerical advantage, two of his light cruisers were destroyer squadron flagships and really ought to be considered more in the category of destroyers. His other four light cruisers could not stand up to Blue's. Blue's light cruisers were almost as tough and rugged as heavy cruisers in terms of side armor as well as main, secondary, and antiaircraft guns. The real advantage that OC had in cruisers lay in their Type G-3 torpedo armament. His heavy cruisers, for instance, carried a total of 128 of these weapons, sixty-four in tubes and the remainder as reloads. His light cruisers carried a total of 192 shorter-range torpedoes, of which ninety-two were in tubes. He recognized what a hazard torpedoes were on deck during a gunfight, but in general they gave him an immense tactical advantage. Blue destroyers, he surmised, had greater life and speed, though his own were of two classes. His *Fubuki*-class ships were superior in terms of gunfire, while his *Kamikaze*-class units were inferior in this category, but collectively his destroyers carried 366 long-range torpedoes, with 183 in the tubes. Blue destroyers had 240 short-range torpedoes but no reloads; Admiral OC's destroyers could attack at long range without exposing themselves to attack. Longer range even gave them the advantage of reloading (where Maneuver Rules permitted), though overall he still took Blue destroyers to be about equal with his. Esprit de corps was his ace in the hole: "I am inferior in air and battleship strength, but I can rely on the fighting spirit of my men to ensure victory."[21]

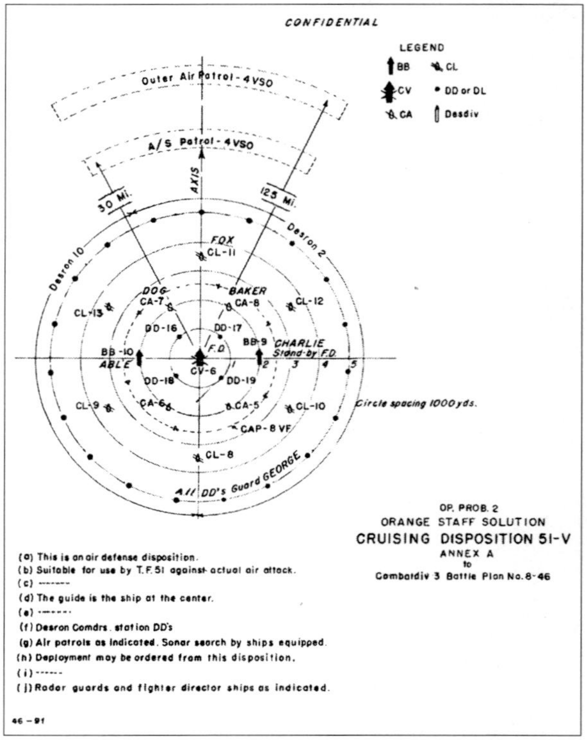

Figure 147
Operations Problem 2, Orange Staff Solution, Cruising Disposition

OC considered that both sides possessed the strategic initiative to an equal degree, but he would proceed as if he had the initiative completely. In terms of doctrine and tasks, destroyers whose primary task was antiaircraft defense could also defend against submarines. Admiral OC thought it unlikely that the Blue commander would detach light forces while the Blue carrier was still operational. He therefore did not think it likely that he would have to deal with a Blue light-force attack in daylight: "Such an attack, unsupported by battleline gunfire, would give me the opportunity of destroying Blue piecemeal." Admiral OC also thought that the chance of encountering a surface attack prior to deployment was minimal. A Blue air attack prior to deployment, however, was not only possible but certain. A Blue air strike could sink or cripple his carrier, damage one or both of his battleships, and damage other ships. If the Blue force did not sustain equivalent damage, the result for Orange could be "disastrous." "If one force loses its carrier air power, the force retaining a carrier and aircraft will have a decisive advantage in the surface engagement."[22]

In addition, OC's battleships had to use their speed to close the range so that his main batteries could fire from their most advantageous position after Blue battleship firepower was reduced. Air attack was the only way to reduce Blue battleship firepower prior to deployment. Defense against Blue air strikes, however, was his major concern in the early stages of approach. Only his carrier, battleships, and heavy cruisers had "effective heavy" antiaircraft batteries, none of them really effective suites overall. His Cruising Disposition, therefore, was formed to give the best "practicable" antiaircraft defense, with priorities of protection being first the carrier, then the battleships, and then mutual support to the rest of the ships. He knew he needed a strong combat air patrol and a good antisubmarine patrol. Scouting against the Blue force would be carried out by the floatplanes of his battleships and cruisers.[23]

In the air strikes themselves, planes assigned to attack roles would relieve the floatplanes over the target. In addition, overcast with a five-thousand-foot ceiling and a cloud thickness of five thousand feet precluded dive-bombing, permitted glide bombing, affected the use of torpedo bombers, reduced the effectiveness of Blue fighter attacks on his own planes, and eliminated any advantages from attacking out of the sun. Admiral OC had advantages of speed and range in his dive-bombers and torpedo planes, and he had nine more of the latter than Blue, but given his weaker antiaircraft defenses, he assumed his force would take as much damage from an initial Blue air strike as it would be able to inflict. If, however, he could effect the early destruction of Blue's carrier, he could prevent further damage to his force. The best way to achieve this was a first strike, but he thought that the Blue commander would have his strike in the air first. This reality necessitated as large a CAP as possible; he had only thirty-six fighters, which meant a smaller escort for his strike force.[24]

In spite of these weaknesses, OC assumed that once the Blue carrier was destroyed, his planes would inflict as much damage as possible on the other surface ships, with the order of priority being battleships, cruisers, and then destroyers. His air-strike force would consist of twenty-seven dive-bombers, twenty-seven torpedo planes, and twenty fighters. Another eight fighters would constitute a CAP; a further eight would have "full ammunition," including rockets; fourteen dive-bombers would have eight rockets and two five-hundred-pound bombs each; and fourteen torpedo planes would be armed with eight rockets and one torpedo each. This initial strike would proceed at low altitude to avoid radar detection until it rendezvoused with the subsequent strike. The latter would consist of twelve fighters, thirteen dive-bombers, and thirteen torpedo planes armed with similar loads. The escort fighters would jettison their rockets if necessary to protect the strike aircraft. In addition, there was to be a "liberal" use of window when enemy radar range was reached. The primary focus for his dive-bombers was to be knocking out the Blue carrier deck, while his torpedo planes damaged or sank at least one battleship. Fighters, however, were to go in first to make strafing and rocket runs in order to reduce antiaircraft fire effectiveness. Battleship and cruiser floatplanes would continue to scout until relieved by the strike force. Four cruiser floatplanes were assigned to this task, and four more were assigned to antisubmarine patrols. When relieved, the cruiser floatplanes would all focus on providing antisubmarine patrols to OC's force and spotting gunfire. His strike aircraft would protect the floatplanes, attack enemy floatplanes, and continue to defend Task Force 51 during the main engagement. During the main engagement, his torpedo bombers would strike again at the Blue battleships, in conjunction with light surface forces.[25]

As concerned the surface engagement itself, Admiral OC did not think a nighttime engagement was likely. Nor did he think it practical, given the conditions, to launch light-force attrition attacks. He did, however, want to use his light forces to the utmost in the gun battle. Given his perceived superiority in heavy cruisers, inferiority in light cruisers, and equality in destroyers, he was going to assign his cruisers and destroyers to certain tasks. Since his heavy cruisers were the main gun strength of his light forces, he was going to place them in the Van on deployment. "The forty 8" guns of these cruisers should also enable me to force Blue center forces back on their battleline and thus hurry or hamper Blue deployment or even to force the direction of his deployment," he wrote. Having two different types of destroyers, of unequal fighting strength, he could not divide them between the flanks except by mixing their units, which he saw as an "abomination." Since he was going to deploy to the south to keep his carrier in company, he wanted to keep his best destroyers on his Right Flank; using his light cruisers to reinforce his Van might so weaken his Rear Flank as to invite a Blue reverse action. If, however, he deployed his light cruisers to the Rear, it could give him an opportunity for a reverse action on Blue. Therefore, his Center was to be composed of the heavy and light cruisers of Cruiser Divisions 5 and 10, respectively, with Destroyer Squadron 10—less two destroyers on his Left Flank—providing a screen for the cruisers. His Right Flank forces would consist of Destroyer Squadron 2, and he would retain the two destroyers of Destroyer Squadron 10 to protect his carrier. Again, in a throwback to interwar American naval doctrine, the student serving as "Admiral OC" was going to keep the carrier close so that it could afford fighter protection to his battle line.[26]

He saw his Type G-3 torpedoes as his "trump card," since they could be used against the Blue battleships by his light forces in the Rear, as well as before visual or even radar contact. In spite of the danger of misses because of dispersion over long ranges, Admiral OC was going to launch his G-3s prior to deployment, as soon as torpedo control data could be obtained from the floatplanes. He was also going to deliver not only early but repeated torpedo attacks and then have his light forces retire in order to, he hoped, draw the Blue battleships into torpedo waters. OC thought the Blue commander would take this bait, in a desire to close the gun range. Reloading could then be carried out during the retirement feint. The light forces in the Center were also going to defend his battle line from enemy light forces; if they were taken under heavy Blue gunfire, they could retreat in the direction of the Orange battle line. Left Flank light forces were also to defend the battle line from Blue light force attacks and, when directed, attack with long-range torpedoes. Torpedo attacks were to be made until all torpedoes had been expended. Given his battleships' weaknesses, he wanted to try to exploit those of the Blue battleships—that is, he wanted to prevent Blue from penetrating his ships while presenting a target angle of forty-five degrees. If Blue's battleships were not damaged, however, he would have to fight at a gun range of more than twenty-eight thousand yards. He hoped that the torpedo attack by his light forces would inflict enough damage to weaken Blue's battleships significantly. A southerly course would allow him to keep his carrier active in flight operations. If his carrier became incapable of flight operations, however, he wanted it to place it on the downwind side, so as to be able to cover it with smoke.[27]

Admiral OC saw the definite possibility of an air strike prior to deployment. If his intelligence was correct, Blue had eighteen torpedo bombers, thirty-six scout bombers, and thirty-seven fighters, some of which—probably about a third—would be detailed as escorts. He also knew that the enemy had a large number of floatplanes on board his battleships and cruisers for scouting and spotting and that Blue ships had "excellent" antiaircraft protection. Additionally assuming that Blue would cruise in a Circular Disposition during the Orange air strike, OC envisioned the Blue carrier dropping behind, with a "suitable" escort but staying within supporting air range, as he intended to do with his carrier. He indicated that the Blue escort would consist of between two to four destroyers, two if the carrier were out of operation and four if it could still launch strikes. He also expected the Blue commander to make maximum use of his light forces at the beginning of the surface engagement. If Blue sent its light forces ahead to stop his own light forces from launching Type G-3s, however, Blue light forces would be out of supporting range of their battleships. Further, to prevent the torpedo attack by his light cruisers, Blue could use its light forces in a reverse action. Similarly, if Blue placed all of its cruisers and half its destroyers in the Van, it would have fire superiority and invite a reverse action. Admiral OC therefore expected Blue to have two heavy cruisers, two light cruisers, and twelve destroyers in the Van, two light cruisers and eight destroyers in the Rear, and two to four destroyers as the Carrier Screen. OC also had to be careful not to allow Blue to use the one-knot speed advantage of its battleships to maneuver those ships into the least vulnerable angles and the most advantageous ranges.[28]

Admiral OC presumed he would lose forces to air strikes, attrition attacks, and perhaps even night attacks, but he thought that he could accomplish his mission with acceptable losses. His Cruising, Approach, and Battle Dispositions took into account his initial air strikes on the Blue force and subsequent move to a surface engagement, which was to be initiated by his light-force torpedo strike. Given the importance of the initial air operations, he wanted a "most capable" air commander for that phase. His most experienced air commander was Captain O-1, but this officer had no flag staff and had no tactical training as the commander of a task force. Therefore, O-1 would be employed to supply advice about the air operations while Admiral OC exercised tactical command, with O-1 then taking separate command of the carrier and its escorts when those ships detached prior to the surface engagement. During the approach, Admiral OC would retain tactical command and then maintain that tactical control during the battle. If the Orange Force was able to engage as expected, Rear Admiral OD, Commander, Cruiser Division 5, would command the forces in the Van, while Rear Admiral OE, Commander, Cruiser Division 10, would command those in the Rear. The battle line would be tactically commanded by Rear Admiral OH, Commander, Battleship Division 3. If OC had to reverse course, the command arrangements would reverse also. If the flagships were damaged, command of the various forces would devolve on the next senior officers.[29]

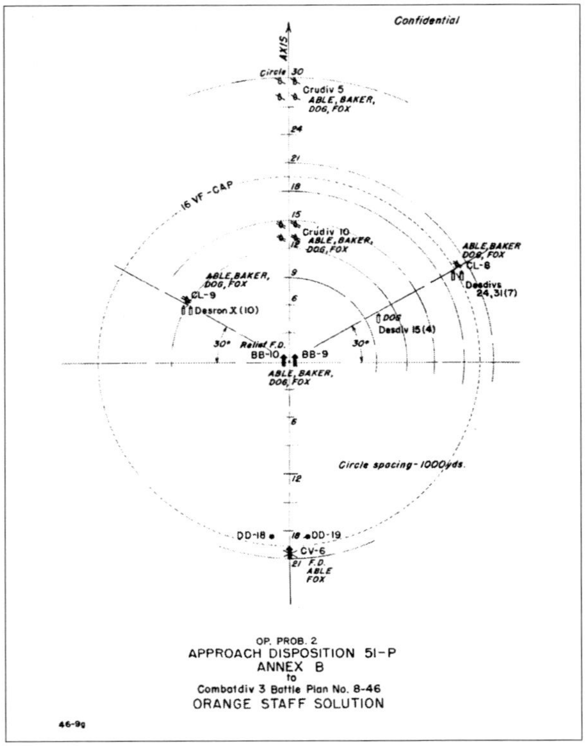

Fig. 148
Operations Problem 2, Orange Staff Solution, Approach Disposition

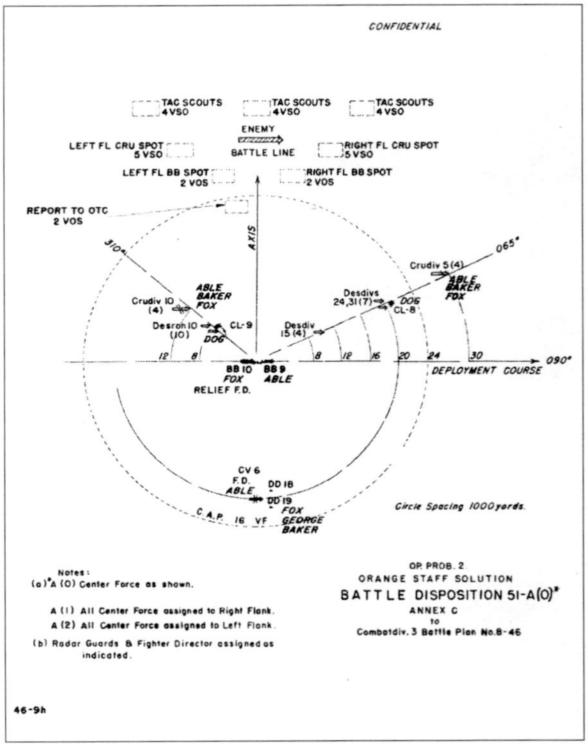

Fig. 149
Operations Problem 2, Orange Staff Solution, Battle Disposition

Conclusion

This Operations Problem is another classic example of the interplay between interwar operational-tactical doctrine and late–Pacific War strategic scenarios. Though the Naval War College staff and students would surely have known by this time that an invasion of the Japanese home islands through the Kuriles and Hokkaido had been rejected in favor of one through Kyushu, they may have been interested in such a northern scenario in case of future hostilities with Japan or, more likely, the Soviet Union. Yet, as Blue and Orange would both show as they carried out their respective board maneuvers, interwar and wartime operational and tactical doctrine here again combined in a way that reflected late wartime U.S. Pacific Fleet practice.[30]

NOTES

1 Command and Staff Class of June 1946, "Operations Problem 2: The Blue Statement," 18 February 1946, pp. 1–2, NHC, folder 2571, box 131, RG 4, NWC.

2 Ibid., pp. 2–4, 7.

3 Ibid., pp. 4–6, 8–9.

4 Ibid., pp. 15–17.

5 Ibid., pp. 17–18.

6 Ibid., pp. 18–21.

7 Ibid., pp. 21–23. For the effects of the Long Lance torpedoes, see Russell Crenshaw, *The Battle of Tassafaronga* (Baltimore: Nautical and Aviation, 1995).

8 Command and Staff Class of June 1946, "Operations Problem 2: Blue Staff Partial Solution, Section B," 19 February 1946, pp. 21–22, NHC, folder 2571, box 131, RG 4, NWC.

9 Ibid., pp. 23–29, 30–34.

10 For accounts of Japanese expertise at night surface warfare in late 1942 and early 1943, see Vincent O'Hara, *The U.S. Navy against the Axis: Surface Combat, 1941–1945* (Annapolis, Md.: Naval Institute Press, 2007), pp. 67–137; Crenshaw, *Battle of Tassafaronga,* pp. 1–206; and Richard Frank, *Guadalcanal: The Definitive Account of the Landmark Battle* (New York: Random House, 1990), pp. 83–123, 428–518. For American naval preponderance in the western Pacific by 1943 and after, see Clark Reynolds, *The Fast Carriers: The Forging of an Air Navy* (Annapolis, Md.: Naval Institute Press, 1968), pp. 113–210. Moreover, there were instances during the latter part of the Pacific War in which orders from American commanders were virtually identical to Admiral BC's. Both Admiral Spruance (during the battle of the Philippine Sea) and Admiral Halsey (during the battle of Leyte Gulf) ordered force dispositions that were very much like those in this war game; see Ivan Musicant, *Battleship at War: The Epic Story of the USS* Washington (New York: Harcourt Brace Jovanovich, 1986), pp. 257–59, 295–303, and see Trent Hone, "U.S. Navy Surface Battle Doctrine and Victory in the Pacific," *NWCR* 62, no. 1 (Winter 2009), pp. 67–105. For the combination of interwar and wartime doctrines, see Hone, "The Evolution of Fleet Tactical Doctrine in the U.S. Navy, 1922–1941," *Journal of Military History* 67 (October 2003), pp. 1107–48; and Hone, e-mails to author, 28 February, 2 March, 3 March, and 14 March 2011.

11 Command and Staff Class of June 1946, "Operations Problem 2: Blue Staff Partial Solution, Section B," pp. 1–9.

12 Ibid., pp. 10–11. For the unique and absolutely vital factor of afloat logistics trains that allowed American naval forces to range across the Pacific between 1943 and 1945, see Reynolds, *Fast Carriers,* pp. 128–30.

13 Command and Staff Class of June 1946, "Operations Problem 2: Blue Staff Partial Solution, Section B," pp. 10–11; and Command and Staff Class of June 1946, "Operations Problem 2: Remainder of Blue Staff Solution, Section B," February 1946, pp. 4–5, 12–14, NHC, folder 2571, box 131, RG 4, NWC.

14 Command and Staff Class of June 1946, "Operations Problem 2: Blue Staff Partial Solution, Section B," p. 15.

15 Command and Staff Class of June 1946, "Operations Problem 2: Orange Statement, Section A," 19 February 1946, pp. 1–3, 6, NHC, folder 2571C, box 131, RG 4, NWC.

16 Ibid., pp. 4–5, 7–9.

17 Command and Staff Class of June 1946, "Operations Problem 2: Orange Staff Solution (Partial), Section A," 19 February 1946, pp. 1–5, NHC, folder 2571D, box 131, RG 4, NWC.

18 Ibid., pp. 6–8.

19 Ibid., pp. 11–12, 8–10.

20 Ibid., pp. 13–16.

21 Ibid.

22 Ibid., pp. 17–18.

23 Ibid., p. 18.

24 Ibid., pp. 18–19.

25 Ibid., pp. 19–21, 34.

26 Ibid., pp. 21–22.

27 Ibid., pp. 22–24, 34.

28 Ibid., pp. 24–26.

29 Ibid., pp. 26–32.

30 Hone, "Evolution of Fleet Tactical Doctrine in the U.S. Navy, 1922–1941," pp. 1123–24; Hone, "U.S. Navy Surface Battle Doctrine and Victory in the Pacific," pp. 67–105; and Hone, e-mails to author, 28 February, 2 March, 3 March, and 14 March 2011.

XV Operations Problem 2

The History of the Maneuver, February 1946

After reviewing both the Blue and Orange strategic and tactical situations, the History of the Maneuver for Operations Problem 2 began by noting that at 0430 Blue Task Force 91 was at a position of latitude 51 degrees north and longitude 166 degrees east on course 260 true, at a speed of eighteen knots, in the Circular Cruising Disposition. At the same time Orange Task Force 51 was at latitude 50 degrees north and longitude 164 degrees east on a course of 070 true at twenty knots. In Move 1, Blue changed course and formation axis to 180 degrees and increased speed to twenty-five knots, while Orange changed course to 180 and increased to twenty-five knots, with a formation axis of 050. At 0435, Blue fleet carrier CV-9 launched a strike of twenty fighters, each armed with a thousand-pound bomb as well as six rockets; twenty-eight scout bombers, each armed with two thousand-pound bombs; nine scout bombers armed with torpedoes; eighteen torpedo bombers, each carrying a torpedo and eight rockets; and eight additional fighters for the combat air patrol. Launching was completed by 0505; the entire group rendezvoused and then proceeded just below the overcast ceiling at five thousand feet. The Blue Strike Group divided its targets: twelve fighters and thirty-seven scout bombers attacked Orange fleet carrier CV-6, while eight fighters and eighteen torpedo bombers attacked one of the Orange heavy cruisers. Four scout planes were also launched from Blue battleships BB-55 and BB-56, with orders to stay clear of the impending air battle and spot for the battle line during the expected surface action. Similarly, Blue Cruiser Division 6 launched its cruiser scouting squadron of eight scout planes, with four planes launched from each of heavy cruisers CA-36 and CA-38 to conduct antisubmarine patrols, while light cruisers CL-55, CL-56, CL-57, and CL-58 each launched an additional two scout planes for gun spotting.*

At the same time, Orange fleet carrier CV-6 launched eighteen fighters armed with eight rockets each; twenty-seven dive-bombers, with one five-hundred-pound bomb and eight rockets each; twenty-seven torpedo planes, with one torpedo and

* Command and Staff Class of June 1946, "Operations Problem 2: History of Maneuver," n.d., pp. 1–4, NHC, folder 2571G, box 131, RG 4, NWC.

eight rockets each; and ten fighter planes for its CAP. The attack group had orders to rendezvous and proceed at two hundred feet to within thirty miles of the Blue force, at which point the fighters and dive-bombers would climb to 4,500 feet for their attack while the torpedo planes dropped to fifty feet for their phase of the strike. Eighteen of the torpedo planes were to attack one of the Blue battleships, while the other nine went after Blue fleet carrier CV-9. The dive-bombers were to glide-bomb CV-9, while the Orange fighters conducted rocket and strafing attacks and engaged the Blue CAP. The Orange Strike Group was launched simultaneously with eight scout observation planes from Orange CRUDIV 5 that were to conduct searches and antisubmarine patrols. Not long after Orange launched its planes, Orange radar picked up the Blue Strike Group at a distance of eighty miles bearing 054 true on a course of 230 at 175 knots at an altitude of four to five thousand feet. Nine Orange fighters were vectored to intercept, while nine others orbited the Orange force at 3,500 feet. Both forces then changed course and slowed slightly; the Blue Air Group commander sighted the Orange air strike at high speed and low altitude. By 0524, Orange TF 51 had gone to General Quarters and worked up to full speed. One minute later, the Orange Strike Group made radar contact with the Blue force, bearing 052 at a distance of thirty miles, on a course of 235 and at a speed of twenty-two knots. At the same time, Blue fleet carrier CV-9 made radar contact with the Orange force, bearing 233 at a distance of thirty-two miles, on a course of 052 and making from twenty-two to 23.5 knots. Blue vectored twelve fighters to intercept the Orange Strike Group. (Pages 4–5.)

Move 2 took place between 0530 and 0600, during which both forces continued on their courses and speeds and both came under air attack. Both forces executed emergency turns, Orange began zigzagging, and both forces took damage. Blue fleet carrier CV-9 was hit by two five-hundred-pound bombs and two torpedoes; it took 70 percent damage but managed to make 16.5 knots. Blue battleship BB-55 was hit by five torpedoes, took 90 percent damage, and could manage only three knots. Orange fleet carrier CV-6 was hit by three thousand-pound bombs and took 60 percent damage but maintained 20.4 knots. Orange heavy cruisers CA-7 and CA-8 were sunk, the former by two torpedoes and the latter by three. Orange also lost two fighters, five dive-bombers, and ten torpedo planes through aerial interception and another four fighters, six dive-bombers, and six torpedo planes to Blue antiaircraft fire. A further two fighters, three dive-bombers, and four torpedo planes were lost afterward; the remaining Orange planes were ordered to make Kamikaze attacks on the Blue light forces. The suicide attacks, however, resulted in only minor damage to Blue forces. The Blue Strike Group attacked the Orange carrier with eleven fighters and twenty-four scout bombers. Another six fighters, six scout bombers, and eleven torpedo bombers attacked Orange heavy cruisers CA-7 and CA-8; it was these attacks that, as noted above, sank them both. Blue lost four fighters, nine scout bombers, and three torpedo bombers to antiaircraft fire. Blue also lost six fighters from its own combat air patrol; only ten fighters, seventeen scout bombers, and six torpedo bombers returned to TF 91. At 0540, the Blue OTC shifted his flag from the damaged battleship BB-55 to the undamaged BB-56, formed his ships into the Approach Disposition at course and axis 250, and slowed to twenty knots. Six minutes later, he ordered the Right and Left Flanks to send destroyer divisions to screen the battle line and then ordered fleet carrier CV-9 to take battleship BB-55 under tow and retire to Attu with two destroyers from Destroyer Division 41. The remaining destroyers of DESDIV 41 were to join the Center force. A few minutes before, the Blue air commander ordered his surviving fighters to remain airborne and attack targets of opportunity until forced to ditch. All other planes were directed to Attu, where they were to rearm and resume attacks as soon as they could. At 0545, Orange fleet carrier CV-6 was ordered to return to base at eighteen knots on course 220. (Pages 5–7.)

In Move 3 during the next half hour, Orange changed to its Approach Disposition, with its Guide on course 052 at a speed of ten knots. By 0603, the Blue battle line was on course 250, but by 0620 it had changed its axis to 230 at twenty knots. At that same time, the Left and Right Flanks were alerted to send a light cruiser to the Van and Rear, respectively, or both to the Van, depending on the signal at execution. At 0610, the Orange OTC was informed by one of his scout planes that Blue battleship BB-55 was under tow at three knots (by 0630 Blue fleet carrier CV-9 was back at sixteen knots, having cut the tow to the battlewagon, which had been lost). The Orange OTC then informed Commander, Submarine Squadron 2 about the Blue battleship and carrier, whereupon Orange forces increased speed to twenty knots. About the same time, the Blue OTC was informed that battleship BB-55 had sunk. The Orange OTC was told that his own carrier had been hit by two torpedoes from a submarine and sunk. Meanwhile, Blue ordered all remaining fighters to return to the vicinity of the Orange force to protect Blue spotter planes. At the same time, Orange Battleship Division 3 launched aircraft as the ships took Battle Disposition stations. Orange by this time was at fifteen knots on a course of 052 and an axis of 050. Move 4 brought both forces to a speed of twenty knots in their Approach Dispositions, at course 230 for Blue and 052 for Orange. Also, since Blue forces had inaccurately figured fuel consumption for the scout planes carrying depth charges on antisubmarine patrol, eight scout planes of CRUDIV 12 were lost; neither the planes nor the crews were recovered. (Pages 7–8.)

In Move 5, from 0645 to 0651, Orange BB-10 was hit by a submarine torpedo, took 10 percent Under Water Damage, and slowed to twenty-three knots. Both forces, however, continued to close at speeds of twenty knots, though Blue had an eight-destroyer screen around its remaining battleship. At 0651, one of the Blue destroyers in the Van made radar contact on two Orange ships of intermediate size at a range of thirty-five thousand yards, bearing 231 to 233. The Blue OTC accordingly ordered the Commander, Center to have the picket retire to avoid Orange cruiser gunfire. In Move 6, both forces continued to close as the Blue picket destroyer retired and the Orange heavy cruisers obtained radar contact on it. Blue and Orange heavy cruisers in their respective Vans simultaneously obtained radar contacts at ranges of 38,500 yards as the Orange Left Flank destroyer division formed an anti-submarine screen around the Orange battle line, which was limited now to twenty-three knots. At this time, the Orange Van commander ordered CRUDIV 5 to attack with torpedoes according to the Operation Plan, and the Orange OTC ordered his ships to launch torpedoes at discretion. In Move 7, with both forces continuing to close and maintaining their respective courses and speeds, Orange CRUDIV 5 increased to thirty knots; the Orange heavy cruisers in the Van were twenty-nine to thirty thousand yards apart. At 0700, Blue heavy cruiser CA-38 made radar contact on two ships closing at 1,500 yards per minute; by the end of Move 7, the targets were only thirty thousand yards away. Orange heavy cruiser CA-6 picked up radar contact, as did Blue light cruisers CL-56 and 58. Orange CRUDIV 5, along with heavy cruisers CA-5 and CA-6, fired thirty-two Type G-3Y torpedoes on a course of 051, with a spread of fifteen degrees, and a speed of forty knots. Twenty-nine of these torpedoes ran "hot, straight, and normal." (Pages 8–10.)

In Move 8, the Blue CAP shot down four Orange scout observation planes as the battle lines continued toward each other. The Blue heavy cruisers opened up on their Orange counterparts; the Orange cruisers turned away after firing their torpedoes as Orange settled on 040 at a speed of fifteen knots. Orange combat air patrols shot down four Blue scout observation planes. Orange heavy cruiser CA-6 continued to develop a radar contact as Blue heavy cruisers CA-36 and CA-38 opened fire with their main batteries on it and Orange heavy cruiser CA-5, at ranges of thirty-five and thirty-one thousand yards, respectively, inflicting 0.6 percent and 0.7 percent damage, respectively. At 0700, the two Blue heavy cruisers changed course to 155 at thirty knots and opened fire again with their main batteries on their Orange counterparts. At 0703, under fire from Orange, Blue changed course to 270 and then 180 to draw the Orange forces toward the Blue Main Body. In Move 9, the Blue battle line continued on the same course and speed while Orange's battle line temporarily slowed to fifteen knots to enable all of its ships to attain Battle Disposition stations. Once that was accomplished, Orange heavy cruisers CA-5 and CA-6 changed course to 155, increased speed to thirty knots, and opened fire on Blue heavy cruisers CA-36 and CA-38. The Orange ships then ceased fire and changed course to 270. After changing course again, they came under fire yet again from their Blue counterparts, though the Blue cruisers could bring only their forward turrets to bear; they were being enfiladed by the Orange ships. Both groups of ships continued firing, meanwhile making radar contact with additional enemy ships.

In this encounter, Orange heavy cruisers CA-5 and CA-6 sustained 10 percent and 20 percent damage, respectively, while each Blue heavy cruiser received 10 percent damage. The move ended with Blue heavy cruiser CA-38 making radar contact on two ships bearing 288 at a range of seventeen miles and Orange light cruiser CL-12 picking up a large blip bearing 049 at a range of seventeen miles. Orange light cruiser CL-8 made radar contact on one ship bearing 054 at a range of seventeen miles, on a course of between 205 and 210, and at a speed of twenty knots. (Pages 10–12.)

Blue heavy cruisers CA-36 and CA-38 then changed course to 210, continuing to fire on Orange heavy cruisers CA-5 and CA-6 as the Blue ships formed a column and increased speed to twenty knots. Orange heavy cruisers CA-5 and CA-6 resumed a speed of twenty knots, on a southwesterly course, while continuing to fire. By the end of this move, the Blue heavy cruisers had sustained a total of 20 percent damage. Orange heavy cruiser CA-5 had sustained 20 percent Above Water Damage, and CA-6 had sustained 30 percent Above Water Damage, had had its search and fire control radars destroyed, and had been slowed by 10 percent. The Orange heavy cruisers changed course to 270 and continued to fire as the other Orange ships continued to deploy. By 0721, all Orange ships were at twenty knots, but not all of them had reached their stations. At twenty-three knots, the Blue heavy cruisers went column left, after which they were no longer enfiladed, and fired with all batteries unmasked. Orange torpedoes at this time passed the Blue battle line without hitting, while the Orange force turned column right to course 110 so that the Orange battle line could bring the Blue heavy cruisers under fire. It did so but without inflicting any damage on the Blue ships, since they were out of gun range. However, by this time, Blue heavy cruiser CA-36 had sustained 30 percent Total Damage. The Blue ships also fired on Orange destroyers but without effect. In the next move, the two Blue heavy cruisers continued firing with six guns each on their Orange counterparts, devoting one turret each to Orange destroyers approaching from the southwest. (This fire-direction decision, however, was considered by the Staff a waste of ammunition since a ship could realistically concentrate on only one target.) Also, by this time the Orange force had reached its new station; there was no sign of such deployment by the Blue force. Blue light forces at this point were concentrated on the Left Flank, and the Orange OTC accordingly thought that the Blue commander was committed to a deployment to the left. Orange, on the other hand, had equal distribution to the right and left, could deploy either way, and was in a favorable position for a torpedo attack by either its Right Flank or Left Flank forces. (Pages 12–14.)

Orange's heavy cruisers CA-5 and CA-6 had been driven in toward the Center, but conditions were nevertheless favorable for an Orange torpedo attack. At this point, Blue light cruisers CL-56 and CL-58 opened fire on Orange light cruisers

CL-12 and CL-13 at twenty-seven thousand yards, but their fire had no effect as they were still out of range. Orange heavy cruiser CA-5 had lost all of its fire-control and search radars by this time and was 10 percent reduced in speed. Blue battleship BB-56 was on a course of 270 at a speed of twenty knots, while the Blue heavy cruisers were on a course of 090 and Blue light cruisers CL-56 and CL-58 were heading north on a course of 200. The Orange heavy cruisers were also on a course of 270 when they started firing at the Blue heavy cruisers at 0727 and receiving fire from the Blue light cruisers. Additionally, the Orange battle line headed north at twenty knots at this time. One minute later, Orange DESDIV 41 fired twenty-four torpedoes to port on a base course of 077 with a twelve-degree spread. Twenty-two of these Type G-3s ran hot, straight, and normal, and a minute later DESDIV 41 fired nine more, eight of which ran, on a nine-degree spread. The targets were the Blue heavy cruisers, which had by now inflicted 30 percent damage on Orange heavy cruiser CA-5. In addition, Blue light cruisers CL-58 and CL-56 had inflicted 10 percent damage on Orange light cruisers CL-12 and CL-13. (Pages 14–15.)

By the time Move 14 opened, the Blue heavy and light cruisers were in column headed south, firing on the Orange heavy cruisers. The Blue light cruisers were dividing their fire between the Orange light cruisers and Orange Destroyer Division 24. The Orange heavy cruisers were on course 140; the Orange destroyers had turned toward the Blue cruisers while launching torpedoes. At the same time, the Orange light cruisers had "countermarched" and were on a northwesterly course. At 0731, Orange heavy cruisers CA-5 and CA-6 fired thirty-two G-3s on a course of 073 with a spread of fifteen degrees, twenty-two of which ran. Orange DESDIV 17 fired another twenty-four torpedoes, of which twenty-two ran, on course 036 with a spread of twelve degrees. Additionally, Orange DESDIV 61 fired twenty-four torpedoes on course 039 with the same spread and with the same number running. The Orange OTC thought that he had inflicted 60 percent damage on the Blue heavy cruisers because of a deliberately false report by the Blue OTC. At the end of this move, Orange heavy cruisers CA-5 and CA-6 had sustained 40 percent damage from Blue heavy cruisers CA-36 and CA-38. Blue light cruisers CL-56 and CL-58 were firing on Orange light cruisers CL-12 and CL-13, respectively, as well as on Orange DESDIVs 24 and 31. In addition, Blue battleship BB-56 had fired on Orange heavy cruiser CA-6 at a range of thirty-seven thousand yards but inflicted no damage, as it was out of range. Orange heavy cruisers CA-5 and CA-6 also maintained fire on Blue heavy cruisers CA-36 and CA-38, without inflicting further damage. (Pages 15–16.)

In Move 15, the Blue heavy cruisers turned right to 270 while still firing at the Orange heavy cruisers. The Blue light cruisers did the same, still dividing their fire between the Orange heavy cruisers and destroyers. Other Blue forces had still not deployed into battle. The Orange heavy cruisers changed course to the southwest in order to open the range and start firing on their Blue counterparts. The Orange light cruisers were now headed northwesterly but were not engaged, and the Orange battle line was on a northerly course at its prior speed. Orange Left Flank forces were doing quite a bit of maneuvering by this time to rectify their positions vis-à-vis the Orange Main Body (though the maneuver looked unnecessary to the umpires). At 0734, Orange DESDIV 24 fired nine Type G3-X torpedoes (six successfully) on a course of 061 with a total spread of eight degrees. Orange DESDIV 31 fired twelve Type G3-Xs on the same course with a spread of sixteen degrees; eight ran. Orange heavy cruisers CA-5 and CA-6 fired on Blue heavy cruiser CA-38 at a range of twenty-four thousand yards; CA-38 now had 40 percent Total Damage. Orange light cruisers CL-12 and CL-13 were under fire at the same range by Blue light cruisers CL-56 and CL-58 and sustained 20 percent Total Damage. (Pages 16–17.)

At 0736, the Blue heavy cruisers were on course to the west, the Blue light cruisers were headed northwest, and the Blue Main Body's course remained unchanged. Orange forces on a northwesterly course were still under fire from the Blue heavy cruisers, Orange destroyers were under fire from the Blue light cruisers, and the entire Orange battle line disposition appeared to be headed northwest. The Blue heavy cruisers were no longer under fire. Instead, Orange fire had been shifted to the Blue light cruisers to try to draw the Blue heavy cruisers into torpedo waters, the Orange OTC having realized that the Blue heavy cruisers were not as badly damaged as previously reported. As Move 16 came to an end, Blue light cruisers CL-56 and CL-58 continued to fire on Orange DESDIVs 24 and 31 at a range of twenty thousand yards, inflicting 10 percent Total Damage, while Blue light cruiser CL-58 received 10 percent damage from Orange DESDIV 24 and Orange light cruiser CL-8. In Move 17, the Orange heavy cruisers resumed fire on their Blue counterparts. The Blue light cruisers had meanwhile turned south under concentrated fire from Orange light cruisers and destroyers, and Blue battleship BB-56 was taken under concentrated fire by both Orange battleships. The Orange light cruisers were no longer under fire, but Orange DESRON 2 was being shot at by Blue light cruisers CL-55, CL-56, and CL-58. (Pages 17–18.)

At 0740, Orange DESDIV 15 fired thirty-six G-3Y torpedoes on a course of 062, with a spread of eighteen degrees, and saw thirty-two of them run. Orange DESDIV 16 fired another thirty-six G-3Ys on the same course, with a twelve-degree spread; twenty-two ran. The Blue heavy cruisers were the targets of all of these torpedoes. At the same time, Orange battleships BB-9 and BB-10 were firing on Blue battleship BB-56, but no damage was inflicted, as it was out of range; BB-56 returned fire on the Orange battlewagons but with the same results and for the same reason. Orange heavy cruisers CA-5 and CA-6 came under fire from Blue heavy cruisers CA-36 and CA-38, respectively. Orange heavy cruiser CA-5 had sustained

40 percent damage by this time, while CA-6 had no damage assessed during this move. (Pages 18–19.)

By Move 18, the Blue heavy cruisers were still under fire from their Orange counterparts, and the Blue light cruisers were still under fire from the Orange light cruisers and destroyers. Blue light cruiser CL-58 was struck by a torpedo and was left moving at five knots in a "sinking" condition. Blue light cruiser CL-56 was struck two minutes later and reduced to a similar state, while Blue light cruiser CL-57 went north to open the range and Blue battleship BB-56 headed south to close with the Orange battle line. The Orange destroyers were still under fire from the Blue light cruisers, the Orange heavy cruisers were still being fired at by their Blue counterparts, and the Orange battleships were being engaged by Blue battleship BB-56 as the move ended. The Blue battleship had still not received damage from its Orange counterparts because of the range, but Blue heavy cruiser CA-36 was now 40 percent damaged by fire from Orange heavy cruisers CA-5 and CA-6, and Blue light cruisers CL-56 and CL-58 had each sustained a torpedo hit that inflicted 80 percent damage. Blue light cruiser CL-55 had fired on Orange light cruiser CL-8 and DESDIV 31, inflicting 20 percent damage on these units generally. (Pages 19–20.)

In Move 19, Blue light cruisers CL-56 and CL-58 were sinking, while Blue heavy cruisers CA-36 and CA-38 and Blue battleship BB-56 were all under fire from Orange battleships BB-9 and BB-10. Blue Right Flank destroyers heading west were taken under fire by the Orange battleships' secondary batteries, and Blue light cruiser CL-57 was engaged by the Orange battleships' main batteries. The Orange heavy cruisers now retired to the southwest under fire from their Blue counterparts, which were also firing on the Orange Right Flank destroyers. The Orange battleships headed north under divided fire from Blue battleship BB-56, and Orange CRUDIV 10—the Left Flank cruisers—was under secondary-battery fire from the Blue Right Flank destroyers. Unknown to the Blue heavy cruisers, they were closing the range to the Orange Right Flank forces, which were in a very favorable position for a torpedo attack. (It was unclear to the umpires why the Blue ships exposed themselves in this way.) At the end of Move 19, Blue DESDIVs 43 and 44 were under fire from the secondary batteries of Orange battleships BB-9 and BB-10 at a range of twenty-four thousand yards but suffered no damage, because of the range. Blue light cruiser CL-57 was also attacked by Orange battleship BB-10, suffering 20 percent damage at a range of thirty thousand yards. In addition, Orange heavy cruisers CA-5 and CA-6 came under fire from Blue heavy cruisers CA-36 and CA-38, respectively. Orange heavy cruiser CA-6 now had 50 percent Total Damage, but there was no damage assessed in this move on Orange heavy cruiser CA-5. Orange light cruiser CL-8 and Orange DESDIV 31 came under fire from Blue light cruiser CL-55, all three Orange units suffering 30 percent Total Damage. Orange battleships BB-9 and BB-10 continued to take fire from Blue battleship BB-56, whose gunnery inflicted 20 percent damage on Orange BB-10. Orange BB-9, however, remained undamaged in this move. (Pages 20–21.)

By the next move, Orange had intercepted a dispatch from the Blue OTC to his Right Flank destroyers to be prepared for a possible right turn by the Orange battleships. Blue Left Flank destroyers were closing on a course of 270 but were not under fire, while the Blue Right Flank destroyers were being shot at by the Orange battleships' secondary batteries, the Orange Left Flank light cruisers CL-10 and CL-11, and Orange DESDIVs 17 and 61. Orange heavy cruisers CA-5 and CA-6 then came around to a northerly course and were no longer under fire, though the Orange Right Flank destroyers—DESDIVs 15 and 16—were being attacked by Blue light cruiser CL-55 and Blue light cruiser CL-57 was being attacked by Orange battleships BB-9 and BB-10. The Orange battleships were in column on a northerly course, still under divided fire from Blue battleship BB-56; the Orange Left Flank destroyers were on a northeasterly course under secondary-battery fire from Blue light cruiser CL-57; and Orange CRUDIV 10 was headed northeast and under secondary-battery fire from Blue light cruiser CL-57. At 0750, Orange light cruiser CL-8 fired eight Type G-1Z torpedoes on course 085 with a one-degree spread and saw six of them run. A minute later, Blue DESDIV 43 fired forty Type G-1Z torpedoes on course 255 with an eighteen-degree spread; twenty-eight ran. Blue DESDIV 44 now fired forty additional torpedoes on a course of 272 with the same degree spread and saw twenty-eight run as well. About this time, Blue heavy cruisers CA-36 and CA-38 were sunk by the torpedoes from Orange light cruiser CL-8 and DESDIVs 24 and 31. (Pages 21–23.)

By the end of Move 20, Orange battleship BB-9 had sustained 10 percent damage from Blue battleship BB-56, Orange light cruisers CL-10 and CL-11 suffered the same percentage of damage from fire by Blue light cruiser CL-57, and Orange DESDIV 17 had also taken 10 percent damage from fire by Blue DESDIVs 43 and 44. In addition to the lost heavy cruisers, Blue light cruiser CL-56 had received 90 percent damage from fire by Orange battleship BB-9. Orange BATDIV 3 also caused 40 percent damage to Blue light cruiser CL-57. Further, Orange BATDIV 3 and light cruisers CL-9 and CL-11 were firing on Blue DESDIV 43 with their secondary batteries, while Orange light cruiser CL-9 and DESDIV 61 were shooting at Blue DESDIV 44. Each Blue DESDIV took 20 percent damage from this gunfire. (Pages 22–23.)

This action was followed in the next move by the sinking of Blue light cruiser CL-57 by torpedoes fired by Orange DESDIVs 17 and 61. This loss caused the Blue Right Flank destroyers to retire to the east under secondary-battery fire from the Orange battle line, Left Flank cruisers, and destroyers. Blue battleship BB-56 was also retiring under fire from Orange battleships BB-9 and BB-10, while the Blue Left Flank destroyers were headed south under fire from the Orange cruisers and destroyers. In addition, Blue DESDIV 48 was headed south under fire from Orange DESDIVs 15 and 16, having lost two of its four ships to Orange torpedoes. The Orange Right Flank destroyers had also lost one ship to torpedoes but were headed southwest clear of Blue fire, while the Orange heavy cruisers were headed northeast, also without taking any fire. Orange DESDIV 15 was headed north under fire from Blue light cruiser CL-55 and DESDIV 48; Orange DESDIV 16 was headed in the same direction under fire from the same Blue light cruiser. The Orange battleships were still receiving divided fire from Blue battleship BB-56 and had turned south; the Orange Left Flank cruisers and destroyers went southeast, and these destroyers were engaged by Blue light cruiser CL-57 until the latter sank. Move 21 resulted in 20 percent damage to Orange DESDIVs 17 and 61 by Blue DESDIVs 43 and 44, while Orange DESDIVs 15 and 16 took fire from Blue light cruiser CL-58; the result was 10 percent damage on DESDIV 15. Blue light cruiser CL-57 had been sunk by torpedoes, Blue battleship BB-56 had a total of 10 percent damage from its Orange adversaries, and Blue DESDIV 48 had received 60 percent damage from DESDIVs 15 and 16. Blue DESDIV 43 received 80 percent Total Damage from the concentrated fire of Orange light cruisers CL-10 and CL-11, the secondary batteries of Orange battleship BB-10, and Orange DESDIV 7, while Blue DESDIV 41 took 10 percent damage from Orange light cruiser CL-8 and DESDIVs 24 and 31. Blue DESDIV 44 came under fire from Orange light cruiser CL-9 as well as DESDIV 61 and secondary-battery fire from Orange battleship BB-9, all resulting in 70 percent Total Damage. (Pages 23–24.)

Move 22 began with Blue DESDIV 47 moving on a westerly course and coming under fire from the Orange heavy cruisers, while Blue DESDIV 41 and the attached destroyer DD-446, moving in the same direction, were sunk by torpedoes from Orange DESDIV 15. Orange battleships BB-9 and BB-10 were still under fire from Blue battleship BB-56, and Orange DESDIV 16, on an easterly course, was under fire from Blue light cruiser CL-55. Orange DESDIV 15 was also on an easterly course and being engaged by the same Blue light cruiser and Blue DESDIV 48. The Orange Left Flank heavy cruisers on a southeasterly course were not under fire, nor were the Right or Left Flank destroyers of either side. Orange was also maintaining its Battle Disposition, while Blue was scattered and all of its units were under fire, though Orange did lose a previously damaged destroyer, DD-46, sunk at 0757. Overall damage to the two sides' units included 30 percent damage to Orange DESDIV 15 from Blue light cruiser CL-55 and DESDIVs 43, 44, and 48; 20 percent damage to Blue battleship BB-56 from Orange battleships BB-9 and BB-10; and 30 percent damage to Blue DESDIV 43 from Orange BB-10's secondary batteries as well as the concentrated fire of Orange light cruisers CL-10 and CL-11 and Orange DESDIV 17. Blue DESDIV 48 also took 20 percent damage from Orange DESDIVs 15, 16, and 17, while Blue DESDIV 44 took 40 percent damage from the secondary-battery fire of Orange battleship BB-9, Orange light cruiser CL-9, and Orange DESDIV 61. (Pages 24–26.)

In the subsequent move, Blue DESDIV 43 came under the concentrated fire of Orange light cruisers CL-9, CL-10, and CL-11, the destroyers of Orange DESDIVs 16, 17, and 61, and the secondary battery of Orange battleship BB-9. Blue DESDIV 44 was also under secondary-battery fire from Orange battleship BB-10, and Blue DESDIV 47 came under fire from, according to the sources, Blue DESDIV 44 and Blue light cruiser CL-57. Blue DESDIV 48 by this time was being fired at by Orange DESDIV 15; Blue DESDIV 42 was under fire from Orange light cruisers CL-8, CL-12, and CL-13, as well as Orange DESDIV 24; and Blue battleship BB-56 was under "double concentration" fire from both Orange battleships. The Orange heavy cruisers were still clear of fire and were now on a southwesterly course. The Orange Right Flank light cruisers were also clear of fire, though the Orange Right Flank destroyers, on a northerly course, were under concentrated fire from Blue DESDIVs 41 and 42. Orange DESDIV 15 now came under fire from Blue light cruiser CL-55 and DESDIV 48; Orange DESDIV 16 was also under fire from the same Blue light cruiser. Orange Left Flank cruisers and destroyers were on a northwesterly course and free of fire, though the Orange Main Body itself was headed southeast by this time and under divided fire from Blue battleship BB-56. Blue DESDIV 43 was also under fire from Orange BB-10, CL-9, CL-10, CL-11, and DESDIV 61. Damage sustained during the move included Blue DESDIV 48's remaining two ships receiving 60 percent damage while Orange DESDIV 16 took 30 percent damage from Blue light cruiser CL-55 and DESDIVs 43 and 44. Orange battleship BB-10 had taken 30 percent damage from Blue battleship BB-56; Orange BB-9 was undamaged. Orange DESDIV 15 had received 70 percent damage from the fire of Blue light cruiser CL-55 and of DESDIVs 43, 44, 47, and 48. As the move ended, Blue DESDIV 48 fired twenty G1-Z torpedoes on a course of 271 with a spread of eight degrees, of which fourteen ran. Blue DESDIV 42 fired another forty G1-Zs on a course of 291 and a spread of eighteen degrees, but only twenty-eight ran. At the same time, Blue DESDIV 47 fired an additional forty G1-Zs in a similar spread on course 290; again, only twenty-eight of these ran. (Pages 26–27.)

In Move 24, starting at 0800, Blue battleship BB-56 was still headed southwest and was still under concentrated fire from both Orange battleships. The Blue Right Flank destroyers were headed southeast, as was Blue light cruiser CL-55, the latter having been taken under fire by the Orange battleships' main batteries. Blue DESDIV 44 was headed north and still under secondary-battery fire from Orange battleship BB-10; Blue DESDIV 42 was headed in the same direction under secondary fire from Orange battleship BB-9 as well as Orange light cruisers CL-8, CL-12, and CL-13, and Orange DESDIV 24. Blue DESDIV 48 was headed west under the concentrated fire of Orange battleship BB-10, the secondary-battery fire of Orange heavy cruiser CA-5, and Orange DESDIV 15. The Orange Right Flank by this time was headed northwest under fire from Blue DESDIV 42; Orange DESDIV 16 was headed south under fire from Blue light cruiser CL-55; and Orange heavy cruisers CA-5 and CA-6 were headed northwest, not under fire. By 0806, however, all ships of Orange DESDIV 15 had been sunk. At the same time, the Orange battleships were headed north under fire from Blue battleship BB-56, and the Orange Left Flank cruisers and destroyers were maneuvering "considerably" and headed east. Damage in this move included 20 percent damage on the two Orange battleships from their Blue adversary; Orange DESDIV 15 sunk in its entirety; Orange DESDIV 24 taking 50 percent damage from fire by Blue DESDIVs 47 and 48; Orange DESDIV 24 receiving 50 percent damage from Blue DESDIV 42; Orange DESDIV 31 receiving 60 percent damage from Blue DESDIV 42; and Orange DESDIV 16 receiving 40 percent damage from Blue DESDIV 43 and Blue light cruiser CL-55. (Pages 27–28.)

The Blue force, in turn, saw light cruiser CL-55 receive 20 percent damage from the main batteries of Orange battleship BB-9. In addition, Blue DESDIV 42 received 10 percent damage from the Orange battleships' secondary batteries as well as fire from light cruisers CL-8, CL-12, and CL-13 and DESDIVs 24 and 31. Further, all the ships of Blue DESDIV 43 received 80 percent damage, were put out of action, and were reduced to five knots by fire from Orange light cruisers CL-9, CL-10, and CL-11. Blue DESDIV 44 was almost as badly damaged by Orange DESDIV 16, with 70 percent damage, while Blue DESDIV 47 received 10 percent damage from the main batteries of Orange heavy cruisers CA-5 and CA-6. Blue DESDIV 48 was 60 percent damaged by Orange DESDIV 15 as well as by secondary-battery fire from Orange battleship BB-10 and heavy cruisers CA-5 and CA-6. Blue battleship BB-56 also took 30 percent damage from the fire of the two Orange battlewagons.

Given the greater damage inflicted by Orange than by Blue, the question arose in the minds of umpires as to why in this move Orange Left Flank forces closed the Blue Main Body in a way that put themselves directly in the path of torpedoes from the Blue Right Flank destroyers instead of taking a course that would have avoided that threat. (Pages 28–29.)

The folly of this action was seen in the next move when Orange light cruisers CL-9 and CL-11 were each struck by a torpedo after the Orange Left Flank ignored an order by the Orange OTC to avoid torpedo waters when closing the Blue ships. At the same time, Blue DESDIVs 43 and 44 were headed southeast, as was the Blue battleship, the latter under fire from Orange battleships BB-9 and BB-10. Blue light cruiser CL-55 was also being shot at by the Orange battleships as it headed west; the Blue Left Flank destroyers were headed in the same direction. Blue DESDIV 47, meanwhile, was under fire from the secondary batteries of Orange battleship BB-9; from Orange light cruisers CL-8, CL-12, and CL-13; and from Orange DESDIV 16.

Blue DESDIV 42 was also under fire from the secondary battery of Orange battleship BB-10; from light cruisers CL-8, CL-12, and CL-13; and from Orange DESDIV 24. Additionally, Blue DESDIV 48 was being engaged by the secondary battery of Orange battleship BB-10 and Orange heavy cruiser CA-5's secondary battery as the Blue ships headed west. The Orange battleships were still under divided fire from Blue battleship BB-56. In addition, Orange Right Flank forces were headed north under fire from Blue DESDIV 42, while Orange DESDIV 16 was headed east and under fire from Blue light cruiser CL-55. Orange heavy cruisers CA-5 and CA-6 were headed east, receiving no fire. It was now that the Left Flank cruisers and destroyers closed the Blue Main Body on an easterly course in torpedo range of the Blue Right Flank destroyers. (Page 29.)

Aside from the loss of light cruisers CL-9 and CL-11, DESDIV 16 received 50 percent damage from Blue light cruiser CL-55, DESDIV 31 was 90 percent damaged by fire from Blue DESDIV 42, and DESDIV 24 suffered 70 percent damage by fire from Blue DESDIV 42. Blue DESDIV 42 was 40 percent damaged by fire from Orange light cruisers CL-12 and CL-8 and Orange DESDIVs 24 and 31, while Blue light cruiser CL-55 took 30 percent damage from the main batteries of Orange battleship BB-9, Orange light cruiser CL-8, and Orange DESDIV 24. Blue DESDIV 48 took fire and 70 percent damage from Orange DESDIV 16, as well as from the secondary battery of Orange battleship BB-10 and from Orange heavy cruiser CA-5. Additionally, Blue DESDIV 42 took 30 percent damage from the fire of Orange light cruiser CL-12, while Blue DESDIV 47 took 20 percent damage from the combined fire of Orange DESDIV 16, the main battery of Orange heavy cruiser CA-6, and the secondary batteries of Orange battleships BB-9 and BB-10. (Page 30.)

The next move saw Blue DESDIV 44 retire to the southeast under the secondary-battery fire of Orange battleship BB-10, the Orange heavy cruisers, and Orange DESDIV 16. At the same time, Blue DESDIV 42 retired southeastward under the fire of Orange DESDIV 24; of Orange light cruisers CL-8, CL-12, and CL-13; and of the secondary battery of Orange battleship BB-9. Blue DESDIV 48's two remaining destroyers came under fire from Orange heavy cruiser CA-5 and the secondary battery of Orange battleship BB-10. The Orange heavy cruisers were still clear of fire, but Orange DESDIV 16 was sunk in its entirety by torpedoes from the Blue Left Flank destroyers, Orange DESDIV 24 was badly damaged (80 percent) by fire from Blue DESDIVs 42 and 47, and all ships of Orange DESDIV 31 were sunk by Blue DESDIV 42 at 0807. The Orange battleships by this time were no longer under fire, and the Orange Left Flank cruisers and destroyers were closing the enemy at high speed on an easterly course. Orange light cruiser CL-13 took 30 percent damage from the main battery of Blue battleship BB-56, and Orange light cruiser CL-8 took 40 percent damage from Blue light cruiser CL-55. Blue DESDIV 47 had taken 40 percent damage from the secondary batteries of BATDIV 3 and Orange heavy cruisers CA-5 and CA-6, while Blue battleship BB-56 had taken 40 percent damage from the divided fire of the two Orange battlewagons. Blue light cruiser CL-55 had also taken 40 percent damage from Orange battleship BB-9, and all ships of Blue DESDIV 48 sank under secondary-battery fire of the same Orange battleship and Orange heavy cruiser CA-5. All ships of Blue DESDIV 42 were also sunk from gunfire by Orange DESDIV 24 and Orange light cruisers CL-8, CL-12, and CL-13. (Pages 30–32.)

In Move 27, Blue took defensive measures that included effectively interfering with Orange radio traffic. In addition, Blue battleship BB-56 continued divided fire on the two Orange battlewagons. Blue DESDIV 43, however, severely damaged, continued to be attacked by Orange light cruiser CL-10 and Orange DESDIVs 17 and 61. Blue DESDIV 44 was also heavily damaged but was retiring to the east to get clear of enemy fire. Blue DESDIV 47 was also retiring, in this case to the southeast, under the fire of the Orange heavy cruisers, light cruiser CL-8, and DESDIV 24. Worse, from the Blue perspective, by 0809 all of its DESDIV 48 had been sunk. The Orange heavy cruisers were not under fire at this point, but Orange DESDIV 24, on a southeasterly course, was heavily damaged and under fire from Blue DESDIVs 42 (perhaps resurrected by umpires after its recent destruction) and 47, while Orange DESRON 2 was under fire from Blue light cruiser CL-55 and DESDIVs 42 and 47. The Orange battleships by this time were free of fire, and the Orange Left Flank was on a northeasterly course and still closing Blue, also free of enemy fire. In this move, Blue DESDIV 47 succeeded in sinking Orange DESDIV 24. Orange light cruiser CL-8 took 60 percent damage from Blue light cruiser CL-55, and Orange light cruiser CL-12 took 40 percent damage from the main battery of Blue battleship BB-56 and from Blue DESDIV 44. Damage to the Blue battleship from Orange battleships BB-9 and BB-10, meanwhile, reached 50 percent. Blue DESDIV 47 also came under fire from the secondary batteries of Orange battleship BB-10 and from Orange heavy cruisers CA-5 and CA-6. The result was 50 percent damage. Blue DESDIV 43 fared even worse, taking 90 percent damage from Orange DESDIV 61. (Pages 32–33.)

Move 28 entailed an unsuccessful attempt by Orange to deceive Blue with false radio traffic. By this time, Blue battleship BB-56 was retiring eastward without escort. The ships of Blue DESDIV 43 were heavily damaged and under fire from the two Orange battleships, light cruiser CL-10, and DESDIVs 17 and 61. Blue DESDIV 44 was also retiring east, heavily damaged and under fire from Orange light cruisers CL-12 and CL-13. Blue light cruiser CL-55 was also retiring to the east under the main-battery fire of the Orange battleships. Blue light cruisers CL-56 and CL-58 were heading east as well, both in a sinking condition. Moreover, DESDIV 47 was still under fire from the Orange heavy cruisers, light cruiser CL-8, and DESDIV 24. Orange DESDIV 41 was sunk in its entirety. Heavy cruisers CA-5 and CA-6, on an easterly course, were under fire, and the Orange battleships were still taking fire from their Blue counterpart. The Orange Left Flank cruisers and destroyers, now in column, headed east at high speed to close with the Blue Main Body. (Pages 33–34.)

All of Blue DESDIV 43 was sunk by Orange light cruiser CL-10 and the secondary-battery fire of Orange battleship BB-9. Blue light cruiser CL-55 sustained 80 percent damage from the main batteries of the two Orange battlewagons, while Blue DESDIV 47, under fire from Orange heavy cruisers CA-5 and CA-6 and Orange light cruiser CL-8, took 80 percent damage. Further, Blue DESDIV 44 received 80 percent damage from the fire of Orange light cruisers CL-12 and CL-13; Orange light cruiser CL-12 in turn took 70 percent damage from Blue DESDIVs 44 and 47. Orange light cruiser CL-13 sustained 60 percent damage from the same two Blue destroyer units. By this time, the remaining Blue forces, including Blue battleship BB-56, were widely scattered and retiring at maximum speed to the east. Orange forces retained their approximate positions in the Battle Disposition. The Orange Right Flank light forces were considerably reduced in numbers but well concentrated and in a good position for pursuit. The Orange battleships were still in company and pursuing the Blue battlewagon at high speed, while Orange Left Flank forces remained intact, less the two light cruisers they had lost. These units were also closing with Blue at high speed and in concentrated formation when the end of the maneuver was called. (Page 34.)

The most intriguing aspect about this Exercise, of course, is that Orange wins, in spite of the fact that it was set late in the Pacific War, when Blue would have held most, if not all, of the advantages. The other surprising aspect is how strongly this maneuver was marked by the combination of interwar and wartime American naval doctrine that has been noted above. Each side only had one aircraft carrier, and those ships, used very early in the Exercise, were able only to knock out each other's airpower and inflict some damage on small portions of the enemy's surface forces. Given the paucity of airpower, the maneuver quickly became essentially a surface gun and torpedo fight. Pedagogical aspects aside, however, there were significant reasons for conducting an Operations Problem like this one. In the book's conclusion, I will look at reasons for the Naval War College constructing such scenarios, as well as the rationale behind other Exercises and Operations Problems recounted in this book.*

* For the pedagogical theory behind war gaming, see Francis McHugh, *A Manual of War Games* (Newport, R.I.: Naval War College, 1960), pp. 1–3; McHugh, *Fundamentals of War Gaming* (Newport, R.I.: Naval War College, 1966); McHugh, "War Gaming and the Navy Electronic Warfare Simulator," *NWCR* 19, no. 7 (March 1967), pp. 22–32; and McHugh, "Eighty Years of War Gaming," *NWCR* 21, no. 7 (March 1969), pp. 88–90. For interwar and wartime U.S. Pacific Fleet doctrine, see Trent Hone, "The Evolution of Fleet Tactical Doctrine in the U.S. Navy, 1922–1941," *Journal of Military History* 67 (October 2003), pp. 1107–48; Hone, "U.S. Navy Surface Battle Doctrine and Victory in the Pacific," *NWCR* 62, no. 1 (Winter 2009), pp. 67–105; and Hone, e-mails to author, 28 February, 2 March, 3 March, and 14 March 2011.

Blue versus Orange
Some Conclusions

Between 1945 and 1946, the U.S. Naval War College continued to focus on Japan as its primary hypothetical enemy in the Pacific as it transitioned from reduced wartime operations back to its peacetime status as the Navy's premier command and staff college. The Naval War College's continued focus on Orange was an understandable result of the suddenness of the end of the Pacific War and the academic-administrative reality that the curriculum for 1945–46 had already been established. In fact, the late spring of 1946 would see the beginning of a shift to Purple (the Soviet Union) as the primary U.S. enemy, and although Orange would appear again in the summer of 1946, the 1946–47 academic year would see Purple become the number-one hypothetical enemy of the U.S. Navy.

Highly noticeable in the war games described in this book are the theoretical underpinnings of that shift as well as more practical applications of interwar and wartime doctrine. The theoretical and pedagogical bases of war gaming were such that the strategic and tactical scenarios themselves were not as important as the simulation of the stressful conditions under which naval commanders would have to make decisions. The games could be set in strategic and tactical contexts that the Naval War College staff and the Director of the Maneuver thought would provide learning opportunities and experience in naval operational decision making. In fact, the games did not even have to be played out to a conclusion; whether one side won or lost might even obviate the learning experience the staff was attempting to afford the student officers (see chapter 1).

Nevertheless, the strategic and geographic foci of the various Exercises and Maneuvers—especially those concerning operations in the South China Sea aimed at severing supply and transportation lines between Southeast Asia and the Orange homeland—reflected what had been, or might have been, wartime experiences. Operations Problems gaming the interdiction of Orange shipping in the South China Sea as a way of getting at heavy surface units are not fundamentally different from Admiral Halsey's attempt in January 1945 to flush the hybrid battleship/carriers *Ise* and *Hyuga* and other survivors of the battle of Leyte Gulf out of Cam Rahn Bay with carrier air strikes in the South China Sea region.[1] Other aspects of the Operations Problems were also very realistic, such as the use of radio deception. In fact, there was at least one case of successful tactical radio deception by the U.S. Navy, during its surface battle near the Komandorski Islands in the North Pacific.[2]

Potential operations toward the southern Kuriles and Hokkaido also made quite a bit of sense in the context of the once-assumed Allied need to invade the home islands in the fall of 1945, though it was probably common knowledge by the time the game was planned that the invasion was actually to have taken place in Kyushu.[3] Operations in Hawaii and Micronesia would have seemed a bit less likely by that time, given the strength of American naval power and its projection into the western Pacific and East Asia. American naval officers, however, were not confident—in fact, were fairly pessimistic—about the prospects for turning the postwar Pacific Basin into an "American lake" and thereby obviating the need for the Navy to fight its way across the Pacific again in some future conflict.[4]

What seems less plausible upon initial examination was the operational and tactical focus on surface warfare. At a time when the prewar gunship-oriented fleet had been firmly transformed into one focused on carrier aviation and submarines, it might seem highly anachronistic—pedagogy aside—for the Naval War College to have created and practiced a series of war games, only one of which placed carrier air forces at the center. With the exception of Operations Problem 5, played in October 1945, the war games all simulated operational and tactical engagements fought seemingly with interwar doctrine; carriers were restricted to fighting other carriers, providing reconnaissance for the surface fleets, and employing their airpower to aid surface ships in their decisive battles. Similarly, submarines were barely seen in the war games, and when they were, they were primarily used as scouting platforms.[5]

Yet upon reexamination of the war games in the context of wartime events, Admiral Spruance's disposition of 17 June 1944 at the Battle of the Philippine Sea can be seen as the model for many of these war games, in that he directed the enemy carriers to be knocked out first, as a prelude to air strikes on Japanese heavy surface forces, which in turn were preparations for a surface engagement. The war games designed by Admiral Pye and his staff were, in fact, highly similar to Spruance's orders off the Marianas as well as to Admiral Halsey's initial preparations for the battle of Cape Engaño in October 1944.[6] Far from anachronistic throwbacks to the interwar period, Naval War College Exercises and Operations Problems represented combinations of interwar and wartime doctrine that reflected some of the most recent experiences of the U.S. Pacific Fleet.[7]

Some of these operational and tactical scenarios also make more sense if viewed in the context of their own historical settings. One might expect to find records of war games in which large numbers of Blue fast carriers—with battleships, cruisers, destroyers, and afloat trains in support—used massed carrier air strikes, and only carrier air strikes, to demolish Orange carriers, surface ships, and shore bases. Instead, the typical scenario included a small number of Blue carriers facing off against an even smaller number of Orange carriers, both Blue and Orange carriers being put out of action early, and a surface battle resulting. It must be remembered, however, that the U.S. Navy was rapidly demobilizing after 1945; the huge fleet planned for the postwar period during the war itself was already being reduced significantly in size, primarily for fiscal reasons.[8] Admiral Pye and his officers in the fall of 1945 might easily have assumed that in the future the Navy might have to fight in the Pacific or even across the globe with a severely limited number of aircraft carriers.

It should also be recalled that at the end of November 1942 the Navy had only two fleet carriers and two fast battleships operational in its entire fleet, because of how many aircraft carriers had been sunk that year and how few fast battleships had yet been commissioned. In addition, there was such a shortage of cruisers after the battles around Guadalcanal in 1942–43 that by the summer of 1943 the Navy was forced to rely heavily on destroyers as its primary combatant platform.[9]

In 1950, at the beginning of the Korean War, the Navy would again find itself with just two carriers deployed in the Pacific.[10] (Additional ships were provided as soon as President Harry Truman made the decision to use American forces to defend South Korea.) To be sure, the situation in 1950 was different from that of 1942; in 1950, most of the American carriers were in the Atlantic and Mediterranean, because of the central focus on Europe in the context of the Containment Doctrine—there were additional carriers to deploy to the Pacific. In late 1942, there were no more fleet carriers at all, until the arrival of the *Essex* class in 1943.[11] The small number of carriers that had been available in 1942 may have been on the minds of Pye and his staff as they envisioned a potential future in which American foreign-policy commitments exceeded military assets. The war games they designed seem less anachronistic against this politico-military-fiscal backdrop.

Even during the war, a number of training exercises in the Pacific pitted only one or two carriers with surface escorts against either a surface force or a force of similar composition. Some of these wartime exercises, conducted even as the fast carrier task force concept was maturing, looked like war games played in Newport, both before the war and in 1945–46.[12]

It is also possible that the Naval War College staff at this time thought future naval engagements would take place in the northern Pacific, where more reliance on surface ships would result from poor weather and the degraded performance of aircraft and other vital equipment, such as radars. These scenarios were a major theme during Admiral Spruance's tenure as President of the College, when carrier operations were being demonstrated in the spring of 1946 to be less than tenable in Arctic waters.[13]

Fear about the future as a driver of war games needs to be taken into account in some additional possible contexts as well. Looking back today, it seems ridiculous that the Naval War College was replaying what appear to be surface-warfare scenarios from war just ended. If, however, we realize that these were really combined-arms scenarios that reflected late–Pacific War U.S. naval practice and that no one knew what the next conflict would entail, the war games again lose some of their anachronistic quality. No one in the American national security establishment at this time had a crystal ball or could predict what the next war would be like. Obviously, there was a good majority who understood the Soviet Union to be the next probable enemy, but even the Joint Chiefs of Staff did not dismiss the possibility of a resurgent Japan until 1946.[14]

In addition, whoever the probable next enemy, there was no guarantee that the next war would be a "brush fire," as the United States would actually encounter in Korea and Vietnam. Between 1945 and 1950, even the Joint Chiefs assumed that the next war would mostly likely be a full-scale conventional conflict, fought in the Atlantic, the Mediterranean, and Europe, almost exactly mirroring World War II in those theaters.[15] If no one in the American national security establishment was postulating differently, why should not the Naval War College assume the next conflict might look very much like the last one? A resurgent Japan—once evacuated by the occupying powers—might rebuild its fleet, or the Soviet Union might build an oceangoing fleet of its own; why then would the U.S. Navy assume in 1945 and 1946 that it would never again have to fight another blue-water navy? If the Navy had to fight such a force in the Pacific, perhaps in northern latitudes, the battles of the late Pacific War needed to be rethought and refought.[16] We can comfortably say it did not happen that way and that they should have known better, but that would not have been so easy for the Naval War College staff and students to see in 1945–46.

Fear of a repetition of the dire strategic situation of the early years of the Pacific War, fiscal uncertainty about future force structure, and immersion in wartime American naval doctrine gave Pye and his staff the basis for many of the chosen scenarios. The result was a series of war games primarily oriented toward combined-arms surface engagements between opposing fleets, their scenarios reflecting the realization that although fast carriers had become the focal point of American naval power, they could not perform all missions and needed to act in conjunction with heavy surface forces, such as fast battleships. The fast battleship was no longer the focal point of postwar naval forces, but it was not yet obsolete.[17]

The bottom line is an appreciation of how difficult a charge the Naval War College faced in this period. In an era of rapid demobilization, domestic reconversion, and a foreign policy that was changing in a revolutionary way at breakneck speed, its staff and students needed to translate the lessons of the war into new strategy, tactics, and procedures for the fleet. Moreover, they had to do all this on a slim budget and in a way that would help American naval forces deter future war, which was now to be especially avoided because of atomic weapons. In the context of all of these transformations—aside from the undeniable bureaucratic inertia of having simulated conflicts against the same enemy for the previous forty years—providing for the security of the Republic and perceiving who its future enemy might be were extraordinarily difficult tasks.

NOTES

1 Ivan Musicant, *Battleship at War: The Epic Story of the USS* Washington (New York: Harcourt Brace Jovanovich, 1986), p. 318.

2 John Lorrelli, *The Battle of the Komandorski Islands, March 1943* (Annapolis, Md.: Naval Institute Press, 1984), p. 139.

3 John Skates, *The Invasion of Japan: Alternative to the Bomb* (Columbia: Univ. of South Carolina Press, 1995).

4 See Hal M. Friedman, *Creating an American Lake: United States Imperialism and Strategic Security in the Pacific Basin, 1945–1947* (Westport, Conn.: Greenwood, 2001), pp. 1–36.

5 These post-1945 war games were in sync with interwar American naval doctrine, which had so evolved by 1941 that the battleships were the strategic reserve, to be used only after auxiliary forces, such as aircraft carriers, had provided for fleet air defense, knocked out enemy carriers, and taken care of enemy land-based air forces. Subsequently the carriers, along with cruisers, destroyers, and submarines, were to whittle down the enemy heavy and light surface forces so that the American battle line could finish the job in a surface gunfight; see John Kuehn, *Agents of Innovation: The General Board and the Design of the Fleet That Defeated the Japanese Navy* (Annapolis, Md.: Naval Institute Press, 2008). For the ascendancy of the naval aviators in the wartime and immediate postwar Navy, see Vincent Davis, *Postwar Defense Policy and the U.S. Navy, 1943–1946* (Chapel Hill: Univ. of North Carolina Press, 1966), pp. 157–270.

6 Musicant, *Battleship at War,* pp. 257–59, 295–303. See "The Conduct of the Maneuver," June 1945, p. 47, NHC, folder 2525-D, box 124, RG 4, NWC.

7 Trent Hone, "The Evolution of Fleet Tactical Doctrine in the U.S. Navy, 1922–1941," *Journal of Military History* 67 (October 2003), pp. 1107–48; Hone, "U.S. Navy Surface Battle Doctrine and Victory in the Pacific," *NWCR* 62, no. 1 (Winter 2009), pp. 67–105; and Hone, e-mails to author, 28 February, 2 March, 3 March, and 14 March 2011.

8 Cdr. Dennis Ringle, USN (Ret.), conversation with author, 30 May 2010; Joel Davidson, *The Unsinkable Fleet: The Politics of U.S. Navy Expansion in World War II* (Annapolis, Md.: Naval Institute Press, 1996), pp. 141–81.

9 Vincent O'Hara, *The U.S. Navy against the Axis: Surface Combat, 1941–1945* (Annapolis, Md.: Naval Institute Press, 2007), pp. 67–137, 162–92.

10 Thomas Cutler, "Sea Power and Defense of the Pusan Pocket, June–September 1950," in *The U.S. Navy in the Korean War,* ed. Edward Marolda (Annapolis, Md.: Naval Institute Press, 2007), pp. 1–51.

11 Dr. Jeffrey Barlow, History and Archives Division, Naval History and Heritage Command, Washington, D.C., e-mail to author, 22 September 2010.

12 Musicant, *Battleship at War,* p. 189.

13 Friedman, *Digesting History: The U.S. Naval War College, the Lessons of World War Two, and Future Naval Warfare, 1945–1947* (Newport, R.I.: NWCP, 2010), pp. 143–49, 156–61, 183–96, 199–207.

14 Friedman, *Creating an American Lake,* pp. 1–36.

15 Steven Ross, *American War Plans, 1945–1950* (London: Frank Cass, 1996), pp. 25–136.

16 Friedman, *Creating an American Lake,* pp. 1–62; Friedman, *Governing the American Lake: The U.S. Defense and Administration of the Pacific, 1945–1947* (East Lansing: Michigan State Univ. Press, 2007), pp. 83–115; Eric Morris, *The Russian Navy: Myth & Reality* (London: Hamish Hamilton, 1977), p. 25; Jürgen Rohwer and Mikhail Monakov, *Stalin's Ocean-Going Fleet: Soviet Naval Strategy and Shipbuilding Programmes, 1935–1953* (London: Frank Cass, 2001), pp. 178–220.

17 Hone, e-mails to author, 28 February, 2 March, 3 March, and 14 March 2011.

APPENDIX: UNIDENTIFIED WAR GAMERS

Records were not available to identify all student officers by their real names. Those who could be identified have been named and their photographs given where they first appear in the book. Those who could not be identified are listed below by their Exercise or Maneuver roles, as well as where they first appear in this account.

Chapter VI, Operations Problem 4 (Command and Staff Class of December 1945)

Commander, Blue 3rd Fleet, p. 121

Admiral OA, Orange 1st Fleet Commander, p. 124

Chapter VII, Operations Problem 5, Blue (Command and Staff Class of December 1945)

Admiral BN, Commander-in-Chief, U.S. Pacific Command–Commander-in-Chief, Pacific Ocean Area, p. 141

General BX, Commander, Southwest Pacific, p. 141

Admiral LMB, Commander, UN Forces, Southeast Asia, p. 141

Captain B-31, Commander, Mine Craft, 7th Fleet, p. 142

Rear Admiral B-7, Commander, Cruiser Division 12, 7th Fleet, p. 143

Rear Admiral B-11, Commander, Cruiser Division 11, 7th Fleet, p. 143

Rear Admiral B-12, Commander, Cruiser Division 15, 7th Fleet, p. 143

Rear Admiral B-9, Commander, Carrier Division 11, 7th Fleet, p. 143

Captain B-9, Commander, Destroyer Squadron 5, 7th Fleet, p. 143

Captain B-23, Commander, Destroyer Squadron 6, 7th Fleet, p. 143

Captain B-24, Commander, Destroyer Squadron 47, 7th Fleet, p. 143

Chapter VIII, Operations Problem 5, Orange (Command and Staff Class of December 1945)

Rear Admiral O-2, Commander, Carrier Division 2, p. 171

Rear Admiral O-4, Commander, Cruiser Division 4, p. 171

Rear Admiral O-11, Commander, Cruiser Division 10, p. 171

Rear Admiral O-10, Commander, Cruiser Division 17, p. 171

Rear Admiral O-15, Commander, Cruiser Division 21, p. 171
Rear Admiral O-20, Commander, Destroyers, Task Force 2, p. 171
Captain O-25, Commander, Destroyer Squadron 10, TF 2, p. 171
Captain O-21, Commander, Destroyer Squadron 2, TF 2, p. 171
Commander O-48, Commander, Patrol Division 30, TF 2, p. 171
Commander O-31, Commander, Patrol Division 31 and Task Unit 2.3.3, Escort Group No. 3, TF 2, p. 171
Commander O-27, Commander, Patrol Division 32, TF 2, p. 171
Vice Admiral OF, Commander, Southwest Area Fleet and Task Group 2.3, p. 180

Chapter X, Operations Problem 6, Blue (Classes of December 1945)

Admiral BB, Commander, Pacific Fleet and 13th Fleet, p. 211
Vice Admiral BZ, Commander, Amphibious Force, p. 211
Fleet Admiral BA, Commander-in-Chief, Pacific Fleet, p. 213
Commanding General, 14th Air Force, p. 214
Commanding General, 20th Air Force, p. 214
Commander, Submarine Force, Pacific Fleet, p. 215
Commander, Northern Pacific Force, p. 215
Commander, Fleet Flagship, p. 215
Commander, Search and Reconnaissance Group, p. 216
Commander, Anti-Submarine Warfare Group, p. 217
Commander, Service Squadron 10, p. 217
Commander, Forward Area, Central Pacific, p. 218
Commander, Service Force, Pacific, p. 218
Commander, Air Forces, Pacific, p. 218
Commander, South Pacific Force, p. 218
Commander-in-Chief, Southwest Pacific Area, p. 218
Commander, Marshall-Gilberts Area, p. 219
Commander, Logistic Support Group, p. 220
Commander, Destroyers, Pacific, p. 220
Commandant, 14th Naval District, p. 221
Vice Admiral CA, Commander, Fast Carrier Task Force, p. 223
Atoll Commander, Ulithi, p. 225

Chapter XI, Exercises (Command and Staff Class of June 1946)

Commander, Blue 5th Fleet and Officer in Tactical Command, p. 233
Commander, Blue Task Force 58 and Commander, Fast Carrier Task Force, p. 233
Vice Admiral BL, Commander, Blue Battleship Squadron 2 and Task Force 59, Striking Force, p. 233
Rear Admiral BA, Commander, Blue Battleline, p. 235
Vice Admiral BC, Commander, Center, Light Forces, p. 235

Rear Admiral BD, Commander, Right Flank, Light Forces, p. 235
Rear Admiral BF, Commander, Left Flank, Light Forces, p. 235
Rear Admiral BG, Commander, Carrier Group 4, p. 235
Rear Admiral OJ, Commander, Orange Striking Force, p. 245

Chapter XII, Demonstrative Problem (Command and Staff Class of June 1946)

Admiral B, Commander, Blue 7th Fleet, p. 256
Vice Admiral BE, Commander, Blue Control Force, p. 256
Vice Admiral BA, Commander, Blue Main Body, p. 257
Vice Admiral BF, Commander, Blue 7th Amphibious Force, p. 257
Rear Admiral BC, Commander, Blue Fast Carrier Task Force, p. 257
Rear Admiral BD, Commander, Blue Fleet Air Wings, p. 257
Captain BL, Commander, Blue Mine Force, p. 258
Captain BJ, Commander, Blue Patrol Force, p. 258
Rear Admiral BH, Commander, Blue Service Force, p. 258
Commanding General, Blue 6th Army, p. 258
Captain BS, Commander, Blue Submarine Force, p. 258
Rear Admiral BG, Commander, Blue Philippine Sea Frontier, p. 258
Rear Admiral BM, Commander, Blue Task Group 73.1, Covering Group, p. 263
Rear Admiral BO, Commander, Blue Task Group 73.2, Escort Group 1, p. 263
Rear Admiral BP, Commander, Blue Task Group 73.3, Escort Group 2, p. 263
Rear Admiral BF, Commander, Blue Task Group 73.4, Escort Group 3, p. 263
Captain BG, Commander, Blue Task Group 73.5, Hunter Killer Group, p. 263
Vice Admiral BF, Commander, Blue Task Force 31, the Striking Group, and Commander, Air Forces, Blue 3rd Fleet, p. 265
Captain B-15, Commander, Blue Task Group 31.2, Air Support Group, and Fleet Air Wings 1 and 10, p. 265
Captain B-14, Commander, Blue Task Group 31.3, Submarine Group, and Submarine Squadron 6, p. 265
Vice Admiral BF, Commander, Blue Task Force 52, p. 265
Rear Admiral B-1, Commander, Blue Battleline and Battleship Division 5, p. 266
Rear Admiral B-3, Commander, Blue Center Force, p. 266
Captain B-20, Commander, Blue Right Flank Force, p. 266
Rear Admiral B-2, Commander, Blue Carrier Group, p. 266
Vice Admiral BZ, Commander, Blue Carrier Division 3 and Task Group 58.4, p. 268

Chapter XIII, Operations Problem 1, Orange (Command and Staff Class of June 1946)

Admiral OA, Commander, Combined Fleet, p. 272
Vice Admiral OB, Commander, 5th Fleet, North East Area Fleet, and Task Force 51, p. 272
Vice Admiral OZ, Commander, 1st Mobile Fleet, p. 272
Rear Admiral OC, Commander, Battleship Division 3, p. 272
Rear Admiral OD, Commander, Cruiser Division 5, p. 272

Rear Admiral OE, Commander, Cruiser Division 10, p. 272
Rear Admiral OF, Commander, Destroyer Squadron 2 and Northern Search Group, p. 272
Rear Admiral OG, Commander, Destroyer Squadron 10 and Southern Search Group, p. 272
Rear Admiral OR, Deputy Task Force Commander and Commander, 12th Air Fleet and Air Group, p. 272
Rear Admiral OS, Commander, Patrol Wing 5, p. 273
Rear Admiral OJ, Commander, Submarine Squadron 1 and Submarine Group, p. 273
Vice Admiral OH, Commandant, Ominato Guard District Forces, p. 273

Chapter XIV, Operations Problem 2, Blue and Orange (Command and Staff Class of June 1946)

Admiral BN, Commander, Blue Pacific Fleet, p. 297
Admiral BA, Commander, Blue 9th Fleet and North Pacific, p. 297
Vice Admiral BC, Commander, Air Forces, 9th Fleet and Task Force 91, the Raiding Force, p. 297
Rear Admiral BE, Commander Blue Van, Center Group, and Cruiser Division 6, p. 298
Rear Admiral BF, Commander, Blue Left Flank Group and Cruiser Division 12, p. 298
Captain B-2, Commander, Blue Destroyer Squadron 21, p. 298
Captain B-3, Commander, Blue Destroyer Squadron 22, p. 298
Captain B-4, Commander, Blue Destroyer Squadron 24, p. 298
Rear Admiral BG, Commander, Blue Carrier Division 1, p. 298
Rear Admiral BD, Commander, Blue Battleline and Battleship Division 6, p. 302
Captain B-10, Commander, Blue Right Flank Group, p. 302
Admiral OA, Commander, Orange 1st Mobile Fleet, p. 305
Vice Admiral OB, Commander, Orange 5th Fleet, p. 305
Rear Admiral OC, Commander, Orange Task Force 51, the Attack Force, p. 305
Rear Admiral OD, Commander, Orange Van and Cruiser Division 5, p. 305
Rear Admiral OE, Commander, Orange Rear and Cruiser Division 10, p. 305
Rear Admiral OF, Commander, Orange Destroyer Squadron 2, p. 305
Rear Admiral OG, Commander, Orange Destroyer Squadron 10, p. 305
Captain O-1, Commander, Orange Air Group, p. 305
Rear Admiral OH, Commander, Orange Battleline and Battleship Division 3, p. 312

BIBLIOGRAPHY

Archival and Manuscript Records

Bates, Richard W., Papers. Naval Historical Collection, Naval War College, Newport, R.I.

Record Group 4: Publications, 1915–1977. Naval Historical Collection, Naval War College, Newport, R.I.

Record Group 19: College Publications, 1897–2000. Naval Historical Collection, Naval War College, Newport, R.I.

"Report of Board to Study the Methods of Educating Naval Officers." January 1944, Naval War College Library.

Spruance, Raymond A., Papers. Naval Historical Collection, Naval War College, Newport, R.I.

Secondary Sources

Asada, Sadao. *From Mahan to Pearl Harbor: The Imperial Japanese Navy and the United States*. Annapolis, Md.: Naval Institute Press, 2006.

Barber, James. "Mahan and Naval Strategy in the Nuclear Age: A Lecture Delivered at the Naval War College." *Naval War College Review* 24, no. 7 (March 1972), pp. 78–88.

———. "The School of Naval Warfare." *Naval War College Review* 21, no. 8 (April 1969), pp. 89–96.

Barlow, Jeffrey. *From Hot War to Cold: The U.S. Navy and National Security Affairs, 1945–1955*. Stanford, Calif.: Stanford Univ. Press, 2009.

Blair, Clay. *Silent Victory: The U.S. Submarine War against Japan*. Annapolis, Md.: Naval Institute Press, 1975.

Bredestege, Joseph. "Limited Nuclear War at Sea." *Naval War College Review* 19, no. 6 (February 1967), pp. 4–31.

Buell, Thomas. *The Quiet Warrior: A Biography of Admiral Raymond A. Spruance*. Boston: Little, Brown, 1974.

Carney, Robert. "Role of the Navy in a Future War." *Naval War College Review* 6, no. 10 (June 1954), pp. 1–12.

Crenshaw, Russell. *The Battle of Tassafaronga*. Baltimore: Nautical and Aviation, 1995.

Crowl, Philip. "Education versus Training at the Naval War College: 1884–1972." *Naval War College Review* 26, no. 3 (November–December 1973), pp. 2–10.

Cutler, Thomas. "Sea Power and Defense of the Pusan Pocket, June–September 1950." In *The U.S. Navy in the Korean War,* edited by Edward Marolda, pp. 1–51. Annapolis, Md.: Naval Institute Press, 2007.

Davidson, Joel. *The Unsinkable Fleet: The Politics of U.S. Navy Expansion in World War II*. Annapolis, Md.: Naval Institute Press, 1996.

Davis, Vincent. *Postwar Defense Policy and the U.S. Navy, 1943–1946*. Chapel Hill: Univ. of North Carolina Press, 1966.

Dyer, George. *The Amphibians Came to Conquer: The Story of Admiral Richmond Kelly Turner.* 2 vols. Washington, D.C.: U.S. Navy Dept., 1972.

Evans, David, and Mark Peattie. *Kaigun: Strategy, Tactics, and Technology in the Imperial Japanese Navy, 1887–1941*. Annapolis, Md.: Naval Institute Press, 1997.

Felker, Craig. *Testing American Sea Power: U.S. Navy Strategic Exercises, 1923–1940*. College Station: Texas A&M Univ. Press, 2007.

Field, James. "The Influence of Sea Power on Modern Strategy." *Naval War College Review* 10, no. 1 (September 1957), pp. 31–52.

———. "Origins of Maritime Strategy and the Development of Sea Power." *Naval War College Review* 7, no. 7 (March 1955), pp. 1–23.

———. "Seapower and Military Strategy Today." *Naval War College Review* 8, no. 8 (April 1956), pp. 21–39.

Fisher, Robert. "The U.S. Navy's Search for a Strategy, 1945–1947." *Naval War College Review* 48, no. 3 (Summer 1995), pp. 73–86.

Frank, Richard. *Guadalcanal: The Definitive Account of the Landmark Battle*. New York: Random House, 1990.

Friedman, Hal M. *Creating an American Lake: United States Imperialism and Strategic Security in the Pacific Basin, 1945–1947*. Westport, Conn.: Greenwood, 2001.

———. *Digesting History: The U.S. Naval War College, the Lessons of World War Two, and Future Naval Warfare, 1945–1947*. Newport, R.I.: Naval War College Press, 2010.

———. *Governing the American Lake: The U.S. Defense and Administration of the Pacific, 1945–1947*. East Lansing: Michigan State Univ. Press, 2007.

Gile, Robert H. *Global War Game: Second Series, 1984–1988*. Newport Paper 20. Newport, R.I.: Naval War College Press, 2004.

Hattendorf, John B. *The Evolution of the U.S. Navy's Maritime Strategy, 1977–1986.* Newport Paper 19. Newport, R.I.: Naval War College Press, 2004.

———, ed. *U.S. Naval Strategy in the 1970s: Selected Documents.* Newport Paper 30. Newport, R.I.: Naval War College Press, 2007.

Hattendorf, John B., B. Mitchell Simpson III, and John Wadleigh. *Sailors and Scholars: The Centennial History of the U.S. Naval War College.* Newport, R.I.: Naval War College Press, 1984.

Hay, Bud, and Robert Gile. *Global War Game: The First Five Years.* Newport Paper 4. Newport, R.I.: Naval War College Press, 1993.

Hegmann, Richard. "Reconsidering the Evolution of the US Maritime Strategy, 1955–1965." *Journal of Strategic Studies* 14 (September 1991), pp. 299–336.

Holwitt, Joel. *"Execute against Japan": The U.S. Decision to Conduct Unrestricted Submarine Warfare.* College Station: Texas A&M Univ. Press, 2009.

Hone, Thomas, and Trent Hone. *Battle Line: The United States Navy, 1919–1939.* Annapolis, Md.: Naval Institute Press, 2006.

Hone, Trent. "The Evolution of Fleet Tactical Doctrine in the U.S. Navy, 1922–1941." *Journal of Military History* 67 (October 2003), pp. 1107–48.

———. "U.S. Navy Surface Battle Doctrine and Victory in the Pacific." *Naval War College Review* 62, no. 1 (Winter 2009), pp. 67–105.

Hurlburt, J. S. "War Gaming at the Naval War College, 1969–1989." *Naval War College Review* 42, no. 3 (Summer 1989), pp. 46–51.

James, Harry. "An Analysis of Limited Maritime War." *Naval War College Review* 19, no. 6 (February 1967), pp. 33–74.

Kuehn, John. *Agents of Innovation: The General Board and the Design of the Fleet That Defeated the Japanese Navy.* Annapolis, Md.: Naval Institute Press, 2008.

Levy, James. "Race for the Decisive Weapon: British, American, and Japanese Carrier Fleets, 1942–1943." *Naval War College Review* 58, no. 1 (Winter 2005), pp. 137–50.

Lorrelli, John. *The Battle of the Komandorski Islands, March 1943.* Annapolis, Md.: Naval Institute Press, 1984.

Mahnken, Thomas. "Asymmetric Warfare at Sea: The Naval Battles off Guadalcanal, 1942–1943." *Naval War College Review* 64, no. 1 (Winter 2011), pp. 95–121.

McBride, William. *Technological Change and the United States Navy, 1865–1945.* Baltimore, Md.: Johns Hopkins Univ. Press, 2000.

McHugh, Francis. "Eighty Years of War Gaming." *Naval War College Review* 21, no. 7 (March 1969), pp. 88–90.

———. *Fundamentals of War Gaming.* Newport, R.I.: Naval War College, 1966.

———. *A Manual of War Games.* Newport, R.I.: Naval War College, 1960.

———. "War Gaming and the Navy Electronic Warfare Simulator." *Naval War College Review* 19, no. 7 (March 1967), pp. 22–32.

McInteer, J. F. "The Significance of Seapower to the United States." *Naval War College Review* 12, no. 1 (September 1959), pp. 1–32.

Miller, Edward. *War Plan Orange: The U.S. Strategy to Defeat Japan, 1897–1945.* Annapolis, Md.: Naval Institute Press, 1991.

Morris, Eric. *The Russian Navy: Myth & Reality*. London: Hamish Hamilton, 1977.

Musicant, Ivan. *Battleship at War: The Epic Story of the USS* Washington. New York: Harcourt Brace Jovanovich, 1986.

Nofi, Albert. *To Train the Fleet for War: The U.S. Navy Fleet Problems, 1923–1940.* Newport, R.I.: Naval War College Press, 2010.

O'Hara, Vincent. *The U.S. Navy against the Axis: Surface Combat, 1941–1945*. Annapolis, Md.: Naval Institute Press, 2007.

Palmer, Michael. *Origins of the Maritime Strategy: The Development of American Naval Strategy, 1945–1955*. Annapolis, Md.: Naval Institute Press, 1990.

Parshall, Jonathan, and Anthony Tully. *Shattered Sword: The Untold Story of the Battle of Midway*. Washington, D.C.: Potomac Books, 2005.

Peattie, Mark. *Sunburst: The Rise of Japanese Naval Air Power, 1909–1941*. Annapolis, Md.: Naval Institute Press, 2001.

Phelan, George. "Sea Power and Strategies for the Control of the Seas." *Naval War College Review* 6, no. 10 (June 1954), pp. 15–36.

Prange, Gordon. *Miracle at Midway*. New York: McGraw-Hill, 1982.

Reynolds, Clark. *The Fast Carriers: The Forging of an Air Navy*. Annapolis, Md.: Naval Institute Press, 1968.

Rohwer, Jurgen, and Mikhail Monakov. *Stalin's Ocean-Going Fleet: Soviet Naval Strategy and Shipbuilding Programmes, 1935–1953*. London: Frank Cass, 2001.

Rosenberg, David. "Being 'Red': The Challenge of Taking the Soviet Side in War Games at the Naval War College." *Naval War College Review* 41, no. 1 (Winter 1988), pp. 81–93.

Skates, John. *The Invasion of Japan: Alternative to the Bomb*. Columbia: Univ. of South Carolina Press, 1995.

Smith, Douglas. *Carrier Battles: Command Decision in Harm's Way*. Annapolis, Md.: Naval Institute Press, 2006.

Smith, Nepier. "Historical Analysis of the Organizational Success of the Naval War College during the Twenty-Five Years following the Second World War." Naval War College Advanced Research Project. Naval Historical Collection, Naval War College, Newport, R.I., 1974.

Snyder, Frank, ed. *Sound Military Decision: U.S. Naval War College*. Annapolis, Md.: Naval Institute Press, 1992.

Spector, Ronald. *Eagle against the Sun: The American War with Japan*. New York: Free Press, 1985.

———. *Professors of War: The Naval War College and the Development of the Naval Profession*. Newport, R.I.: Naval War College Press, 1977.

U.S. Navy Dept. *The Battle of Guadalcanal, 11–15 November 1942*. Washington, D.C.: Naval Historical Center, 1994.

Vlahos, Michael. *The Blue Sword: The Naval War College and the American Mission, 1919–1941*. Newport, R.I.: Naval War College Press, 1980.

Warner, Dennis, and Peggy Warner. *Disaster in the Pacific: New Light on the Battle of Savo Island*. Annapolis, Md.: Naval Institute Press, 1992.

Watman, Kenneth. "Global 2000." *Naval War College Review* 54, no. 2 (Spring 2001), pp. 75–88.

White, Donald. "Admiral Richard L. Conolly: A Perspective on His Notions of Strategy." *Naval War College Review* 24, no. 3 (November 1971), pp. 73–79.

Wildenberg, Thomas. *Destined for Glory: Dive Bombing, Midway, and the Evolution of Carrier Airpower*. Annapolis, Md.: Naval Institute Press, 1998.

Willmott, H. P. *The Battle of Leyte Gulf: The Last Fleet Action*. Bloomington: Indiana Univ. Press, 2005.

Wood, Robert S. "The Conceptual Framework for Strategic Development at the Naval War College." *Naval War College Review* 40, no. 2 (Spring 1987), pp. 4–16.

ABOUT THE AUTHOR

Hal M. Friedman is Professor of Modern History in the History and Political Science Department of the Humanities and Social Science Division at Henry Ford Community College, Dearborn, Michigan. Educated at Eastern Michigan University with a bachelor of science in political science and history (magna cum laude) in 1987, Professor Friedman took his master of arts and doctor of philosophy degrees in the history of international relations at Michigan State University in 1991 and 1995, respectively. Additionally, he teaches part-time as an adjunct faculty member of the Command and Staff College Distance Education Program within the U.S. Marine Corps University's College of Continuing Education; a section instructor in Norwich University's online Master of Arts in Military History Program; and an adjunct lecturer of strategy and policy in the U.S. Naval War College's Non-Resident Strategy and Policy Seminar Program. Professor Friedman has formerly taught for Central Michigan University-Metro Detroit, Eastern Michigan University, Michigan State University, the University of Detroit-Mercy, the University of Michigan-Dearborn, and Washtenaw Community College.

Professor Friedman is the author of *Creating an American Lake: United States Imperialism and Strategic Security in the Pacific Basin, 1945–1947* (Westport, Conn.: Greenwood, 2001), *Governing the American Lake: The US Defense and Administration of the Pacific, 1945–1947* (East Lansing: Michigan State Univ. Press, 2007), *Arguing over the American Lake: Bureaucracy and Rivalry in the U.S. Pacific, 1945–1947* (College Station: Texas A&M Univ. Press, 2009), and *Digesting History: The U.S. Naval War College, the Lessons of World War II, and Future Naval Warfare, 1945–1947* (Newport, R.I.: Naval War College Press, 2010). Professor Friedman is also editing a collection of essays about the United States military in the Pacific and East Asia in the 1940s entitled *Establishing Hegemony: The American Military Presence in the Pacific Basin, 1940–1950.* He has additionally published numerous articles in *The Hawaiian Journal of History, Intelligence and National Security, ISLA: A Journal of Micronesian Studies, The Journal of Pacific History, Pacific Historical Review,* and

Pacific Studies. Professor Friedman has also written book reviews for *The Historian, The History Teacher*, H-Net, the *International Journal of Naval History, The Journal of American History, The Journal of Military History, Michigan Historical Review, Michigan War Studies Review, National Identities, Pacific Historical Review*, the *Journal of World History*, the *World History Bulletin*, and *World War II*. Professor Friedman was an article abstractor for ABC-CLIO; an article referee for *Michigan Historical Review, The Michigan Academician*, and *Pacific Historical Review;* a monograph referee for Texas A&M University Press; and a textbook reviewer for Houghton Mifflin, Longman Publishers, McGraw-Hill, and Prentice Hall.

Professor Friedman has been a conference program committee member for the Organization of American Historians, the Liberal Arts Network for Development, and the Ohio Academy of History; both a consulting historian and summer scholar with the National Endowment for the Humanities' Landmarks of American History and Culture Program; and a consulting historian for the Department of Education's Teaching American History Program. He is also a cofounder of the Metro Detroit Historians Collegium, a member of the *Michigan War Studies Review* Editorial Board, and a member of the Ohio Academy of History's Professional Development Committee.

Professor Friedman has also been honored a number of times, including with the Major General Thomas S. Jones, USMC, Distance Learning Instructor of the Year Award, Fort Worth Region, Command and Staff College Distance Education Program, Marine Corps College of Continuing Education; with an Edward S. Miller Fellowship in Naval History for the Naval War College's Naval Historical Collection; with an Alumni Award for Outstanding Academic and Professional Accomplishments from Eastern Michigan University's Department of History and Philosophy; and as an honorary member in Phi Theta Kappa, the Two-Year College International Honor Society.

Heavily involved with Henry Ford Community College's shared-governance system, Professor Friedman has served on a variety of committees for HFCC and his union local, the last as the secretary of the Henry Ford Community College Federation of Teachers, American Federation of Teachers Local 1650.

Professor Friedman served as a personnelman (PN) in the U.S. Naval Reserve, primarily as a pay and personnel clerk in several Naval Reserve Intelligence Command units. His service awards include the Naval Reserve Meritorious Service Medal, the National Defense Service Medal, and the Joint Meritorious Unit Award. He was also chosen as the Fleet Intelligence Rapid Support Team 1773 Junior Sailor of the Year for 1991.

INDEX

Note: Page numbers in *italics* indicate photographs and illustrations.

NAVAL WAR COLLEGE HISTORICAL MONOGRAPH SERIES

1. *The Writings of Stephen B. Luce,* edited by John D. Hayes and John B. Hattendorf (1975).
3. *Professors of War: The Naval War College and the Development of the Naval Profession,* Ronald Spector (1977).
4. *The Blue Sword: The Naval War College and the American Mission, 1919–1941,* Michael Vlahos (1980).
5. *On His Majesty's Service: Observations of the British Home Fleet from the Diary, Reports, and Letters of Joseph H. Wellings, Assistant U.S. Naval Attaché, London, 1940–1941,* edited by John B. Hattendorf (1983).
7. *A Bibliography of the Works of Alfred Thayer Mahan,* compiled by John B. Hattendorf and Lynn C. Hattendorf (1986).
8. *The Fraternity of the Blue Uniform: Admiral Richard G. Colbert, U.S. Navy and Allied Naval Cooperation,* Joel J. Sokolsky (1991).
9. *The Influence of History on Mahan: The Proceedings of a Conference Marking the Centenary of Alfred Thayer Mahan's* The Influence of Sea Power upon History, 1660–1783, edited by John B. Hattendorf (1991).
10. *Mahan Is Not Enough: The Proceedings of a Conference on the Works of Sir Julian Corbett and Admiral Sir Herbert Richmond,* edited by James Goldrick and John B. Hattendorf (1993).
11. *Ubi Sumus? The State of Naval and Maritime History,* edited by John B. Hattendorf (1994).
12. *The Queenstown Patrol, 1917: The Diary of Commander Joseph Knefler Taussig, U.S. Navy,* edited by William N. Still, Jr. (1996).
13. *Doing Naval History: Essays toward Improvement,* edited by John B. Hattendorf (1995).

14. *An Admiral's Yarn,* edited by Mark R. Shulman (1999).
15. *The Memoirs of Admiral H. Kent Hewitt,* edited by Evelyn Cherpak (2004).
16. *Three Splendid Little Wars: The Diary of Joseph K. Taussig, 1898–1901,* edited by Evelyn Cherpak (2009).
17. *Digesting History: The U.S. Naval War College, the Lessons of World War Two, and Future Naval Warfare, 1945–1947,* by Hal. M. Friedman (2010).
18. *To Train the Fleet for War: The U.S. Navy Fleet Problems, 1923–1940,* by Albert A. Nofi (2010).
19. *Talking about Naval History: A Collection of Essays,* John B. Hattendorf (2011).
20. *New Interpretations in Naval History: Selected Papers from the Sixteenth Naval History Symposium Held at the United States Naval Academy 10–11 September 2009,* edited by Craig C. Felker and Marcus O. Jones (2012).